Methods for Computational Gene

Inferring the precise locations and splicing patterns of genes in DNA is a difficult but important task, with broad applications to biomedicine. The mathematical and statistical techniques that have been applied to this problem are surveyed here and organized into a logical framework based on the theory of parsing. Both established approaches and methods at the forefront of current research are discussed. Numerous case studies of existing software systems are provided, in addition to detailed examples that work through the actual implementation of effective gene predictors using hidden Markov models and other machine-learning techniques. Background material on probability theory, discrete mathematics, computer science, and molecular biology is provided, making the book accessible to students and researchers from across the life and computational sciences. This book is ideal for use in a first course in bioinformatics at graduate or advanced undergraduate level, and for anyone wanting to keep pace with this rapidly advancing field.

W. H. MAJOROS is Staff Scientist at the Center for Bioinformatics and Computational Biology, in the Institute for Genome Sciences and Policy at Duke University. He has worked as a research scientist for over a decade in the fields of computational biology, natural language processing, and information retrieval. He was part of the Human Genome Project at Celera Genomics and has taken part in the sequencing and analysis of the genomes of numerous organisms including human, mouse, fly, and mosquito.

Methods for Computational Gene Prediction

William H. Majoros
Duke University

CAMBRIDGE UNIVERSITY PRESS
Cambridge, New York, Melbourne, Madrid, Cape Town, Singapore, São Paulo

Cambridge University Press
The Edinburgh Building, Cambridge CB2 8RU, UK

Published in the United States of America by Cambridge University Press, New York

www.cambridge.org
Information on this title: www.cambridge.org/9780521877510

First published 2007

Printed in the United Kingdom at the University Press, Cambridge

A catalog record for this publication is available from the British Library

ISBN 978-0-521-87751-0 hardback
ISBN 978-0-521-70694-0 paperback

Contents

Foreword by Steven Salzberg *page* xi
Preface xiii
Acknowledgements xvi

1 Introduction 1

1.1 The central dogma of molecular biology 1
1.2 Evolution 13
1.3 Genome sequencing and assembly 15
1.4 Genomic annotation 19
1.5 The problem of computational gene prediction 25
Exercises 26

2 Mathematical preliminaries 28

2.1 Numbers and functions 28
2.2 Logic and boolean algebra 33
2.3 Sets 34
2.4 Algorithms and pseudocode 35
2.5 Optimization 38
2.6 Probability 40
2.7 Some important distributions 48
2.8 Parameter estimation 54
2.9 Statistical hypothesis testing 55
2.10 Information 58
2.11 Computational complexity 62
2.12 Dynamic programming 63
2.13 Searching and sorting 66
2.14 Graphs 68
2.15 Languages and parsing 74
Exercises 80

3 Overview of computational gene prediction 83

3.1 Genes, exons, and coding segments 83
3.2 Orientation 86
3.3 Phase and frame 89
3.4 Gene finding as parsing 92
3.5 Common assumptions in gene prediction 97
3.5.1 No overlapping genes 98
3.5.2 No nested genes 98
3.5.3 No partial genes 98
3.5.4 No noncanonical signal consensuses 99
3.5.5 No frameshifts or sequencing errors 99
3.5.6 Optimal parse only 100
3.5.7 Constraints on feature lengths 100
3.5.8 No split start codons 100
3.5.9 No split stop codons 101
3.5.10 No alternative splicing 101
3.5.11 No selenocysteine codons 101
3.5.12 No ambiguity codes 101
3.5.13 One haplotype only 102
Exercises 102

4 Gene finder evaluation 104

4.1 Testing protocols 104
4.2 Evaluation metrics 113
Exercises 118

5 A toy exon finder 120

5.1 The toy genome and its toy genes 120
5.2 Random exon prediction as a baseline 121
5.3 Predicting exons based on {G,C} bias 126
5.4 Predicting exons based on codon bias 127
5.5 Predicting exons based on codon bias and WMM score 130
5.6 Summary 134
Exercises 135

6 Hidden Markov models 136

6.1 Introduction to HMMs 136
6.1.1 An illustrative example 138
6.2.1 Representing HMMs 139
6.2 Decoding and similar problems 140
6.2.1 Finding the most probable path 140
6.2.2 Computing the probability of a sequence 143
6.3 Training with labeled sequences 145

6.4 Example: Building an HMM for gene finding 147
6.5 Case study: VEIL and UNVEIL 157
6.6 Using ambiguous models 159
6.6.1 Viterbi training 159
6.6.2 Merging submodels 161
6.6.3 Baum–Welch training 162
6.6.3.1 Naive Baum–Welch algorithm 163
6.6.3.2 Baum–Welch with scaling 165
6.7 Higher-order HMMs 169
6.7.1 Labeled sequence training for higher-order HMMs 169
6.7.2 Decoding with higher-order HMMs 170
6.8 Variable-order HMMs 171
6.8.1 Back-off models 171
6.8.2 Example: Incorporating variable-order emissions 172
6.8.3 Interpolated Markov models 173
6.9 Discriminative training of HMMs 174
6.10 Posterior decoding of HMMs 177
Exercises 179

7 Signal and content sensors 184
7.1 Overview of feature sensing 184
7.2 Content sensors 185
7.2.1 Markov chains 185
7.2.2 Markov chain implementation 188
7.2.3 Improved Markov chain implementation 188
7.2.4 Three-periodic Markov chains 190
7.2.5 Interpolated Markov chains 191
7.2.6 Nonstationary Markov chains 192
7.3 Signal sensors 193
7.3.1 Weight matrices 195
7.3.2 Weight array matrices 197
7.3.3 Windowed weight array matrices 198
7.3.4 Local optimality criterion 198
7.3.5 Coding–noncoding boundaries 200
7.3.6 Case study: GeneSplicer 201
7.3.7 Maximal dependence decomposition 201
7.3.8 Probabilistic tree models 205
7.3.9 Case study: Signal sensing in GENSCAN 207
7.4 Other methods of feature sensing 208
7.6 Case study: Bacterial gene finding 209
Exercises 211

8 Generalized hidden Markov models 214
8.1 Generalization and its advantages 214
8.2 Typical model topologies 218
8.2.1 One exon model or four? 221
8.2.2 One strand or two? 222
8.3 Decoding with a GHMM 223
8.3.1 PSA decoding 228
8.3.2 DSP decoding 237
8.3.3 Equivalence of DSP and PSA 242
8.3.4 A DSP example 244
8.3.5 Shortcomings of DSP and PSA 246
8.4 Higher-fidelity modeling 247
8.4.1 Modeling isochores 247
8.4.2 Explicit modeling of noncoding lengths 249
8.5 Prediction with an ORF graph 251
8.5.1 Building the graph 252
8.5.2 Decoding with a graph 253
8.5.3 Extracting suboptimal parses 254
8.5.4 Posterior decoding for GHMMs 255
8.5.5 The ORF graph as a data interchange format 256
8.6 Training a GHMM 259
8.6.1 Maximum likelihood training for GHMMs 260
8.6.2 Discriminative training for GHMMs 260
8.7 Case study: GHMM versus HMM 263
Exercises 264

9 Comparative gene finding 267
9.1 Informant techniques 269
9.1.1 Case study: TWINSCAN 269
9.1.2 Case study: GenomeScan 271
9.1.3 Case study: SGP-2 272
9.1.4 Case study: HMMgene 273
9.1.5 Case study: GENIE 274
9.2 Combiners 275
9.2.1 Case study: JIGSAW 275
9.2.2 Case study: GAZE 277
9.3 Alignment-based prediction 277
9.3.1 Case study: ROSETTA 278
9.3.2 Case study: SGP-1 278
9.3.3 Case study: CEM 279
9.4 Pair HMMs 281
9.4.1 Case study: DoubleScan 285
9.5 Generalized pair HMMs 287
9.5.1 Case study: TWAIN 289

9.6 Phylogenomic gene finding 299
9.6.1 Phylogenetic HMMs 299
9.6.2 Decoding with a PhyloHMM 305
9.6.3 Evolution models 306
9.6.4 Parameterization of rate matrices 310
9.6.5 Estimation of evolutionary parameters 312
9.6.6 Modeling higher-order dependencies 316
9.6.7 Enhancing discriminative power 318
9.6.8 Selection of informants 318
9.7 Auto-annotation pipelines 319
9.8 Looking toward the future 320
Exercises 321

10 Machine-learning methods 325
10.1 Overview of automatic classification 325
10.2 *K*-nearest neighbors 328
10.3 Naive Bayes models 329
10.4 Bayesian networks 330
10.5 Neural networks 332
10.5.1 Case study: GRAIL 336
10.6 Decision trees 337
10.6.1 Case study: GlimmerM 339
10.7 Linear discriminant analysis 340
10.8 Quadratic discriminant analysis 342
10.9 Multivariate regression 343
10.10 Logistic regression 343
10.11 Regularized logistic regression 345
10.12 Genetic programming 346
10.13 Simulated annealing 349
10.14 Support vector machines 349
10.15 Hill-climbing with the GSL 351
10.16 Feature selection and dimensionality reduction 352
10.17 Applications 354
Exercises 355

11 Tips and tricks 358
11.1 Boosting 358
11.2 Bootstrapping 359
11.3 Modeling additional gene features 361
11.4 Masking repeats 366
Exercises 367

12 Advanced topics 369

12.1 Alternative splicing and transcription 369
12.2 Prediction of noncoding genes 373
12.3 Promoter prediction 379
12.4 Generative versus discriminative modeling 382
12.5 Parallelization and grid computing 384
Exercises 386

Appendix 388
A.1 Official book website 388
A.2 Open-source gene finders 388
A.3 Gene-finding websites 389
A.4 Gene-finding bibliographies 389

References 390

Index 408

Foreword

When Frederick Sanger sequenced the very first genome – the bacteriophage ϕ-X174 – in 1977, it was clear that DNA sequencing offered a dramatically faster way to find genes than earlier, traditional mapping methods. The small phage genome spans just 5386 bases, about 95% of which is used to encode 11 genes. For this and other viruses, gene finding is fast and easy: the proteins are encoded virtually end-to-end, sometimes even overlapping one another. The early days of DNA sequencing proceeded slowly but with great excitement, as the new technology was applied to small fragments of DNA from many different species. By the mid-1980s, scientists were attempting to automate the sequencing process, which soon led to larger sequencing projects, and in 1989 the Human Genome Project was launched, with the ambitious goal of sequencing the entire 3 billion base pairs of the 24 human chromosomes over the course of the next 15 years. In 1995 a team at The Institute for Genomic Research (TIGR) sequenced the first genome of a free-living organism, the bacterium *Haemophilus influenzae*, which at 1.8 million base pairs was considerably larger than any genome that had been sequenced before. The *H. influenzae* genome was the beginning of an enormous outpouring of DNA sequences, fueled by ever-lower sequencing costs and an ever-increasing thirst for new discoveries, that has now produced hundreds of genomes, both large and small.

As more sequences emerged, it became clear that finding genes – the primary goal of most early sequencing projects – was much more difficult in eukaryotic genomes. The human genome was believed in 1989 to contain about 100 000 genes – an estimate that was later downgraded dramatically and now stands at around 25 000 – and each gene encodes a protein of around 350 amino acids. This means that the entire protein-coding portion of the human genome spans only about 25–30 million base pairs, or 1% of the genome. It was clear from the beginning that finding genes might be a bit like finding a needle in a haystack. There was also tremendous interest in what the remaining 99% of the genome contained: was it "junk DNA" as some called it? Or did it serve somehow to regulate the genes?

Greatly complicating matters is the inconvenient fact that genes in most eukaryotes are interrupted by long stretches of DNA that do not encode proteins. This astonishing discovery was made by Richard Roberts and Phillip Sharp in 1977, who discovered and named "introns" and began studying the process by which introns in messenger RNA are spliced out and removed. (Roberts and Sharp won the Nobel Prize in 1993 for their discovery.)

As exciting as these biological disocoveries were, they presented major new challenges to those who wanted to use DNA sequence to identify genes. Not only did we have to find the protein-coding portions of the genome, we also had to find and remove the introns in order to predict the correct protein. Gene-finding programs became a critical part of annotation "pipelines" that the major sequencing centers used to identify the genes for each project. As the technology improved, the sequencers kept throwing new and more difficult challenges at the bioinformaticists: for example, many of the mammalian genomes sequenced since the human draft genome was announced (in 2001) have been low-coverage sequencing projects, meaning that the genomes emerge in thousands of fragments, and the genes are often split among multiple fragments of DNA.

Bill Majoros participated in some of the most important projects of the first ten years of the genome revolution, first at Celera Genomics and then at TIGR, where I had the privilege to work closely with him for several years. While at TIGR, he developed an expertise on virtually all of the major gene finding algorithms in use today. He built several gene finders of his own, notably the hidden Markov model gene finders TIGRscan (for one species) and TWAIN (for two species). He worked closely with genome sequencers and contributed his gene-finding expertise to multiple genome projects, including the genomes of the human, the mouse, the malaria-carrying mosquito *Anopheles gambiae*, the pathogenic fungus *Aspergillus fumigatus*, and the strange but wonderful ciliate *Tetrahymena thermophila*. Along the way, he learned about the shortcomings of past gene finders and about the difficulties faced by a genome project for which no gene finder is available.

Today, with the pace of genomics still increasing, there is a growing need for bioinformatics experts trained in the computational intricacies of gene finding. I can think of no one better suited to put together a book that explains these methods than Bill Majoros, and this book does not disappoint. Bill has created an elegantly written text that covers it all, first providing a mathematical context and then describing all the major techniques, providing details and tips that will prove invaluable to anyone wishing to gain a better understanding of how these algorithms work.

Steven L. Salzberg
Center for Bioinformatics and Computational Biology
University of Maryland
College Park, MD

Preface

This book grew out of a number of conversations between Dr. Ian Korf and myself, in May of 2004, in which we jointly lamented the rather large number of small but important details which one is required to know when building a practical gene-finding system from scratch, and which were at that time not fully documented in the gene-finding literature. Page limits in traditional print journals invariably force authors of research reports to omit details, and although online journals have relaxed this constraint to some degree, the larger impact factors of the more venerable print journals still attract many researchers to those more constrained venues. Of course, even the online journals urge authors to be brief, so that implementation details are either relegated to supplementary online documents, or – more often than not – simply omitted entirely. Because practical gene-finding software typically consists of many thousands of lines of source code, including all the details necessary for replicating any of these programs is out of the question for any journal article. Placing one's gene finder into the public domain, or making it *open source*, can help to alleviate this problem, since the public is then free to peruse the source code. Of course, perusing a 30 000-line program can be tedious, to say the least, and ideally one would like a format more geared toward human consumption, in which the essential concepts have been distilled from the implementation and provided in a rigorous but accessible manner. That was the ambitious goal which Ian and I set out to accomplish with this book when we jointly crafted the original outline in the latter half of 2004. Unfortunately, excessive demands on the time of my respected colleague prevented his participating in this project beyond the outlining phase, so I must duly accept responsibility for any and all defects which the reader may perceive in the final product. I do hope, however, that the monograph will be of some use to the enthusiastic reader, and that the goals of genomic and genetic research may be positively, if only very marginally, impacted by the availability within a single volume of this information.

Prerequisites and audience

This text comes with very few prerequisites, chief among them being merely an interest in understanding state-of-the-art methods in computational prediction of protein-coding genes. Because all of the relevant biological concepts are defined in Chapter 1, computer scientists and computer science students at both the graduate and undergraduate levels should have little difficulty following the material, if digested in a progressive manner. Likewise, in Chapter 2 the reader lacking a background in computer programming will find much of the necessary primer material on discrete mathematics and formal algorithms, though the prior attainment of some practical programming experience on the part of the novice reader will of course be strongly encouraged. Algorithms are presented in a uniform pseudocode throughout the text, so that knowledge of a specific programming language is not required. Because experience in writing and debugging software takes some dedication of time and effort, the non-programmer may take longer to progress through the material, but with sufficient perseverance and by following the suggested reading links and references, the task should be rendered quite feasible.

Structure of the text

Following the primer materials on molecular biology (Chapter 1) and mathematics and computer science (Chapter 2), we give an overview of the gene prediction problem and define many of the concepts which will be central to the rest of the book (Chapter 3). Because gene-finding methods are imperfect, we describe common methods for evaluating the accuracy of a gene finder (Chapter 4). In Chapter 5 we construct a number of "toy" gene finders for a simulated genome, with the intention being to help the reader to rapidly develop some intuition for the methods of gene identification via statistical methods of sequence analysis. Chapters 6, 7, and 8 constitute the main body of the work, as they describe the most common probabilistic frameworks (namely, HMMs and GHMMs) for *ab initio* gene finding, and Chapter 9 extends these frameworks by considering the incorporation of homology and expression evidence. In Chapter 10 we review a number of machine-learning algorithms and describe possible ways of incorporating these methods into hybrid systems for the maximum discrimination of gene elements from nonfunctional DNA. Chapter 11 reveals several "tips and tricks" which we have found useful in practice, and Chapter 12 concludes with a small number of advanced topics representing the current frontier of gene-finding research. The Appendix enumerates several useful online resources.

Errata and updates

Though it is my hope that the number of typographical and/or factual errors in this text will be found to be small, I have greater hopes that readers of this first edition will send me their corrections so that they may be incorporated as appropriate into the next edition. Instructions for submitting errata can be found on the book's official website at www.geneprediction.org/book

Acknowledgements

Quite a sizeable number of people deserve my sincerest gratitude for patiently aiding my induction into the dizzying world of genomics. Chief among them are Mark Yandell, Steven Salzberg, Ian Korf, Mani Subramanian, and Uwe Ohler, who have at their own expense voluntarily given over sizeable portions of their very precious time to act as mentors and in some cases as supervisors. Many other fond acquaintances were made during my time at Celera during the sequencing of the fruitfly, human, mouse, and mosquito genomes, and subsequently during the pleasant three years I spent at TIGR and the year I have been at Duke, yet space permits me to name only a few of the many talented individuals with whom I have been fortunate enough to interact, and who have helped me in one way or another in forming an understanding of the material covered in this book: Jennifer Wortman, Pawel Gajer, Ela Pertea, Jonathan Eisen, Brian Haas, Art Delcher, Mark Adams, Peter Li, Richard Mural, Andy Clark, George Miklos, Rob Holt, Sam Levy, Jonathan Allen, Jason Stajich, Razvan Sultana, Corina Antonescu, Adam Phillippy, Mihai Pop, Jonathan Badger, Tom Heiman, Qing Zhang, Deborah Nusskern, Chris Desjardins, Rosane Charlab, Valentina Di Francesco, Linda Hannick, Chris Shue, Jian Wang, Vineet Bafna, Karen Eilbeck, Ashwin Naik, Sridhar Hannenhalli, Sayan Mukherjee, Natesh Pillai, Terry Furey, Maryam Farzad, Paul Thomas, Owen White, Martin Wu, Mike Schatz, Saul Kravitz, Kabir Chaturvedi, Barbara Methé, Andrei Gabrielian, Roger Smith, Vivien Bonazzi, Ron Wides, Anne Deslattes-Mays, Robin Buell, Joe McDaniel, Matt Newman, Jim Jordan, Sue Pan, Jeff Hoover, and Christine Carter.

I also thank those who have graciously provided technical aid and/or scientific guidance (genomic or otherwise) during my various other terms of employment over the past 12 years, including but not limited to: Steve Finch, Mark Turner, Gerry Perham, Ravi Sundaar, Andy Doyle, and Dave Vause.

Much of this text describes work carried out by others, and to the extent possible, I have sought to have the authors of particular software systems and/or scientific reports review the sections describing their respective works. For their time in this endeavor and their resulting comments, criticisms, and suggestions, and more generally for their varied and valuable contributions to the field, I owe great thanks; any errors that remain in the current edition are of course my fault alone.

To the students of the spring 2006 section of CMSC828N at UMD and of the fall 2006 section of CBB231/COMPSCI261 at Duke I also offer my thanks for their comments and suggestions, and for their patience in dealing with a rough draft of the text. I especially thank Justin Guinney, Alan Boyle, and Laura Kavanaugh for their many useful comments and corrections. Other reviewers who provided invaluable comments and suggestions included Jeffrey Chang, Pawel Gajer, Jonathan Badger, Uwe Ohler, Steven Salzberg, Tom Heiman, Brian Haas, Jonathan Eisen, Art Delcher, Jason Stajich, and Jennifer Wortman.

And finally, I am grateful to all of those highly respected and distinguished colleagues who at one conference or another have patiently allowed me to "pick their brains" on various topics related to gene finding, including: Richard Durbin, Lior Pachter, Chris Burge, Adam Siepel, David Haussler, David Kulp, Mario Stanke, Michael Brent, Simon Kasif, Suzanna Lewis, Michael Ashburner, Roderic Guigó, Ed Orias, Eric Snyder, and Martin Reese.

To all of you I offer my sincerest thanks.

William H. Majoros
Duke University
September, 2006

Notes

1. The individual animals shown on the front cover were drawn by Mateus Zica and were contributed by him to Wikipedia's article "Timeline_of_human_evolution" (http://en.wikipedia.org/wiki/Timeline_of_human_evolution).
2. Celera Genomics and Celera are trademarks of Applera Corporation and its subsidiaries in the U.S. and/or certain other countries.

1 Introduction

The problem that we wish to address in this book is that of predicting computationally the one-dimensional structure of eukaryotic protein-coding genes. Our first order of business will be to define this problem more precisely, and to circumscribe the issue so as to reflect the set of assumptions and constraints which typically apply in the case of practical gene-finding systems. As we recognize that not all readers will be familiar with the relevant facts and theories from molecular biology, we begin with the so-called *central dogma of molecular biology*, in which the significance of genes and their genomic structure are defined in relation to current biological understanding and the goals of modern medicine.

1.1 The central dogma of molecular biology

Life on Earth began approximately 3.5 billion years ago, and since that time it has advanced through a number of stages of increasing complexity.[1] Beginning with the first replicating molecules and unicellular organisms, our own evolutionary trajectory has taken us along an epic journey progressing from microscopic invertebrates to bony fishes, to the amphibians and reptiles who first established the vertebrate kingdom on dry land, to the basal mammals who hid from the dinosaurs among the primitive trees and bushes of the Mesozoic era, and then on to our closest living relatives, the great apes. Through all that time, one fact has remained constant: that our animal selves, both physically and cognitively (with the latter following, of course, the advent of that organ which we call the brain) have been extensively shaped by the action of our *genes*, through their influence both on our bodily development (our *ontogeny*) and also on our ongoing biological processes, both of which are overwhelmingly mediated through the action of *proteins*, the primarily biochemical products of gene expression.

[1] But see, e.g., Gould (1994), Dawkins (1997), and Heylighen (1999).

It is this fundamental influence of the genes and their protein products on human health which is largely responsible for the success and continued momentum of the biological revolution and the tremendous insight into human genetics which has materialized in the late twentieth and early twenty-first centuries, and which at present drives the search for cures to such maladies as cancer, heart disease, diabetes, and dementia, all of which continue to afflict a significant portion of our population, despite the best efforts of modern medicine. It is precisely because of their direct influence on our development and on our physical and mental health that the academic and medical communities have focused so much of their attention on identifying and characterizing human genes. In addition, because many organisms act as pathogens either to ourselves or to animals or plants of economic importance to us, there is also a great interest in identifying and understanding the genes of many non-human species.

Each of the trillion cells in the human body contains in its nucleus the complete human *genome* – 23 pairs of chromosomes – a collection of DNA molecules encoding the information necessary for building and maintaining a healthy human being. It is this information which we pass on to our children in the form of resemblances and inherited traits, and which, when unfavorably perturbed, can lead to disease.

As depicted in Figure 1.1, the DNA in our chromosomes forms a double helix, and in its native state is tightly bound into a complex involving extra-genetic elements called *histones*, about which, though they may have substantial influence on the expression of genes, we will have relatively little to say except in relation to the methylation status of certain elements upstream of mammalian genes (i.e., *CpG islands* – see section 11.3).

In *eukaryotic* organisms (those having nucleated cells), the chromosomes are contained within the *nucleus* of the cell, though their influence can extend quite far beyond the nuclear walls, and indeed, even beyond the boundaries of the organism (e.g., Dawkins, 1982). Different organisms have different numbers of chromosomes, with the fruitfly *Drosophila melanogaster* having only four pairs as compared to our 23. Each member of a pair constitutes a single *haplotype*, which in *diploid* organisms (those whose chromosomes normally occur in pairs) is normally inherited from one of the two parents. These pairs of chromosomes should not be confused with the two strands of DNA; each of the two haplotypes in a diploid genome consists of a set of double-stranded DNA chromosomes.

Figure 1.2 gives a closer look at the structure of the DNA molecule. DNA, or *deoxyribonucleic acid*, consists of a series of paired *nucleotides*, or *bases*, joined at their margins by a sugar–phosphate backbone. The four nucleotides occurring in DNA are *adenine* (denoted chemically as $C_5H_5N_5$), *thymine* ($C_5H_6N_2O_2$), *cytosine* ($C_4H_5N_3O$), and *guanine* ($C_5H_5N_5O$). Adenine and guanine are known as *purines*, while cytosine and thymine are called *pyrimidines*. An important variant of DNA – *ribonucleic acid*,

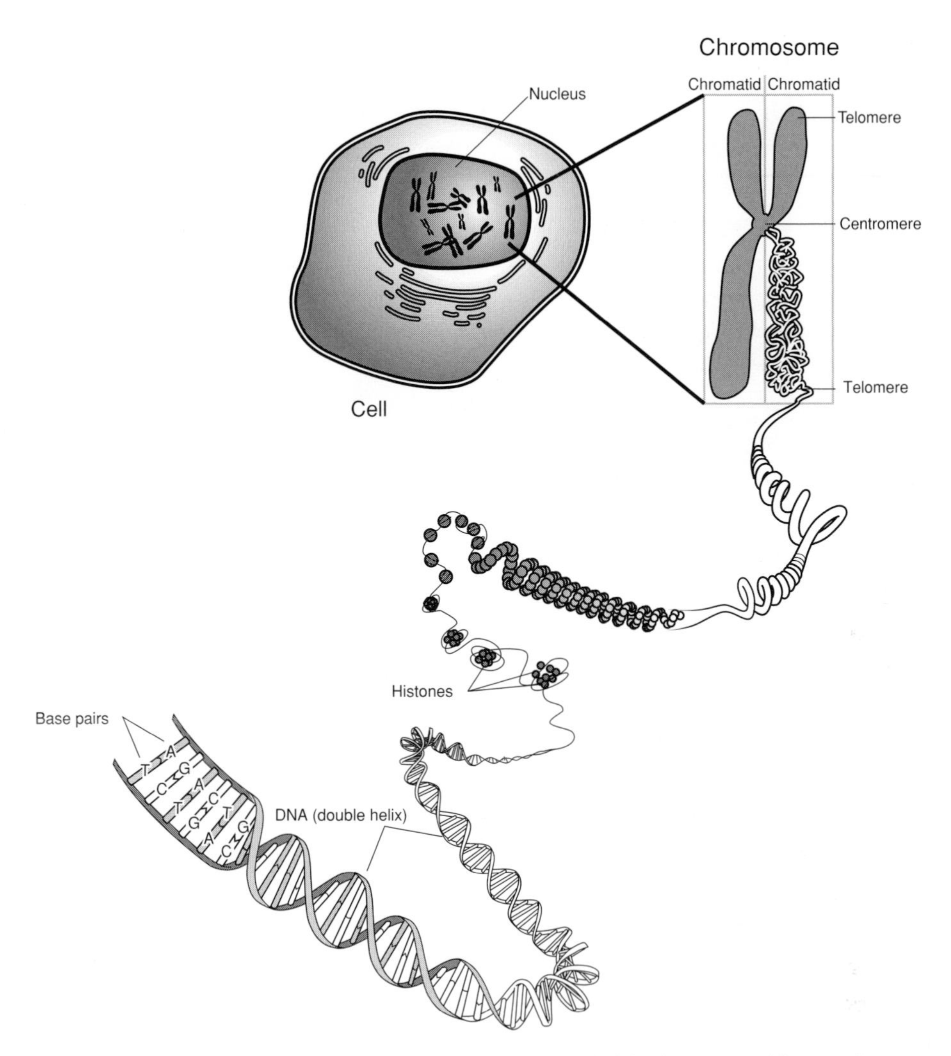

Figure 1.1 The eukaryotic cell. Genetic information is encoded in the DNA making up the chromosomes in the nucleus. (Courtesy: National Human Genome Research Institute).

or *RNA* – substitutes *uracil* ($C_4H_4N_2O_2$) for thymine. The pairing of nucleotides in DNA and RNA follows a strict rule, in which adenine and thymine (or uracil) may bind to one another, and cytosine and guanine may bind to one another, but all other pairings are strictly avoided and occur only in extremely rare circumstances. This pairing is known as *Watson–Crick complementarity*, in honor of the two men who first discovered the structure of DNA and who correctly hypothesized its great importance to genetics (Watson and Crick, 1953). Abbreviating adenine to A, cytosine to C, guanine to G, and thymine to T, we can represent the Watson–Crick complementarity rule as: C $\Leftrightarrow$ G, A $\Leftrightarrow$ T (or in the case of RNA, A $\Leftrightarrow$ U). Figure 1.3 shows the chemical structure of the individual nucleotides.

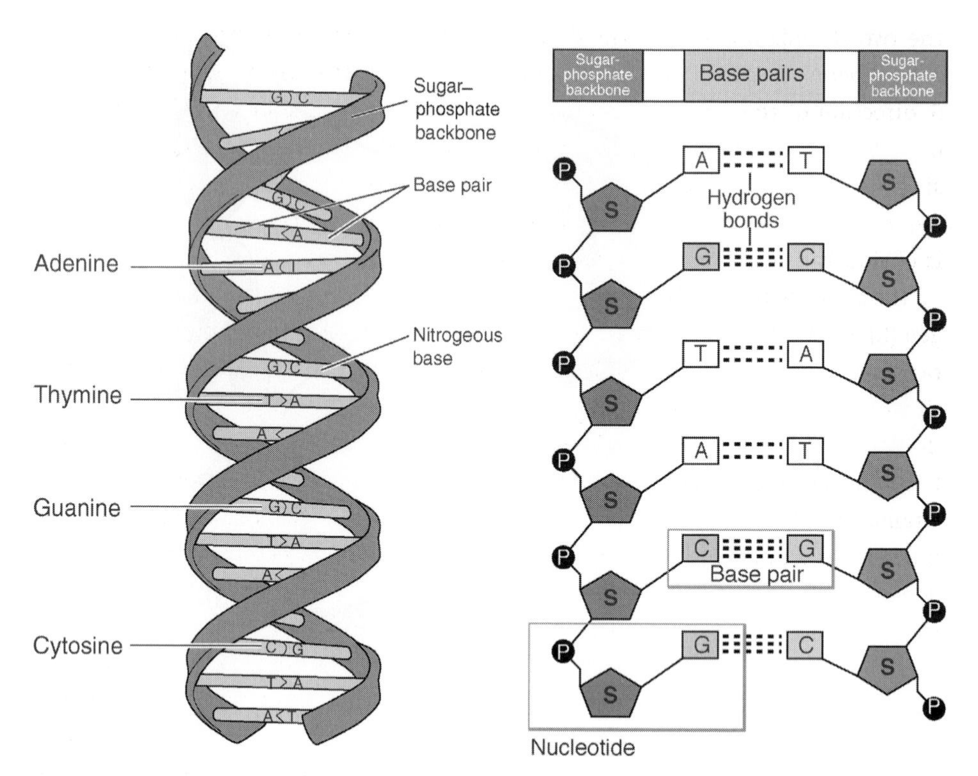

Figure 1.2 Chemical structure of DNA. Each half of the double helix is composed of a sequence of nucleotides linked by a sugar–phosphate backbone. (Courtesy: National Human Genome Research Institute.)

adenine cytosine guanine thymine

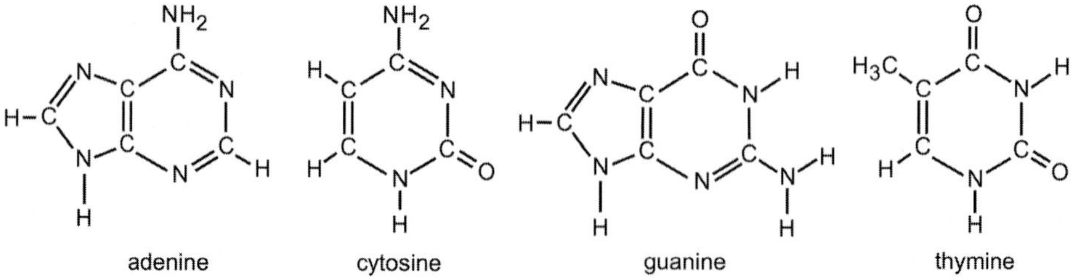

Figure 1.3 Chemical structure of the nucleotides making up DNA and RNA. The thymine in DNA is replaced by uracil in RNA. (Courtesy: National Human Genome Research Institute.)

It is important to note that the nucleotides are not symmetric, so that they have a preferred orientation along the sugar–phosphate backbone. In particular, the *5′-phosphate* group of a nucleotide orients toward the "upstream" direction of the strand, while the *3′-hydroxyl* group orients toward the "downstream" direction, where *upstream* and *downstream* refer to the direction in which the DNA replicates during cell division (i.e., upstream-to-downstream). Thus, we say that DNA and RNA strands are synthesized in the 5′-to-3′ direction. Furthermore, the orientations of the two strands making up the double helix run in opposite directions (i.e., they are *antiparallel*), so that the 5′-to-3′ direction of one strand corresponds to the 3′-to-5′ direction of the opposite strand. As a convention, whenever describing features of a DNA or RNA sequence, one generally assumes that the sequence under consideration is given in the 5′-to-3′ direction along the denoted molecule, and this sequence is always referred to as the *forward strand* (or *sense* strand). The other strand is obviously referred to as the *reverse strand* (or *antisense* strand).

The bonds which join complementary bases in a DNA or RNA molecule are hydrogen (H) bonds. G ⇔ C has three H-bonds, A ⇔ T has two H-bonds, and the extremely rare G ⇔ T base-pairing has only one H-bond. For this reason, when DNA is heated, the A–T pairings will disintegrate (or *denature*) at lower temperatures than will the C–G pairings, since more energy is required in order to break all of their bonds. It should be noted that the bonds joining complementary bases across the two strands of a DNA molecule are chemically different from those that join successive nucleotides along a DNA strand. The latter are termed *phosphodiester bonds*, and are denoted, e.g., CpG, for nucleotides C and G and phosphodiester bond p. When referring to arbitrary DNA sequences, we will generally omit the p and give only the sequence of nucleotides along one of the two DNA strands (most often the forward strand): e.g., ACTAGCTAGCTCTTGATCG for DNA, or ACUAGCUAGCUCUUGAUCG for the corresponding RNA sequence. Note, however, that we will rarely give explicit RNA sequences in this book, opting instead to give the corresponding DNA sequence whenever discussing RNA, for notational simplicity. Hence, we will typically substitute T for U in what would otherwise be RNA sequences, and will therefore confine our discussions to sequences of letters drawn from the set {A,C,G,T}.

Because of the strict Watson–Crick base-pairing observed in normal DNA and RNA, it is possible to infer the precise sequence of nucleotides along one DNA strand when given the sequence of the other strand. For example, given the sequence ATCTAGGCA, the *reverse complement* sequence (i.e., the complementary bases, in reverse order) making up the opposite strand can be confidently deduced (except in extremely rare circumstances) to be TGCCTAGAT, where both of these sequences are given in the 5′-to-3′ order for the respective strands (remembering that 5′-to-3′ for one strand is 3′-to-5′ for the other – hence our convention of describing the opposite strand using the *reverse complement*, rather than just the *complement*). Thus, all the information encoded in a complete DNA molecule is present in one strand of that molecule (with a few exceptions, which we will not consider), and as a general practice, we will consider only one strand or the other, whichever suits our present needs. We shall at this juncture make only a brief remark on the profundity of this equating of sequence to information content. Although we will delay to subsequent chapters a precise definition of sequence information content, the astute reader will perceive that the very potential for DNA to assume arbitrary sequences of As, Cs, Gs, and Ts has important and far-reaching implications for the

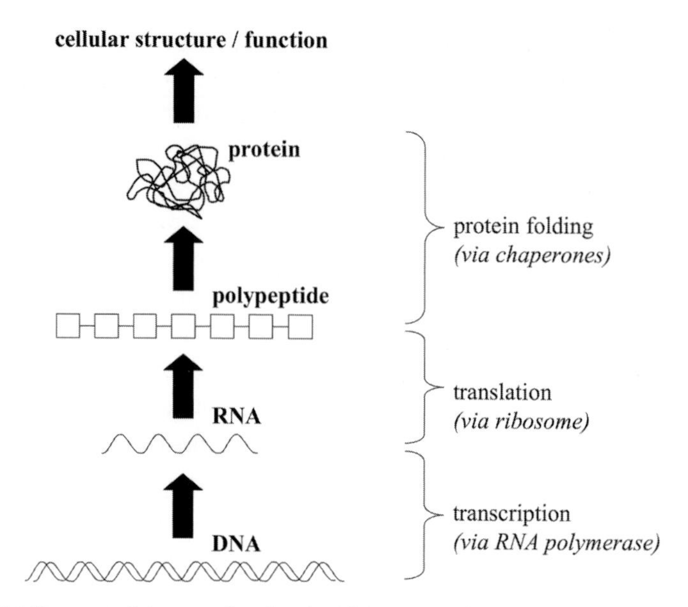

Figure 1.4 The central dogma of molecular biology. DNA is transcribed via RNA polymerase into messenger RNA; RNA is translated via a ribosome complex into a polypeptide; finally, the polypeptide is folded into a completed protein, one of the fundamental building blocks of living organisms.

robustness of the genetic code and of the evolutionary process and its associated transfer of genetic information between generations, even over millions of years of biological evolution.

It is the very existence of a discrete genetic code which allows us to treat in so straightforward a fashion the relations between genes and their protein products using rigorous mathematical techniques. We now come to the *central dogma of molecular biology* (Figure 1.4), which stipulates that DNA sequences give rise to *messenger RNAs (mRNAs)*, which are then translated into the *polypeptides* (chains of *amino acids*) that fold into functional proteins. These protein products perform most of the vital functions of the cell, and therefore of the organism; hence the great importance attached to this particular molecular pathway. Indeed, much of modern biomedical research is founded on the hope that a better understanding of the genes (both singly and in combination) will enable more successful intervention in the development of disease states, through the reshaping or altered regulation of relevant proteins.

Let us now consider in greater detail the processes involved in gene expression as illustrated in Figure 1.4. The first step involves the *transcription* of DNA into RNA. It is a curious fact that much of eukaryotic DNA is non-genic in many organisms; in all the vertebrate genomes sequenced to date, long stretches of DNA can be readily found which do not appear to stimulate the production of proteins or other biologically active molecules in the cell. Punctuating these *noncoding regions* (or *intergenic* regions) are the actual genes, or *loci* (singular: *locus*), which individually encode the functional units of hereditary information, and which in many species constitute only a small percentage of the organism's genome. Because a particular locus may take slightly different forms in the different individuals of a species (accounting for, e.g., some individuals having blue eyes and others having green or brown eyes), we may differentiate between these forms – called *alleles* – and their different effects on the biology of a particular organism. The differential effects and expression patterns of multi-allelic loci are of primary concern in the field of *quantitative genetics* (e.g., Falconer, 1996), in which *single-nucleotide polymorphisms (SNPs)* are often used as surrogates for the full complement of allelic variations in a population. For the purposes of computational gene prediction, however, these issues of individual variation are generally ignored in practice. Instead, given a nucleotide sequence along one DNA strand, the task is to identify the loci that are present in the DNA, and to delimit precisely the boundaries separating those loci from the noncoding regions which surround them, as well as the internal structure of each locus as defined by the *splicing* process (see below). It should also be noted that not all transcribed loci in a genome are protein-coding genes – that is, not all transcribed genes give rise to mRNAs which are translated into functional proteins. We will consider noncoding genes very briefly in Chapter 12.

The process of transcription is carried out by an *RNA polymerase* molecule, which scans one DNA strand in the 5′-to-3′ direction, pairing off each DNA nucleotide (A,C,G,T) on the antisense strand with a complementary RNA nucleotide (U,G,C,A). In this way, the DNA of a gene acts as a template for the formation of an RNA sequence from the free *ribonucleotides* (RNA nucleotides) which are present individually in the nucleus. The RNA nucleotides are joined together with phosphodiester bonds to produce an RNA molecule known as a *messenger RNA* – abbreviated *mRNA* – which will later migrate out of the nucleus into the cytoplasm of the cell, where it then acts as a template for protein synthesis. First, however, the emergent mRNA – known as a *pre-mRNA*, or a *transcript* – is generally spliced by a molecule known as the *spliceosome*, which removes intervals of RNA known as *introns* from the sequence. This process is illustrated in Figure 1.5.

Each intron begins with a *donor site* (typically GT) and ends with an *acceptor site* (typically AG). The entire intron, including the donor and acceptor *splice sites*, is excised from the mRNA by the spliceosome via a two-step process, in which the donor site is first cleaved from the preceding ribonucleotide and brought into association with a region upstream from the acceptor site known as the *branch point*, to produce a loop-like structure known as a *lariat*, and then in the second step, the acceptor site is cleaved from the following ribonucleotide, thereby completing the separation of the intron from the mRNA. The excised intron is discarded, to be degraded by enzymes in the cell back into individual ribonucleotides for use in future transcription events, and the remaining portions of the mRNA are *ligated*

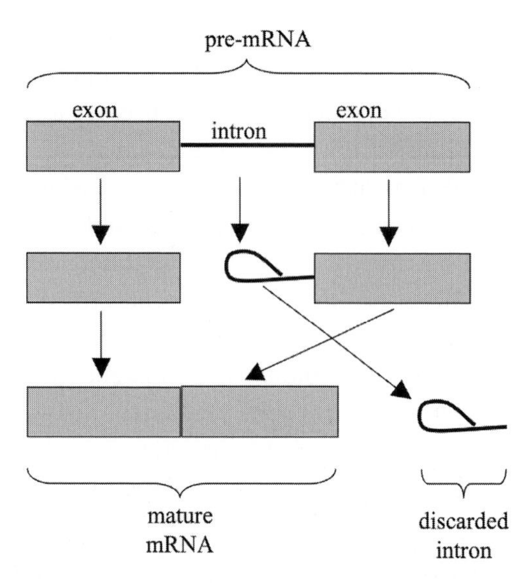

Figure 1.5 Splicing of a two-exon pre-mRNA to produce a mature mRNA. Splicing removes the introns and ligates the remaining exons together to close the gap where the intron used to reside.

(i.e., joined together by newly created chemical bonds – in this case, phosphodiester bonds) so as to close the gap created by the excised intron. The regions separated by introns are known as *exons*; these are the portions of the transcript that remain after all splicing of a particular mRNA has been completed, and which generally influence the resulting structure of the encoded protein.

Although every protein-coding gene contains at least one exon, introns are not always present. In *prokaryotes* (organisms whose cells lack a nucleus), introns do not occur, so that the identification of functional genes is synonymous with the identification of coding exons. Although we will largely limit our discussion to eukaryotes, it should be noted that even in the latter organisms, genes lacking introns can sometimes be found (though what may appear to be an intronless gene may often be a *retrotransposed pseudogene* – a mature mRNA which has been reverse-transcribed back into the chromosome in a random location and subsequently rendered nonfunctional through the accumulation of mutations; see Lewin, 2003). Another exceptional case in the biology of introns occurs in the case of *ribozymes* – introns that autocatalytically splice themselves out of a transcript – of which the interested reader may learn more elsewhere (e.g., Doudna and Cech, 2002)

The ends of the spliced transcript are additionally processed while still within the nucleus, as follows. The 5′ end of the mRNA is *capped* by the addition and methylation of a *guanosine*, to protect the mRNA from being destroyed by the cell's RNA degradation processes (which are carried out by a molecule called an *exonuclease*). At the 3′ end of the mRNA a *poly-A tail* (a long string of adenine residues) is appended after the transcript is cleaved at a point roughly 20–40 nucleotides

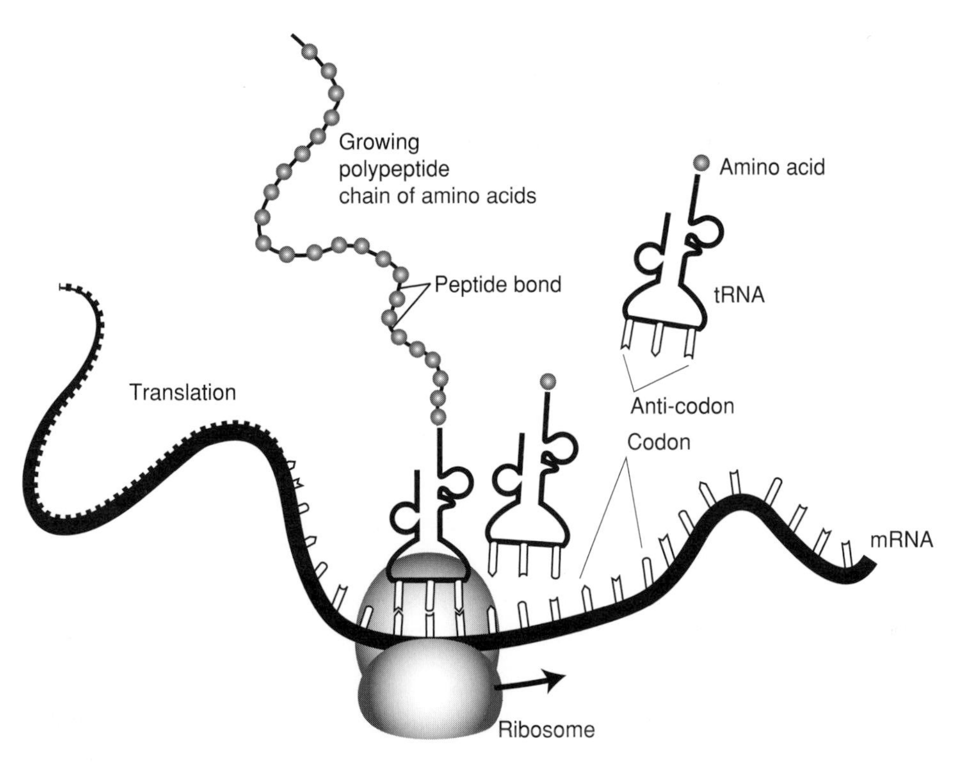

Figure 1.6 Protein synthesis. RNA is translated by a ribosome complex into a polypeptide chain of amino acids. The polypeptide will then fold into a protein structure. (Courtesy: National Human Genome Research Institute.)

downstream from a special *polyadenylation signal* (typically ATTAAA or AATAAA). The poly-A tail likewise protects the 3′ end of the mRNA from exonuclease activity, and also aids in the export of the mRNA out of the nucleus.

Once the splicing, 5′ capping, and polyadenylation processes are complete, the processed transcript – now known as a *mature mRNA* – can be exported from the nucleus in preparation for protein synthesis.

Figure 1.6 illustrates the process of *translation*, whereby an mRNA is translated into a polypeptide. At the bottom of the figure can be seen the *ribosome*, a molecular complex which scans along the mRNA, much as the RNA polymerase scans the DNA during transcription. Whereas in transcription a series of free RNA nucleotides are polymerized into a contiguous RNA molecule, the ribosome instead attracts molecules known as *transfer RNAs* (*tRNAs*) which aid in the formation of the polypeptide product. Each tRNA possesses at one end an *amino acid*, and at the other end an *anti-codon* which will bind only to a particular combination of three nucleotides. Such a sequence of three nucleotides is called a *codon*. Although there are 64 possible nucleotide triplets ($4 \times 4 \times 4 = 64$), there are only 20 amino acids (A = *alanine*, R = *arginine*, N = *asparagine*, D = *aspartic acid*, C = *cysteine*, Q = *glutamine*, E = *glutamic*

Table 1.1 The genetic code: beside each amino acid are listed the codons that are translated into that acid

Amino acid	Codons	Amino acid	Codons	Amino acid	Codons	Amino acid	Codons
A	GCA	G	GGA	M	ATG	S	AGC
	GCC		GGC				AGT
	GCG		GGG				TCA
	GCT		GGT				TCC
							TCG
							TCT
C	TGC	H	CAC	N	AAC	T	ACA
	TGT		CAT		AAT		ACC
							ACG
							ACT
D	GAC	I	ATA	P	CCA	V	GTA
	GAT		ATC		CCC		GTC
			ATT		CCG		GTG
					CCT		GTT
E	GAA	K	AAA	Q	CAA	W	TGG
	GAG		AAG		CAG		
	TTC						
	TTT						
F	TTC	L	CTA	R	AGA	Y	TAC
	TTT		CTC		AGG		TAT
			CTG		CGA		
			CTT		CGC		
			TTA		CGG		
					CGT		

acid, G = *glycine*, H = *histidine*, I = *isoleucine*, L = *leucine*, K = *lysine*, F = *phenylalanine*, P = *proline*, S = *serine*, T = *threonine*, W = *tryptophan*, Y = *tyrosine*, V = *valine*, and lastly M = *methionine* which is encoded by the *start codon* – see below), so that the mapping from codons to amino acids is *degenerate*, in the sense that two codons may map to the same amino acid (though one codon always maps unambiguously to a single amino acid; see Table 1.1). Furthermore, the frequencies of the various codons within protein-coding genes tend to be significantly nonuniform; this *codon bias* (i.e., profile of codon usage statistics) is an essential component of nearly all gene-finding programs in use today.

The operation of the ribosome involves advancing by threes along the mRNA sequence, allowing a free tRNA having the appropriate anti-codon (as determined by Watson–Crick complementarity) to bind momentarily to the currently exposed codon in the mRNA. The amino acid at the other end of this tRNA is then ligated to those amino acids that have already been donated by tRNAs for earlier codons

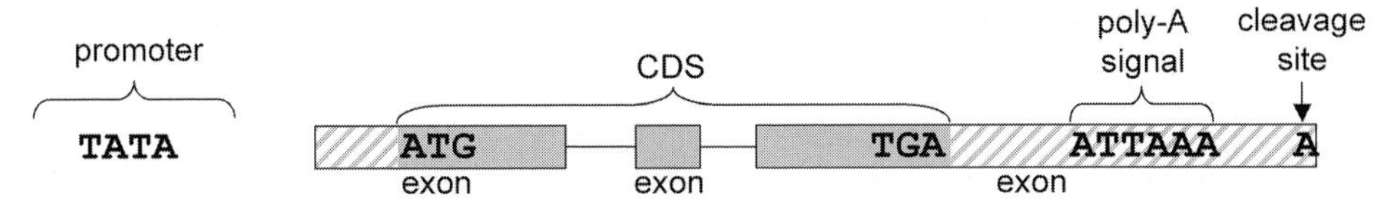

Figure 1.7 Structure of a typical human gene. A promoter sequence (including a TATA-box element in this example) is followed by the first exon of the gene, which may have an untranslated region (hatching). The coding region (solid gray) begins with a start codon (ATG) and ends with a stop codon; both the coding and noncoding regions of a gene may be segmented into multiple exons. The coding segment (CDS) ends at the stop codon (TGA in this example) and is followed at some distance by a polyadenylation signal (ATTAAA in this example). A short but variable-length interval follows, ending at a cleavage site (typically A).

in the mRNA sequence. These ligated amino acids form a *polypeptide chain*, which is released by the ribosome when the latter reaches the end of the mRNA. The actual translation process is terminated before the end of the mRNA, when the ribosome reaches a *stop codon* or *termination site* (in most eukaryotic genomes, one of TAA, TGA, or TAG – or in the alphabet of mRNAs, UAA, UGA, or UAG). Similarly, translation begins not at the very beginning of the mRNA, but at a special codon called a *start codon* or *translation initiation site* (ATG in eukaryotic DNA, or AUG in the mRNA (see Figure 1.7); in prokaryotic DNA, start codons can appear as ATG, GTG, or TTG – see section 7.5). We call the upstream end (i.e., the region translated earliest) of a polypeptide the *N-terminus*, and the downstream end we call the *C-terminus*. As stated above, ATG codes for the amino acid *methionine*; stop codons normally do not code for any amino acid (but see section 3.5.11 for a rare exception).

The final process in protein synthesis is the folding up of the polypeptide chain into its final three-dimensional shape. This process is typically aided by one or more other proteins known as *chaperones*. Although the folding process is thought to be highly reproducible (i.e., deterministic) for a given polypeptide, presence or absence of the necessary chaperones (and possibly other environmental variables, such as temperature or particular chemical concentrations) can potentially influence the final shape (and the stability of that shape) for a given polypeptide.

The final three-dimensional shape of a protein, as well as other properties such as *hydrophobicity* (tendency for avoidance of water), molecular weight, and electric charge, can all contribute to the biological function of the protein within a living organism. From the foregoing it should now be reasonably clear – in particular, from the fact that DNA acts as a template for mRNA, which in turn acts as a template for protein – that *mutations* (changes to the sequence) in genomic DNA can strongly influence the health of the organism. It is therefore of great importance for the advancement of modern medicine that the precise relationship between DNA sequence features and their contribution to specific aspects of organismal biology be elucidated.

Unfortunately, the relation between individual genetic features and their *phenotypic effects* (i.e., effects on the molecular and biological operation of the cell, and more generally on the form and behavior of the complete organism – collectively referred to as the *phenotype*) is exceedingly difficult to predict in advance, due to the enormous complexity of the cell. A key contributor to this complexity is the sheer number of molecules and potential types of molecular interactions which variously present themselves along the route from a particular gene to the ultimate fate of an encoded protein. Predicting these molecular interactions can be very difficult when only pairs of putative proteins are considered; indeed, simply predicting the three-dimensional structure of individual proteins from their amino acid sequences is still a very challenging computational problem, despite considerable and sustained attention from computational scientists over a number of years. Beyond this, given correct three-dimensional structures for a pair of proteins, the challenge remains to predict the set of possible molecular interactions (if any) that can occur between two proteins under normal cellular conditions, and the resulting biological significance of such interactions. In living cells, these interactions are often mediated by one or more additional molecules, as in the case of the chaperones which aid a protein to fold properly.

The very expression patterns of genes – i.e., patterns describing whether a gene is expressed in a given cell type, or during a certain phase of the cell cycle or of the organism's ontogeny – are themselves subject to regulation by other proteins (and other types of molecules) present in the cell. Specific regions of a gene (such as the *promoter* region 5′ of its start codon) can act as binding sites for various regulatory proteins which may chemically bind to the DNA in order to promote or repress the transcription of the gene. The propensity of these regulatory proteins to affect their target genes can vary between cell types and other conditions, as indicated above, so that the prediction of expression patterns from local sequence *motifs* (i.e., specific patterns of As, Cs, Gs, and Ts) can be highly unreliable. Although some forms of protein interactions can be predicted by identifying *protein domains* (motifs at the amino acid level), the reliability of these predictions can vary widely.

Thus, the problem of *systems biology* – of elucidating the large-scale operation of networks of interacting genes and their protein products – while essential to our understanding of molecular and organismal biology, promises to challenge scientists for some time yet. In the meantime, research into methods for reliably predicting gene regulatory mechanisms and small-scale metabolic pathways will depend to a great extent on our ability to first identify the precise locations and exonic structure (i.e., splicing patterns) of individual genes, as well as to catalog the set of transcribed genes (the *transcriptome*) and their protein products (the *proteome*) of individual organisms.

1.2 Evolution

Looking beyond the individual organism, much useful information can be gleaned from studying the evolutionary patterns between the genomes of related species. As first suggested over a century ago by Charles Darwin (Darwin, 1859), it is very likely the case that all life arose from a single ancestor, and that all extant organisms thus share a common genetic history via their lines of evolutionary descent. This common ancestry promises to become an enormously useful source of information for the biological interpretation of genetic patterns, once the processes of molecular evolution are better characterized (and more species' genomes are sequenced). In particular, because we expect each successive generation in an evolutionary lineage to retain to a high degree both the genetic information inherited from its parents as well as the molecular and biological mechanisms associated with those genes, we can posit that regions of elevated sequence similarity between a genome of interest and that of some suitably related organism are likely to indicate *coding potential* (i.e., to contain a corresponding density of exons of protein-coding genes). In this way, concepts from evolutionary biology can guide us in using the evidence provided by *sequence conservation* (i.e., sequence similarity) between related genomes for more accurate gene structure prediction.

We will elaborate on the topic of sequence conservation within the framework of *phylogenetics* (the study of evolutionary relationships between species) and more specifically *phylogenomics* (the study of evolutionary relationships between genomes) in Chapter 9. In the meantime, we will have recourse to use the following definitions. An individual species is referred to as a *taxon* (plural: *taxa*), and a "family tree" showing relationships between taxa is known as a *phylogeny* or a *phylogenetic tree* (Figure 1.8) A subtree of a phylogeny (i.e., some parent taxon and all its descendents) is called a *clade*. A pair of genomic features which are descended from a corresponding feature in some common ancestor are known as *homologs*, and their relation is termed one of *homology*. If the genomic features (not including the ancestor) reside in the same genome (i.e., if one resulted from the phenomenon of *gene duplication* – see Lewin, 2003), we call them *paralogs*, and say that they are *paralogous*, whereas if they occur in different genomes we call them *orthologs* and deem them rather to be *orthologous* (the more complex possibility, namely that the relation includes both orthology and paralogy, is an important one that we will, unfortunately, not have the space to address). In this way it should be seen that individual loci, like complete genomes, can be represented using phylogenetic models of descent, with the parent of a given node in the phylogeny representing the immediate ancestor of the child locus (subject to the resolution of the known ancestral relationships). Clearly, the sibling of any given locus may be an ortholog (in the case of *speciation* – see below) or a paralog (in the case of gene duplication).

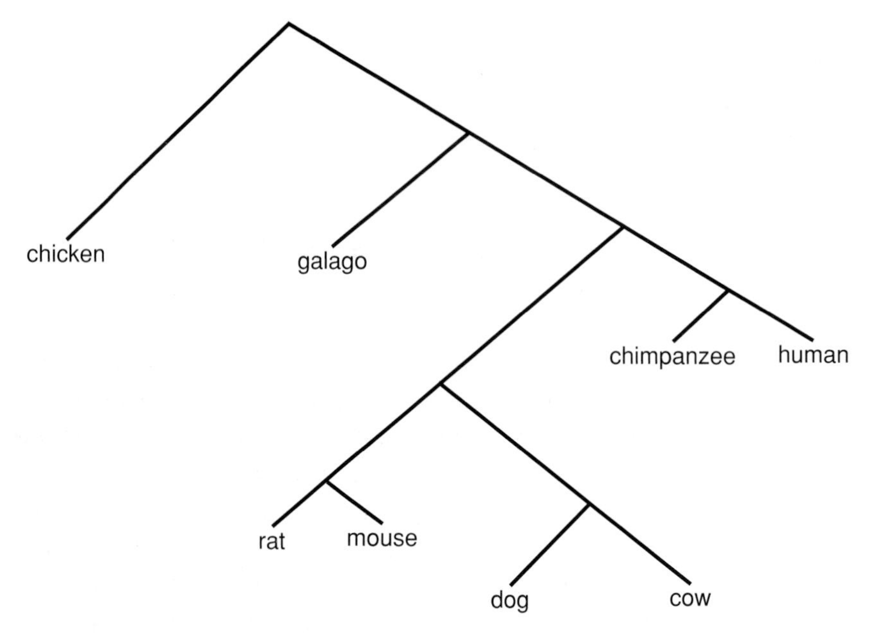

Figure 1.8 A phylogenetic tree inferred via sequence conservation. Specific taxa are shown at the bottom of the tree. Each branching point in the tree denotes the root of a clade. The species occurring in a clade share a common ancestor implicitly positioned at the root of the clade.

The phenomenon of *evolution* (or *descent with modification* – Darwin, 1859) can be understood as the result of two processes, both occurring in the context of naturally arising genetic variations: *genetic drift*, and *natural selection*. Whereas the former is caused by random fluctuations and *sampling error* (discussed in Chapter 2) between generations, effectively resulting in random, undirected evolution, the latter – natural selection – is by no means purely random (contrary to popular misinformation promoted by the anti-science establishment). Natural selection operates by permitting beneficial genetic mutations, when they arise by chance, to enjoy elevated transmission to succeeding generations (in terms of the proportion of the future population having the mutant allele), with the magnitude of that elevation in transmission fidelity varying roughly with the degree of benefit which the mutant phenotype confers on the organism, either in terms of enhanced survival tendencies or improved reproductive success (i.e., the degree of elevation in the organism's *fitness*). These phenomena all operate stochastically – that is, they are subject to some random fluctuations – but they are not purely random, since natural selection imposes a definite bias on allele frequencies which, given a sufficient number of generations, can result in observable changes in the average population phenotype.

Subpopulations of a species which are isolated (geographically, for example) for a number of years may eventually accumulate so many mutations that they are no longer reproductively compatible with other subpopulations, so that a new species is effectively formed – i.e., a *speciation* event has occurred. Comparison of the genomes of the newly divergent species would reveal differences corresponding to the effects of selection (i.e., due to *selective pressures* for or against specific mutations), as well as those resulting from mere genetic drift – i.e., mutations in nonfunctional regions of the genome, or even coding mutations which are *selectively neutral* due to their not conferring any *selective advantage* or disadvantage (or at least not a strong one) on the organism. Regarding the latter, the reader will recall that due to the degenerate nature of the genetic code – the mapping from codons to amino acids – most amino acids are encoded by more than one codon. Furthermore, in many cases the codons that code for a given amino acid will differ in only one of their bases (typically in the third position of the codon – the so-called *third-position wobble* phenomenon), so that some mutations in coding regions will change the resulting protein product, whereas others will not. The former we call *nonsynonymous* mutations, and the latter *synonymous* or *silent mutations*. The ratio K_s/K_a of synonymous mutations to nonsynonymous mutations in a putative coding segment should therefore reflect (to some degree) the likelihood that the segment is a coding region, since we would expect most mutations in coding regions to be silent. This of course assumes, however, that the homologous regions have been correctly identified in the two genomes, and that the homologs have not significantly diverged in function since the speciation event.

We will elaborate on some of these issues in Chapter 9 when discussing phylogenetic methods for comparative gene finding; the reader may also find the discussion of *genetic programming* in section 10.12 to be illuminating as an illustration of *in silico* evolution – that is, evolution in the computer.

1.3 Genome sequencing and assembly

Practical genomic applications depend on our being able to determine the precise sequence of *residues* (e.g., nucleotides, amino acids) comprising the chromosomes, genes, transcripts, and proteins of the target organism. This is most visibly manifested in the process of *genome sequencing*, by which we attempt to ascertain the precise nucleotide sequence comprising the chromosomes of a target organism (Pop *et al.*, 2002). Although DNA is natively a double-stranded molecule, the uniqueness of the base-complementarity map (ignoring the exotic G–T pairings) allows us to simplify the problem to that of determining one strand only. Thus, the ultimate goal of genome sequencing is to produce a string of As, Cs, Gs, and Ts denoting the nucleotide sequence of one strand of each chromosome.

The range of options for whole-genome sequencing is generally dichotomized into the *BAC-by-BAC* method versus *whole-genome shotgun* (*WGS*) sequencing (Venter *et al.*, 1996); we will describe the latter, since it is arguably more efficient and because it serves our present purposes as well as any other. In the WGS method, cells of the target organism are harvested and their DNA is extracted using various technologies which need not concern us here. Because eukaryotic chromosomes can be very large, and because current sequencing technologies are applicable only to relatively short sequences, the target genome must be fragmented into short pieces which can be individually sequenced and then reassembled (by computer) into a complete genomic sequence.

The method is necessarily complicated by the fact that current biotechnology techniques are still quite primitive, relying largely on the cellular reproductive machinery of bacteria and the gross electromagnetic properties of organic molecules. The first task is to shear the chromosomal DNA into much smaller pieces which can be reliably sequenced. One common method of shearing DNA is by forcing it at high pressure through a nozzle or syringe. Ideally, the resulting DNA fragments would have random lengths. Next, the fragments must be cloned. For this purpose we generally utilize a *cloning vector* such as a bacterium or yeast. Bacterial vectors, which are circular, are cut open using a *restriction enzyme*, and then the fragments from our genome of interest are inserted into the vectors via an enzyme known as a *DNA ligase*. The vectors are reinserted into living cells and the cells are allowed to multiply, thereby cloning the fragments which have been inserted into the vectors.

When billions of copies have been made, the cells are killed and the vectors are extracted and their two strands denatured into single-stranded DNA. Special fluorescently dyed nucleotides (as well as nonfluorescent nucleotides) are added to the mixture and allowed to form a complementary strand via Watson–Crick base-pairing. The new fragments are then stimulated by an electric current to migrate through a gel-filled capillary. Steps are taken to ensure that all of the molecules passing through the capillary during a single run will have originated from the same point in the genome, but due to the stochastic nature in which the complementary strands were formed, they will tend to differ in length.

Furthermore, because of the chemical properties of the specially engineered fluorescent nucleotides, each fragment contains exactly one fluorescent nucleotide, which is positioned only at the end of the fragment. Because the differential migration velocity of the fragments of different lengths along the electric current is very precisely determined, the fluorescent bases which reach the end of the capillary will be effectively sorted by their position in the genome, so that the time order of their arrival at the end of the capillary is correlated in a predictable way with their position along the genome. A laser directed at the end of the capillary causes the terminal nucleotides to fluoresce in one of four colors specific to each of the four nucleotides, and the color of each florescence is captured by a detector and transferred to a printout called a *chromatogram*, or *trace file* (Figure 1.9).

Figure 1.9 Trace image from a sequencing machine. Colored peaks (shown here in shades of gray) at the bottom are used to infer identities of individual nucleotides. (Courtesy: U.S. Department of Energy Human Genome Program.)

Using a program known as a *base-caller*, we can then infer the original sequence of nucleotides from the peaks in the trace diagram. In this way we infer the sequence of each segment of the fragmented genome; through appropriate oversampling and appropriately randomized fragmentation of the genome, one can hope that these fragments can then be reassembled to produce the original genomic sequence. A number of practical issues arise at this stage, however. The first is the fact that the sequencing reaction works only for relatively short sequences. This is in fact the reason we fragmented the genome in the first place, rather than simply sequencing each chromosome from end to end. The second complication arises out of the fact that the ends of the reads (where a *read* is a sequence of base-calls for a given fragment) tend to be less reliable than the central portion of the read. Thus, a *clear range* – interval of high confidence – is assigned to each read to delimit the low-confidence peripheral regions. Areas outside the clear range may be permanently trimmed from the read or down-weighted in importance as the reads are later assembled (see below) into a consensus sequence. Other issues related to the chemistry of DNA may interfere, such as the existence of *heterochromatic regions* of chromosomes, about which the interested reader can read elsewhere (e.g., Hoskins *et al.*, 2002).

It should be intuitively obvious that the fragments need to overlap in order for the later assembly process to infer correctly the order of the reads. Thus, it is necessary that more than one copy of the genome be included into the assay; the random nature of the fragmentation process will tend to produce a degree of overlap among fragments from different genome copies, with a larger number of genome copies giving rise to a higher overall degree of overlapping. It is thus possible to derive a formula for the expected percentage of the genome covered by at least one read, given the number of copies of the genome which have been sequenced, according to a nonlinear relation (Lander and Waterman, 1988).

At the resolution of current sequencing technologies, reads tend to be roughly 500 bp (*bp* = base pairs) long on average. Because the fragments are often longer than this, reads are generally taken from the ends of the fragments, and the pairing of fragment ends is utilized later to infer the order and orientation of *contigs* in a *scaffold* (see below).

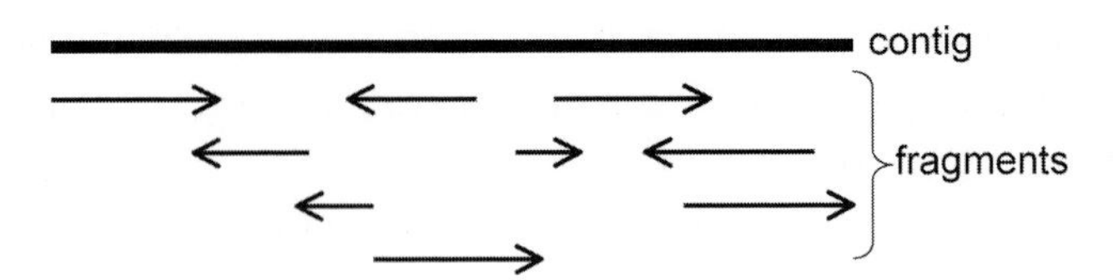

Figure 1.10 Assembling overlapping fragments into a contig. Fragments are shown as directed arrows to indicate which strand of the DNA they represent. The sequence comprising the contig represents the consensus of the fragments.

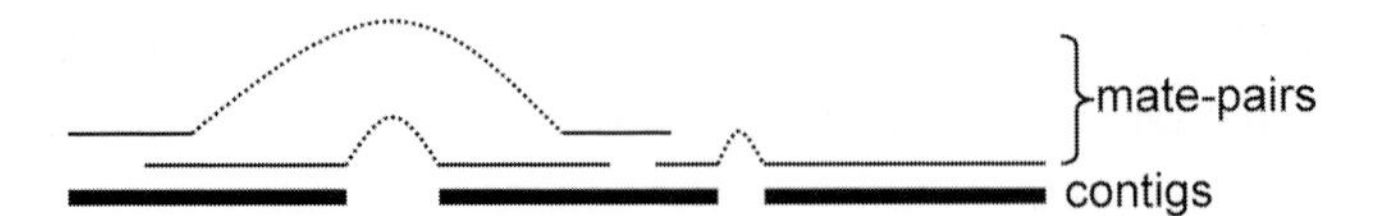

Figure 1.11 A single scaffold consisting of three contigs. The order and orientation of the contigs in the scaffold are inferred through the connectivity of the mate-pairs.

Once the reads have been completely sequenced, *in silico* reassembly of the fragments into a complete genomic sequence can be attempted using computational methods. The first step is to combine overlapping reads into a contiguous consensus sequence known as a *contig*. Because the orientation of the vector inserts cannot be easily controlled, the sequenced fragments may individually come from either DNA strand. This is illustrated in Figure 1.10, in which the fragments are denoted by directed arrows denoting *forward strand* (rightward-directed arrows) or *reverse strand* (leftward-directed arrows) fragments, where one of the two strands of the template DNA is arbitrarily designated the forward strand.

A fundamental problem in the *in silico* reassembly of genomic reads is that of *genomic repeats* – multiple regions of the genome that are nearly identical, such as those resulting from the movement of *transposons* within the genome (Lewin, 2003). The sequence similarity of repeats creates an ambiguity in the assembly process which often cannot be easily resolved, and may result in assembly errors. In the case of the human genome, repeats may account for as much as 35% of the sequence (Venter *et al.*, 2001). A further difficulty is the occurrence of *gaps* – regions of the genome which, through sampling error and other technical difficulties, receive no coverage by any of the sequenced reads. The locations of gaps can often be inferred by constructing *scaffolds*, through the utilization of *mate-pair* information, as illustrated in Figure 1.11.

Recall that for long fragments, both ends of the fragment may be sequenced, so that the order and orientation of such mate-pairs can be used as additional information in the assembly process. In particular, the order and orientation of any mate-pair straddling a gap between contigs can be used to infer the order and relative orientation of the corresponding contigs, resulting in a *scaffold* structure. The remaining gaps in the scaffolds (and between scaffolds) may be resolved through a process referred to as *gap closure*, in which gap regions are specifically targeted during a resequencing phase. For *draft genome sequencing* projects, gap closure is typically not performed, due to budget constraints.

Genome sequencing methods can also be applied to mRNAs culled from cells, producing what are known as *ESTs* (*expressed sequence tags*) if only part of the mRNA is sequenced, or *cDNAs* (*complementary DNAs*) if the entire mRNA is sequenced. ESTs and cDNAs can be used to great effect to improve *genome annotation* accuracy (see section 1.4), though some care must be taken in the use of ESTs since EST databases tend to contain many errors. Because ESTs generally originate from mature mRNAs, their introns usually have been spliced out, so that a *spliced alignment* program such as *sim4* (Florea *et al.*, 1998) or *BLAT* (Kent, 2002) must be employed before a set of ESTs can be used as evidence during gene prediction; we describe the use of ESTs for comparative gene finding in Chapter 9.

For the purposes of gene finding, it is important to note that current sequencing technologies occasionally result in the omission from the emitted sequence of an individual base in the sequenced DNA; these deletions of single bases can cause *frameshifts* in coding segments, in which one of the positions of a codon is inadvertently skipped and the remaining portions of the coding sequence are thereby misinterpreted as coding sequences residing in an incorrect *phase*. The relevance of this fact will become clear when we consider *phase-tracking* in putative coding sequences. The notions of *frame* and *phase* will be more precisely defined in section 3.3.

1.4 Genomic annotation

Once a genome has been sequenced and assembled into a set of consensus contigs and/or scaffolds, we can then contemplate the problem of searching the resulting sequences for evidence of functional exons and genes. This is precisely the problem which will occupy the remainder of this book, though we will be confining our attention in subsequent chapters to fully automated means of approaching this task, and we will primarily focus on eukaryotic genes (the special case of prokaryotic gene finding is considered in section 7.5). For the present, however, it is instructive to consider the task of *manual annotation* (also called *curation*), since the ultimate consumers of computational gene predictions are quite often the human annotators who are charged with preparing the best possible set of gene identifications given the available evidence. Thus, we distinguish between the *predicted genes* obtained from a gene-finding program and the *curated genes* which have undergone manual curation by a human annotator.

Unfortunately, the problem of eukaryotic annotation is sorely plagued with ambiguity. As we explained previously, the protein-coding portions of eukaryotic genes follow a fairly strict pattern beginning with a start codon (ATG), ending with a stop codon (one of TAG, TAA, or TGA in most eukaryotes), and having embedded within them zero or more introns delimited by donor sites (usually GT) at their 5′ ends and acceptor sites (usually AG) at their 3′ ends. Yet, as the reader may have guessed, not all ATG triplets occurring in a DNA sequence constitute functional start codons, nor do all GT and AG dinucleotides in an mRNA denote functional splice sites as recognized in practice by the eukaryotic spliceosome. For this reason, the problem of accurately identifying the precise series of *signals* (e.g., start codons, splice sites, and stop codons) comprising real genes residing on eukaryotic chromosomes is a highly ambiguous one. In practice, the most accurate *annotations* (identifications of genes and their exon–intron structures) are still produced by human annotators with access to multiple forms of evidence, rather than by purely computational prediction methods based only on the DNA sequence.

The process of manual genomic annotation, in which gene boundaries are delimited, their exon–intron structures are identified, and the likely biological functions of the gene products are stated, is more formally called *functional annotation*. Functional annotation is a black art, governed by at best a set of rules of thumb. For our purposes we will be largely concerned with the role that computational predictions of gene structure play in the manual annotation process.

Proper functional annotation is based on the full body of available evidence, with the types of potential evidence ranging from computational predictions based only on sequence patterns (termed *ab initio* predictions), to similarities of the DNA sequence to previously described genes, to protein-level similarities with known proteins from the same or another organism. In addition, we may consider more generally the patterns of evolutionary conservation between the target sequence and the genomes of related organisms, as well as the patterns of *genomic repeats* (section 1.3). Indeed, the seasoned genome annotator sees a given genome not as a static entity, but as a snapshot in time of a dynamical system evolving endlessly in response to the selective pressures – both internally within the genome and externally at the level of organism phenotypes – which constrain the content and structure of the genome. As such, the annotation process must ultimately be an integrative one in which an array of biological information – as much of it gleaned from the published biological literature as from sequence databases – is combined within the framework of current biological understanding to produce annotated gene structures which best fit the available data.

An indispensable tool for manual curation is the *genome browser*, a software framework in which multiple forms of evidence may be visualized simultaneously in the context of a particular stretch of the target genome. A number of genome browsers exist, though most of them have recognizably similar interfaces. Most of them feature a view of the data in which a region of target DNA is shown as a horizontal *track* within the workspace, with other tracks above (representing the forward strand) or below (representing the reverse strand) denoting similar genes,

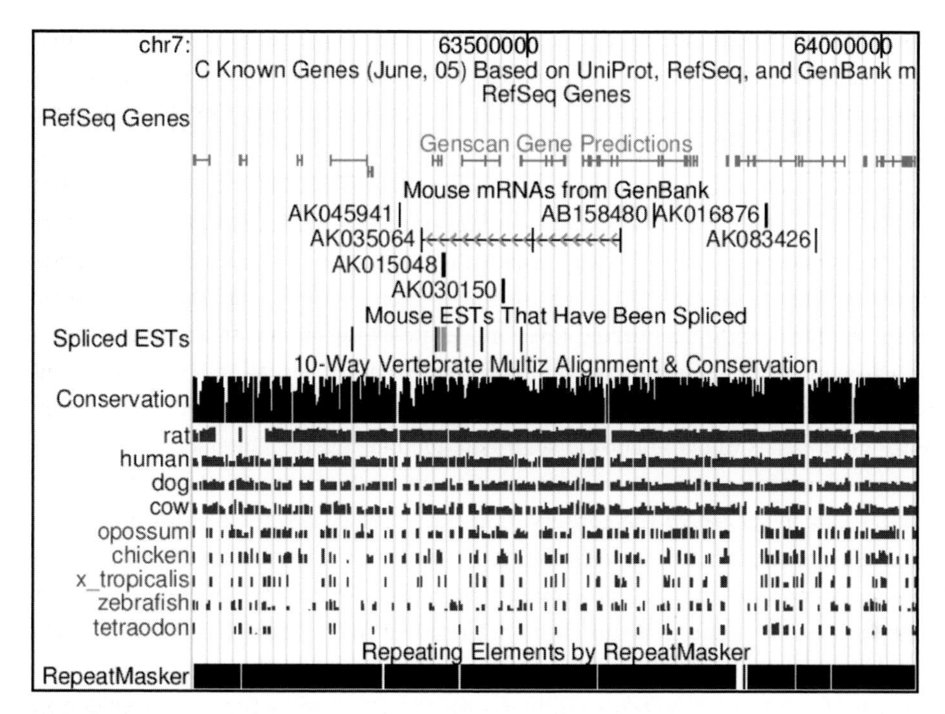

Figure 1.12 Snapshot of the UCSC Genome Browser (http://genome.ucsc.edu) on a region of mouse chromosome 7.

ESTs, cDNAs, proteins, and repeats, using some system of rectangular boxes and icons to show the boundaries of the denoted *features* (i.e., the similar genes and proteins, etc.) relative to the target genome.

Figure 1.12 shows a snapshot of a popular internet-based program known as the *UCSC Genome Browser* (Kent *et al.*, 2002). At the top of the screen are shown the genomic coordinates, which for this example range around the 63 Mb (*Mb* = *megabase*: one million base pairs) point along mouse chromosome 7. Below this are a number of tracks drawn from the *RefSeq* database (Pruitt *et al.*, 2005). The first is a set of computational gene predictions from the popular program *GENSCAN* (Burge and Karlin, 1997). The predictions are shown as tall, narrow boxes (representing putative exons) connected by horizontal line segments (representing putative introns). It should be noted that GENSCAN predictions, as with most other *gene finders* (computational gene-prediction programs), include only the *coding segments* (*CDSs*) of the putative genes; that is, they denote only the region extending from a putative start codon to a putative stop codon, and do not explicitly indicate the *untranslated regions* (*UTRs*) of the putative mRNA that flank the coding regions at the 5′ and 3′ ends of the gene. Though UTRs are known to be of great biological relevance (e.g., they are often targeted by regulatory proteins and *micro RNAs* (*miRNAs*) which can modulate the transcription and translation of the gene), they are more difficult to predict than coding segments, and are therefore rarely annotated (with any confidence). Beneath the GENSCAN predictions are shown a set of similar mouse mRNA sequences (i.e., cDNAs) and ESTs drawn from the *GenBank* database (Benson *et al.*, 2005).

The lower half of the display is taken up by the *conservation tracks*, which reflect the level of sequence similarity between the target genome (mouse in this case) and the sequenced genomes of various other organisms. Because of their great importance to the health of the organism, the protein-coding regions of a genome tend to be most highly conserved (over moderate evolutionary distances), through the action of natural selection, so that the patterns of conservation between related species along a stretch of homologous DNA can be very good indicators of the locations of genes and the coding segments of their exons. This fact is central to the operation of *comparative gene-finding* techniques, which we will examine in great detail in Chapter 9. Below the conservation tracks are shown the predictions of the *RepeatMasker* program (Smit and Green, 1996), which identifies sequence motifs having significant similarity to known genomic repeats. Although repeats may occur within genes, they tend to be much more prevalent in intergenic regions, and thus provide another line of evidence for the identification of likely gene boundaries.

Much of the similarity information displayed in a genome browser is produced via a popular family of programs known collectively as *BLAST* (*Basic Local Alignment Search Tool* – Altschul *et al.*, 1997; Korf *et al.*, 2003). Given an interval of DNA sequence, BLAST quickly searches through large databases of known DNA, RNA, or protein features, and returns the set of features with *statistically significant* similarities (see section 2.9 for a discussion of statistical significance) – i.e., levels of similarity which, under some statistical model, are deemed likely to indicate a common ancestry for the two sequence features. That is, the features identified by BLAST may be presumed to represent *evolutionarily conserved* (and therefore possibly biologically significant) genomic elements, such as exons or regulatory regions of genes. Some common nucleotide and/or protein databases searchable by BLAST are: GenBank, RefSeq, *dbEST* (Boguski *et al.*, 1993), *GenPept* (Burks *et al.*, 1991), *SwissProt* (Bairoch and Apweiler, 1996), *TrEMBL* (Bairoch and Apweiler, 1996), and *PIR* (Wu *et al.*, 2003).

The major programs in the BLAST family are *BLASTN*, *BLASTP*, *BLASTX*, *TBLASTN*, and *TBLASTX*. Each program takes a query sequence as input and produces a set of *hits*, or *HSPs* (*high-scoring segment pairs*) which document the similarity between the *query* and the matching *subject sequences* found in the database. BLASTN performs this search at the nucleotide level – i.e., it searches a nucleotide sequence database for matches to a nucleotide query sequence. BLASTP instead searches a protein database for matches to a protein query. BLASTX allows a protein database to be searched using a nucleotide query sequence; the program translates the query in all six reading frames (i.e., three possible codon positions on two strands, resulting in six combinations) into possible protein products and then uses these hypothetical protein fragments to search the protein database. TBLASTN takes the opposite approach: it searches for matches to a query protein in a nucleotide database by performing six-frame translations of the subject sequences. Finally, TBLASTX searches a nucleotide database using a nucleotide query sequence, by performing six-frame translation on both the query and the subject sequences, and then assessing match quality by comparing the hypothetical proteins from the six-frame translations. The scoring of putative protein matches utilizes a *substitution matrix* (see section 9.6.4) to assess similarities between amino acids; these similarity measures attempt to take into account the different propensities for amino acids of one type to be modified over evolutionary time into an amino acid of another type.

Various other tools and databases are available for searching for known patterns in DNA and proteins. Prominent among these are the *Prosite* (Bucher and Bairoch, 1994), *Prints* (Attwood *et al.*, 2003), and *Blocks* (Henikoff *et al.*, 2000) protein motif databases (now unified under *InterPro*: Apweiler *et al.*, 2001), and the *Pfam* (Bateman *et al.*, 2004), *SMART* (Schultz *et al.*, 1998), and *ProDom* (Servant *et al.*, 2002) protein *domain* databases. Another important resource is *TRANSFAC* (Wingender *et al.*, 1997), which consists of a number of motifs for well-characterized *transcription factor binding sites* – locations where regulatory molecules known as *transcription factors* can bind to a gene or a gene's *promoter* (a region upstream from a gene which regulates its transcription by serving as the binding site for the polymerase or its intermediaries such as transcription factors). All of these are useful in identifying potential functional sites in a putative locus.

An essential task in gene annotation is ensuring that the coding region of a gene consists of an *open reading frame*. Recall that the CDS (*coding segment*) of a gene consists of a series of codons, beginning with a start codon and ending with a stop codon. Although not all ATGs in a sequence constitute true start codons as recognized by the ribosome, in most eukaryotic organisms the codons TAG, TGA, and TAA will terminate translation of the current transcript wherever they occur (as long as they occur at a position which is a multiple of 3, relative to the translation initiation site in the mature mRNA). Thus, a CDS must contain a stop codon at its end and *only at its end*, since a stop codon within the interior of the CDS – at a distance from the start codon which is a multiple of 3 – would terminate the CDS at that point. Thus, we say that the CDS portion of a gene may not contain *in-frame stop codons*; we will formalize the notions of in-frame stop codons and open reading frames in Chapter 3.

Other sources of information during the annotation process include genomic markers, such as the *cytogenetic bands* which are visible when viewing a chromosome through a high-powered microscope, and clusters of ESTs, such as those contained in the *UniGene* (Pontius *et al.*, 2003) and *HomoloGene* (Wheeler *et al.*, 2005) databases, or provided by such programs as *PASA* (Haas *et al.*, 2003) or *CRASA/PSEP* (Chuang *et al.*, 2004).

Each of these forms of evidence has its own special place in the hierarchy of reliable annotation evidence. The experienced genome annotator takes this reliability ranking into consideration when considering the body of evidence for a particular gene annotation. For example, full-length cDNAs aligned with high stringency to the genome are generally taken to be much more reliable than imperfect matches to distant protein homologs. Computational gene predictions (especially *ab initio* predictions, which are not based on any expression evidence such as ESTs or cDNAs) are generally considered the *least* reliable form of evidence by experienced annotators, and this very fact should strike a sobering chord in any computational gene-finding researcher, as it is a strong indication of the need for improvement in this field.

A number of biological phenomena often conspire to mislead the unwary annotator. One of these is the existence of *pseudogenes*, which are genomic constructs that appear in many respects to be real genes, but that are in fact (in most cases) the genomic fossils of genes which presumably were once functional but now are defunct. Pseudogenes tend to have codon statistics that resemble those of real genes, but they often contain in-frame stop codons or invalid or rare splice sites. At present the very definition of a pseudogene is subject to some debate, and additional work is needed to define more rigorous methods for their identification. A different but equally misleading phenomenon is that of gene *shadows*, which occur when a functional gene residing in one reading frame induces a statistical pattern in one of the other two frames which resembles the statistical pattern of a real gene, though no genes actually exist in these other frames. This latter phenomenon is particularly relevant to the problem of accurately detecting short coding segments via statistical methods.

Another problem stems from the phenomenon of *alternative splicing*, in which a single locus may give rise to multiple (mature) mRNAs and multiple distinct protein products, depending on precisely how the spliceosome splices the transcripts of the locus. Although we will, unfortunately, be able to give little guidance with respect to this phenomenon in this edition (see section 12.1), much attention is at present focused on this very problem, since the phenomenon of alternative splicing is now known to be very prevalent in human genes (possibly affecting as many as 80% of human loci – Matlin *et al.*, 2005), and also because current state-of-the-art gene prediction methods disregard this issue and generally predict only one *isoform* (i.e., one possible splicing pattern) for each predicted gene.

Once the intron–exon structure of a putative gene has been annotated by a human annotator, some attempt is usually made to assign a probable function to the gene, usually based on the most similar BLAST hit to the nucleotide or amino acid sequence of the annotated gene. If a novel gene appears to be significantly similar to some well-characterized gene in a related genome, the annotator may feel confident in assigning to the novel gene the same function as the known gene from the related organism. Annotation of function typically takes the form of assigning one or more categories from the *gene ontology* (*GO* – Gene Ontology Consortium, 2000). In cases where a specific function cannot be confidently assigned, putative gene isoforms are often characterized into one of many *protein families* which effectively circumscribe the possible protein functions without explicitly indicating a particular function for a particular molecule.

Finally, a symbol must be assigned to each putative gene by the annotator. If the gene appears to have been previously identified (perhaps at the protein level but not at the level of genomic DNA), then the assignment is simply one of synonymy to an existing database entity. If instead the gene appears to be uncharacterized, the annotator may assign a unique name and/or a unique database identifier (ID). Database links may also be created to link the new gene to other known genes or to suspected homologs or protein families. Gene identifiers are collected into databases such as *LocusLink* (Maglott *et al.*, 2000) and *HUGO* (Wain *et al.*, 2002).

A short but useful overview of genome annotation may be found in Stein (2001).

1.5 The problem of computational gene prediction

Although we will postpone a rigorous definition of the task of gene structure *parsing* to Chapter 3, from the foregoing it should be clear that the task is in no way a trivial one. As current estimates put the proportion of the human genome comprising actual coding sequences at roughly 1%, it is perhaps not terribly inaccurate to describe the problem as one of finding a needle – by which we mean precisely defining its coordinates, orientation, and internal structure – in the proverbial haystack of genomic DNA (which for humans consists of 2.9 *billion* nucleotides). While the earlier portions of this book will focus on the task of *ab initio* prediction – i.e., prediction based solely on the sequence of As, Cs, Gs, and Ts making up a genomic sequence – we will in due course introduce additional forms of evidence and show how these can substantially increase our ability to accurately demarcate the boundaries of genes and exons in genomic DNA.

As will be seen in later chapters, most of the techniques which we will be employing for this task are *statistical* in nature – that is, we will be observing the frequencies of various patterns that occur in confirmed coding segments (particularly the frequencies of individual codons), and using these to estimate the probability that a particular interval of a novel sequence contains an exon or some other gene feature. Because mRNAs are transcribed from the reverse strand of the DNA where a gene is located (i.e., the "template" is reverse complementary to the actual gene), and because transcription forms an mRNA which is the reverse complement of the template DNA, the resulting mRNA will be in the same sense as the DNA strand containing the gene (i.e., the reverse complement of the reverse complement of a sequence is the same as the original sequence), so that all of our computations can be performed on the original DNA strand. That is, the DNA sequence forms the primary substrate for our computations, even though it is implicitly acknowledged that some of the phenomena which we will be modeling (e.g., splicing, translation) actually occur not at the level of DNA, but rather at the level of mRNAs and proteins.

EXERCISES

1.1 Give the reverse complement (i.e., the Watson–Crick complementary sequence, in reverse order) of the following sequences:
  (a) TAGCTGATCTATCTCGCGCATTATCGTATTCTCAGTGATCTCTTTTACCCAG
  (b) ATGGATTTATCTGCTCTTCGCGTTGAAGAAGTACAAAATGTCATTAATGCTATG
  (c) GCGATCTCGATCTATCTCTCTCTTAGGCGCTAGCCGATTCTAGCTGCTATCTAA
  (d) GCGCGATCTATATATCGCGATCGTACTGACTGACTGACTGACGCTATCTAGCCG
  (e) ATATAGCTGCAGGAGCTTATTATCTCTATATAGCGTATCTGATCGATGCTAGCT
  (f) CTCTCTAGATTATATAGCTACGTAGCTGACTGACTGCGTGCTATAGCTACTATT

1.2 At the NCBI website (www.ncbi.nlm.nih.gov/) perform BLASTN searches for the sequences given in Exercise 1.1. Familiarize yourself with the BLAST reports which result.

1.3 Write a parser module in your favorite programming language to parse BLAST output. Your parser should retain the significant HSPs and their scores.

1.4 The human genome consists of 23 pairs of chromosomes. Perform a literature search to find out how many chromosomes there are in the genomes of the following organisms. Also provide an estimate of the genome sizes, in bp.
  (a) Chimpanzee (*Pan troglodytes*)
  (b) Dog (*Canis familiaris*)
  (c) Mouse (*Mus musculus*)
  (d) Chicken (*Gallus gallus*)
  (e) Cow (*Bos taurus*)
  (f) Zebrafish (*Danio rerio*)
  (g) Yeast (*Saccharomyces cerevisiae*)
  (h) Rice (*Oryza sativa*)
  (i) Fruitfly (*Drosophila melanogaster*)
  (j) Worm (*Caenorhabditis elegans*)
  (k) Thale cress (*Arabidopsis thaliana*)

1.5 Find a site on the internet allowing you to run the RepeatMasker program on an uploaded sequence. Use it to mask low-complexity repeats from the sequences given in Exercise 1.1.

1.6 Visit the UCSC genome browser website (http://genome.ucsc.edu). Navigate the genome browser to the region shown in Figure 1.12. Turn on all available tracks in the browser. List the tracks which are currently available which are not shown in Figure 1.12.

1.7 In the sequences given in Exercise 1.1, find all possible start codons, stop codons, and splice sites. Count the number of possible introns in each sequence.

1.8 Repeat the previous exercise for the reverse-complementary sequences that you computed in Exercise 1.1. Compare the number of start codons, stop codons, and splice sites in the original sequences versus in their corresponding reverse-complementary sequences. Group the results by sequence.

1.9 Extend the phylogeny shown in Figure 1.8 to include all of the species in Exercise 1.4, based on:
   (a) your intuition;
   (b) a search of recent literature;
   (c) sequence similarity, via the PHYLIP program (Felsenstein, 1989) or a similar program approved by your instructor.

2

Mathematical preliminaries

In this chapter we will briefly review the basic mathematical machinery and notational conventions that it will be necessary to know in order to most effectively absorb the rest of the material in this book. It will be of use primarily to those lacking a strong computational background. We assume only that the reader has knowledge of algebra and has at least some minimal facility with a structured programming language. In section 2.4 we will summarize the syntax and semantics of the pseudocode which we will be using when describing algorithms, and which should be largely intuitive to anyone familiar with a modern programming language such as C/C++, Java, Perl, or Pascal.

The sections that follow will be, we believe, appropriately brief; readers requiring further elaboration of any of the topics in this chapter are encouraged to pursue the cited references. Conversely, readers with a strong computational background may skim this chapter, noting only the typographical conventions that we have opted to employ.

2.1 Numbers and functions

A number of standard constants and functions will be utilized in the chapters ahead. The two infinities, ∞ and $-\infty$, should be familiar enough; we will use them to denote the largest and smallest values, respectively, that can be represented in a particular computer system. As computer architectures advance, these respective values may grow, and so it is advisable when writing a gene finder to use an appropriate portable means of referring to these values in source code.

The Euler number, $e \approx 2.718\ldots$, will be useful when converting to and from logarithms of base e. The latter will be denoted either $\log_e(x)$ or $\log(x)$, with or without the parentheses, so that the default base will be taken to be e. In many languages one can obtain an approximation to e by using a function call of the form $exp(1)$. The base of the logarithm function, though for some purposes arbitrary, is in other cases rather important. Different programming languages provide different defaults for this parameter. The following identity is useful for converting between bases of logs:

$$\log_b(x) = \log_a(x)/\log_a(b). \tag{2.1}$$

Because the log transformation (i.e., replacing x with $\log(x)$) will prove so indispensable to us later, it is worth noting several other identities involving logs:

$$\log(a^b) = b\log(a), \tag{2.2}$$

$$\log(ab) = \log(a) + \log(b), \tag{2.3}$$

$$\log(a/b) = \log(a) - \log(b), \tag{2.4}$$

$$b^{\log_b x} = x, \tag{2.5}$$

$$\log_b(b) = 1. \tag{2.6}$$

The log of zero is undefined; in many languages log(0) evaluates to $-\infty$.

When formally describing a function $f(x)$, we will on occasion specify the *domain* A and *range* B of the function using the notation:[1]

$$f : A \mapsto B. \tag{2.7}$$

In this way, if we agree that $\mathbb{Z}$ is the set of all integers, $\mathbb{N}$ is the set of nonnegative integers (including zero), and $\mathbb{R}$ is the set of real numbers, then $f : \mathbb{Z} \mapsto \mathbb{R}$ (for example) specifies that f is a function taking an integer argument (or *pre-image*) and returning a real value (or *image*). In the case of $y = f(x)$, x is the pre-image and y the image. We can also denote the domain and range of f as *dom*(f) and *range*(f), respectively. When considering an invocation of a function without regard to the particular arguments which are provided, we may use the notation $f(\bullet)$, where $\bullet$ denotes an arbitrary argument value or list.

We will only rarely have need to refer to the *inverse* of a function. The *inverse* of f, denoted f^{-1}, is that function having the property $f^{-1}(f(x)) = x$ and $f(f^{-1}(x)) = x$; not all functions are *invertible*, and hence not all functions have an inverse.

The non-negative integers $\mathbb{N}$ will prove especially useful in algorithmic descriptions, where we will almost always assume the use of *zero-based* indices for arrays and sequences, so that, e.g., the very first element of a list will have index 0. Although 1-based indices are common in some biological domains, our use of zero-based coordinates will facilitate easier translation into working software implementations for most programming languages.

Rounding a real value x up or down to the nearest integer will be specified using $\lceil x \rceil$ (the *ceiling* of x) or $\lfloor x \rfloor$ (the *floor* of x), respectively. Rounding to the

[1] Technically, B is known as the *co-domain*, which may be a superset of the range; we will gloss over this distinction since it will not be overly important for our purposes. Note also that we use the $\mapsto$ symbol instead of $\rightarrow$ for specifying functions, to avoid confusion with the other uses to which we will put the $\rightarrow$ operator.

nearest integer will be denoted *round*(x), and may be implemented as $round(x) = \lfloor x + 5.0/9.0 \rfloor$. Caution is advised to the novice programmer in implementing this for negative numbers.

The *Kronecker delta* function $\delta(x, y)$ evaluates to 1 if $x = y$, and to 0 otherwise.

Generalized summation and multiplication will be denoted with the standard sigma and pi notations, $\sum_i f(x_i)$ and $\prod_i f(x_i)$, respectively. Regarding the latter, it will be seen when we consider probabilistic models that products of large numbers of probabilities will have a tendency to cause *underflow* in the computer by producing values too small (i.e., too close to zero) to represent in the number of bits available on the computer. Underflow occurs when either a nonzero value is rounded down to zero or when two values which differ by a small amount are rounded to the same value, eliminating our ability to distinguish between those values in the computer. This problem can often be avoided by using the log transformation when dealing with probabilities, so that the logarithms of products become sums of logarithms, according to Eq. (2.3), giving us:

$$\log\left(\prod_i f(x_i)\right) = \sum_i \log(f(x_i)). \tag{2.8}$$

Logarithms of sums are more difficult to manipulate algebraically. A well-known solution is given by:

$$\begin{aligned} \log(p+q) &= \log(e^{\log p} + e^{\log q}) \\ &= \log(e^{\log p}(1 + e^{\log q - \log p})) \\ &= \log(e^{\log p}) + \log(1 + e^{\log q - \log p}) \\ &= \log p + \log\left(1 + e^{\log \frac{q}{p}}\right) \\ &= \log p + \log\left(1 + \frac{q}{p}\right) \end{aligned} \tag{2.9}$$

(Kingsbury and Rayner, 1971; Durbin *et al.*, 1998). Thus, if we factor $\log(p+q)$ as $\log(p) + \log(1 + e^{\log q - \log p})$, with $p \geq q$, then the term $e^{\log q - \log p} = q/p$ should, for probabilities of similar events (see section 2.6), remain large enough in most cases to avoid underflow, since it is a ratio in which the numerator and denominator will have similar magnitudes. That is, given that q and p are already represented in log space as q' and p', respectively, we can safely compute $z = \log(p+q)$ by taking $d' = q' - p'$, then $d = e^{d'}$, then $x = 1 + d$, and finally $z = p' + \log(x)$. While taking p or q out of log space individually may cause underflow if either of these are extremely small, taking d' out of log space typically should not, as long as p and q are probabilities of similar events (i.e., their probabilities do not differ by many orders of magnitude), such as the occurrence of an A versus a C in a typical nucleotide sequence. The case of $p = q = 0$ can be handled separately.

The tracking of phase and frame (section 3.3) will necessitate the use of *modular arithmetic*. By a mod b we mean the remainder after integer division of a by b; e.g.,

$7 \div 3 = 2$ with a remainder of 1. Languages such as C/C++ and Perl provide this operation via the % operator, but the reader should note that different languages can give different results for a mod b when the dividend a is negative. For example, the expression $-7\%3$ will evaluate to 2 in Perl and -1 in C/C++; in the case of the latter, adding b to the result of a mod b when a mod b returns a negative result will ensure that only nonnegative values are produced. We will adopt the convention that the remainder after division should always be positive, and so therefore should the result of mod.

It should be obvious that a sequence of integers such as (0, 1, 2, 3, 4, 5, 6, 7, 8) is cyclic under the mod 3 operation: the former sequence subjected to the mod 3 operation becomes (0, 1, 2, 0, 1, 2, 0, 1, 2). Thus we say that, for example, $2 \equiv 5 (\text{mod } 3)$, read "two is congruent to five, modulo three," because 2 and 5 produce the same remainder after division by 3. Addition and subtraction are well-behaved under modular arithmetic: $a \equiv b (\text{mod } 3)$ implies that $(a + x) \equiv (b + x)(\text{mod } 3)$ for any $x \in \mathbb{Z}$. Also, $a \equiv b(\text{mod } 3)$ implies that $(a + 3) \equiv b(\text{mod } 3)$. Congruence is symmetric, in that $a \equiv b(\text{mod } 3)$ is equivalent to $b \equiv a(\text{mod } 3)$.

We will have occasion to use the *binomial coefficient*:

$$\binom{n}{r} = \frac{n!}{r!(n-r)!} \tag{2.10}$$

for nonnegative integers n and r, $0 \leq r \leq n$. The binomial coefficient, also denoted ${}_nC_r$, counts the number of r-element subsets of an n-element set. The notation $x!$ is the *factorial of* x, which is given by:

$$x! = \begin{cases} \prod_{i=1}^{x} i & \text{for } x > 0 \\ 1 & \text{for } x = 0. \end{cases} \tag{2.11}$$

The factorial is best computed in log space, as is the binomial coefficient. We give an algorithm for computing the binomial coefficient in section 2.4.

An ordered set of numbers can be represented using a *vector*; i.e., $v = (1.3, 8.9, 2.7)$ is a 3-element vector. The *dot product*, $v \cdot w$, of two vectors (also known as the *inner product* and sometimes denoted $<v, w>$) having the same number of elements is defined as:

$$v \cdot w = \sum_{i=0}^{|v|-1} v_i w_i, \tag{2.12}$$

where $|v|$ denotes the number of elements in v, and v_i is the $(i + 1)$th element (v_0 being the first). If v and w are interpreted geometrically as directed arrows in an n-dimensional space ($n = |v|$), with the bases of the arrows jointly positioned at the *origin* $(0, 0, 0, 0, \ldots, 0)$ and the arrowheads positioned at the respective points given by the vector tuples $(v_0, v_1, \ldots, v_{n-1})$ and $(w_0, w_1, \ldots, w_{n-1})$, then the dot product can be defined in terms of the angle θ between v and w:

$$v \cdot w = \|v\| \cdot \|w\| \cdot \cos\theta, \tag{2.13}$$

where the *Euclidean norm* $\|v\|$ of a vector v is defined as:

$$\|v\| = \sqrt{\sum_{i=0}^{|v|-1} v_i^2}. \tag{2.14}$$

Thus, the angle between two vectors is given by $\theta = \cos^{-1}(v \cdot w/(\|v\| \cdot \|w\|))$, for $\cos^{-1}$, the inverse of the *cosine* function.

A two-dimensional array of numbers (or other objects) is known as a *matrix*, and is denoted in boldface: **M**. The element residing at the intersection of the ith row and the jth column of a matrix can be denoted $\mathbf{M}_{i,j}$. We will sometimes denote a matrix using the notation $\mathbf{M} = [w_{i,j}]$, where the function $w_{i,j} = w(i, j)$ specifies the element $\mathbf{M}_{i,j}$.

A *scalar* – in contrast to a vector or matrix – is simply a number. Although in linear algebra matrices typically consist of numbers, we will eventually consider matrices in which the elements are *n-tuples* – i.e., *n*-element arrays of objects – or *pointers* to objects, in the sense of addresses in a computer's memory space where such objects reside.

Matrix multiplication, denoted $\mathbf{M} \cdot \mathbf{W}$, is defined as:

$$r_{i,j} = \sum_{k=0}^{n-1} \mathbf{M}_{i,k}\mathbf{W}_{k,j}, \tag{2.15}$$

for matrix **M** of size $m \times n$ and **W** of size $n \times p$, where the result $\mathbf{R} = [r_{i,j}]$ is of size $m \times p$. A *square matrix* is one having the same number of rows and columns. A *diagonal matrix* is a square matrix in which all elements not on the main diagonal are zero – i.e., $\mathbf{M}_{i,j} = 0$ if $i \neq j$. The *identity matrix* $\mathbf{I}_m$ is defined as a diagonal matrix in which the diagonal elements are all set to 1. The *transpose* of an $m \times n$ matrix **M**, denoted $\mathbf{M}^T$, is the $n \times m$ matrix defined by $\mathbf{M}^T_{i,j} = \mathbf{M}_{j,i}$. The *inverse* of a square matrix **M**, denoted $\mathbf{M}^{-1}$, is defined as that matrix which satisfies the relation $\mathbf{M} \cdot \mathbf{M}^{-1} = \mathbf{I}_m$, where **M** is $m \times m$. Not all square matrices are invertible. Methods for computing a matrix inverse can be found in any linear algebra textbook (e.g., Anton, 1987).

The product of a scalar c and a matrix **M** – denoted $c\mathbf{M}$, or $\mathbf{M}c$ – is simply the matrix which results by multiplying all entries in **M** by c. The product of a matrix and a vector, $\mathbf{M}v$, is defined trivially by considering the vector v as an $n \times 1$ matrix, where **M** is $m \times n$; the result is obviously an $m \times 1$ matrix, which we can again interpret as a vector, and we say that **M** has performed a *coordinate transformation* on v into m-dimensional space. This will be useful when we consider *eigenvectors*, which are those vectors v for which $\mathbf{M}v = \lambda v$ for matrix **M** and scalar (or possibly *complex*) value λ; the latter is known as an *eigenvalue*. We will have use for such entities when we consider *phylogenetic modeling* in section 9.6.

2.2 Logic and boolean algebra

The two *boolean* values are *true* and *false*. A logical *proposition* or *predicate* is a sentence or formula which, when the values of all the variables appearing in the proposition have been specified, or *bound*, evaluates to true or false. A proposition which evaluates to true regardless of the values of its variables is a *tautology*; one that always evaluates to false is a *contradiction*. If two propositions $f(x)$ and $g(x)$ always evaluate to the same value for any and all x, we say they are *logically equivalent*, and we denote this fact by $f(x) \Leftrightarrow g(x)$. The $\Leftrightarrow$ operator is read "if and only if," and is sometimes abbreviated *iff*.

Logical *conjunction* is denoted by $\wedge$, the "and" operator. Denoting true by T and false by F, we have $\mathrm{T} \wedge \mathrm{T} \Leftrightarrow \mathrm{T}$, $\mathrm{T} \wedge \mathrm{F} \Leftrightarrow \mathrm{F}$, $\mathrm{F} \wedge \mathrm{T} \Leftrightarrow \mathrm{F}$, and $\mathrm{F} \wedge \mathrm{F} \Leftrightarrow \mathrm{F}$. Note that in expressions involving probabilities we will sometimes use a comma rather than $\wedge$ to denote conjunction. Logical *disjunction* is denoted by $\vee$, the "or" operator, which obeys $\mathrm{T} \vee \mathrm{T} \Leftrightarrow \mathrm{T}$, $\mathrm{T} \vee \mathrm{F} \Leftrightarrow \mathrm{T}$, $\mathrm{F} \vee \mathrm{T} \Leftrightarrow \mathrm{T}$, and $\mathrm{F} \vee \mathrm{F} \Leftrightarrow \mathrm{F}$. Logical *negation* or *complementation* is indicated by $\neg$, the "not" operator; $\neg\mathrm{T} \Leftrightarrow \mathrm{F}$, and $\neg\mathrm{F} \Leftrightarrow \mathrm{T}$. Of these three operators, $\neg$ has the highest precedence and $\vee$ the lowest, so that $\mathrm{A} \wedge \neg\mathrm{B} \vee \mathrm{C} \Leftrightarrow (\mathrm{A} \wedge (\neg\mathrm{B})) \vee \mathrm{C}$.

$\forall$ is the *universal quantifier*, and is used to parameterize a proposition over a set of variable bindings. $\forall_{x \in S} f(x)$, read "for all x in S, $f(x)$," asserts that the proposition $f(x)$ is true for all values of x in set S (sets are described in the next section). We will use the shorthand $\forall_{x \in S, y \in T} f(x, y)$ to denote $\forall_{x \in S} \forall_{y \in T} f(x, y)$. The *existential quantifier*, $\exists$, has a similar meaning: $\exists_{x \in S} f(x)$, read "there exists an x in S such that $f(x)$," asserts that $f(x)$ is true for at least one value of x in S. The symbol $\ni$ means *such that*, so that $\exists_{x \in S} f(x)$ could also be written "there exists an $x \in S \ni f(x)$."

In the few mathematical proofs occurring in this book we will use the $\rightarrow$ operator to denote logical *implication*; $A \rightarrow B$, read "A implies B," means that if A is true, then B must be true also. It is important to note that $A \rightarrow B$ is itself only a proposition; if A does indeed imply B, then the proposition evaluates to true; otherwise, the proposition evaluates to false. The *converse* of an implication, denoted *converse* $(A \rightarrow B)$, is given by $B \rightarrow A$, which may or may not be true, regardless of the truthfulness of $A \rightarrow B$. The *contrapositive*, on the other hand, defined as *contrapositive* $(A \rightarrow B) = ((\neg B) \rightarrow (\neg A))$, is logically equivalent to the original implication, $A \rightarrow B$. The $\therefore$ ("therefore") operator is very similar to the implication operator, except that its first operand is generally taken to be the set of facts already set forth in a proof; it can be read "the foregoing arguments imply the following . . .". The end of a proof will be denoted with the symbol ❑.

A special kind of proof is that which is based on the notion of *mathematical induction*. Suppose we have a proposition $f(x)$ which we wish to prove for all $x \in \mathbb{N}$. Then it suffices for us to show that $f(0)$ is true and to also show that $\forall_{x \geq 0} f(x) \rightarrow f(x+1)$. A proof that utilizes an argument of this form is called an *inductive proof* or a *proof by induction*. The part of an inductive proof that establishes the truth of $f(0)$ is called the *basis step*, and the part of the proof that establishes $\forall_{x\geq 0} f(x) \rightarrow f(x+1)$ is known as the *inductive step*. When establishing the inductive step, we refer to $f(x)$ as the *inductive hypothesis*.

To illustrate some of this notation, we will give the definition of a *limit at infinity* from elementary calculus. Let $f : \mathbb{R} \mapsto \mathbb{R}$ be a function. Then *the limit of f(x) as x approaches infinity*, denoted $\lim_{x\rightarrow\infty} f(x)$, can be said to equal real value L iff we can make f as close to L as we please simply by choosing higher values of x:

$$\left(\lim_{x\rightarrow\infty} f(x) = L\right) \Leftrightarrow (\forall_{\varepsilon>0}\exists_c\forall_{x>c} x \in dom(f) \rightarrow |f(x) - L| < \varepsilon). \tag{2.16}$$

This can be read: "the limit of $f(x)$ as x approaches infinity equals L if and only if for any ε greater than zero there exists a c such that for all $x > c$, x being in the domain of f implies that the absolute difference between $f(x)$ and L is less than ε."

2.3 Sets

We will use sets extensively. We denote by $\{x|f(x)\}$ the set of all x such that the proposition $f(x)$ is true. If a set has only a few elements we may list them explicitly, as in $\{\texttt{a},\texttt{b},\texttt{c}\}$. Note that the set $\{\texttt{a},\texttt{b},\texttt{c},\texttt{d}\}$ and the *4-tuple* $(\texttt{a},\texttt{b},\texttt{c},\texttt{d})$ are distinct; the latter is an ordered sequence of four elements, whereas order is undefined for the former. Furthermore, a tuple may contain repeated elements, whereas a set may not. The *empty set* is denoted $\{\}$ or $\emptyset$. The set consisting of the first n whole numbers – i.e., $\{x|0 \leq x < n\}$ – may be denoted $\mathbb{N}_n$. The expression $|S|$ denotes the *cardinality* of, or number of elements in, set S. Note that while the vertical bars will also be used to denote absolute value and string length (section 2.15), it should always be clear from the context which meaning we intend.

Set membership is tested with the $\in$ operator: $x \in A$ iff x is an element of set A; $x \notin A$ iff x is *not* an element of A. Note that the meaning which we adopt for universal and existential quantification over empty sets may be somewhat counterintuitive: the statement "$\forall_{x\in\emptyset} f(x)$" is true for any function $f(x)$, and the statement "$\exists_{x\in\emptyset} f(x)$" is false for any function $f(x)$. Summation over an empty set will be taken to equal 0: $\sum_{i\in\emptyset} f(x_i) = 0$. Multiplication over the empty set will be taken to be 1: $\prod_{i\in\emptyset} f(x_i) = 1$.

The *intersection*, $A \cap B$, and *union*, $A \cup B$, should be familiar from elementary school. These operators can be generalized as $\cap_{A\in S} A$ and $\cup_{A\in S} A$, where S is a collection (or "family") of sets to be jointly intersected or united, respectively. Note that when $A \cap B = \phi$ we say that A and B are *disjoint*, or have no members in common. Set *subtraction*, denoted, $A - B$ can be formalized as $A - B = \{x \in A | x \notin B\}$.

The *cartesian product* of two sets, denoted $A \times B$, is the set of all pairs (x, y) such that x and y are elements of A and B, respectively. Formally, $A \times B = \{(a, b)|a \in A \wedge b \in B\}$. This definition can be generalized so that the product of n sets denotes a set of *n*-tuples rather than a nested set of pairs: e.g., $A \times B \times C = \{(a, b, c) | a \in A \wedge b \in B \wedge c \in C\}$, rather than $\{((a, b), c) | a \in A \wedge b \in B \wedge c \in C\}$. Similarly, $f : A \times B \times C \mapsto D \times E$ will be read as a function $f(a, b, c)$ taking parameters $a \in A$, $b \in B$, and $c \in C$, and returning a pair $(d, e) \ni d \in D, e \in E$.

The *subset* and *superset* operators should likewise be familiar from elementary school: $A \subseteq B$ and $A \supseteq B$ denote the subset and superset relations, respectively. $A \subset B$ denotes that A is a *proper subset* of B; i.e., $A \subseteq B$ but $A \neq B$. Similarly, $A \supset B$ denotes the *proper superset* relation.

The *power set* of a set S, denoted 2^S, is the set of all subsets of S; i.e., $2^S = \{A | A \subseteq S\}$. Note that the result of a power set operation is a set in which each element is itself a set. The cardinality of a (finite) power set is given by $|2^S| = 2^{|S|}$.

An interval over the integers or reals is denoted (a, b) or $[a, b]$, the latter (a *closed interval*) including its endpoints and the former (an *open interval*) not. These notations can be mixed, as in $(a, b]$, which includes b but not a, and $[a, b)$, which includes a but not b. Whether an interval is over the integers or the reals will be clear from the context.

A set can sometimes be interpreted as a function, and vice versa. If a set $S = \{(x_i, y_i)\}$ has the property that no x occurs in two or more distinct pairs (x, y_i) and (x, y_j) with $y_i \neq y_j$, then we can define the function $f_S(x)$ such that $\forall_{(x,y) \in S}\, f_S(x) = y$. Conversely, given a function $f(x)$ we can without ambiguity refer to f as a set of pairs $S_f = \{(x, y) | x \in dom(f) \wedge y = f(x)\}$.

A function $f : X \mapsto Y$ for which every element in Y has a unique pre-image in X (and assuming f is defined on its entire domain), is called a *one-to-one correspondence*. A *countable set* is any set X for which there exists a one-to-one correspondence between X and a subset $Y \subseteq \mathbb{N}$ of the natural numbers.

A common graphical illustration used to show set relations is the *Venn diagram*, in which an outermost rectangle denotes the *universal set* of all possible elements for the problem at hand, and various closed shapes within that outer rectangle denote subsets of the universal set. Areas of overlap between these closed shapes represent intersections of the corresponding subsets; various other set relations may obviously be represented as well. An extended example utilizing a Venn diagram is given in section 2.6.

2.4 Algorithms and pseudocode

All algorithms, with few exceptions, will be presented in a relatively formal pseudocode, which will afford us language independence while still achieving a high degree of rigor in our descriptions. The pseudocode should be familiar to users of structured programming languages; our constructs and conventions (i.e., scoping rules, operator precedence, etc.) are similar to those used in Pascal, C/C++, Java, and Perl, so that there should be little room for ambiguity.

Assignment is indicated by:

```
x ← expression;
```

in which the value of `expression` is assigned to variable `x`; in this case we say that `x` is *bound* to the value of `expression`. If `x` is an array, we can access a location in that array using `x[index]`, where `index` is zero-based, so that the first element of the array has index zero. `x[a..b]` denotes the sub-array of `x` beginning at index `a` and ending at index `b`, inclusive (so that the sub-array has $b - a + 1$ elements). Note that the symbol = in an algorithm listing indicates a test for equality, not assignment.

An array literal can be expressed as a comma-separated list, so that:

```
(a,b,c) ← array;
```

causes `a`, `b`, and `c` to be assigned the 0th, 1st, and 2nd values of the array, respectively. Similarly,

```
array ← (a,b,c);
```

causes the values of `a`, `b`, and `c` to be assigned into the 0th, 1st, and 2nd positions of the array, respectively. The number of elements in an array `A` will be denoted $|A|$.

Conditional execution is indicated by **if**:

```
if boolean_expression then
  body_1;
else
  body_2;
```

where `body_1` and `body_2` are arbitrary blocks of pseudocode, with the block boundaries indicated by indentation rather than braces or `begin..end` pairs.

Iteration (i.e., "looping") is achieved using the familiar **while** and **for** constructs:

```
while boolean_expression do
  block_of_code;

for x←v0 to vf by stepsize do
  block_of_code;
```

where the **by** `stepsize` is optional; when absent, `stepsize` defaults to 1. Reversing the order of iteration is achieved either by negating `stepsize` or replacing **to** with **down to**; we will generally substitute **up to** in place of **to** as a reminder that the default increment is 1. The `boolean_expression` (the "test") of the **while** construct is evaluated before the execution of the body of the loop, so that the body may be executed zero or more times. In the case of the **for** statement, the body is executed once for each value of `x` in the integer interval $[v_0, v_f]$, though it should be noted that the expression v_f is evaluated anew for each iteration. As before, we indicate block structure using indentation rather than braces or other punctuation, as will be true for all block constructs.

A variant of the **while** loop is the **do** loop, which executes the block_of_code, then checks whether the boolean_expression is true, and if so, executes the block_of_code again, etc.:

```
do
  block_of_code
while boolean_expression;
```

The difference between a **while** loop and a **do** loop is that in the **do** loop the block_of_code is always executed at least once, regardless of the value of the boolean_expression.

The **next** keyword causes execution to prematurely jump to the next iteration of the innermost loop containing that **next** statement. The **break** statement causes a loop to terminate, with execution resuming at the first statement following the loop (i.e., the next instruction which would be executed after the loop if the loop had terminated normally).

The **foreach** statement causes a block of code to execute once for each element in a given set:

```
foreach x∈S do
  body
```

In this example the body of the loop would execute |S| times, with x bound to a different element of S each time; in cases where S has a natural ordering (e.g., when S is a set of numbers), we may assume the execution order is defined similarly.

String and character literals appearing in pseudocode will be enclosed in single or double quotes; e.g., "ATTAAA"; the quotes will generally be omitted within descriptive text when there is no chance of ambiguity. String concatenation is indicated with the & operator, so that "AT"&"G" evaluates to the string "ATG". |s| denotes the length of string s. The empty string is denoted by ε in text, or "" within pseudocode. A string can be indexed just like an array, using zero-based indices, to produce a single element of the string. Substrings are denoted using the notation S[i..j] within pseudocode or $S_{i..j}$ in text, where i and j are the zero-based indices of the first and last elements of the substring, respectively. Assigning to a substring causes the assigned value to be substituted in for that substring, thereby changing the contents (and potentially the length, as a consequence) of the original string.

Function calls are indicated in the usual way; i.e., f(x). Parameters are assumed to be passed *by value*, unless preceded by the keyword **ref** in the function specification, to indicate passing *by reference*. If a parameter is passed by value, then assignments to that variable within the function have no effect on any variable passed as a corresponding argument to the function. If a parameter is passed by reference, then assignments to that parameter within the function apply to the variable given in the function call.

Associative arrays generalize the notion of an array to allow strings or other types of objects to be used as indices. In this way, `A[s]` denotes an element of any type (i.e., number, string, array, etc.) which has been associated with object `s` in associative array `A`. If no value has been associated with `s`, then `A[s]` evaluates to the undefined value, `NIL`. Association is established by assignment; e.g., `A[s]←x` associates `x` with `s` in array `A`.

Algorithm 2.1 shows the computation of the binomial coefficient as a concrete example of our notational conventions.

Algorithm 2.1 Computing the binomial coefficient ${}_nC_r$ using logs.

```
1.  procedure binCoeff(n,r)
2.     c←0;
3.     for x←r+1 up to n do c←c+log(x);
4.     for x←1 up to n-r do c←c-log(x);
5.     return exp(c);
```

2.5 Optimization

We will often need to maximize a given function. If $f(x)$ is a real-valued function, then $max_x\, f(x)$ denotes the largest value y to which $f(x)$ can evaluate; i.e., that $y \in range(f)$ such that $y = f(x) \wedge \neg\exists_{x'}[f(x') > y]$, for some $x \in dom(f)$.

A related but distinct maximization operation is the *argmax*, which denotes a domain value x such that $f(x)$ is maximal – i.e., an $x \in dom(f)$ such that $\neg\exists_{x'}\, f(x') > f(x)$. Unlike the *max* of a function, the *argmax* is not necessarily unique. For our purposes this will seldom matter; when the *argmax* is non-unique we will allow that the *argmax* arbitrarily evaluates to any one of the domain values of f which maximize the function. The *max* and *argmax* are illustrated in Figure 2.1, in which it can be seen that the former produces the optimal y-value and the latter produces the optimal x-value.

It is also worth noting that the *max* and *argmax* operators apply to any real-valued or integer-valued function, regardless of the domain of the function. In particular, we will find it useful to consider functions of the form $f : G \mapsto \mathbb{R}$ for G the set of all possible gene structures for a given sequence. In this case the *argmax* will evaluate to a particular gene structure and the *max* will evaluate to a real value associated with that gene model through f.

Actually finding the *max* or the *argmax* of a function can be a rather difficult problem (especially if we wish to do so quickly), and it is one to which we will

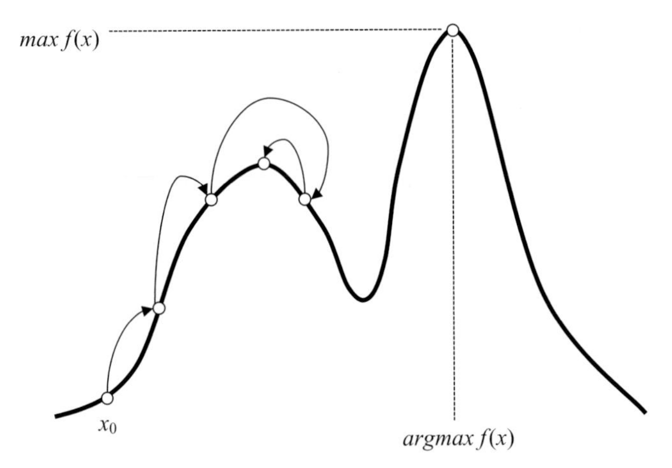

Figure 2.1 Finding a (local) maximum by hill-climbing. In this example, the second-to-last step overshot the peak, requiring a reversal for the final step. The (globally) optimal x-value is shown as *argmax* $f(x)$, and the optimal y-value is shown as *max* $f(x)$.

dedicate quite a number of pages in the chapters ahead. For now we will consider a simple heuristic which can be used to find an *optimum* (maximum or minimum) for a real-valued function of one variable. This method, variously called *gradient ascent*, *gradient descent*, or *hill-climbing*, is the underlying basis for many optimization procedures, though we will present it in its most basic form. We will consider the problem of finding $argmax_x f(x)$ for a continuous function $f : \mathbb{R} \mapsto \mathbb{R}$.

The method involves picking a domain value x_0 at random, observing the *gradient* of the function at $f(x_0)$ – i.e., whether the function is increasing or decreasing (or neither) at that point – and then selecting $x_{i+1} = x_i + \Delta$, where Δ is a small positive number if f is increasing at x_i and is negative if f is decreasing at x_i. In this way, we always attempt to move "uphill," as illustrated in Figure 2.1. As the search progresses, the absolute value of the *step size* Δ may be decreased so that we do not leap right past the peak and into the valley beyond, as we would be more likely to do if an overly large step size is used. This procedure is described in Algorithm 2.2.

Algorithm 2.2 Optimizing a function f by performing N iterations of gradient ascent, starting at position x and using initial stepsize Δ. The domain value for the selected point is returned.

```
1. procedure gradientAscent(f,x,Δ,N)
2.   for i←1 to N do
3.     if f(x+Δ)>f(x) then s←Δ;
4.     else s←-Δ;
5.     while f(x+s)≥f(x) do x←x+s;
6.     Δ←Δ/2;
7.   return x;
```

As suggested by the figure, the method is not guaranteed to find the global maximum of the function. If a function has one global maximum and many smaller local maxima, the method is often more likely to find a suboptimal peak than to find the globally optimal one. For this reason, it is common practice to run the procedure many times, starting at a different x_0 each time and keeping track of the best value over all runs.

Algorithm 2.2 operates by iterating N times, with each iteration taking a variable number of steps in the selected direction until no improvement can be found, and then reducing the step size Δ before commencing the next iteration. Note that rather than iterating a fixed number of times, we could instead have specified a small value ε and decided to terminate the search when $\Delta < \varepsilon$, or perhaps when $|f(x+s) - f(x)| < \varepsilon$. More sophisticated methods for numerical optimization are well characterized in the literature (e.g., Press *et al.*, 1992; Galassi *et al.*, 2005), which the reader is encouraged to peruse; the procedure given in Algorithm 2.2 is intended only to illustrate the general idea, and is not designed for optimal efficiency. Section 10.15 describes an efficient open-source software package for multidimensional hill-climbing.

As noted above, the gradient ascent method is not in general guaranteed to find the global maximum of a function. We will see that there are cases in which it is possible to efficiently find the global maximum of a function, even when the domain is excessively large. For these cases we will apply a technique known as *dynamic programming*, which is described in section 2.12.

2.6 Probability

We will be relying very heavily on basic probability theory in our modeling of the gene prediction problem. It will be seen in later chapters that many gene finders model genomes as having been produced by a *stochastic* (probabilistic) process in which key elements of a gene structure can be assigned probabilities. Probabilities provide a principled way to rank alternative gene structures. The probability of a gene structure ϕ can be thought of loosely as a measure of how confident we are that ϕ is the correct gene structure for a given sequence S.

A *sample space* is the set of all possible outcomes (or *sample points*) of an *experiment* (the latter also being called a *trial* or an *observation*). For example, if we have a stochastic process R_{nuc} which generates nucleotides and we consider one trial to be the act of sampling one nucleotide emitted by the process, then the sample space of that experiment is $\{\texttt{A,C,G,T}\}$. We can think of an experiment as the assignment (not by ourselves, but by some natural phenomenon) of an element from the sample space to a special kind of variable, called a *random variable*. If X is a random variable for the R_{nuc} process, then $X \in \{\texttt{A,C,G,T}\}$. Another way of thinking of a random variable is as a function with an implicit time parameter:

$X = X(t)$, so that observing X at successive times may clearly result in different values.

An *event* is any subset of the sample space. For example, for the process R_{nuc} we may be interested in the occurrences of Cs and Gs. Let us suppose we consider the emission of any C or G to be an event of particular importance, but that we are unconcerned with the actual identity of the nucleotide (i.e., we care whether it was a C or G, but we do not care which). Then we can formulate the event E as the assertion $X \in \{\mathtt{C},\mathtt{G}\}$, where X is the random variable which observes the output of the process. Equivalently, we can instead say that event E *is* the set $\{\mathtt{C},\mathtt{G}\}$. In this way, an event is both a boolean proposition *and* a subset of the sample space.

We can define the *probability* of an event E, denoted $P(E)$ or $P(X \in E)$, as the limiting proportion of times the experiment results in a sample point in E:

$$P(X \in E) = \lim_{n\to\infty} \frac{|\{i|x_i \in E, 0 \leq i < n\}|}{n}, \tag{2.17}$$

where the x_i are obtained by repeated observation of random variable X (i.e., as a result of conducting n experiments, resulting in outcomes $\{x_0, x_1, \ldots, x_{n-1}\}$). A simpler formulation is possible if the points in the sample space are all equally likely:

$$P(X \in E) = \frac{|E|}{|S|} \tag{2.18}$$

for a finite, nonempty sample space S. The numerator of Eq. (2.18) is known as a *frequency* (i.e., an integer count), and the entire ratio constitutes a *relative frequency* (i.e., a frequency normalized to lie in the interval [0, 1]). Note that both relative frequencies and probabilities can be denoted using either a real number in [0, 1], such as 0.25, or using a percentage such as 25%. When speaking of probabilities, it should be clear that, e.g., 0.25 = 25%.

If events A and B are *mutually exclusive* (i.e., cannot occur simultaneously), then the probability of either A or B occurring is obtained through simple addition:

$$[A \cap B = \emptyset] \rightarrow [P(A \vee B) = P(A) + P(B)]. \tag{2.19}$$

Thus, since the sample points comprising an event are by definition mutually exclusive,

$$P(E) = \sum_{a\in E} P(a). \tag{2.20}$$

If events A and B are not known to be mutually exclusive we can resort to:

$$\begin{aligned} P(A \vee B) &= P(A \cup B) \\ &= \sum_{a\in A} P(a) + \sum_{b\in B} P(b) - \sum_{c\in A\cap B} P(c) \\ &= P(A) + P(B) - P(A \wedge B), \end{aligned} \tag{2.21}$$

where we have made use of the fact that an event is both a boolean variable and a subset of the sample space, so that $A \wedge B = A \cap B$, and $A \vee B = A \cup B$. Thus,

$$P(A \wedge B) = P(A \cap B) = \sum_{c \in A \cap B} P(c). \tag{2.22}$$

As noted previously, an alternative notation for $P(A \wedge B)$ is $P(A, B)$.

If $P(B) > 0$, then the probability that event A will occur during the next trial, given that B also occurs, denoted $P(A|B)$, is given by:

$$P(A|B) = \frac{P(A, B)}{P(B)}. \tag{2.23}$$

$P(A|B)$ is the *conditional probability of A given B*, also called the *posterior probability* of A (given B), and is the basis for *Bayes' rule*:

$$P(A|B) = \frac{P(B|A)P(A)}{P(B)} = \frac{P(B|A)P(A)}{\sum_{\text{disjoint } C} P(B|C)P(C)}, \tag{2.24}$$

where we have utilized the fact that

$$P(B) = \sum_{\text{disjoint } C} P(B, C), \tag{2.25}$$

for a *partitioning* of the sample space S into a set of disjoint sets C whose union is S (i.e., a set of mutually exclusive events having probabilities that sum to 1). This construction is referred to as *marginalization*, and $P(B)$ as a *marginal* (or *prior*) *probability*.

If the occurrence of event B gives no information as to the probability of A also occurring, then we say that A and B are *independent*, and denote this fact $A \perp B$:

$$\begin{aligned} A \perp B &\Leftrightarrow P(A|B) = P(A) \\ &\Leftrightarrow P(A, B) = P(A)P(B). \end{aligned} \tag{2.26}$$

Note that $A \perp B$ is symmetric, so that $(A \perp B) \Leftrightarrow (B \perp A)$. If $P(A|B, C) = P(A|B)$, we say that A is *conditionally independent of C, given B*.

The notion of conditional probability is central to probabilistic gene modeling techniques, and is utilized heavily in this text. It is thus essential that the novice reader fully master this concept. The following simple example will aid in the intuitive understanding of the idea of conditional probability.

As shown in Figure 2.2, the set of flying vertebrate species (left oval) intersects with the set of all bird species (class *Aves*; right oval), though the two do not perfectly coincide, due to the existence of flightless birds (e.g., penguins, ostriches) as well as non-avian flying vertebrates (e.g., bats, pterosaurs). The outer rectangle encloses all animal species, and represents the *universal set*. Let us consider

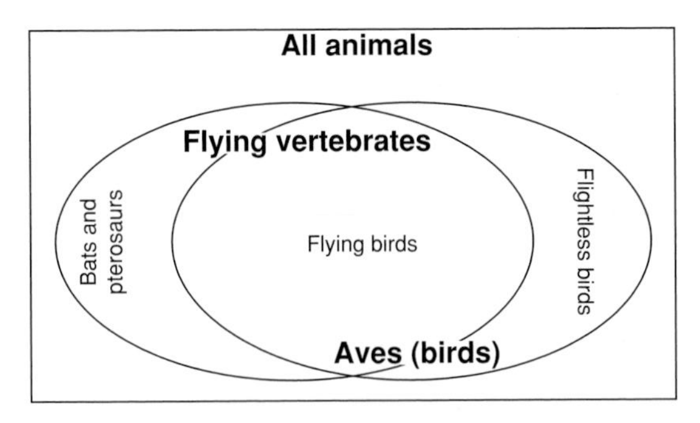

Figure 2.2 A Venn diagram illustrating the idea of conditional probability. The left oval represents all flying vertebrates and the right oval represents all species of birds. Conditional probabilities among these classes may be conceptualized as ratios of areas of subclasses and their enclosing superclasses (assuming sample points are equally likely).

a sampling experiment in which all animals are equally likely to be selected. If we let X denote the species of a randomly selected animal, we may compute the probability *P*(*bird*) that $X \in Aves$, the probability *P*(*flying*) that X flies, and all combinations of these predicates, based on the population sizes of these groups. In particular, the value *P*(*flying bird*) that X falls in the intersection of the two ovals shown in the figure is equal to the joint probability *P*(*flying, bird*). This joint probability may be assessed by dividing the number of animals in the intersection of the two ovals by the total number of animals: $N(flying \wedge bird)/N_{animals}$. The conditional probabilities *P*(*flying*|*bird*) and *P*(*bird*|*flying*) are distinct from this joint probability. The former, *P*(*flying*|*bird*), can be assessed via $N(flying \wedge bird)/N_{bird}$ and may be conceptualized (if the areas in the diagram were scaled proportional to their probabilities) as the ratio formed by taking the area in the overlap region of the two ovals in the figure divided by the area of the right oval. In contrast, the probability *P*(*bird*|*flying*) is given by $N(flying \wedge bird)/N_{flying}$. Recalling that the joint probability *P*(*flying, bird*) is given by $N(flying \wedge bird)/N_{animals}$, it can be seen that these three quantities are not generally equal.

As a second example, let us suppose that we are planning to randomly select a single *codon* (i.e., a nucleotide triplet coding for an amino acid – see Chapter 1) from a randomly selected protein-coding gene in the human genome, and that we wish to determine the probability P(ATG) that the codon is $x_0x_1x_2 = \text{ATG}$. Because the probability of a given nucleotide in a given codon position is known to be dependent on both the codon position and also on the identities of the nucleotides in the other codon positions, we need to model P(ATG) using conditional probabilities.

We can first decompose the composite event P(ATG) as a joint probability:

$$P(x_0x_1x_2 = \text{ATG}) = P(x_0 = \text{A} \wedge x_1 = \text{T} \wedge x_2 = \text{G})$$

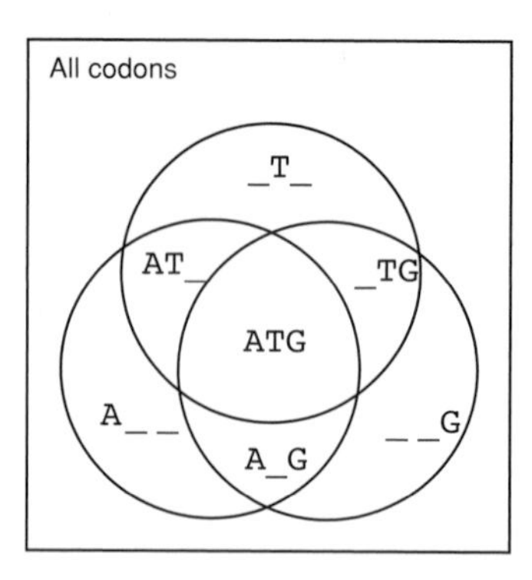

Figure 2.3 A Venn diagram for start codons. The lower-left circle represents all codons beginning with A, the lower-right circle represents all codons ending with G, and the top circle represents all codons having a T in the middle position.

which may then be factored into:

$$P(x_0 = \mathtt{A}) \cdot P(x_1 = \mathtt{T}|x_0 = \mathtt{A}) \cdot P(x_2 = \mathtt{G}|x_1 = \mathtt{T}, x_0 = \mathtt{A})$$

using repeated applications of Eq. (2.23). We may estimate these probabilities by counting various patterns in the set of codons comprising all (known) human genes:

$$P(x_0 = \mathtt{A}) \approx N(x_0 = \mathtt{A})/N_{codons}$$
$$P(x_1 = \mathtt{T}|x_0 = \mathtt{A}) \approx N(x_0 = \mathtt{A} \wedge x_1 = \mathtt{T})/N(x_0 = \mathtt{A})$$
$$P(x_2 = \mathtt{G}|x_0 = \mathtt{A}, x_1 = \mathtt{T}) \approx N(x_0 = \mathtt{A} \wedge x_1 = \mathtt{T} \wedge x_2 = \mathtt{G})/ N(x_0 = \mathtt{A} \wedge x_1 = \mathtt{T})$$

where we have applied the same reasoning as in the flying vertebrate example, but with three top-level classes ($x_0 = \mathtt{A}$, $x_1 = \mathtt{T}$, and $x_2 = \mathtt{G}$) instead of two, as illustrated in Figure 2.3. Thus, for example, the probability $P(x_1 = \mathtt{T} \mid x_0 = \mathtt{A})$ may be conceptualized as taking the set $U_A = \{x_0x_1x_2|x_0 = \mathtt{A}\}$ of all codons beginning with A to be the universal set, and then computing the probability $P(x_1 = \mathtt{T}|x_0 = \mathtt{A})$ as the simpler probability $P(x_1 = \mathtt{T})$ relative to this modified universal set U_A, rather than the default universal set $U = \{x_0, x_1, x_2\}$ consisting of all codons. The reader can easily verify that $P(x_0 = \mathtt{A}) \cdot P(x_1 = \mathtt{T}|x_0 = \mathtt{A}) \cdot P(x_2 = \mathtt{G}|x_1 = \mathtt{T}, x_0 = \mathtt{A}) = N(x_0 = \mathtt{A} \wedge x_1 = \mathtt{T} \wedge x_2 = \mathtt{G})/N_{codons}$, which is consistent with our definition of probability in Eq. (2.17) when the sample size N_{codons} is allowed to increase without bound. We will utilize a decomposition much like the one employed in this example when we consider higher-order *Markov* models in Chapters 6 and 7.

A *discrete probability distribution* P_X is a function $P_X : S \mapsto \mathbb{R}$ defined on a countable set S such that $\forall_{x\in S} P_X(x) \geq 0$ and $\sum_{x\in S} P_X(x) = 1$, where $P_X(x)$ is interpreted as the probability $P(X = x)$ for random variable X. We write $X \sim P_X$ to denote that random variable X follows distribution P_X.

The *expected value* of a function $f(X)$ for random variable X with discrete probability distribution P_X is given by:

$$E(f(X)) = \sum_{x \in S \cap dom(f)} P_X(x) f(x), \tag{2.27}$$

where S is the sample space for X. For continuous random variables we obviously have:

$$E(f(X)) = \int_{x \in S \cap dom(f)} P_X(x) f(x) dx. \tag{2.28}$$

As a simple example, if $dom(f) = \{0, 1\}$, $f(0) = 1$, $f(1) = 2$, $P_X(0) = 0.3$, and $P_X(1) = 0.7$, then $E(f(X)) = 0.3 \times f(0) + 0.7 \times f(1) = 0.3 \times 1 + 0.7 \times 2 = 1.7$. An expected value is also referred to as an *expectation*. Some useful identities regarding the expected value are:

$$E(X + Y) = E(X) + E(Y), \tag{2.29}$$

$$E(cX) = cE(X), \tag{2.30}$$

for random variable X and constant c. In the case where X and Y are independent,

$$E(XY) = E(X)E(Y). \tag{2.31}$$

The expected value of a random variable X, denoted $\mu_X = E(X)$, is known as the *mean* of X. For a sample $\{x_i | 0 \leq i < n, n > 0\}$ we can estimate the mean of the underlying distribution by computing the *sample mean* $\bar{x}$, or *average*:

$$\bar{x} = \frac{\sum_{i=0}^{n-1} x_i}{n}. \tag{2.32}$$

Somewhat similar to the mean are the *median* and *mode*: the *median* of a sample is the middle value when the sample is sorted numerically, and the *mode* is simply the most commonly occurring value in a sample (i.e., the one occurring at the "peak" of the distribution). The relative ordering of the median and mean can sometimes be used to infer *skewness* (roughly, asymmetry – see below) in a distribution, though care must be taken in doing so (e.g., von Hippel, 2005). A *unimodal* distribution is one having only one local maximum or "peak."

The *variance* of a random variable X, denoted $Var(X)$ or σ_X^2, is defined as:

$$\begin{aligned} \sigma_X^2 &= E[(X - \mu_X)^2] \\ &= E(X^2) - E(X)^2. \end{aligned} \tag{2.33}$$

As an example, if X denotes the length of a randomly sampled coding exon from the human genome, the mean and variance of X are roughly 200 bp and 89 000 bp, respectively, showing that human coding exons, while fairly small on average, can nonetheless vary quite substantially in length. Some useful identities regarding the variance are: $Var(cX) = c^2Var(X)$ for constant c, and in the case where X and Y are independent, $Var(X + Y) = Var(X) + Var(Y)$.

We can estimate σ_X^2 by collecting a sample $\{x_i | 0 \leq i < n, n > 1\}$ and computing the *sample variance*, s^2:

$$s^2 = \frac{\sum_{i=0}^{n-1} x_i^2 - \frac{1}{n}\left(\sum_{i=0}^{n-1} x_i\right)^2}{n-1}. \tag{2.34}$$

The *standard deviation* of X, denoted σ_X, is defined as the positive square root of σ_X^2, and can be estimated by computing the square root of Eq. (2.34). It is typical to report the mean and standard deviation together, using the notation $\mu \pm \sigma$ – e.g., 1.4 ± 0.7 – where the sample mean $\bar{x}$ and sample standard deviation s are obviously substituted for the theoretical values when the latter are not known. As an example, the $\bar{x} \pm s$ values for the lengths of human single-exon genes (i.e., those not interrupted by any introns) are 833 ± 581 bp, whereas the exons in spliced human genes have lengths averaging to 178 ± 267.

The *standard error*, $\sigma_{\overline{X}}$, or *standard deviation of the mean* – which measures the precision of our estimate of the mean – is given by:

$$\sigma_{\overline{X}} = \frac{\sigma_X}{\sqrt{n}} \tag{2.35}$$

for sample size n. The *law of large numbers* states that $\sigma_{\overline{X}}$ should approach zero as our sample size approaches infinity:

$$\lim_{n \to \infty} \sigma_{\overline{X}} = 0, \tag{2.36}$$

so that with larger volumes of *training data* (i.e., data from which to estimate a distribution's parameters), our estimate of the mean should get closer to the true mean of the underlying distribution.

A useful rule of thumb for "mound-shaped" or "bell-shaped" distributions (i.e., those which, when graphed, produce a smooth, symmetrical, single-peaked curve) is the so-called *empirical rule*, which states that roughly 68% of all observations should be within one standard deviation of the mean (i.e., within the interval $\mu \pm \sigma$), roughly 95% should be within two standard deviations of the mean ($\mu \pm 2\sigma$), and roughly 99% should be within three standard deviations ($\mu \pm 3\sigma$). Thus, for an experiment which is run only a single time, observing a value more than two or three standard deviations away from the mean on that single run of the experiment may be interpreted as evidence that the random variable does not in fact follow the supposed distribution, and may therefore be used as justification for rejecting the hypothesis that the random variable follows that distribution. This is only a rule of thumb, however; more rigorous methods for *statistical hypothesis testing* are described in section 2.9.

A closely related quantity to the variance of a single variable is the *covariance* between two random variables, denoted $Cov(X, Y)$ or $\sigma^2_{X,Y}$:

$$\begin{aligned} Cov(X, Y) &= E[(X - \mu_X)(Y - \mu_Y)] \\ &= E(XY) - E(X)E(Y). \end{aligned} \tag{2.37}$$

It should be readily apparent that $Cov(X, X) = Var(X)$. Furthermore, we have:

$$Var(X + Y) = Var(X) + Var(Y) + 2Cov(X, Y), \tag{2.38}$$

$$Cov(X, cY) = cCov(X, Y), \tag{2.39}$$

$$Cov(X, Y + Z) = Cov(X, Y) + Cov(X, Z), \tag{2.40}$$

and for the *sample covariance*, $s^2_{X,Y}$:

$$s^2_{X,Y} = \frac{\sum_{i=0}^{n-1} x_i y_i - \frac{\sum_{i=0}^{n-1} x_i \sum_{i=0}^{n-1} y_i}{n}}{n-1} \tag{2.41}$$

for a paired sample $\{(x_i, y_i) | 0 \leq i < n, n > 1\}$.

The *correlation* between two variables (also called the *Pearson correlation coefficient*, $\rho_{X,Y}$) is defined as:

$$\rho_{X,Y} = Cor(X, Y) = \frac{\sigma^2_{X,Y}}{\sqrt{\sigma^2_X \sigma^2_Y}} \tag{2.42}$$

As an example, if we generate random nucleotide sequences of varying {A,T} density, we find that the number of randomly occurring stop codons in these sequences is highly correlated ($\rho \approx 0.994$) to the {A,T} density of the generated sequences (see Exercise 2.45).

The *coefficient of variation* (*CV*) of a random variable X is simply:

$$CV(X) = \frac{\sigma_X}{\mu_X} \tag{2.43}$$

The *skew* and *kurtosis* of a distribution are general characteristics of the shape of the distribution when it is graphed in two dimensions. A distribution which is perfectly symmetric (left-to-right) has no *skew*, whereas a peaked distribution having more area to one side of the peak than the other (i.e., having a prominent *tail*) is said to be *skewed*. The *kurtosis* of a distribution is a characterization of its peakedness; a distribution which is perfectly flat has no kurtosis (or is *platykurtotic*), whereas one with a very high peak has much kurtosis (or is *leptokurtotic*). We will have occasion to consider modifying the skew and kurtosis of a distribution, and by this we mean lengthening or shortening the left or right tail of a distribution (skew) or causing the distribution to have more or less of a dominant peak (kurtosis).

The *log odds* of a probability $p < 1$ is:

$$\log odds = \log\left(\frac{p}{1-p}\right). \tag{2.44}$$

A similar concept to the log odds is the *log-likelihood ratio*:

$$\log\left(\frac{P(C\,|A)}{P(C\,|B)}\right), \tag{2.45}$$

for events A, B, and C; when we refer to a *likelihood* we mean merely a conditional probability. The log-likelihood ratio gives a measure of how much more likely the event C is under hypothesis A than under hypothesis B, with positive values indicating that C is more likely under A, and negative values instead favoring B. We will see a number of cases in which this simple construction can prove highly useful (e.g., classification of random genomic intervals as coding vs. noncoding – see section 7.2.5).

2.7 Some important distributions

Consider an experiment that can have one of only two possible outcomes, such as heads or tails, or true or false. Such an experiment is called a *Bernoulli trial*. Without loss of generality, we can arbitrarily designate one of the two outcomes a *success* and the other a *failure*. If a series of trials are independent and follow the same probability distribution, we say that they are *independent and identically distributed*, or *i.i.d.*

Suppose we perform n Bernoulli trials and denote by random variable X the number of successes in those n trials. Then X obeys a probability distribution called the *binomial distribution*, in which:

$$P_X(x) = \binom{n}{x} p^x (1-p)^{n-x} \tag{2.46}$$

for probability of success p. The mean and variance of a binomial distribution are $\mu = np$ and $\sigma^2 = np(1-p)$, respectively. An example of a binomial distribution is the number of heads in n flips of a coin: if the coin is fair, then the mean number of heads should be $np = n/2$.

If X instead represents the number of trials needed to obtain the first success, then X obeys a *geometric distribution*:

$$P_X(x) = (1-p)^{x-1} p. \tag{2.47}$$

An example is the number of coin flips before the first head is observed (plus 1, for the head). The mean and variance of this distribution are $\mu = 1/p$ and $\sigma^2 = (1-p)/p^2$, respectively, where p is still the probability of success. For the coin flip example, the expected number of flips needed to observe a head is $\mu = 1/p = 2$.

Note that a geometric distribution – which is a *discrete distribution* (see below) – may be approximated with a (continuous) *exponential distribution*, for which the cumulative distribution function is more easily computed: $P(X \leq x) = 1 - e^{-px}$. The *cumulative distribution function (cdf)* for a distribution is defined as the probability mass lying to the left of a given point in the probability density function for the distribution.

The *Poisson* distribution gives the probability of some rare event occurring k times in a given continuous interval of s units:

$$P_X(k; \lambda) = \frac{e^{-\lambda}\lambda^k}{k!}, \tag{2.48}$$

where λ is the expected number of occurrences of the event in the s-unit interval under consideration. For the Poisson distribution, $\mu = \sigma^2 = \lambda$.

The *Gaussian*, or *normal*, distribution is given by:

$$P_X(x) = \frac{1}{\sqrt{2\pi}\sigma} e^{-\frac{1}{2}\left(\frac{x-\mu}{\sigma}\right)^2}, \tag{2.49}$$

for $x \in \mathbb{R}$. Many natural phenomena follow Gaussian (or approximately Gaussian) distributions. It is important to note that the function defined in Eq. (2.49) gives *probability densities*, rather than probabilities, owing to the fact that the Gaussian distribution is a *continuous distribution* rather than a *discrete distribution*. In a discrete distribution, the domain of the probability function is a discrete, countable set, such as the integers. For such a distribution, the probability of any given sample point x can be given as $P_X(x)$. In contrast, a continuous distribution is one which is defined on an uncountably infinite set, such as $\mathbb{R}$, and in such a case, $P(X = x)$ for any particular sample point x is so small that it is taken to be effectively zero. Thus, for continuous distributions, probabilistic questions are posed in terms of intervals, such as $P(a \leq x \leq b)$, the probability that x lies in the interval $[a, b]$. Computing this probability involves the integration of the probability density function over the interval of interest (i.e., $\int_{[a,b]} P_X(x)dx$), and therefore requires some knowledge of calculus. In this book, however, we will be using discrete distributions only (including discrete approximations to continuous distributions).

Another important (if trivial) distribution is the *uniform distribution*, in which every sample point is equally likely; we have already made allusion to this case.

It is often the case that the probability distribution for a random variable of interest is not known analytically, and must be estimated from empirical observations. Given a *sample* $M = \{x_0, x_1, \ldots, x_{n-1}\}$ of repeated observations taken from a (discrete) random variable X, we can estimate $P_X(s)$ for any sample point s by simply computing the relative frequency of s in M:

$$\frac{|\{i | x_i = s\}|}{n}. \tag{2.50}$$

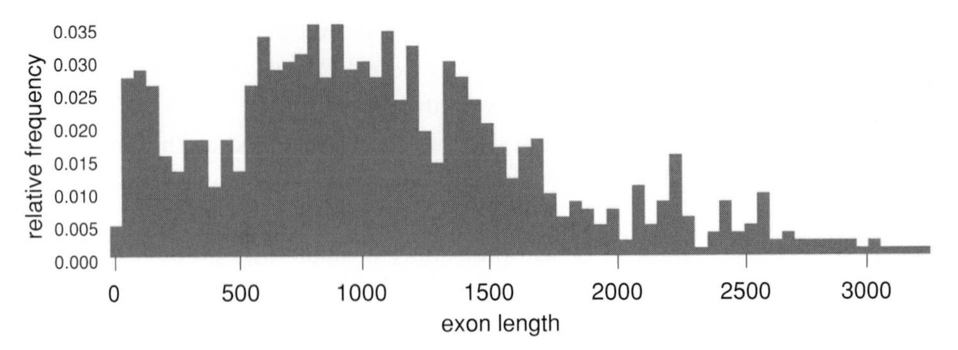

Figure 2.4 A histogram of exon lengths (bp). Interval size was 50.

According to Eq. (2.17), the limit of this quantity as our sample size approaches infinity is precisely the probability we seek. Unfortunately, sample sizes are often too small in practice to provide reliable estimates using this simple method. There are three popular ways of circumventing this problem: *binning*, *smoothing*, and *regression*.

Binning refers to the use of a *histogram* to represent the unknown distribution. A histogram provides a partitioning of the domain of a distribution into a number of discrete, nonoverlapping intervals. Each interval can then be treated as an event, with all the sample points falling within an interval being treated as mutually indistinguishable instances of that event. Figure 2.4 shows an example of a histogram. Each vertical bar in the figure corresponds to an interval, or *bin*, with the bar height representing relative frequency.

The purpose of using a histogram is to *conflate*, or collapse together, those sample points which are similar enough that they can be safely treated as having the same probability of occurrence. In this way, the counts for individual sample points can be *pooled* together to produce larger and more reliable *sample sizes*. The difficulty is in choosing an interval size for the histogram which is large enough to produce reasonable sample sizes for all of the intervals, yet small enough that significant differences between sample points will not be lost (i.e., such that we retain sufficient *resolution* for the task at hand).

Algorithm 2.3 gives a procedure for constructing a histogram for a collection of integers. The input to this algorithm is a series S of observations from a discrete random variable X, and a parameter N specifying the desired number of intervals. A reasonable way of choosing N for a given problem is to iteratively construct a number of histograms for a set of sample data and for a number of values for N, graphing each histogram to assess its smoothness. For excessively large N it may be observed that the contour of the histogram alternately rises and falls to produce many peaks and valleys – i.e., is very *rugged*. Decreasing N generally tends to reduce the ruggedness of the histogram by averaging together adjacent peaks and valleys. We recommend selecting the largest N which produces a sufficiently smooth histogram, where the definition of "sufficiently smooth" will naturally depend on

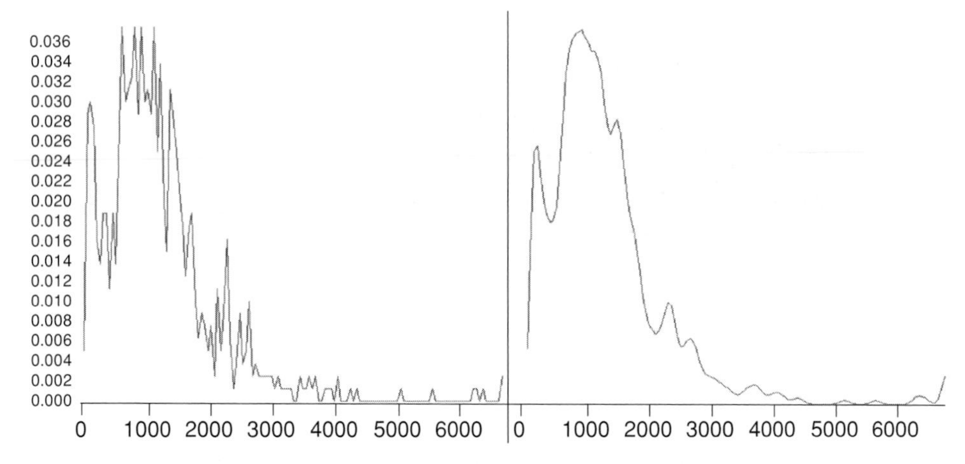

Figure 2.5 A histogram of exon lengths (bp), before (left) and after (right) three iterations of smoothing with a window size of three bins. The bin size was 50 nucleotides.

the phenomena being modeled and on one's preconceived notion of how smooth the *true* distribution is likely to be.

Algorithm 2.3 Constructing an *N*-interval histogram for a collection *S* of integers. *max*(*S*) and *min*(*S*) return the largest and smallest elements of *S*, respectively.

```
1.  procedure constructHistogram(S,N)
2.    Δ←(max(S)-min(S)+1)/N;
3.    for b←0 up to N-1 do H[b]←0;
4.    foreach x∈S do
5.      b←⌊(x-min(S))/Δ⌋;
6.      H[b]←H[b]+1/(|S|*Δ);
7.    return H;
```

Note that we divide by Δ on line 6 so that the probabilities are normalized to the level of individual sample points; this will be important when we model feature lengths in *generalized hidden Markov models* (*GHMMs* – Chapter 8), in which the granularity of state duration events must be comparable between all states of the model, in order to avoid unintended biases.

Once a histogram has been constructed it may (or may not) also be desirable to apply a *smoothing* procedure to increase the smoothness of the histogram at the selected resolution. In this way, one can attempt to obtain a relatively smooth histogram without sacrificing too much intrinsic information.

Figure 2.5 illustrates the effect of this procedure by showing a histogram before and after smoothing.

Algorithm 2.4 Smoothing an N-interval histogram H using a window of size W, over T iterations. Returns the smoothed histogram.

```
1.  procedure smoothHistogram(H,N,W,T)
2.    G←H;
3.    for i←1 up to T do
4.      for b←0 up to N-W do
5.        c←b+⌊(W-1)/2⌋;
6.        G[c]←0;
7.        for j←b up to b+W-1 do
8.          G[c]←G[c]+H[j]/W;
9.      for b←0 up to ⌊(W-1)/2⌋-1 do
10.       G[b]←G[⌊(W-1)/2⌋];
11.     for b←N-W+1+⌊(W-1)/2⌋ up to N-1 do
12.       G[b]←G[N-W+⌊(W-1)/2⌋];
13.     H←G;
14.   for b←0 up to N-1 do sum←sum+G[b];
15.   for b←0 up to N-1 do G[b]←G[b]/sum;
16.   return G;
```

There is a large literature on smoothing (e.g., Simonoff, 1996), including many descriptions of rather sophisticated techniques. We have opted to provide a variant on a simple heuristic which is commonly used and often works reasonably well in practice. Algorithm 2.4 describes a procedure for sliding a fixed-length window along the domain of the histogram, averaging together the probabilities of all the sample points falling within the window and assigning the resulting probability to the sample point nearest the center of the window.

It is worth emphasizing that smoothing should be applied only to the extent that one believes the true distribution underlying the phenomenon being modeled is indeed smooth. The desire to smooth a particular histogram should be driven not by esthetics, but by a belief that the level of detail observed in the contour of the histogram is due in larger part to sampling error (described below) than to the intricacies of the biological phenomena that generated the observed data; only then are we justified in obliterating that detail by smoothing.

An alternative to the use of histograms and smoothing is *regression*, the fitting of a parameterized curve to a set of points. Regression is useful if, after generating a histogram and displaying it graphically, one happens to notice a strong similarity to a familiar curve with a known algebraic formula. As with smoothing, there is a fairly large body of literature relating to regression, which the interested reader may wish to peruse. Here we will outline a simple *hill-climbing* procedure for fitting a curve $f(x, \theta)$ with parameters $\theta = (\alpha_1, \alpha_2, \ldots, \alpha_n)$ to a histogram H. If after performing the regression the curve is indeed found to approximate the observed data well, the function $f(x, \theta)$ may then be used as an estimate for the unknown probability function $P_X(x)$ represented by the histogram.

Let $H(x)$ denote our histogram, where $y_i = H(x_i)$ is the relative frequency associated with interval x_i. Then we can define an error measure, $\delta_{f,H}(\theta) = \sum_{x \in dom(H)} |f(x, \theta) - H(x)|^2$, the sum of the squared differences between the probability assigned by the curve $f(x, \theta)$ and that assigned by the histogram H. The problem can then be approached by applying gradient descent to the error function to find the minimum-error parameterization $\theta_{min} = argmin_{\theta} \delta_{f,H}(\theta)$. Except for the fact that we are taking the *argmin* rather than the *argmax*, and that $\delta_{f,H}(\theta)$ is effectively a function of n variables $\theta = (\alpha_1, \alpha_2, \ldots, \alpha_n)$, this is essentially the same problem addressed by Algorithm 2.2. Though we do not specify the algorithm, it should be reasonably easy for the reader to see how Algorithm 2.2 can be generalized to perform n-dimensional (rather than one-dimensional) gradient descent, and so we leave this as an exercise for the reader (see Exercises 2.8 and 2.9). Numerical optimization methods are well described in the literature (e.g., Press *et al.*, 1992; Galassi *et al.*, 2005; see also section 10.15).

An issue of particular importance when performing parameter estimation for a gene finder is that of *sampling error*. We will see in the chapters ahead that in the training of probabilistic gene finders, a very large number of distributions are typically estimated from empirical data. Broadly speaking, the purpose of using these probability distributions is to encode biases of biological significance relating to the base composition at particular locations within and around genes, and to reward putative gene predictions in proportion to the degree to which they affirm, or conform to, those biases.

As we have noted previously, the amount of training data available for training a gene finder is generally limited, resulting in sample sizes which are rarely ideal. Due to the random nature of sampling, the natural variation which is present in the sampled *population* (i.e., sample space) can cause a sample to exhibit statistical biases that are not present (or are present, but in weaker form) in the population as a whole. These spurious biases misrepresent the patterns actually present in the underlying biological phenomena being modeled, thereby reducing the accuracy of the model. These spurious patterns, which we refer to collectively as *sampling error*, are generally present in a sample in inverse proportion to the (square root of the) sample size (i.e., the *law of large numbers* – section 2.6), so that smaller samples are likely to contain more sampling error than larger samples. Thus, it is always preferable to obtain the largest samples practicable when training a gene finder.

In practice, however, one often has little or no control over the amount of training data which is made available for training a gene finder. Thus, given a set of training data, the emphasis should be on trying to assess how much sampling error is likely to be present, and on trying to mitigate the effects of that sampling error. We have already presented two very crude approaches to the latter, in the form of histogram smoothing and regression, both of which tend to eliminate small-scale variations present in a sample when recreating a probability distribution from empirical data. These methods are very crude because they make no attempt to objectively determine the degree of sampling error present, but merely smooth out the contour of the histogram under the practitioner's subjective eye. Thus, these methods should always be applied with a care for the meaning of the underlying distribution.

2.8 Parameter estimation

In the previous section we described a method based on gradient descent of a squared error function $|f(x, \theta) - H(x)|^2$ for fitting an arbitrary curve $f(\bullet, \theta)$ to a histogram H of observed relative frequencies. This method for estimating a set of parameters for a model is referred to as *least-squares estimation* (*LSE*). In the case of probabilistic modeling, a more popular alternative is *maximum likelihood estimation* (*MLE*). Maximum likelihood estimation involves finding the set of parameters θ which renders the training observations T maximally likely:

$$\theta_{\text{MLE}} = \underset{\theta}{argmax}\; P(T|\theta). \tag{2.51}$$

Precisely how this maximization can be performed efficiently in practice depends on the problem at hand; we will see a number of examples in Chapters 6–8, some of them fairly nontrivial to implement.

A somewhat less popular (but in some cases preferable) alternative to MLE is *maximum a posteriori* (*MAP*) estimation. The MAP estimator θ_{MAP} is the one which is most probable, given the training observations:

$$\begin{aligned}\theta_{\text{MAP}} &= \underset{\theta}{argmax}\; P(\theta|T) \\ &= \underset{\theta}{argmax}\; P(\theta, T),\end{aligned} \tag{2.52}$$

where the second equality follows from the invariance of $P(T)$ over the *argmax*.

We will see a number of uses for these several formulations, both for parameter estimation and also for finding the optimal *parse* of a sequence (the notion of parsing will be introduced in section 2.15), in which case θ in the foregoing equations would be replaced with a parse ϕ, and we instead speak of the *maximum likelihood* parse $\phi_{\text{ML}} = argmax_{\phi} P(S|\phi)$ of a given sequence S, or the *maximum a posteriori* parse $\phi_{\text{MAP}} = argmax_{\phi} P(\phi|S)$. In section 6.9 we will consider optimization criteria based on the goal of maximal *prediction accuracy* rather than maximal probability.

2.9 Statistical hypothesis testing

We will occasionally have use for statistical hypothesis testing, such as when assessing the significance of nucleotide composition patterns at key locations within a gene. Many statistical tests have been devised and are described in the statistics literature. Here we will limit ourselves to a brief description of the philosophy of hypothesis testing, and provide a few examples of useful tests. A number of excellent textbooks and monographs exist on the subject, including those by Zar (1996), Sokal and Rohlf (1995), and Edwards (1992).

The idea behind hypothesis testing is conceptually simple. First, we form a hypothesis H_0, called the "null" hypothesis to differentiate it from the "alternative" or "research" hypothesis H which we favor; H_0 is generally the logical negation of H. H_0 is typically expressed in the form "p is not significantly different from p_0," for some parameter p and value p_0. Given a sample S, we may then compute a *test statistic* $x = f(S)$, based on the measurements in S. The assumption of hypothesis H_0 (i.e., that $p \approx p_0$) generally induces some probability distribution D on the test statistic x, so that we can assess $P = P_D(x)$, or, more typically, $P = \sum_k P_D(k)$, where the summation is over all values k which are considered as extreme as x. If the resulting probability is small (say, not more than 5%), then we can make the argument that the value observed for the test statistic x should be unlikely to occur if hypothesis H_0 was true. We can therefore reject the null hypothesis, under our conviction that the observed test statistic would most likely not have been observed if H_0 was true. Alternatively, if the P value was *not* especially small (i.e., it was $>5\%$, for example), then we might conclude that we have insufficient justification for rejecting H_0. It should be noted that the test statistic is a surrogate for the sample; assessing the likelihood of the test statistic is merely a substitute for assessing directly the likelihood of the sample.

It should be clear from the foregoing description that hypothesis testing does not in general allow one to draw conclusions with complete certainty regarding the validity of a given hypothesis. What it can do, if practiced properly, is allow us to reduce the expected number of false hypotheses that we will inadvertently accept. If we set our rejection criterion for $P_D(x)$ at $\alpha = 5\%$, then we can expect that of the null hypotheses which we reject, roughly 5% were in fact true and should not have been rejected, thereby constituting what are known as *Type I errors*. Unfortunately, while the expected percentage α of erroneously rejected null hypotheses (i.e., Type I errors) is typically fairly simple to assess, the expected percentage β of null hypotheses which should have been rejected but were not (termed *Type II errors*) is often very difficult to estimate, and so therefore is $1 - \beta$, the *power of the test* to reject a false null hypothesis.

Let us consider a test based on the binomial distribution, which was defined in Eq. (2.46). Suppose we are given a sample consisting of N sequences in a *multiple alignment* (i.e., a table in which each row represents a sequence and each column contains residues or *gap characters* making up the aligned sequences – see Figure 9.23 in Chapter 9), and we wish to assess whether the observed frequency $X = f_G$ of Gs occurring at a particular position differs significantly from the average frequency b_G of Gs at all other positions. If we believe that the bases occurring in the N sequences at any given position can be reasonably modeled as having resulted from N i.i.d. Bernoulli trials (almost certainly not a valid assumption – see section 9.6), then the random variable X representing the frequencies of Gs observed in samples such as this one should follow a binomial distribution with parameter p, the probability of occurrence of a G. Therefore, the hypothesis H_0 which we wish to test for this example is that the frequency of Gs follows a binomial distribution $D(p)$ with parameter $p = b_G/N$. By applying Eq. (2.46) with $n = N$ we can assess the probability $P_D(X = x)$ of observing x Gs for any given x. Once we have settled on a rejection criterion α (typically 0.05 or less) we can delimit a region $R \subset dom(X)$ for which $P(R) = \sum_{x \in R} P_D(x) \approx \alpha$; for small values of α, the event R should be rare. We call R the *rejection region*, and impose the rule that if an observed X falls in region R then we will reject the hypothesis H_0.

Exactly how we identify the rejection region depends on the form of the hypothesis being tested. If the alternative hypotheses being considered are of the form $H_0 : x \geq x_0$ versus $H : x < x_0$, then we need only consider those values at the extreme right of the distribution. This is termed a *one-tailed test*. For hypotheses of the form $H_0 : x = x_0$ versus $H : x \neq x_0$, we will generally perform a *two-tailed test*, in which the rejection region R will consist of all sample points less than some x_L or greater than some x_R, with x_L and x_R typically being chosen so that $P_D(X < x_L) = P_D(X > x_R) = \alpha/2$; i.e., so that $P(R) = \alpha$.

As a second example we will consider the χ^2 *goodness-of-fit* test. Suppose that we are given a multiple sequence alignment as in the previous example, but rather than assessing whether the frequency of Gs at a given position differs significantly from the background probability, we instead wish to test whether the entire distribution of As, Cs, Gs, and Ts at that position differs significantly from the background distribution. That is, our null hypothesis is $H_0 : D \approx D_{background}$, that the distribution D of bases at that position is not different from the pooled distribution $D_{background}$ of bases at all other positions. We can test this hypothesis by computing a χ^2 test statistic according to:

$$\chi^2 = \sum_{i=0}^{k-1} \frac{(c_i - c_i')^2}{c_i'} \tag{2.53}$$

(for observed counts c_i and expected counts c_i' – see below) and then finding the corresponding *P*-value in a χ^2 table (such a table and the C++ source code for utilizing it are available for download at the book's website; see Appendix). Values of $P < 0.05$ suggest that the background distribution is *not* a good fit for the sample,

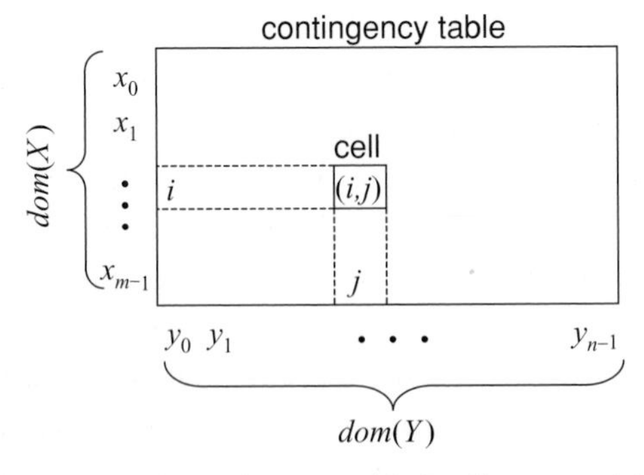

Figure 2.6 An example contingency table for discrete variables $X \in \{x_i | 0 \leq i < m\}$ and $Y \in \{y_j | 0 \leq j < n\}$.

so that in this example we would conclude that the distribution of residues at this position is significantly different from the others. In our example, $k = 4$, c_i is the observed frequency (i.e., count) of nucleotide i in the sample, and c_i' is the expected count of nucleotide i based on the background distribution and the sample size – i.e., $N \cdot P_{background}(x_i)$. Note that the χ^2 test may be unreliable if any individual expected count is extremely small. When the background distribution exhibits only moderate departure from uniformity, requiring that all $c_i' \geq 1$ may be sufficient to ensure a reliable test at the $\alpha = 0.05$ level; for larger departures from uniformity or for smaller α values it may be advisable to require all $c_i' \geq 4$ or $c_i' \geq 5$. For more guidelines on the application of the test and the avoidance of unwanted bias (see, e.g., Zar, 1996).

A related test is the χ^2 *test of independence*. Suppose that we are given a set of observations for which variables X and Y each take on a set of discrete values – i.e., $X \in \{x_i | 0 \leq i < m\}, Y \in \{y_j | 0 \leq j < n\}$. Then we can form a *contingency table*, as illustrated in Figure 2.6, in which cell (i, j) contains the number of observations for which $X = x_i$ and $Y = y_j$.

Given a contingency table of observed counts c_{ij}, we can compute a table of expected counts c_{ij}' where:

$$c_{ij}' = \frac{\left(\sum_{h=0}^{m-1} c_{hj}\right)\left(\sum_{k=0}^{n-1} c_{ik}\right)}{\sum_{h=0}^{m-1}\sum_{k=0}^{n-1} c_{hk}}. \tag{2.54}$$

This latter table gives the expected counts under the assumption of independence. From these two tables we can compute a χ^2 statistic similar to Eq. (2.53):

$$\chi^2 = \sum_{\substack{0 \leq j < n \\ 0 \leq i < m}} \frac{(c_{ij} - c_{ij}')^2}{c_{ij}'}. \tag{2.55}$$

Looking this value up in a χ^2 table again gives us the probability P of observing this table under the assumption of independence, so that a value of $P < \alpha$ suggests that we reject the hypothesis of independence and instead favor the conclusion that the variables are in some way dependent. The reader is again referred to a standard statistics reference such as Zar (1996) for suggested guidelines on minimum sample sizes for this test (i.e., a common stipulation is no zero entries and few entries smaller than 5, though not all texts agree).

Note that classical hypothesis testing is in some ways analogous to *binary classification* (see section 10.1) based on likelihood ratios – i.e., given the test statistic and a rejection region, the foregoing outlines a decision procedure for classifying the result of an experiment as supporting either the research hypothesis or the null hypothesis.

A sample of additional readings related to statistical significance in biosequence analysis includes those by Mitrophanov and Borodovsky (2006), Eddy (2005), Pearson and Wood (2001), and Karlin and Altschul (1990).

2.10 Information

We will require several concepts from *information theory*, which is a mathematical discipline concerned with measuring uncertainty and the degree to which an event or a quantity of information reduces uncertainty.

The basis of information theory is the *bit*, or *binary digit*, which is a variable that can assume the value 0 or 1. If we consider a set of symbols $S = \{\mathtt{A}, \mathtt{C}, \mathtt{G}, \mathtt{T}\}$, we can encode the symbols into strings of bits using an encoding such as $E = \{\mathtt{A} \leftrightarrow 00, \mathtt{C} \leftrightarrow 01, \mathtt{G} \leftrightarrow 10, \mathtt{T} \leftrightarrow 11\}$. In this encoding each symbol in S requires two bits. Other encodings are possible, however, which represent each symbol with different numbers of bits, such as $E = \{\mathtt{A} \leftrightarrow 110, \mathtt{C} \leftrightarrow 10, \mathtt{G} \leftrightarrow 0, \mathtt{T} \leftrightarrow 111\}$. If we wish to encode a sequence of symbols from S using a small number of bits, we would be wise in general to choose an encoding E in which the most frequent symbols from S require fewer bits than the rarer ones, so that for typical strings containing mostly common symbols, the average encoding lengths for the sequences will be minimized.

Although we will not give the proof, it can be shown that if the symbols drawn from some countable set S occur in sequences with frequencies given by discrete probability distribution D, then the optimal encoding of such sequences into bit strings can be achieved in no fewer than H_D bits per symbol on average, where H_D is given by:

$$H_D = - \sum_{\substack{x \in S, \\ P_D(x) > 0}} P_D(x) \log_2 P_D(x) \tag{2.56}$$

(Shannon, 1948). H_D is called the *entropy of D*, and it is maximized when D is the uniform distribution, in which case the entropy is simply $H_{max} = \log_2 |S|$. Thus, entropy reflects the *evenness* (formally defined as H/H_{max}) of a distribution. Entropy can also be seen as a measure of the uncertainty of a random variable, since a random variable which is overwhelmingly likely to assume one particular value out of many will have a lower entropy than one which can assume any value with equal probability (assuming the number of possible outcomes, $|S|$, is constant); clearly, the latter, higher-entropy random variable is more uncertain than the former. Inasmuch as entropy measures the uncertainty about which value will be assumed by a random variable, it is also a fitting measure of the amount of information which is gained upon observing the variable – in this sense, entropy is a measure of the amount of uncertainty which has been eliminated as a result of making the observation.

As an example, consider the three nucleotides comprising stop codons in human genes. Roughly 50% of human stop codons are of the TGA variety, 27% are TAAs, and 23% are TAGs. Given a randomly sampled stop codon from the collection of all human protein-coding genes, the first position can be seen to have an entropy of 0 bits (since this position is always occupied by a T, so there is no uncertainty), the second position has an entropy of 1 bit (since a G occupies this position 50% of the time and an A occurs here the other 50%), and the third position has an entropy of 0.78 bits (since the relative frequencies of Gs and As in this position are 23% and 77%, respectively). Thus, the first position of human stop codons is the most predictable and the second position the least.

There are a number of functions related to the entropy which are useful. The first is *relative entropy*, which can be thought of as the number of wasted bits resulting from our modeling a discrete random variable as following some distribution C when in fact it follows distribution D:

$$H_{D,C} = \sum_{\substack{x\in S,\\ P_D(x)>0,\\ P_C(x)>0}} P_D(x)\log_2\frac{P_D(x)}{P_C(x)}. \tag{2.57}$$

The relative entropy is also called the *Kullback–Leibler divergence*, and is used as a measure of how severely two distributions differ (Kullback, 1997). The value is never negative, is zero only when the distributions are identical, and is asymmetric, so that $H_{D,C}$ is not guaranteed to equal $H_{C,D}$. Equation (2.57) can be factored as:

$$\sum_{\substack{x\in S,\\ P_D(x)>0}} P_D(x)\log_2 P_D(x) - \sum_{\substack{x\in S,\\ P_C(x)>0}} P_D(x)\log_2 P_C(x), \tag{2.58}$$

and the second term:

$$-\sum_{\substack{x\in S,\\ P_C(x)>0}} P_D(x)\log_2 P_C(x) \tag{2.59}$$

is called the *cross entropy*. For the purposes of modeling a discrete random variable which follows some unknown distribution D, improving our model C will correspond to minimizing the cross entropy between D and C, or, equivalently, minimizing the relative entropy. Other entropic measures include the *joint entropy*:

$$-\sum_{x \in dom(X)} \sum_{\substack{y \in dom(Y), \\ P(x,y)>0}} P(x,y) \log_2 P(x,y) \tag{2.60}$$

and the *conditional entropy*:

$$-\sum_{x \in dom(X)} \sum_{\substack{y \in dom(Y), \\ P(x|y)>0}} P(x,y) \log_2 P(x|y), \tag{2.61}$$

which, unlike the previous entropic measures, are applied to pairs of discrete random variables X and Y. In the case of sequence analysis, X and Y are typically adjacent positions in a sequence (with Y preceding X), so that the conditional entropy gives us a measure of the information content or "complexity" of a sequence when taking *dinucleotide* patterns (i.e., pairs of adjacent residues) into account.

Higher-order joint and conditional entropies may also be computed by taking Y to be several consecutive symbols in a sequence rather than just the single symbol preceding each X. For example, given the sequence:

AACCAAGGTT

the 2nd-order conditional and/or joint entropies may be computed by taking successive pairs (X, Y) to be:

(C, AA), (C, AC), (A, CC), (A, CA), (G, AA), (G, AG), (T, GG), (T, GT)

and computing as in Eqs. (2.60) and (2.61) to produce a joint entropy of 3 bits and a conditional entropy of 0.25 bits.

Figure 2.7 graphs the nth-order conditional entropies of exons, introns, and intergenic regions in the human genome, for successively higher values of n. From this figure we may perceive several things. First, the 0th-order entropies are close to 2 bits per symbol, which is the expected entropy for random sequences over four symbols of equal frequency:

$$-\sum_{i=1}^{4} \frac{1}{4} \log_2 \frac{1}{4} = -\log_2 \frac{1}{4} = -\log_2 1 + \log_2 4 = \log_2 4 = \log_2 2^2 = 2$$

The fact that the 0th-order entropies are slightly less than 2 bits per symbol merely reflects the fact that the overall {G,C} density of the human genome is not 50%, but is closer to 41%, so that the optimal encoding of the entire genome would not allocate the same number of bits to all four nucleotides. Second, as the order n increases, the conditional entropies decrease. This is true for all three classes of sequences, indicating that both coding and noncoding sequence contains systematic patterns which deviate from pure randomness and which increase the predictability of the sequence when longer contexts are observed (see Exercise 2.46). Third, the entropies of the coding regions are consistently lower than those of the noncoding regions, reflecting the fact that natural selection has acted more

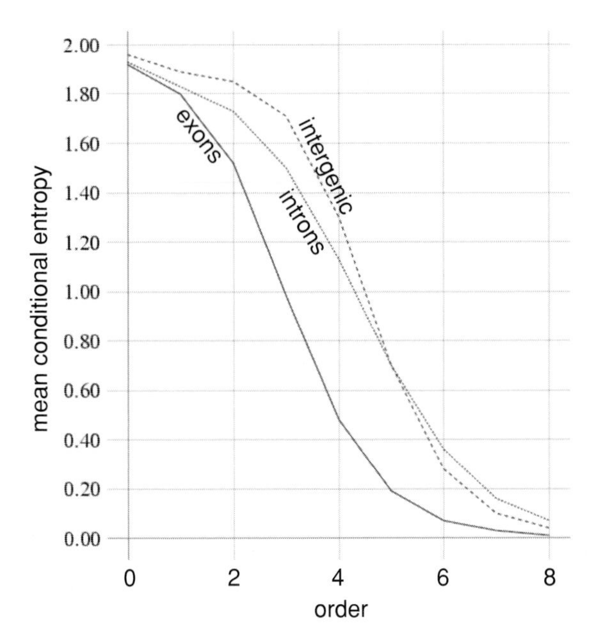

Figure 2.7 Average conditional entropies (y-axis) of human exons, introns, and intergenic regions, for successively higher orders (values of n) (x-axis). Exon coordinates were obtained from RefSeq (Pruitt et al., 2005). Intergenic regions were obtained from the 5′ and 3′ regions in the vicinity of RefSeq genes.

strongly on average in the coding regions to maintain particular patterns necessary for the integrity and functionality of the genome (or, more specifically, of the *proteome*); unfortunately, the differential effects of natural selection as measured by entropy are too crude to be used as a solitary means of discrimination between coding and noncoding intervals, though we will see in later chapters that these patterns may be employed in concert with other information to achieve respectable gene-finding accuracy.

Finally, it can be seen that all three curves follow an S-shape, suggesting that each of the three classes of genomic elements has an intrinsic *word length*, or optimal predictive context length, beyond which no (or very little) additional information may be gleaned (on average). This will become relevant when we consider higher-order emission probabilities for Markov models in Chapters 6 and 7.

The *mutual information* is another useful measure applicable to discrete random variables:

$$\begin{aligned} M(X,Y) &= \sum_{\substack{x \in dom(X), \\ y \in dom(Y), \\ P(x,y)>0}} P(x,y)\log_2 \frac{P(x,y)}{P(x)P(y)} \\ &= H(X) + H(Y) - H(X \times Y) \end{aligned} \tag{2.62}$$

for $H(X)$ the entropy function for random variable X. The mutual information $M(X,Y)$ measures the amount of information about one random variable that we can glean from another random variable. The log term in this formula when applied to specific values for X and Y is called the *pointwise mutual information*. Mutual information is measured for *paired observations*; i.e., when the random variables X and Y are observed at the same time. $M(X, Y)$ evaluates to 0 if X and Y are independent, and to the entropy of X (or Y) if the random variables always behave identically.

2.11 Computational complexity

The *computational complexity* of an algorithm refers to the amount of time and/or space that is consumed during a run of the algorithm. When devising a new algorithm or selecting from among a set of known algorithms for performing a given task, it is important to take into consideration the time and space complexity of a prospective procedure so that the resulting program will run as efficiently as possible given the available processing and memory resources. As we present the algorithms in this book we will often characterize the complexity of our procedures using the standard $\mathcal{O}$ (read "big-Oh") notation, which provides a (largely) hardware-independent means of assessing the resource requirements of an algorithm.

Suppose we wish to characterize the expected execution time of an algorithm A as a function of x, where x is a parameter to A. An empirical study of A on a set of parameters $\{x_i\}$ could be performed on a particular computing system to produce a set of $(x_i, time_i)$ data points, to which we could then fit a suitable curve f using the regression techniques mentioned previously. Then $f(x)$ would provide a reasonable prediction of the amount of time required to run $A(x)$ on this particular computing system. Unfortunately, $f(x)$ would be of questionable utility if we wished to run A on a different computing architecture, due to differences in hardware speed. Furthermore, because $f(x)$ was inferred empirically, the accuracy of our predictions can be quite low for values of x outside the range of our study set, $\{x_i\}$, even for an identical computing system.

The $\mathcal{O}$ formalism addresses the first of these problems by expressing computational complexity in terms of an unknown constant c which controls for (among other things) the differences between computing systems. When assessing expected execution time, the constant c is typically chosen to represent the time required to perform one processor cycle, or one instruction, or one memory access, on the target computing system. The second problem, that of $f(x)$ being an empirically observed relation, can be addressed by instead inferring $f(x)$ analytically from the structure of the algorithm. Methods for performing such an analysis are beyond the scope of this book, and are adequately described elsewhere (e.g., Cormen *et al.*, 1992). For our purposes it is more important that the reader know how to interpret the results of such an analysis.

Formally, let $f(x)$ denote the amount of some computing resource which is consumed, as a function of x, where x is a parameter to algorithm A, and let $g(x)$

Table 2.1 Common characterizations of algorithm efficiency, in terms of parameter n and constant b

Complexity	Description
$O(1)$	constant
$O(\log n)$	logarithmic
$O(n)$	linear
$O(n \log n)$	"n log n"
$O(n^2)$	quadratic
$O(n^b)$	polynomial
$O(b^n)$	exponential
$O(n!)$	factorial

be a real-valued function. Then we define O such that:

$$[f(x) = O(g(x))] \Leftrightarrow [\exists_{c,k} \forall_{x>k} f(x) \leq c \cdot g(x)]. \tag{2.63}$$

This states that for sufficiently large values of x, $f(x)$ can be bounded above by some fixed multiple c of $g(x)$. Given such a characterization, we can then assess the relative efficiency of A with respect to the chosen resource by observing the form of function $g(x)$. Table 2.1 lists a number of commonly used forms for $g(x)$, and provides a useful hierarchy into which an algorithm can be categorized to assess its relative efficiency.

To give the reader some perspective on the asymptotic run-time of algorithms:

- adding two numbers requires constant time
- searching a sorted array can be done in logarithmic time (see Algorithm 2.9)
- iterating through all the elements of such an array is linear
- sorting an array of random numbers is "n log n" on average (see Algorithm 2.8)
- iterating through all the elements of a square matrix is quadratic
- anything slower than polynomial is generally considered impractical when n is large.

Often we will find for a given problem that the choice of algorithm for solving the problem involves a trade-off between time and space complexity, so that improving the run-time of a program may require the use of more memory.

2.12 Dynamic programming

A fundamental concept in bioinformatics is *dynamic programming*, a technique which can greatly improve the efficiency of some classes of algorithms by identifying overlapping subproblems to be solved by the algorithm and ensuring that no work will be needlessly duplicated as the algorithm progresses through the subproblems. We will illustrate the issue using the notion of recurrence relations from discrete mathematics.

A *recurrence relation* is a function which is defined recursively in terms of itself. Perhaps the most celebrated example is that which defines the Fibonacci sequence:

$$F(x) = \begin{cases} 0 & \text{if } x = 0 \\ 1 & \text{if } x = 1 \\ F(x-1) + F(x-2) & \text{otherwise} \end{cases} \tag{2.64}$$

where $x \geq 0$. Note that the "otherwise" case of this definition causes $F(x)$ to be defined in terms of itself – i.e., *recursively*. The parts of a recurrence relation which contain calls to the function being defined (F in the present example) constitute the *recursion* of the definition, and the remaining parts (those not making further calls to the function being defined) are called the *basis*. Recursion and induction (section 2.2) are very closely related ideas, and indeed, the values returned by recursive calls to a function in a recurrence relation are often referred to as *inductive scores*.

The relation shown in Eq. (2.64) generates the sequence 0, 1, 1, 2, 3, 5, 8, 13, 21, 34, 55, ... Computing the nth Fibonacci number can be done simply by evaluating $F(n)$ as specified in Eq. (2.64), using a recursive procedure, as shown in Algorithm 2.5.

Algorithm 2.5 Computing the nth number in the Fibonacci sequence, using recursion.

```
1.  procedure fibonacci_recurs(n)
2.    if n<2 then return n;
3.    return fibonacci_recurs(n-1)+
4.      fibonacci_recurs(n-2);
```

However, this approach turns out to be very wasteful due to the phenomenon of *overlapping subproblems*. Consider the computation of $F(10)$:

$$\begin{aligned} F(10) &= F(9) + F(8) \\ &= F(8) + F(7) + F(7) + F(6) \\ &= F(7) + F(6) + F(6) + F(5) + F(6) + F(5) + F(5) + F(4) \\ &\ \ \dots \end{aligned}$$

By the third line it is clear that a significant amount of wasted computation will be spent in re-computing terms: $F(6)$ and $F(5)$ both appear three times in the third line, and as each of these are expanded recursively according to Eq. (2.64), the problem of redundant evaluation of subproblems manifests itself yet further. The result is that a recursive implementation of $F(n)$ requires exponential time – i.e., $time(F_{recurs}, n) = O(b^n)$, where $b \approx 1.6$ (Cormen *et al.*, 1992).

A more efficient solution is to compute all values of $F(x)$ starting at 2 and working up to n, keeping all intermediate values stored in an array for fast look-up, as shown in Algorithm 2.6. We will call this a *bottom-up dynamic programming* approach. Such an approach is linear in n – a significant improvement over the naive, exponential-time approach.

Algorithm 2.6 Computing the nth number in the Fibonacci sequence, using bottom-up dynamic programming. Run time is $O(n)$.

```
1.  procedure fibonacci_dp(n)
2.     for i←0 up to 1 do array[i]←i;
3.     for x←2 up to n do
4.       array[x]←array[x-1]+array[x-2];
5.     return array[n];
```

The reason Algorithm 2.6 is so much faster than the recursive approach is that the dynamic programming version computes each $F(x)$ for $0 \leq x \leq n$ only once, whereas the recursive version will typically compute each $F(x)$ many times. Our remedy was to ensure that each $F(x)$ was computed exactly once, and this goal was achieved by changing the *evaluation order* of the recurrence so that the smaller values of x were handled before the larger values which depended upon them. This issue is typically described in terms of the *bottom-up* versus *top-down* approach to evaluation, but more generally it is one of evaluation order, where the optimal order for a given recurrence might be simply described as bottom-up, or left-to-right, or might instead be defined by some complex sequence of elements in the parameter space, ordered according to the particulars of the problem at hand. We will see later (section 2.14) that the notion of a *directed graph* is sometimes useful for this purpose; an analysis of the dependency structure of the recurrence can guide the construction of an acyclic network of links that are then sorted so as to provide an evaluation order in which the values of subexpressions are always available by the time they are needed by expressions containing them.

It is not entirely accurate to attribute the value of dynamic programming to evaluation order alone, however. An alternate form of dynamic programming called *memoization* is sometimes practiced, in which a recursive procedure is combined with a global variable or *cache* into which the values of subproblems can be stored for later reuse. In this way, we can adopt a top-down approach while still ensuring that no particular subproblem will ever be evaluated twice. A *memoized* version of Algorithm 2.5 is given in Algorithm 2.7.

Memoization brings with it two advantages: (1) it permits the use of a recursive implementation of an algorithm, which for some problems is most intuitive, and

(2) it ensures that only subproblems that are actually needed will be computed. For example, if the Fibonacci numbers were instead defined such that $F(x) = F(x-2) + F(x-5)$, it might not be immediately obvious which values of $F(x)$ need to be evaluated during a bottom-up procedure and which others can be skipped, so that an analysis of the recursive structure of the recurrence would be necessary in order to determine the evaluation order (or, more precisely, the *evaluation set* – that is, the set of parameters $\{x_i\}$ for which the dynamic programming array needs to be initialized in order to evaluate $F(n)$). With the memoization approach, this analysis is automatically performed by the recursive procedure, making algorithm development simpler. The cost for this simplicity is the additional computation in terms of cache look-ups and recursive function calls, and depending on the level of efficiency desired, this additional cost may or may not be acceptable. Indeed, in cases involving very deep recursion, there is the danger that an excessive number of recursive calls will overflow the program's internal *call stack*, causing the program to terminate abnormally (i.e., *crash*). For this reason, it is wise to apply some caution when using memoization.

Algorithm 2.7 Computing the nth number in the Fibonacci sequence, using memoization. The associative array G is global, and is maintained between calls. Run-time is again $O(n)$.

```
1.  procedure fibonacci_memo(n)
2.     if G[n]≠NIL then return G[n];
3.     if n<2 then G[n]←n;
4.     else G[n]←fibonacci_memo(n-1)+
5.        fibonacci_memo(n-2);
6.     return G[n];
```

It is worthwhile to point out that while dynamic programming algorithms are often used to solve optimization problems, the use of a dynamic programming approach does not automatically ensure that the resulting algorithm will find the globally optimal value of the function to be optimized; this may seem an obvious point, but as it is not always obvious whether a given dynamic programming implementation will produce a globally optimal solution to an optimization problem, the task of verifying the correctness and optimality of a proposed algorithm is a worthwhile one.

2.13 Searching and sorting

Standard algorithms for searching and sorting data of various types are described in great detail in any number of textbooks (e.g., Cormen *et al.*, 1992). We therefore treat the subject very briefly.

Sorting an array A of numbers can be carried out efficiently – $O(N\log_2 N)$ on average – using the *quick sort* algorithm, shown in Algorithm 2.8. This procedure recursively partitions an interval in the array into two smaller intervals, rearranging elements as necessary to ensure that all the elements in the first interval are smaller than the elements in the second interval, before recursively applying the same procedure to each of the intervals in turn.

Algorithm 2.8 Sorting an array A of size N using *quick sort*. The procedure `swap(A,i,j)` exchanges elements i and j in array A.

```
1.  procedure quickSort(ref A,N)
2.    quickSort(A,0,N-1);
3.  procedure quickSort(ref A,begin,end)
4.    if begin≥end then return;
5.    pivot←partition(A,begin,end);
6.    quickSort(A,begin,pivot);
7.    quickSort(A,pivot+1,end);
8.  procedure partition(ref A,begin,end)
9.        pivot←begin+random(0..end-begin);
10.       temp←A[pivot];
11.       swap(A,pivot,begin);
12.       i←begin-1; j←end+1;
13.       while true do
14.         do j←j-1 while A[j]>temp;
15.         do i←i+1 while A[i]<temp;
16.         if i≥j then return j;
17.         else swap(A,i,j);
```

Given a sorted array of N numbers, we can find an element in the array in time $O(\log_2 N)$ by performing a *binary search*.

Algorithm 2.9 Binary search for element E in array A of size N.

```
1.  procedure binarySearch(E,A,N)
2.    begin←0; end←N;
3.    while begin<end do
4.      mid←⌊(begin+end)/2⌋;
5.      m←A[mid];
6.      if E>m then begin←mid+1 else end←mid;
7.    return begin;
```

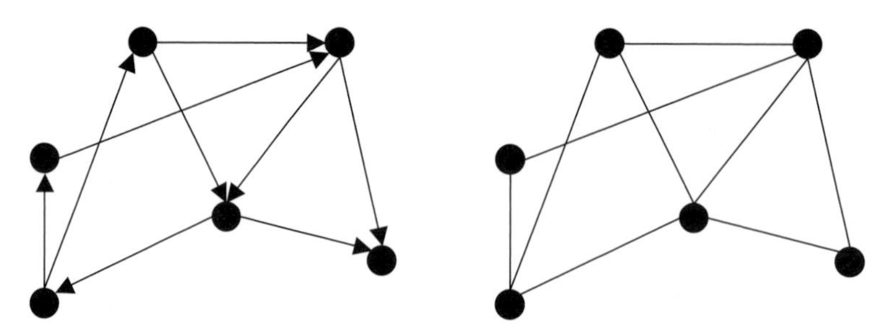

Figure 2.8 An example directed graph (left) and its underlying undirected graph (right).

The binary search algorithm recursively examines the middle element of an interval of an array, and if the sought element is less than the examined element, it recurses onto the first half of the interval; otherwise, it recurses onto the second half of the interval. The procedure is shown in Algorithm 2.9.

2.14 Graphs

A *network* or *graph* is a formalism which allows us to study patterns of connectivity among discrete elements. Let $G = (V, E)$ be defined by a set of elements V together with a set of *edges* $E \subseteq V \times V$ specifying the pairwise connections between those elements. G is called a *graph*, and the elements of V are referred to as the *vertices* or *nodes* of the graph. Vertices are typically depicted visually as small circles, and each edge $e = (v_i, v_j)$ is typically depicted as an arrow pointing from vertex v_i to vertex v_j. We will sometimes denote an edge $e = (v_i, v_j)$ using the notation $v_i \rightarrow v_j$ to reinforce the notion that e is a directed edge pointing from v_i to v_j; this should not be confused with other uses of the arrow operator as defined previously. An *undirected graph* is a special type of graph in which edges have no inherent directionality, so that (v, w) and (w, v) denote the same edge. A *directed graph* is simply a graph which is not undirected. The *underlying undirected graph* of a directed graph $G_{dir} = (V, E)$ is an undirected graph $G_{un} = (V, U)$ in which $E \subseteq U$ and $\forall_{(w,v) \in U} [(w, v) \in E \vee (v, w) \in E]$. Figure 2.8 illustrates this relation.

A *weighted graph* $G = (V, E, f)$ is merely a graph (V, E) that has been augmented with a *weighting function* $f : E \mapsto \mathbb{R}$ which associates a *weight* or *score* with each edge. We will sometimes use the shorthand $G = (V, E)$ to denote a weighted graph $G = (V, E, f)$ when the weighting function f is obvious from the context. The *topology* of a weighted graph $G = (V, E, f)$ is simply the corresponding graph (V, E) having no weighting function.

An edge $e = (w, v)$ is said to be *incident* on vertex w, as well as on v. In a directed graph, such an edge is said to *enter* or *lead into* v, and to *leave* or *lead out of* w. The

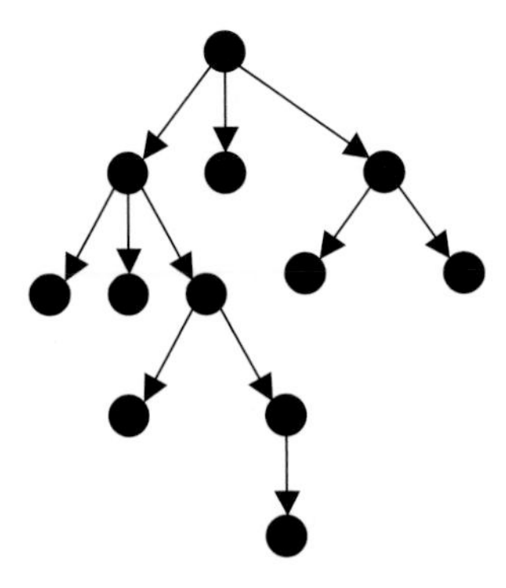

Figure 2.9 An example of a tree. The root is shown at the top. This tree has five internal nodes and seven leaves.

in-degree of a vertex v in a directed graph is the number of edges leading into v, and the *out-degree* is the number of edges leading out of v. In an undirected graph, the *degree* of vertex v is the number of edges incident on v. Given the existence of an edge (w, v) in a graph G, we can say that w and v are *adjacent* in G.

If $G_1 = (V_1, E_1)$ and $G_2 = (V_2, E_2)$ are graphs, then G_1 is a *subgraph* of G_2 iff $V_1 \subseteq V_2 \wedge E_1 \subseteq E_2$. A *path* of length $L > 0$ is a series $P = (v_0, v_1, \ldots, v_L)$ of vertices such that $\forall_{0 \leq i < L}(v_i, v_{i+1}) \in E$; we say that v_i is *reachable* from v_j for any $i > j$ in such a path. A path P of length L beginning and ending at the same vertex is called a *cycle* or a *circuit*. A directed graph having no cycles is called a *directed acyclic graph*, or *DAG*. If every vertex in an undirected graph is reachable from every other vertex, we say that the graph is *connected*; a directed graph is considered connected if its underlying undirected graph is connected. A connected subgraph S of an undirected graph G is called a *connected component* of G, as long as S is not contained within any other connected component having more vertices or edges than S. If $\forall_{i \neq j}(v_i, v_j) \in E$, we say that the graph is a *complete graph* or a *fully connected graph*.

Two graphs $G_1 = (V_1, E_1)$ and $G_2 = (V_2, E_2)$ are said to be *isomorphic* if $|V_1| = |V_2|$ and there exists a function $f : V_1 \mapsto V_2$ having the property that $(u, v) \in E_1 \Leftrightarrow (f(u), f(v)) \in E_2$; in this case f is said to be an *isomorphism* between G_1 and G_2. We may also consider isomorphisms between subgraphs, which are a straightforward generalization of the above definition.

If a connected, directed acyclic graph has the property that every vertex has an in-degree of 1 or less, with only one vertex having an in-degree of zero, then we call the graph a *tree*. Vertices having a nonzero out-degree are called *internal nodes*, while those with zero out-degrees are called *leaves*. The single vertex with in-degree of zero is called the *root*. If a tree has an edge (w, v), then we say that w is the *parent* of v, and v is a *child* of w. Trees are useful in implementing a variety of data structures. An example tree is shown in Figure 2.9.

Tree traversal is a common operation on trees. One method of tree traversal is *postorder* traversal, in which a vertex's children are visited before the vertex itself is visited (see, e.g., Cormen *et al.* (1992) for elaboration). Tree algorithms are often specified in terms of *subtrees*, where a subtree is any subgraph of the original tree such that the subgraph is itself a tree.

A particularly useful construct is the *decision tree*, in which every internal node v_i has associated with it an integer-valued decision function $\zeta_i(\hat{y})$, where $\hat{y}$ is a *test case*, or vector of parameters, and where the value of the function for any given input determines which child of v_i should be visited next. Given a test case $\hat{y}$, we begin at the root of the tree, applying the decision function at each successively chosen child node as we work our way to a leaf node. Once at a leaf node we may use the identity of the particular leaf to classify the test case, or we may evaluate a function specific to that leaf, depending on the problem to which the decision tree is being applied. Decision trees are described in greater detail in section 10.6.

Another useful tree-based data structure is the *binary tree*, which can be used to represent a sorted collection of data. Each vertex in a binary tree has at most two children, referred to as the *left child* and the *right child*. A single datum is stored at each vertex. A valid binary tree has the property that for any given vertex v_i with left child v_{left} and right child v_{right}, the datum d_i stored in v_i is greater than all of the data in the subtree rooted at v_{left} and less than all of the data in the subtree rooted at v_{right}, where *greater than* and *less than* are defined specific to the data type being stored in the tree. Algorithms for building, searching, and otherwise manipulating binary trees and related data structures are well described in a number of texts (e.g., Cormen *et al.*, 1992).

Given a set of common utility routines for binary trees, we can easily implement a useful data structure known as a *priority queue*. A priority queue Q is merely a sorted collection which (efficiently) supports two operations: *insert* (x, Q) and *extract-Min*(Q). The first operation inserts datum x into Q without violating the sorted ordering of Q, and the second operation removes the smallest element of Q and returns that element. Priority queues may be implemented using a variety of techniques, most notably the binary tree and the *heap*, which we will not describe (see Cormen *et al.*, 1992). Exercises 2.36–2.38 explore the implementation of a priority queue using a binary tree.

Given a connected, weighted, undirected graph $G = (V, E, f)$, it is sometimes useful to extract a subgraph called a *minimal spanning tree* (*MST*), from G. An MST of graph $G = (V, E, f)$ is an acyclic, connected, undirected subgraph $M = (V, B, f)$, $B \subseteq E$, for which the sum of the edge weights in the tree is minimal. Extracting an MST from a graph can be done fairly simply, as shown in Algorithm 2.10; an example MST is illustrated in Figure 2.10.

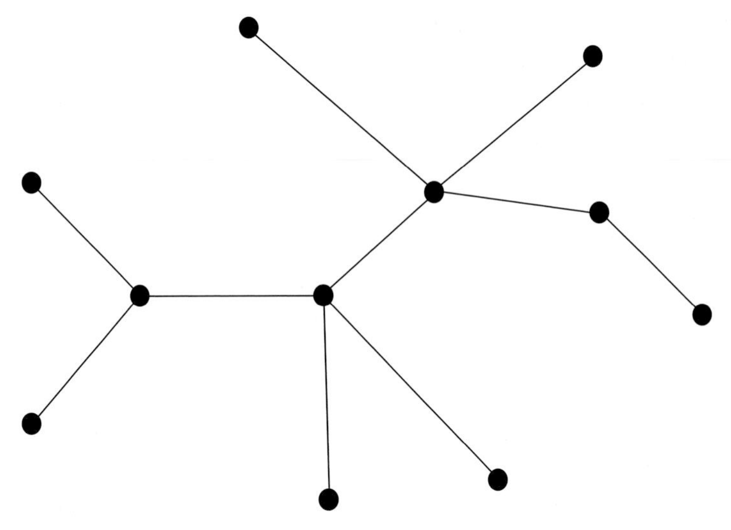

Figure 2.10 An example of a minimal spanning tree (MST). This tree is unrooted. Distances are omitted for clarity.

Algorithm 2.10 Finding a minimal spanning tree for a connected, weighted, undirected graph $G = (V, E, f)$. Returns the edges of the MST. The `sort(E,f)` procedure sorts a collection E in order of increasing value as defined by function $f : E \mapsto \mathbb{R}$.

```
1.  procedure MST(V,E,f)
2.    EMST←{};
3.    foreach x∈V do S[x]←{x};
4.    Esorted←sort(E,f);
5.    i←0;
6.    while i<|E| do
7.      (v,w)←Esorted[i];
8.      if S[v]≠S[w] then
9.         EMST←EMST∪{(v,w)};
10.        Y←S[v]∪S[w];
11.        foreach x∈Y do S[x]←Y;
12.      i←i+1;
13.    return EMST;
```

Several different schemes are possible for the representation of graphs in a computer. These include the *incidence matrix*, the *adjacency matrix*, and the *pointer representation*. An *adjacency matrix* **M** for a graph $G = (V, E)$ is a $|V| \times |V|$ array which has been initialized so that $\mathbf{M}_{ij}$ is 1 if there exists an edge (v_i, v_j), and is 0 otherwise.

For a weighted graph, $\mathbf{M}_{ij}$ may instead store the weight associated with edge (v_i, v_j). An *incidence matrix* $\mathbf{I}$ is an $|E| \times |V|$ array which has been initialized so that $\mathbf{I}_{ij} = 1$ if edge e_i is incident on vertex v_j, and $\mathbf{I}_{ij} = \mathbf{0}$ otherwise. Finally, in the *pointer representation* of a graph, each vertex v_i with out-degree D is represented by an object in memory having associated with it a D-element array P of pointers representing the outgoing edges from v_i (i.e., so that for each edge $e = (v_i, v_j)$ there is a nonnegative integer $k < D$ such that $P[k]$ is the address in memory of the object representing vertex v_j).

Several additional algorithms from graph theory will prove useful later for illustrating methods of computational gene prediction; for the present they will also provide further concrete examples of the invaluable dynamic programming technique introduced earlier. We first consider the *topological sort*.

Suppose we have a directed acyclic graph $G = (V, E)$ and we wish to obtain an ordered, nonredundant list T of all the vertices in V such that if $(v, w) \in E$, then the index of v in T is less than the index of w; i.e., $index_T(v) < index_T(w)$. We call T a *topological sort* of V. T has the property that if we were to draw the graph with the vertices lined up left-to-right according to their ordering in T, all of the edges would appear as arcs pointing left-to-right (see Figure 2.11). Algorithm 2.11 gives a memoized procedure for finding a topological sort of a graph G in pointer representation.

Algorithm 2.11 Topological sort of a DAG with vertex array V. Returns a list representing the left-to-right ordering of the vertices after sorting. The function `prepend(t,T)` inserts `t` into the front of list `T`, and $\texttt{index}_\texttt{V}\texttt{(w)}$ returns the index of `w` in array `V`. $\texttt{E}_\texttt{i}$ is the array of (outgoing) edge pointers for vertex $\texttt{v}_\texttt{i}$.

```
1.  procedure topologicalSort(V)
2.    for i←0 up to |V|-1 do seen[i]←false;
3.    for i←0 up to |V|-1 do
4.      if ¬ seen[i] then
5.        prepend(topSortRecurs(V,i,seen),T);
6.    return T;
7.  procedure topSortRecurs(V,i,ref seen)
8.    seen[i]←true;
9.    for j←0 up to |Ei|-1 do
10.     if ¬ seen[indexV(Ei[j])] then
11.       prepend(topSortRecurs(V,indexV(Ei[j]),
12.         seen),L);
13.   prepend(V[i],L);
14.   return L;
```

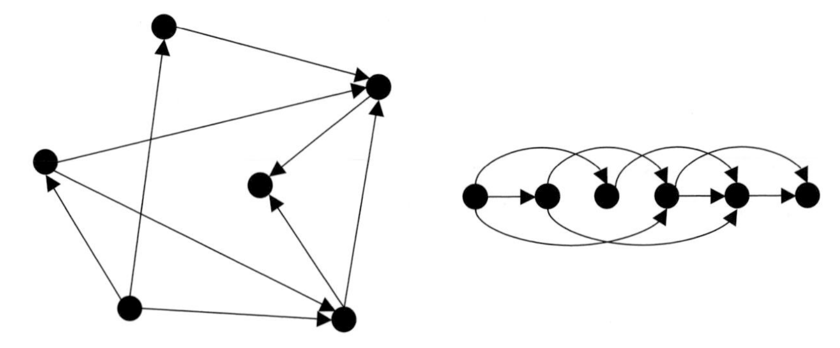

Figure 2.11 An example of a topological sort. The original, unsorted directed acyclic graph (DAG) is on the left; the sorted graph is on the right.

Given a topological sort of a weighted DAG, we can then very easily find the highest-scoring path between two given vertices in the graph, as long as we also have an array of incoming edges stored in each vertex object. As this problem has direct applicability to gene finding, we provide the solution in Algorithm 2.12.

Algorithm 2.12 Finding the highest-scoring path between two vertices in a weighted DAG, given topological sort T. It is assumed that at least one such path exists. The two vertices are given by indices b (the beginning vertex) and e (the ending vertex) into T. The weighting function is given by f.

```
1.  procedure highestScoringPath(T,f,b,e)
2.    for i←0 up to |T|-1 do s[i]←-∞;
3.    for i←0 up to |T|-1 do π[i]←NIL;
4.    s[b]←0;
5.    for i←b+1 up to e do
6.      v←T[i];
7.      I←edgesInto(v);
8.      for j←0 up to |I|-1 do
9.        pred←I[j];
10.       w←f(pred,v);
11.       if s[index_T(pred)]+w>s[i] then
12.          s[i]←s[index_T(pred)]+w;
13.          π[i]←pred;
14.    v←T[e];
15.    while v≠T[b] do
16.      prepend(v,P);
17.      v←π[index_T(v)];
18.    prepend(v,P);
19.    return P;
```

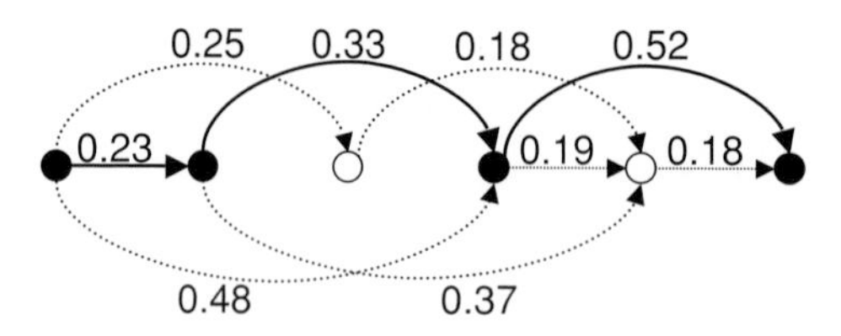

Figure 2.12 Example of a topologically sorted, weighted DAG. The highest-scoring path between the leftmost and rightmost vertices is rendered in bold; all other edges are dotted. Vertices included in the path are shown in black.

The algorithm begins by initializing a score array s to $-\infty$ for all entries except that corresponding to the origin vertex, b. We also initialize a predecessor array (or *trellis*) π to all *NIL* pointers. These two arrays are referred to as the *dynamic programming arrays*, since they store the *inductive score* and pointer to the best predecessor at each vertex in the topological sort. In the loop starting on line 5 we iterate from the origin vertex b up to the target vertex e, filling out the dynamic programming arrays along the way. The conditional construct comprising lines 11–13 detects when a prospective predecessor is found which can improve the current vertex's score. The inductive score for a vertex consists of the inductive score of its best predecessor plus the weight associated with the edge connecting the two vertices. Lines 14 through 18 then walk backward through the predecessor array π, starting at the target vertex e and ending at the initial vertex b, prepending each successive vertex in the walk onto the front of a list P. This list will constitute the highest-scoring path from b to e, and is thus returned by the procedure on line 19.

The run-time of this algorithm depends on the degree of connectivity in the graph, but if the underlying problem allows us to place a constant upper bound on the number of edges per vertex, then the run-time can be described as $O(|T|)$; i.e., linear in the number of vertices.

Figure 2.12 depicts a topologically sorted, weighted DAG, with its highest-scoring path shown in bold. Note that a *greedy algorithm* which processed the graph left-to-right, always selecting the highest-scoring edge, would have selected a suboptimal path, whereas the dynamic programming algorithm finds the optimal path. For a more in-depth discussion of dynamic programming and greedy algorithms, see Cormen *et al.* (1992).

It will be seen in Chapter 3 that the problem solved by Algorithm 2.12 can be reformulated as that of finding the highest-scoring *parse* of a DNA sequence into a valid gene structure.

2.15 Languages and parsing

Because gene finding is often described in terms of *parsing* a DNA sequence, it will be useful for us to present the most rudimentary basics of *formal language theory*, on which the theory of parsing is built (see, e.g., Hopcroft and Ullman, 1979).

Formal language theory is formulated in terms of sets of *strings* or *sentences*, which will correspond to DNA (and possibly protein) sequences in our particular domain; thus, the terms *string* and *sequence* will be considered synonymous throughout this book, except where noted otherwise. Strings are composed of individual *symbols* chosen from a finite set called an *alphabet*. We will generally denote an alphabet with the Greek letter α, and will typically assume $\alpha = \{\mathtt{A,C,G,T}\}$ in the case of DNA sequence, or $\alpha = \{\mathtt{A,R,N,D,C,Q,E,G,H,I,L,K,M,F,P,S,T,W,Y,V}\}$ in the case of proteins; it will always be clear from the context which we mean. The length of a string s will be denoted $|s|$. The *empty string*, which has length zero, will be denoted ε.

Concatenation of two strings x and y will be denoted $x\&y$, so that, e.g., if $x = \mathtt{AT}$ and $y = \mathtt{G}$, then $x\&y = \mathtt{ATG}$. Concatenation with an empty string evaluates to the original string: $x\&\varepsilon = \varepsilon\&x = x$. If s and t are sets of strings, then $s\&t$ denotes the set that results from forming the concatenation of all pairs of strings from s and t; i.e., $s\&t = \{w|w = x\&y, x \in s, y \in t\}$. Exponentiation of a string denotes repeated concatenation of the string with itself; i.e., $x^0 = \varepsilon$, $x^n = x\&x^{n-1}$ for $n > 0$. Exponentiation of a set of strings is defined similarly: $s^0 = \{\varepsilon\}$, $s^n = s\&s^{n-1}$ for $n > 0$. The *Kleene closure* of a string x is denoted x^*, and evaluates to the set of all strings formed by any finite number of concatenations of x; i.e., $x^* = \{x^n | n \geq 0\}$. *Positive closure* is similar to Kleene closure, but omits the empty string: $x^+ = x^* - \{\varepsilon\}$. Kleene and positive closure are defined for sets of strings as well: $s^* = \cup_{n\geq 0} s^n$, and $s^+ = s^* - \{\varepsilon\}$. Thus, α^* (interpreting the elements of α as strings of length 1) denotes the set of all strings over alphabet α. Substrings will be denoted using subscripts specifying the (zero-based) indices of the first and last (inclusive) symbols of the substring: $S_{i..j} = x_i \,\&\, x_{i+1} \& \ldots \& x_j$ for $i \leq j$, x_i being the ith letter in string S; $S_{i..j} = \varepsilon$ if $i > j$. *Alternation* will be denoted with the | operator: $a|b$ for strings a and b denotes the set $\{a, b\}$, and $s_1|s_2$ for sets of strings s_1 and s_2 denotes the set $s_1 \cup s_2$. Expressions composed of variables, Kleene closure, alternation, concatenation, and parentheses (for grouping subexpressions) are referred to as *regular expressions*; adjacent subexpressions in a regular expression are assumed to be joined by an implicit concatenation operator. We will also allow that regular expressions may include positive closure and exponentiation.

Note that these conventions apply to expressions appearing in the main text of the book; section 2.4 describes the corresponding syntax for these operations when appearing within pseudocode.

To give a few concrete examples, the regular expression $\mathtt{T}^*$ matches any string of Ts of any length (including zero), whereas $(\mathtt{TA})^+$ matches any non-empty string of TA dinucleotides, such as TATA or TATATATATA. The expression $\mathtt{ATG}\alpha^*(\mathtt{TGA}|\mathtt{TAG}|\mathtt{TAA})$ denotes the set of strings beginning with a eukaryotic start codon and ending with one of the three stop codons TGA, TAG, or TAA. The expression $\mathtt{GT}\alpha^*\mathtt{AG}$ matches any sequence beginning with GT and ending with AG. The expression $(\mathtt{TCA}|\mathtt{CTA}|\mathtt{TTA})\alpha^*\mathtt{CAT}$ will match any single-exon CDS on the reverse strand

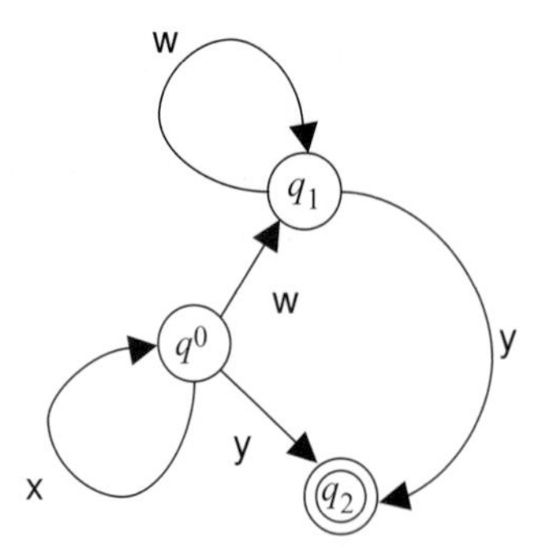

Figure 2.13 State-transition graph of a deterministic finite automaton (DFA) for the language x^*w^*y. States are shown as circles and transitions as labeled arrows. The single final state is shown as a double circle.

beginning with a reverse-strand stop codon in {TCA,CTA,TTA} and ending in the reverse-strand start codon CAT, though it should be clear that not all DNA sequences matching this regular expression will be recognized by the cell as true protein-coding segments, due to the underspecified nature of gene syntax at the level of simple DNA motifs such as start and stop codons. Furthermore, the preceding regular expressions do not account for phase periodicity (i.e., the requirement that coding segments have lengths divisible by 3 – see Exercise 2.47). Thus, regular expressions alone will not suffice to accurately predict the locations of genes in genomic DNA.

A number of different formalisms have been developed for modeling the syntactic structure of strings, most notably those that comprise the so-called *Chomsky hierarchy* (Hopcroft and Ullman, 1979):

unrestricted grammars
context-sensitive grammars
context-free grammars
finite automata

Of these, we will be most interested in the *finite automata* and (more especially) their variants, which it can be shown are, in their simplest form, equivalent to regular expressions in their expressive power (see, e.g., Hopcroft and Ullman, 1979). A *deterministic finite automaton (DFA)* is a state-based machine denoted by a 5-tuple $M = (Q, \alpha, \delta, q^0, F)$, where Q is a finite set of discrete elements called *states*, α is an input alphabet, $q^0 \in Q$ is a special *start state*, $F \subseteq Q$ is a set of *final states*, and $\delta : Q \times \alpha \mapsto Q$ is a transition function mapping (*state, symbol*) pairs to states.

Finite automata are often depicted using *state-transition* graphs such as that shown in Figure 2.13, in which vertices represent states and directed edges represent transitions. The formal representation of this automaton is $M = (\{q^0, q_1, q_2\}, \{\mathrm{w}, \mathrm{x}, \mathrm{y}\}, \{(q^0, \mathrm{x}, q^0), (q^0, \mathrm{w}, q_1), (q^0, \mathrm{y}, q_2), (q_1, \mathrm{w}, q_1), (q_1, \mathrm{y}, q_2)\}, q^0, \{q_2\})$.

An automaton M begins its operation in the start state q^0, reads symbols sequentially from its input, and changes its state according to δ as a result of each input symbol. When the end of the input is reached, if the automaton is in a final state $q \in F$, then we say that the automaton has *accepted* the input string; if instead the machine ends in a nonfinal state $q \notin F$, or if the machine cannot process the entire input because a particular symbol invokes an undefined transition in some state, then we say that the automaton *rejects* the input string. In this way, an automaton M defines a set of strings, $L(M)$, called the *language of* M: $L(M) = \{s | M \text{ accepts } s\}$. For each of the strings in $L(M)$, for a deterministic automaton M, there is a unique path through the state-transition graph of the machine, beginning with the start state q^0 and ending in some final state $q \in F$.

This path can be used to *parse* the input string by noting which regions of the string cause the machine to stay in certain designated states (or sets of states) within the automaton. For example, in order for a compiler for a particular programming language to parse a computer program into a set of keywords, operators, and other language tokens, the sequence of ASCII characters comprising the source code for the program is fed into a DFA called a *scanner*. A series of states corresponding to the string "while" will be active when the keyword **`while`** is read; consequently, when a compiler notes that its scanner has just visited this sequence of states, it concludes that the program contains a **`while`** loop starting at this position in the source code, and acts accordingly.

The act of observing the states visited by an automaton for the purpose of parsing the input sequence is called *decoding*; in the case of a DFA, the decoding problem is a trivial one, since the deterministic nature of the DFA ensures that the state sequence visited by the automaton for a given input string is unique. For the purposes of parsing DNA, however, the problem is not nearly so unambiguous. For example, any given AG dinucleotide observed in a DNA sequence may conceivably be part of an acceptor site, or may instead comprise the last two bases of a TAG stop codon, or may be neither of these.

For this reason, we will be more interested in a family of stochastic models of syntax called *hidden Markov models* (HMMs). Hidden Markov models are superficially similar to finite automata in that they operate by transitioning between discrete states. There is a fundamental difference between the two types of model, however. A DFA operates by reading a string of symbols and changing state in response to the identities of those symbols. In contrast, an HMM operates by choosing its state sequence at random (according to fixed probabilities) and emitting a random symbol upon reaching each state (again, according to fixed probabilities). Thus, a DFA is a deterministic string *acceptor*, whereas an HMM is a nondeterministic string *generator*. Because of this fundamental difference, the problem of parsing a sequence using an HMM is rather different from that of parsing using a DFA. In the case of HMMs, this decoding problem is most commonly solved using the *Viterbi*

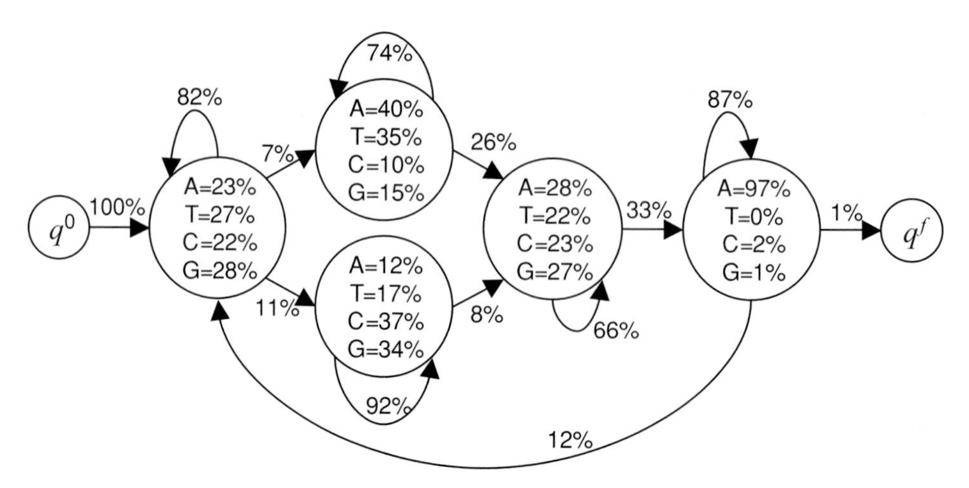

Figure 2.14 An example hidden Markov model (HMM). Emission probabilities are shown inside each state. Each edge is shown with its transition probability (zero-probability edges are omitted for clarity).

algorithm, which we will discuss in detail in Chapter 6. For now we will give the formal definition of an HMM and remark on several of its variants; these will form the basis for much of what follows in this book.

A hidden Markov model is denoted by a 6-tuple $M = (Q, \alpha, P_t, q^0, q^f, P_e)$, where Q, α, and q^0 are as defined for a DFA; q^f is the single final state, $P_t : Q \times Q \mapsto \mathbb{R}$ is the *transition distribution*, and $P_e : Q \times \alpha \mapsto \mathbb{R}$ is the *emission distribution*.

The machine operates by starting in the start state q^0 and transitioning stochastically from state y_{i-1} to y_i according to probability distribution $P_t(y_i|y_{i-1})$, where we have used the subscript i to denote the ith state visited by the machine (i.e., so that y_i and y_j for $i \neq j$ may in fact be the same state q_k, since the machine is generally free to enter a state any number of times). Upon entering a state q, the machine emits a random symbol s according to probability distribution $P_e(s|q)$. When the machine enters state q^f it halts, terminating its output string; after this, the machine cannot perform any other operations until the machine is reset by an external event. Note that there are no transitions into q^0, and none out of q^f, and neither state emits any symbols; these are referred to as *silent* states, and all other states are referred to as *non-silent*, or *emitting* states. Figure 2.14 shows an example HMM.

The initial and final states are sometimes collapsed into a single state, with the convention that a single run of the machine always begins in the combined initial/final state and ends as soon as the machine re-enters that state. Although the topology of an HMM is assumed to be fully connected (i.e., with transitions connecting every pair of states), particular transitions are often assigned zero probability in order to prohibit their use; these zero-probability transitions are generally omitted from HMM diagrams for clarity.

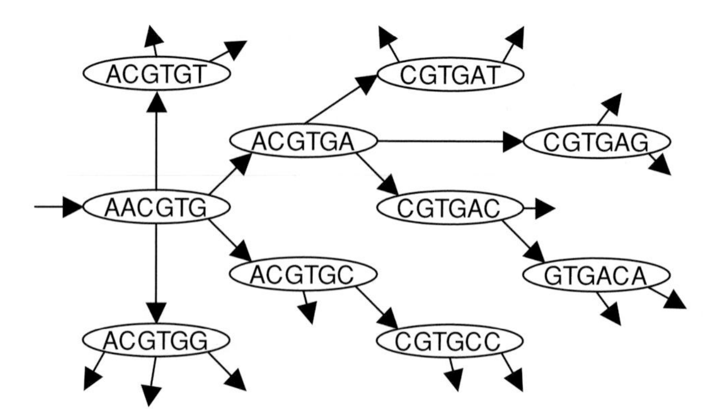

Figure 2.15 A portion of a 6th-order Markov chain. Transition probabilities are omitted for clarity. Successive states have labels shifted by one symbol.

This definition of an HMM can be extended so as to condition the emission (and possibly transition) probabilities on the identities of the n symbols most recently emitted by the machine, for some n; i.e., $P_e(x_i|x_{i-n}, \ldots, x_{i-1})$ for sequence $S = x_0, x_1, \ldots, x_{|S|-1}$. Such a machine is called an *nth-order hidden Markov model*. HMMs are described in detail in Chapter 6.

A *1st-order Markov chain* is a variant of the HMM in which each emitting state can emit only one particular symbol, and no two states emit the same symbol, so that the chain has exactly $|\alpha|$ nonsilent states. Higher-order Markov chains are also possible. For an *nth-order Markov chain*, each emitting state is labeled with an n-letter string called an *n-gram* (or *n-mer*) encoding both the symbol emitted by the state and the $n-1$ most recent symbols emitted by the machine (see Figure 2.15). Although two states are permitted to emit the same symbol, no two states may have the same label, so that knowledge of the n most recent symbols to be emitted is sufficient to determine exactly in which state the machine currently resides. Thus, an nth-order chain with alphabet α requires $|\alpha|^n + 1$ states (assuming a single collapsed initial/final state).

As a special case we can also define a 0th-order Markov chain as a 2-state HMM with a single collapsed initial/final state and an emitting state with an arbitrary emission distribution. It should be clear from this definition that a 0th-order model cannot encode dependencies between serial emission events during a run of the machine – that is, the emission probabilities of the 0th-order model are not conditioned on any prior emissions. More generally, for an nth-order model the emission probabilities will be conditioned on the n most recently emitted symbols.

A distinguishing feature of Markov chains is that for any given sequence s generated by a chain M, we can determine unambiguously which state path the machine followed in generating the sequence. Such is not true in general with a hidden Markov model, in which the state path is "hidden" from us and can only be inferred probabilistically. Unfortunately, the ambiguous nature of *ab initio* gene prediction will preclude us from relying solely on Markov chains as we attempt to model the syntactic structure of whole genes in DNA. Nevertheless, we will be encountering a number of sub-tasks for which Markov chains will be quite useful; a more thorough treatment of Markov chains is postponed until Chapter 7.

EXERCISES

2.1 Use Eq. (2.10) to compute the binomial coefficients ${}_nC_r$ for $0 \leq n \leq 5$ and $0 \leq r \leq n$. If you arrange these values into a triangle with higher n positioned higher on the page and r values arranged left-to-right, you will obtain a portion of *Pascal's triangle*. What pattern do you observe in the left-to-right structure of the triangle? Express this as a theorem regarding binomial coefficients.

2.2 Prove that ${}_nC_r = {}_{n-1}C_{r-1} + {}_{n-1}C_r$ for $n > 0$ and $0 < r < n$.

2.3 Denoting true by T and false by F, evaluate the expression: $(\mathtt{T} \wedge \mathtt{F} \vee \mathtt{T}) \wedge \neg(\mathtt{F} \vee ((\mathtt{T} \wedge \mathtt{T} \wedge \mathtt{F}) \vee (\mathtt{T} \vee \mathtt{F})) \wedge \mathtt{F}) \vee \neg (\mathtt{F} \vee \mathtt{T} \wedge \mathtt{F} \vee \mathtt{T}) \wedge ((\mathtt{T} \wedge \mathtt{T} \vee \mathtt{F}) \wedge \mathtt{T})$.

2.4 Prove that $\neg\forall_{x \in S} f(x) \Leftrightarrow \exists_{x \in S} \neg f(x)$.

2.5 Prove that $\neg\exists_{x \in S} f(x) \Leftrightarrow \forall_{x \in S} \neg f(x)$.

2.6 Prove that $(A \rightarrow B) \Leftrightarrow (\neg B \rightarrow \neg A)$.

2.7 Compute the power set of $S = \{\mathtt{A}, \mathtt{C}, \mathtt{G}, \mathtt{T}\}$. Show that the cardinality of the power set of a set S is $2^{|S|}$.

2.8 Generalize the hill-climbing procedure shown in Algorithm 2.2 to perform n-dimensional gradient ascent of an n-parameter function.

2.9 Modify the procedure shown in Algorithm 2.2 to allow it to perform either gradient ascent or gradient descent, depending on the value of an additional parameter (i.e., -1 = gradient descent, 1 = gradient ascent).

2.10 In section 2.9 it is stated that in performing a statistical test we typically sum over all values as extreme as the observed value. What is the rationale for doing this? You may consult a statistics book for help with this question, but explain the answer in your own words.

2.11 Flip a penny 20 times, recording the number h of heads observed. Using the definition of the binomial distribution, compute the probability of observing that many heads, assuming your penny is unbiased (i.e., $P(heads) = 0.5$).

2.12 Implement Algorithm 2.3 in your favorite programming language.

2.13 Implement Algorithm 2.4 in your favorite programming language.

2.14 Write an algorithm to compute the entropy of a DNA sequence. Implement the algorithm in your favorite programming language. Obtain a DNA sequence and use your program to compute the entropy of the sequence. Report the results.

2.15 Prove that the Kullback–Leibler divergence is always nonnegative.

2.16 Prove that the cross entropy is minimal for identical distributions.

2.17 Prove that the mutual information $M(X, Y)$ is zero whenever X and Y are independent.

2.18 Prove that the mutual information $M(X, Y)$ equals the entropy H_X when X and Y are perfectly dependent (i.e., when they always evaluate to the same value).

2.19 Explain the practical differences between relative entropy and mutual information. Give an example where one is applicable and the other is not, and vice versa.

2.20 Prove that $n^b = O(b^n)$, for variable n and constant b.

2.21 Prove that $\log_2 n = O(n)$.

2.22 Prove that $n \log_2 n = O(n^2)$.

2.23 Prove that $b^n = O(n!)$.

2.24 Implement the `fibonacci_recurs` and `fibonacci_dp` procedures in your favorite programming language. Run them on a number of different parameters, replicating the run for each parameter five times. Measure the execution times (averaged over replicates) and report the results.

2.25 Implement the `fibonacci_dp` and `fibonacci_memo` procedures in your favorite programming language. Run them on a number of different parameters, replicating the run for each parameter five times. Measure the execution times (averaged over replicates) and report the results.

2.26 Devise an algorithm to count the number of connected components in an undirected graph.

2.27 Write a dynamic programming algorithm to compute *Ackermann's function*, which is defined for nonnegative integers m and n as follows: $A(0, n) = n + 1$; $A(m, n) = A(m - 1, 1)$ if $m > 0$ and $n = 0$; otherwise, $A(m, n) = A(m - 1, A(m, n - 1))$. Do not use memoization.

2.28 Derive an equation relating $|V|$ and $|E|$ in a complete graph, $G = (V, E)$.

2.29 Count the number of paths in the weighted graph shown in Figure 2.12. Report the score of each path. List the scores in increasing order.

2.30 Explain the purpose of the loop comprising lines 3–5 of the `topological-Sort` procedure.

2.31 Draw a state-transition diagram for a DFA accepting the language denoted by: `(a|b)+c((a|b)c)+a*d(a|b|c)+`.

2.32 In Eq. (2.9), why do we recommend factoring such that $p \geq q$?

2.33 For the HMM shown in Figure 2.14, which path is most likely to have generated the sequence ATCAGGGCGCTATATCGATCA, the top path or the bottom path? Give the probabilities of the two paths. Ignore the transition labeled "12%."

2.34 Draw the complete state-transition diagram for a 1st-order Markov chain, omitting the transition probabilities.

2.35 Draw the complete state-transition diagram for a 2nd-order Markov chain, omitting the transition probabilities.

2.36 Implement a *find*(x, T) procedure for finding the vertex containing datum x in a binary tree T. Hint: try writing a recursive procedure which, starting at the root of the tree, compares x to the datum at the current vertex and recursively proceeds on to the left or right child of the current vertex based on the results of that comparison.

2.37 Implement the *insert*(x, T) function to insert a datum x into a binary tree T. Hint: use a similar strategy as for *find*(x, T) in the previous exercise, attaching x as a left or right child as appropriate when a leaf node is reached.

2.38 Implement the *extractMin*(Q) function for a binary tree. Hint: find the minimum by searching leftward in the tree; then remove the target vertex v from the tree, splicing in one of the child subtrees of v as a new child for v's parent, and inserting v's other child subtree into the binary tree using *insert*(x, T).

2.39 Prove that the *binary search* algorithm has logarithmic time complexity.

2.40 Prove that the *quick sort* algorithm results in a sorted array.

2.41 Prove that $P(X, Y)P(Z|Y) = P(X|Y)P(Y, Z)$. Give the probability or logic rule that justifies each step of the proof.

2.42 What additional assumption is necessary and sufficient to conclude that $P(A, B)P(C|B) = P(A, B, C)$? Provide a proof for your answer.

2.43 Prove that $P(A, B|C) = P(A|B, C)P(B|C)$.

2.44 Prove that $P(A, D, B|C) = P(A, D|B, C)P(B|C)$.

2.45 Write a program to generate random DNA sequences. The program should take as input the desired {A,T} density of the generated sequence. Write another program to count the total number of stop codons in the three frames of an input sequence. Use these programs to graph a curve showing the average number of stop codons per 1000 bp as a function of the {A,T} density of a random sequence.

2.46 Prove that for random DNA the conditional entropy equals the entropy (for all orders n).

2.47 Give a regular expression for single-exon eukaryotic CDSs with lengths divisible by 3.

3

Overview of computational gene prediction

In this chapter we will develop a conceptual framework describing the gene prediction problem from a computational perspective. Our goal will be to expose the reader to the overall problem from a high level, but in a very concrete way, so that the necessities and compromises of the computational methods which we will introduce in the chapters ahead can be seen in light of the practical realities of the problem. A comparison of the material in this chapter with the description of the underlying biology given in Chapter 1 should highlight the gulf which yet needs to be crossed between the goals of genome annotation and the current state of the art in computational gene prediction.

3.1 Genes, exons, and coding segments

The common substrate for gene finding is the DNA sequence produced by the genome sequencing and assembly processes. As described in Chapter 1, the raw *trace files* produced by the sequencing machines are subjected to a *base-caller* program which infers the most likely nucleotide at each position in a fragment, given the levels of the fluorescent dyes measured by the sequencing machine. The nucleotide sequence fragments produced by the base-caller are then fed to an *assembler*, a program that combines fragments into longer DNA sequences called *contigs*. Contigs are generally stored in *FASTA* files. Figure 3.1 shows an example FASTA file.

A FASTA file contains one or more sequence entries, with each entry consisting of two parts: (1) a *defline*, which consists of a greater-than sign (>) followed by a textual description of the entry, and (2) a sequence of letters specifying the inferred composition of a single read, contig, scaffold, chromosome, EST, cDNA, protein, or other biosequence. The first field after the > on the defline is usually taken to be the unique identifier of the sequence; this is generally the only part of the defline that is utilized by gene-finding programs. Thus, the input to an *ab initio* gene finder (i.e., a gene finder not incorporating any additional evidence, such as homology) effectively consists of a finite sequence over the four-letter alphabet, $\alpha = \{\mathtt{A}, \mathtt{C}, \mathtt{G}, \mathtt{T}\}$.

```
>7832 contig assembled on Nov-19-04 by chASM v3.2
ATCGATCGATCGGCGATGCTAGCTACTAGCTGATTCTCTCTAGAGAGCTAGCTGAC
GGCGTAGCTAGCTAGCTGCGATTCAGCGTACGTAGCTAGCTATCTACTTCGATCGT
AGCTATTCGATCTAGCTAGTCGATGTCAGCGCGCGATTATATCGTGCTATCGTGCG
TATCATATATATAGCGCGCGATCGTCGGCGCATGCGAGAGAGTCGTAGTAGTCGTA
GCGCTAGCTGATGCTGTCGTAGCTATCTTTCAGTAG
```

Figure 3.1 A portion of a FASTA file. The line beginning ">" is the defline, and gives a description of the sequence that follows. This particular sequence has the unique identifier 7832.

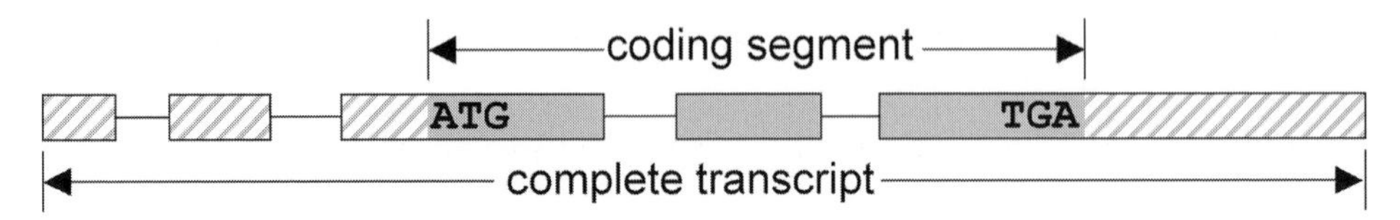

Figure 3.2 An example of eukaryotic gene structure. Rectangles represent exons and the line segments joining them represent introns. The solid shaded region is the coding segment (CDS) of the gene, and the hatched area is untranslated region (UTR). The ATG and TGA mark the effective start and stop codons, respectively.

The goal of computational gene prediction is not only to find any and all genes in the input sequence, but also to precisely identify the coding segments of the exons comprising each gene. Although we would ideally like to identify a number of other features of genes (such as untranslated regions, promoters, splice enhancers, etc.), the problem of accurately identifying the coding regions has still not been completely solved, and is therefore the prime focus of current protein-coding gene finders.

As explained in Chapter 1, eukaryotic genes generally consist of a number of exons separated by introns. During transcription, one or more contiguous genes are transcribed into an mRNA, which we refer to as a *transcript*. For the purpose of gene prediction we will assume that a transcript contains only one gene. Splicing of the transcript then removes introns, so that the mature mRNA consists only of contiguous exons. Translation of the mature transcript into a protein then occurs, beginning at a *start codon* and ending at a *stop codon*. Figure 3.2 shows the intron/exon structure of a sample eukaryotic gene. The case of prokaryotic genes is simpler, since they do not contain introns; we will therefore concern ourselves primarily with the combinatorically more complex problem of eukaryotic gene prediction (see section 7.5 for a summary of the issues encountered specifically in prokaryotic gene finding).

As indicated by Figure 3.2, not all exons in a gene are translated into protein, and even those that are translated may not be translated in full. In particular, only the exonic regions lying between the *translation start site* (denoted by a start codon) and the *translation stop site* (denoted by a stop codon) are translated into protein. These regions collectively comprise the *coding segment* (*CDS*) of the gene. Since the

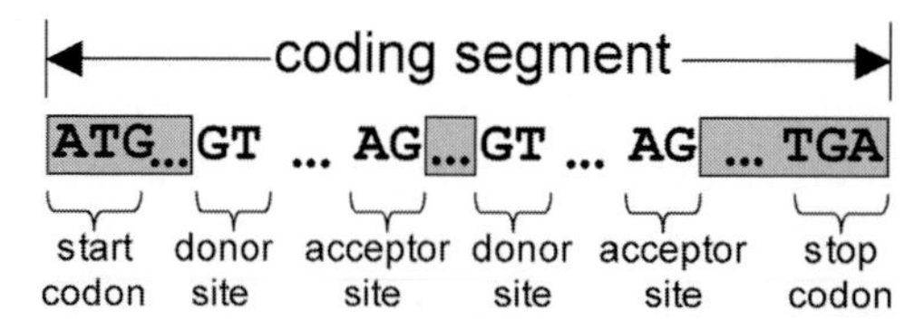

Figure 3.3 Signals delimiting the parts of a CDS. The CDS begins with a start codon and ends with a stop codon. Each intron interrupting the CDS begins with a donor splice site and ends with an acceptor splice site. Exons are shown as shaded rectangles.

current focus of gene finding is the reliable identification of coding segments of genes, the computational problem that we are trying to solve can be described as the accurate identification of the following *signals* in a DNA sequence:

- start codons – typically ATG
- stop codons – typically TAG, TGA, or TAA
- splice donor sites – typically GT
- splice acceptor sites – typically AG

where we have indicated the typical *consensus sequences* of each signal type as they occur in eukaryotes.

As illustrated in Figure 3.3, the locations of donor and acceptor sites are desired because they delimit the beginnings and ends of introns, respectively. Note that while the leftmost exon is marked in the figure as extending from the start codon to the first donor site, in practice the first coding exon may begin some distance upstream from the start codon, resulting in an untranslated portion of the exon. Similarly, the last coding exon of a gene may extend past the end of the CDS. For convenience, we will often ignore this distinction by treating the coding portion of an exon as if it were the entire exon, and referring to the noncoding portions of the spliced gene collectively as *untranslated regions*, or *UTRs*.

The locations of the four signal types described above can thus be seen to jointly specify the CDS of the gene, from which we can then deduce quite easily which protein is produced when the gene is translated (as described in Chapter 1).

Unfortunately, accurate identification of these signals in a DNA sequence is a distinctly nontrivial problem, due to the fact that not every ATG, GT, or AG occurring in a sequence denotes a true start codon, donor site, or acceptor site, respectively, and because current biological knowledge does not yet provide enough clues to disambiguate potential signals with complete certainty. Thus, we are left with the task of making educated guesses based on other cues embedded in the sequence surrounding putative signals, together with a few additional constraints on intron/exon structure that we will describe shortly.

One of the most useful cues in predicting coding exons is the *codon bias* which is commonly present within functional coding segments. Recall from Chapter 1 that the *codons* (nucleotide triplets specifying individual amino acids in a resulting protein product for a gene) tend to occur in such a way that some codons are significantly more common than others within coding segments. In some organisms, this bias is so strong that simply observing the *trinucleotide* (i.e., three successive nucleotides, $x_i x_{i+1} x_{i+2}$) frequencies in local regions of a sequence can provide a reasonably accurate prediction of coding segments. In other cases the bias is more subtle, requiring more sophisticated methods such as those which we will be describing in later chapters.

For the present, we wish merely to point out that although the biological processes of gene regulation, transcription, splicing, and translation are all conceptualized biologically as (more or less) separate phenomena occurring within the cell, current gene-finding techniques almost invariably conflate these processes when modeling the structure of genes. Thus, sequence biases associated with these several phenomena are generally combined into a single mathematical model for the purpose of assessing the likely boundaries of coding segments, with the emphasis being on maximizing the accuracy of the resulting gene predictions. In this way, it will be seen that the models employed in *ab initio* gene finders generally have more to do with the observed statistical properties of gene structures than with the evolutionary and cellular processes giving rise to those structures, and to the extent that this is the case, it can perhaps be said that current *ab initio* gene finders are somewhat better suited to the task of finding and parsing genes than of providing explanatory models of their biology. We will return to this issue when we consider discriminative training of Markov models in Chapters 6 and 8 (see also section 12.4).

3.2 Orientation

As described in Chapter 1, the DNA molecule comprises two strands, each consisting of a sequence of nucleotides, which we will designate using the DNA alphabet $\alpha = \{\mathtt{A}, \mathtt{C}, \mathtt{G}, \mathtt{T}\}$ as mentioned previously. The nucleotides on one strand are (except in relatively rare cases) strictly paired off in reverse-complementary fashion, so that an A on one strand is always paired off with a T on the other strand, and likewise, each C is paired off opposite a G. Thus, we have the rule:

$$\begin{aligned} compl(\mathrm{A}) = T \quad & compl(\mathrm{C}) = \mathrm{G} \\ compl(\mathrm{T}) = A \quad & compl(\mathrm{G}) = \mathrm{C} \end{aligned} \tag{3.1}$$

where *compl*(x) denotes the complementation operator for a single base x. Given one strand of a DNA molecule, we can easily deduce the other strand by simply reversing the order of bases in the sequence for the given strand and then complementing each base in the resulting sequence. This is the *reverse complement* operation

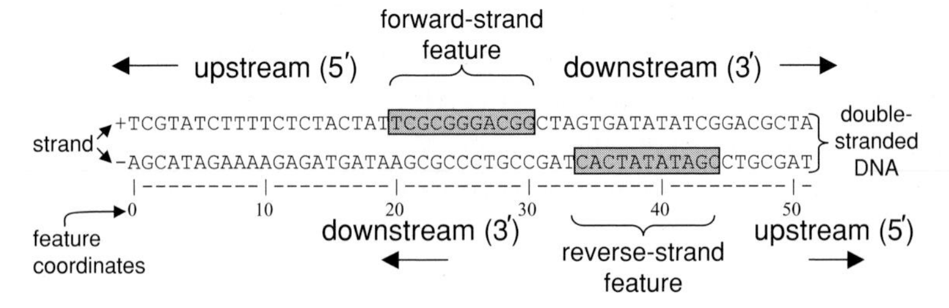

Figure 3.4 Orientation of features in DNA. The forward strand is labeled with a + and the reverse strand with a –. The definitions of upstream (5′) and downstream (3′) are reversed for the two strands: relative to a forward-strand feature, 5′ is to the left and 3′ to the right, and relative to a reverse-strand feature, 5′ is to the right and 3′ to the left. Feature coordinates increase left-to-right, regardless of strand, unless stated otherwise.

introduced in Chapter 1, which we will denote using *revcomp*(*S*) for sequence *S*. Thus we have, for example, *revcomp*(ATG) = CAT.

Without loss of generality, we will arbitrarily designate the input sequence to a gene finder as the *forward strand* sequence and its reverse complement as the *reverse strand* sequence. In illustrations we will often denote the forward strand using the "+" sign and the reverse strand using the "−" sign. Coordinates of all DNA features, regardless of the strand on which the features reside, will also be given relative to the forward strand – that is, relative to the beginning of the input sequence – unless noted otherwise.

Furthermore, we will adopt the convention that terms describing relative direction with respect to a given position in a sequence should be interpreted relative to the forward strand, unless we are explicitly discussing a feature located on the reverse strand. Thus, if we are given a sequence $S = x_0 x_1 \ldots x_n$ where each x_k is a single nucleotide, then we will say that x_i is *upstream from* x_j and that x_j is *downstream from* x_i if $i < j$. As noted in Chapter 1, the biological terms for these directions are 5′ (denoting the upstream direction) and 3′ (denoting the downstream direction). It is very important to understand that when we are giving directions relative to a reverse-strand feature, the terms are reversed, so that 5′ will denote bases further to the right and 3′ will denote bases further to the left. Note that we always use zero-based coordinates, so that the very first base in a sequence is at index 0. These ideas are illustrated in Figure 3.4.

To reinforce these ideas, let us consider an example, and let us be brutally explicit in order to avoid any possibility of confusion. Suppose we are given an input sequence $S = x_0 x_1 \ldots x_n$. By convention, we refer to S as the forward strand of the represented molecule and to *revcomp*(*S*) as the reverse strand. When describing a feature E on the forward strand, if E extends from position i to position j in S, then we will say that all positions k in S such that $k < i$ are *upstream of*, or *more* 5′ *than*, feature E, and that all positions $k > j$ are *downstream from*, or *more* 3′ *than*, feature E.

Now suppose instead that feature E happens to lie on the reverse strand. Then if we say that some feature G lies upstream from (or, equivalently, is more 5′ than) feature E, we mean that the coordinates of G are strictly greater than those of E. Note that this is true regardless of the strand on which feature G lies, since we are describing the location of G relative to E, not the other way around. By convention, the coordinates $[i, j]$ of feature E are still given relative to the beginning of sequence S – that is, the coordinates of E are given relative to the forward strand, even though E resides on the reverse strand. In this way, we can always determine the coordinates of a feature if we know the length of the feature and its distance from the beginning of the *substrate* (the input sequence), even if we do not know on which strand the feature lies.

Finally, continuing with the example of feature $E = [i, j]$ on the reverse strand, if we wish to obtain the subsequence representing E as seen by the cellular machinery, we must do either one of two things: (1) extract the subsequence $S_{i..j}$ and then compute $revcomp(S_{i..j})$, or (2) compute $R = revcomp(S)$ and then extract the subsequence $R_{L-1-j..L-1-i}$ for $L = |S|$ the length of the sequence. In the case of option (2), we have employed the coordinate transformation:

$$revcoord(x, L) = L - 1 - x \tag{3.2}$$

to transform coordinate x on the forward strand to $L - 1 - x$ on the reverse strand; though we will always give coordinates relative to the forward strand by default, reverse-strand coordinates are necessary when indexing into the reverse-complementary sequence, as we just did. This should not cause undue confusion, since our use of reverse-strand coordinates will be exceedingly rare.

Given an input sequence S, the term *left* will always refer to the beginning of the sequence and *right* will always refer to the end of the sequence. Thus, if we say that feature E is left of feature G, then it can be assumed that the coordinates of E are strictly less than those of feature G, regardless of the strands on which E and G lie. In this way, *left* and *right* will always be used relative to the forward strand, without exception.

In genes having multiple coding exons, we will refer to the most 5′ of those exons as the *initial exon* and the most 3′ as the *final exon*; if the gene has three or more coding exons, then all exons lying between the initial and final exons will be referred to as *internal exons*. In genes having only one coding exon, the coding portion of that exon will be referred to as a *single exon*. Figure 3.5 summarizes this terminology.

As indicated earlier, we will refer to the coding portions of exons as if they comprised the entire exon; this is merely a convention of the gene-finding literature which reflects the current emphasis on prediction of coding segments, due to the practical difficulties of identifying untranslated regions with any precision.

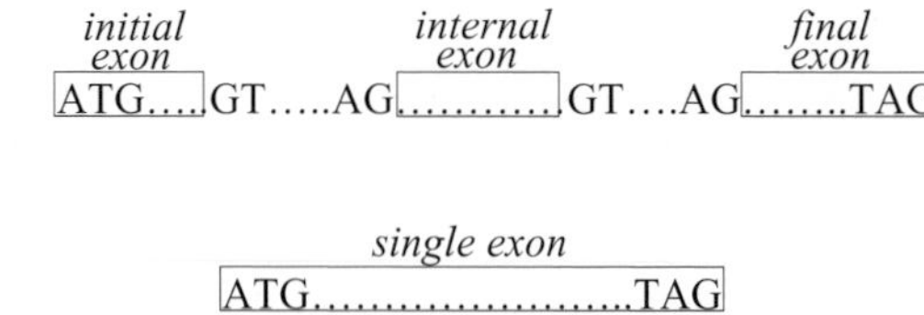

Figure 3.5 Distinctions between exon types. An "initial exon" is one that begins with a start codon and ends at a donor site; an "internal exon" begins after an acceptor site and ends at a donor site; a "final exon" begins after an acceptor site and ends with a stop codon; and a "single exon" denotes a single-exon gene, having no introns in the CDS.

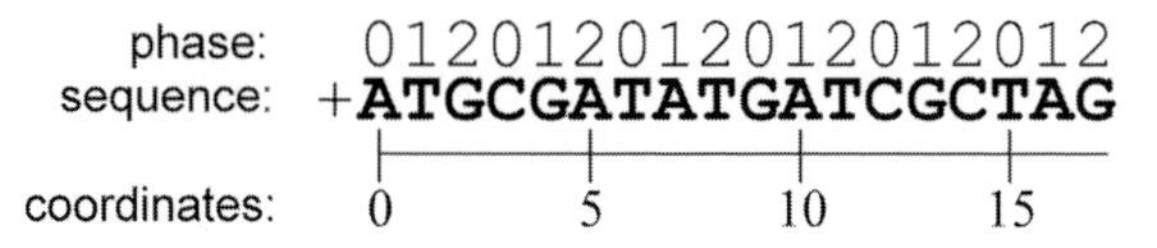

Figure 3.6 Phase periodicity of a forward-strand coding segment. Phase increases (mod 3) left to right. No in-phase stop codons are permitted except at the 3′ end of the CDS.

3.3 Phase and frame

As explained in Chapter 1, the CDS of a gene consists of a series of triples known as *codons*. Each codon specifies an amino acid to be appended to the protein product during translation. Because codons are always three bases long and because a coding segment consists of a series of nonoverlapping codons, it should be clear that the length of a CDS is therefore always divisible by 3 (assuming no errors in the sequencing or assembly processes).

The three positions within a codon are referred to as *phase 0* (the most 5′ base), *phase 1* (the middle base), and *phase 2* (the most 3′ base). Thus, as we move along a coding segment in the 5′-to-3′ direction, we cycle through phases 0, 1, and 2, in order, starting in phase 0.

Figure 3.6 shows a six-codon CDS consisting of the codons ATG, CGA, TAT, GAT, CGC, and TAG. As described in Chapter 1, the translation machinery of the cell will read through these codons in this order, translating each codon into an amino acid to be appended to the end of the growing protein product. When the stop codon is reached (TAG in this case), translation terminates, thereby terminating the protein product. Note that (except in the rare case of *selenocysteine codons* – see section 3.5.11) no stop codons may occur before the final stop codon of the CDS, or the translation machinery will terminate prematurely, manufacturing only a part of the full protein. Thus, if the third codon had been TAG rather than TAT, then the CDS would effectively be 9 base pairs long rather than 18 base pairs as shown in the figure.

The alert reader will have noticed that occupying positions 8 through 10 in this sequence is the triple TGA, which in many eukaryotic genomes denotes a stop

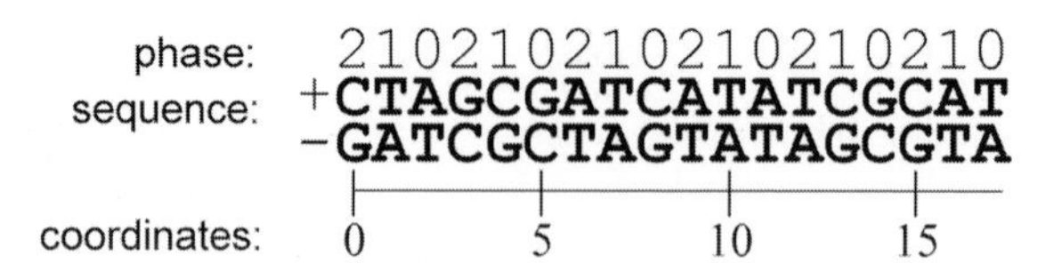

Figure 3.7 Phase periodicity of a reverse-strand coding segment. The segment shown is identical to that given in Figure 3.6, but now occurs on the reverse strand. As a result, phase now increases (mod 3) right to left, rather than left to right.

codon. The reason this triple is permitted within the CDS is that the triple begins in phase 2; only stop codons occurring in phase 0 of the CDS will cause termination of the translation process, because it is only these triples that are actually read together as a codon by the *transfer RNA* (*tRNA*; see Chapter 1). Thus, we say that no *in-phase* stop codons may occur in a coding segment except at the terminal 3′ end. When we say *in-phase* (or *in-frame*) we will always mean *beginning in phase* 0, whether on the forward strand or the reverse strand.

On the reverse strand, phase increases in a right-to-left fashion, as shown in Figure 3.7. In the figure we have shown the same CDS as before, but now on the reverse strand. Thus it can be seen that the A of the start codon ATG at the 5′ end of the CDS occurs in phase 0, as do each of the subsequent codons, just as before, but in right-to-left order rather than left-to-right. Translation of this CDS will therefore result in the same protein product as before. It should be clear that our definition of upstream/downstream for the two strands is consistent with the strand-specific cycling of phase in coding segments, as both are defined consistent with the direction of transcription as it occurs in the cell.

Given that a base x_i on the forward strand is known (or presumed) to be in some phase ω_i, we can compute the phase ω_j of any other base x_j within the same coding exon as follows:

$$\omega_j = (\omega_i + \Delta_{i,j}) \bmod 3, \tag{3.3}$$

where $\Delta_{i,j} = j - i$. For coding segments on the reverse strand, the relation is:

$$\omega_j = (\omega_i - \Delta_{i,j}) \bmod 3. \tag{3.4}$$

It should be noted that the definition we have given for *phase* is often confused with a related concept, the *frame*. In some texts these terms are used interchangeably, but we will opt to exert greater care in discriminating between these easily confused concepts. The frame f_i associated with a position i in substrate S is defined as the distance from the beginning of the substrate to position i, mod 3. Thus, if i is a zero-based coordinate, then:

$$f_i = i \bmod 3. \tag{3.5}$$

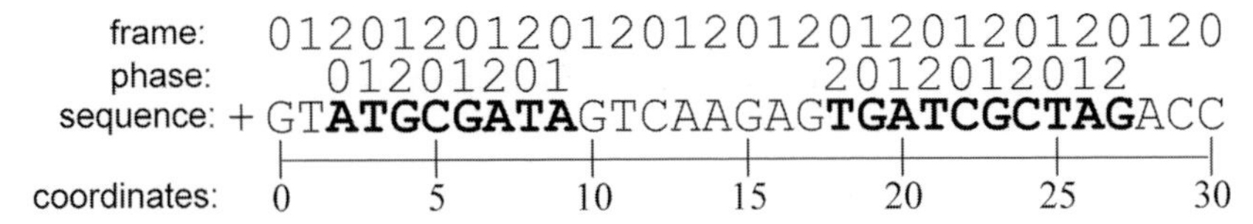

Figure 3.8 Phase and frame progression of a two-exon coding segment on the forward strand. The coding bases are shown in bold. The phase progression of the CDS stops at position 10, where the intron begins, and resumes with the next phase at position 18, after the intron ends.

In this way, the frame provides a mod 3 frame of reference independent of any particular coding segment. This is useful, for example, when a number of prospective exons are being considered for inclusion into a hypothetical gene model. Whereas a putative intron will interrupt the periodic phase progression of a prospective CDS, the progression of frames is unaffected by all such hypothetical features, as illustrated in Figure 3.8.

The difference between phase and frame can be further illustrated by considering the notion of an *open reading frame*, or *ORF*. Let us consider a portion of the second exon in Figure 3.8, beginning at position 18 and ending at position 26 (just one base short of the end of the CDS). This exon begins in frame 0, since 18 mod $3 = 0$. It is clear that this exon cannot be interpreted as beginning in phase 0, since that would entail beginning with the stop codon TGA. Thus, we say that this segment beginning at position 18 and ending at position 26 *is not open* in frame 0 (on the forward strand) because of the stop codon at position 18 – that is, *frame 0* cannot be in *phase 0* for this interval, and so frame 0 is *not open* (in this interval). By contrast, this segment is open in frame 2 because the sequence beginning at position 20 (ATC GCT A) does not contain a stop codon. Likewise, the sequence GAT CGC TA corresponding to frame 1 in this interval does not contain any stop codon, and so frame 1 is also open.

Thus we say that this region contains two *open reading frames*, or *ORFs*, namely, frames 1 and 2. These frames are considered open precisely because *placing their first positions in phase 0 allows uninterrupted translation over the full interval* (not including partial codons at the end). Note that if we had considered the full interval [18,27], then frame 1 would technically not have been open, due to the stop codon at the very end. Because we were working in the context of a putative final exon, we intentionally shortened the interval by 1 base to account for the fact that we knew a priori that the interval ended in an in-phase stop codon in frame 1 and we wished simply to determine whether any other in-phase stop codons occurred in that frame.

Formally,

$$open(i, j, f, S, +) \Leftrightarrow \neg\exists_{\substack{k\in[i, j-2], \\ k \bmod 3 = f}} S_{k..k+2} \in \Gamma \tag{3.6}$$

for $\Gamma = \{\texttt{TAG, TGA, TAA,}\}$ where $open(i, j, f, S, +)$ denotes the proposition that the forward-strand interval $[i, j]$ of sequence S is open in frame f. On the reverse strand:

$$
\begin{aligned}
open(i, j, f, S, -) &\Leftrightarrow \neg\exists_{\substack{k\in[i,j-2],\\ (k+2) \bmod 3=f}} \quad revcomp(S_{k..k+2}) \in \Gamma \\
&\Leftrightarrow \neg\exists_{\substack{k\in[i,j-2],\\ (k+2) \bmod 3=f}} \quad S_{k..k+2} \in \overline{\Gamma}
\end{aligned}
\tag{3.7}
$$

for $\overline{\Gamma} = \{\texttt{CTA,TCA,TTA}\}$.

3.4 Gene finding as parsing

As described earlier in this chapter, a gene (or, more precisely, a CDS) consists of a sequence beginning with a start codon (typically ATG) and ending with a stop codon (typically TAG, TGA, or TAA), possibly interrupted by a number of introns, each of which begins with a donor splice site (typically GT) and ends with an acceptor splice site (typically AG). If we let $\Gamma = \{\texttt{TAG, TGA, TAA}\}$ denote the valid stop codons for a particular organism, then the syntax of genes can be loosely described by the following *regular expression* (see section 2.15) over the DNA alphabet, α:

$$
\alpha^*(\underline{\mathrm{ATG}}\alpha^*(\underline{\mathrm{GT}}\alpha^*\underline{\mathrm{AG}}\alpha^*)^*\underline{\Gamma}\alpha^*)^*, \tag{3.8}
$$

where we have underlined the *signals* for clarity. This expression will match any nonoverlapping set of zero or more genes lying on the forward strand of a contig. A number of well-established methods are available for parsing strings of letters according to a given regular expression, and in the case of languages having unambiguous syntax, these methods can be very efficient (e.g., Aho *et al.*, 1986; Fischer and LeBlanc, 1991).

Unfortunately, the language of genes is highly ambiguous, as we noted several times previously. Although the regular expression given in Eq. (3.8) will match any set of valid gene structures in a sequence, it will also match many invalid gene structures, due to the fact that it does not enforce phase constraints as described in the previous section, nor does it prohibit the occurrence of in-phase stop codons within a coding segment. More importantly, even those gene structures matched by Eq. (3.8) which do satisfy phase constraints and contain valid ORFs will not all turn out to be functional genes, since many (perhaps most) of them will simply fail to be properly transcribed, spliced, and/or expressed by the cellular machinery, for reasons that are not entirely known, and it is this latter source of uncertainty which most severely complicates the gene prediction task.

As a result, the problem of computational gene finding is more akin to the problem of *natural-language parsing* and *speech recognition* than to that of constructing a compiler or interpreter for a computer programming language, since programming languages are generally constructed so as to have relatively unambiguous syntax, while natural (i.e., spoken and/or written) language tends often to be rather ambiguous. In the case of natural-language parsing (e.g., Manning and Schütze, 1999) and speech processing (e.g., Jelinek, 1997), a number of heuristic parsing techniques have been developed, of which we will consider several in due course; these latter primarily take the form of hidden Markov models, which were introduced in Chapter 2 and which are described more fully in Chapter 6. Various other machine-learning methods have been applied to the problem of natural-language processing, including neural networks, decision trees, and a number of other techniques; we will consider these and their use for gene finding in Chapter 10.

For now we will introduce a number of basic formalisms which will prove useful later as we delve into each of these more specialized techniques in greater detail. First we introduce the notion of an *ORF graph* (also called in the literature a *parse graph* or an *induced graph* – e.g, Kulp *et al.*, 1996; Majoros *et al.*, 2004, 2005b), which is simply a directed acyclic graph (DAG) representing all possible coding segments within a sequence. Using the notation $A \rightarrow B$ to denote an edge from a vertex of type A to a vertex of type B, we can define an ORF graph more explicitly as an ordered DAG having the following edge types:

$$
\begin{aligned}
ATG &\rightarrow TAG \\
ATG &\rightarrow GT \\
GT &\rightarrow AG \\
AG &\rightarrow GT \\
AG &\rightarrow TAG \\
TAG &\rightarrow ATG
\end{aligned} \tag{3.9}
$$

where *ATG*, *GT*, *AG*, and *TAG* represent putative start codons, donor sites, acceptor sites, and stop codons, respectively; it should be understood that these are merely identifiers for classes of signals, so that, e.g., a signal of type *TAG* may in fact consist of the bases TAG, TGA, or TAA (or conceivably some other triple, depending on the particular organism under consideration). The vertices in an ORF graph are strictly ordered according to the positions of their corresponding signals in the underlying substrate. A simplistic example of an ORF graph is shown in Figure 3.9.

Given an ORF graph G, any path traversing the graph from left to right outlines an individual parse of the underlying sequence, where a *parse* is defined as a series of zero or more nonoverlapping putative gene structures (exons and introns) and the intergenic regions separating them. In order to ensure that our definition of a parse covers the entire substrate, we introduce two virtual *anchor signals*, *LT* for "left terminus" and *RT* for "right terminus," which reside at the extreme left and

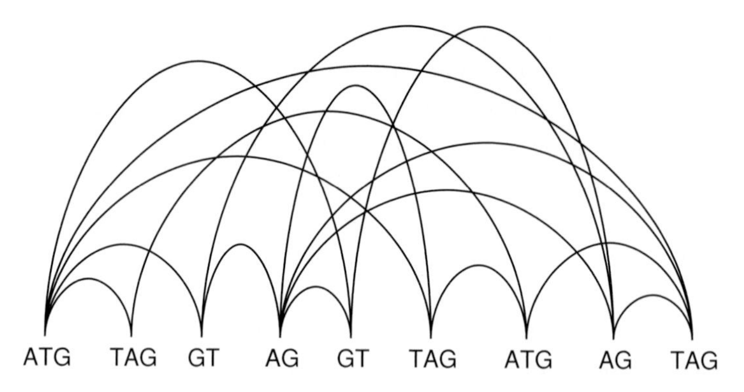

Figure 3.9 A hypothetical ORF graph for a contig. Vertices are shown at the bottom with labels indicating the type of signal represented. Each edge represents a putative exon, intron, or intergenic region, according to the edge type. Edges are implicitly directed from left to right. Vertices are associated with specific positions in a DNA sequence (omitted for clarity).

extreme right ends of the graph, respectively, and are associated with the following additional edge types:

$$\begin{aligned} LT &\rightarrow ATG \\ TAG &\rightarrow RT \\ LT &\rightarrow RT \end{aligned} \tag{3.10}$$

In this way, an ORF graph represents all possible parses of an input sequence into zero or more gene structures, thereby providing a formal framework for dealing concretely with syntactic structures in DNA; it will be seen once we have augmented this formalism with phase constraints and several other relevant attributes that it will serve as a very useful representation for both describing and implementing gene-finding strategies.

First we need to incorporate additional vertex and edge types to allow for genes on the reverse strand:

$$\begin{aligned} \overline{TAG} &\rightarrow \overline{ATG} \\ \overline{GT} &\rightarrow \overline{ATG} \\ \overline{AG} &\rightarrow \overline{GT} \\ \overline{GT} &\rightarrow \overline{AG} \\ \overline{TAG} &\rightarrow \overline{AG} \end{aligned} \tag{3.11}$$

We have used the notation $\overline{X}$ to denote a signal on the reverse strand, so that, e.g., the signal denoted as $\overline{ATG}$ will appear in the input sequence as CAT, assuming the usual consensus for start codons. A further set of additional edge types will enable the linking together of genes on the forward and reverse strand, and to the left and right termini:

$$\overline{ATG} \rightarrow \overline{TAG}$$
$$\overline{ATG} \rightarrow ATG$$
$$TAG \rightarrow \overline{TAG}$$
$$LT \rightarrow \overline{TAG}$$
$$\overline{ATG} \rightarrow RT \tag{3.12}$$

Additional signal types may be introduced to represent features such as *promoters, cap sites, branch points, polyadenylation signals, exon splice enhancers, signal peptides, protein domains*, and any other features known to occur in DNA (see section 11.3 for a case study involving the modeling of several of these features). Given these additional feature types, corresponding vertex and edge types may then be introduced to represent these features and the regions surrounding them, thereby increasing the complexity and representational power of the ORF graph. Though we will be exploring the utility of some of these additional features in later chapters, we will omit them for the present in order to avoid unduly complicating the discussion.

A particularly useful enhancement is to augment both the vertices and the edges of an ORF graph with real numbers representing scores of various types, so that the ORF graph can then be subjected to the various algorithms on weighted graphs, such as those described in Chapter 2. Some particularly promising uses for ORF graphs include the investigation of patterns in suboptimal parses as evidence for *alternative splicing* (see section 12.1).

Incorporating phase constraints into our graph-theoretic framework may now be done fairly simply, by stipulating that each edge e must have associated with it three scores, $score(e, \omega)$ – one for each phase ω – where the special score $-\infty$ will be used to indicate which phases of an edge are unavailable for use in a valid parse. A parse $\phi = (e_0, e_1, \ldots, e_n)$ will consist of a series of edges e_i spanning the input sequence from LT to RT, where the parse ϕ acts to single out a particular phase $\lambda_\phi(e_i)$ for each edge. The score of a prospective parse ϕ can then be defined as:

$$score(\phi) = \sum_{i=0}^{n} score(e_i, \lambda_\phi(e_i)) + score(v_i)$$

for $e_i = (v_i, w_i)$, $score(LT, \omega) = score(RT, \omega) = 0$.

Formally, we stipulate the following:

(1) For a valid parse ϕ, if e_i is an intron edge (GT→AG), then:

$$\lambda_\phi(e_i) = (\lambda_\phi(e_{i-1}) + length(e_{i-1})) \bmod 3$$

(assuming only complete gene structures are predicted – see Exercise 3.17).

1. Given an input sequence S, induce an (explicit or implicit) ORF graph $\mathcal{G}$ for S.
2. Score the vertices and edges in $\mathcal{G}$ using some scoring strategy, f.
3. Extract the highest-scoring valid parse ϕ from $\mathcal{G}$.
4. Interpret ϕ as a series of zero or more gene structure predictions according to f.

Figure 3.10 A conceptual framework for the gene-parsing problem.

(2) If e_i is an initial exon (ATG→GT), single exon (ATG→TAG), or intergenic edge (LT→ATG, TAG→ATG, TAG→RT), then:

$$\lambda_\phi(e_i) = 0.$$

(3) If e_i is an internal exon (AG→GT) or final exon edge (AG→TAG), then:

$$\lambda_\phi(e_i) = \lambda_\phi(e_{i-1}).$$

(4) For a predicted CDS of length L:

$$open(0, L-1, 0, C, +) \Leftrightarrow true$$

where C is the (spliced) sequence of the predicted CDS.

(5) For final exon edge e_i:

$$\lambda_\phi(e_i) = (3 - length(e_i)) \bmod 3.$$

A parse ϕ is valid only if all of these requirements are met. These restrictions apply in the forward-strand cases; a set of similar restrictions may be derived for reverse-strand features (see Exercise 3.16). Note that we arbitrarily adopt the convention that the phase of an edge refers to the phase of the leftmost base in the interval covered by that edge.

Given the above, we can use the notion of an ORF graph to formulate a generic gene-finding strategy, as outlined in Figure 3.10. We first induce an ORF graph for the target sequence, then score the vertices and edges of the graph using techniques which we will describe in subsequent chapters, then extract the single best-scoring path from the graph using a dynamic programming algorithm, and finally we interpret the extracted path as a parse of the target sequence – i.e., as a series of predicted gene features.

Inasmuch as the various gene-finding techniques which we will be describing in this book can be mapped into this conceptual framework, it should become apparent that gene finding can indeed be usefully described in terms of parsing DNA according to the ambiguous syntax of genes, so long as one acknowledges that the problem can be more appropriately characterized as a *heuristic parsing* problem, rather than a simple deterministic one. Furthermore, it will be seen that, as in natural-language processing, gene prediction accuracy can often be improved by informing this parsing process of additional "semantic" information, such as sequence conservation and homology, and other information of biological significance; in exploring several methods for incorporating such information into *ab initio* gene finders, we will see that the above framework readily admits such enhancements in a fairly elegant fashion and without undue difficulty.

The above formulation also brings into focus the distinction between the engineering problems involved in efficiently enumerating and representing alternative gene predictions on the one hand (steps 1 and 3 in Figure 3.10), and the more biologically motivated problem of finding the most effective scoring function f for evaluating those alternative parses on the other hand (step 2). Whereas the former has perhaps been adequately addressed in the literature, the latter clearly remains an ongoing effort in gene-finding research. We will see in the chapters ahead that a broad array of methods have been considered for scoring features in DNA, including several types of Markov model, neural networks, decision trees, Bayesian networks, and various statistical estimation techniques. As the field advances, we obviously expect to see the greatest amount of progress in this particular arena, including more effective integration of external "semantic" information derived from improved understanding of molecular and evolutionary phenomena.

It should be clear from our treatment in this section that the notion of "parsing" involves the partitioning of an input sequence into a series of nonoverlapping intervals of different types (exon, intron, etc.) which collectively span the entire length of the sequence, such that every position in the sequence falls into exactly one interval, or *feature*, in the parse. This is in contrast to the problem of *exon finding* – which we will consider briefly in Chapter 5 – in which the goal is instead to identify likely exon candidates as isolated (though possibly overlapping) intervals in a sequence. Gene finding as parsing (i.e., *gene parsing*) is thus a global sequence *partitioning* problem, whereas exon finding may instead be formulated as a local *classification* problem (see Chapter 10) in which intervals are classified as either belonging to a particular class (such as exons, or *promoters* – see section 12.3) or not belonging to that class. These are fundamentally distinct problems with traditionally different computational solutions. We will revisit the problem of gene feature classification when we consider the phenomenon of *alternative splicing* (section 12.1), in which a single gene may have multiple, overlapping parses, or *isoforms*.

3.5 Common assumptions in gene prediction

We have already made reference to several conceptual simplifications that are commonly employed in the computational modeling of genes (such as the emphasis on coding versus noncoding exons, and the conflating of the transcription, splicing, and translation processes into a single model of sequence composition). These simplifications are generally made in response to the overwhelming practical difficulties involved in trying to faithfully model the full gamut of molecular and evolutionary processes shaping gene structure.

These difficulties can be manifested in intractable computational problems, lack of adequate training data for empirical parameter estimation, or simply in the many gaps in current biological understanding of the relevant causal phenomena. In other cases, compromises in gene finder implementation have been made out of momentary exigencies which implementers have not yet had an opportunity to revisit; these instances can generally be attributed to the complexities of the software engineering process, of which we will have little to say in this book.

The following list is an attempt to acknowledge these various simplifications, assumptions, and deviations from current biological dogma, which are commonly made in the computational modeling of gene structure. We hope that progress in the field will allow us to remove many of these from future editions of this book.

3.5.1 No overlapping genes

The possibility that two or more genes might overlap on the chromosome is generally not considered in most gene-finding paradigms for eukaryotes (see section 7.5 for a discussion of the prokaryotic case). It is, however, a phenomenon that is known to occur (e.g., Normark *et al.*, 1983; Pavesi *et al.*, 1997). Because many of the known cases of overlapping genes involve loci on opposite strands, one solution to this problem is to run the gene finder on the two strands separately, predicting genes on either strand independently of what may have already been predicted on the other strand. Most gene finders do not follow this policy (though some do, one example being *SNAP* – Korf, 2004), opting instead to identify the most likely series of nonoverlapping genes on either strand. To our knowledge, a rigorous comparison of the relative merits of these two approaches has not been performed. Such a study would of course have to take into account the fact that overlapping genes may occur more frequently in some genomes than others, so that an optimal computational strategy may very well be species-specific.

3.5.2 No nested genes

Similar to the case of overlapping genes is that of *nested genes*, in which a locus occurs within an intron of another locus (possibly on the opposite strand). This phenomenon is also known to occur (Yu *et al.*, 2005), and although it appears to be relatively rare in most of the genomes that have been sequenced to date, the prevalence of this phenomenon may also be species-specific.

3.5.3 No partial genes

A *partial gene* occurs when only part of a gene is present on a contig, due to a missing sequence in the genome assembly. Gaps tend to be more common in draft genomes, and those with low sequencing coverage (Chapter 1); thus, many low-budget sequencing projects result in gap-ridden assemblies that may contain many partial genes. Although not all gene finders currently allow the prediction of partial genes, we will see that allowing for partial genes is fairly simple in the case of hidden Markov models (HMMs) and generalized hidden Markov models (GHMMs). In our graph-theoretic framework, we can accommodate partial genes by adding the edge types $LT \rightarrow AG$, $LT \rightarrow \overline{GT}$, $GT \rightarrow RT$, and $\overline{AG} \rightarrow RT$, so that a predicted CDS may begin with an internal or final exon, or end with an internal or initial exon.

3.5.4 *No noncanonical signal consensuses*

In identifying putative signals (e.g., start and stop codons, donor sites, acceptor sites) in DNA, it is common practice to assume that the consensuses for those signals strictly match the commonly seen forms, such as `ATG` for start codons (in eukaryotes) and `GT` for donor sites. There are a number of well-documented cases, however, of signals in functional eukaryotic genes that do not match the majority consensus sequences for signals of that type. The most common examples are the rare `GC-AG` and `AT-AC` introns in human genes (e.g, Burset *et al.*, 2000).

When rare signal consensuses are known beforehand, it is often possible to configure a given gene finder so as to recognize both the common form and the rarer forms of each signal type, thereby increasing the sensitivity of signal identification, where *sensitivity* is defined as the percentage of functional signals in a sequence that are found by the gene finder. However, doing so will necessarily decrease the specificity of signal detection, where *specificity* is defined as the percentage of putative signals identified by the gene finder that are actually functional in the cell (see Chapter 4 for formal definitions of sensitivity and specificity). Although no systematic studies have been published to date regarding the effect on overall gene prediction accuracy of reducing signal specificity in this way, anecdotal evidence suggests that gene prediction accuracy can more often than not be adversely affected by allowing the gene finder to recognize extremely rare signal consensuses.

3.5.5 *No frameshifts or sequencing errors*

Due to the imperfect nature of the sequencing and assembly processes (Chapter 1), the sequences comprising the resulting contigs, which are provided as inputs to a gene finder, may not accurately reflect the exact composition of nucleotides in the DNA molecule under study. As a result, constraints on gene syntax, such as phase constraints, may appear to the gene finder to be violated in the "correct" annotation when in fact they are not. In such cases, a gene will generally be missed or its structure incorrectly predicted as a result. Little attention seems to have been paid to these issues in the construction of the current generation of gene finders, though it is clearly an important problem, given the number of genomes that are being sequenced to low coverage.

3.5.6 *Optimal parse only*

The default behavior of almost all eukaryotic gene finders is to produce a single set of zero or more nonoverlapping gene predictions. Unfortunately, the average accuracy of most *ab initio* gene finders at the level of whole exons and whole genes is generally not much greater than 50% on complex eukaryotic genomes. Ironically, in many cases it could be shown by observing the internal workings of these gene finders that they often come very close to predicting the correct gene structure, only to discard the correct parse in favor of another, incorrect parse which just happens to have scored slightly higher (under the gene finder's scoring function) than the correct one. For this reason, several gene finders now provide the optional feature of reporting both the optimal parse (meaning "optimal" according to the program's scoring function) and a number of suboptimal parses, in hopes that the correct prediction, if it is among those emitted by the program, will be identified later during a manual genome annotation process. We will address this issue when we describe decoding algorithms for HMMs and GHMMs in Chapters 6 and 8.

3.5.7 *Constraints on feature lengths*

In order to place strict upper bounds on computational complexity, some gene finders impose artificial limits on the lengths of the individual features (such as exons) comprising a parse. We will show in Chapter 8 that this practice is not necessary for GHMM-based gene finders, at least for typical eukaryotic genomes. For some other gene-finding architectures, such restrictions may or may not be necessary in order to reduce memory and/or time requirements to a practical level. Unfortunately, the decision to impose such restrictions is often made in the context of a particular computing environment (i.e., on a particular model of computer), so that the actual bounds imposed may be less than ideal when the gene finder is deployed in other computing environments. At the very least, we recommend that such constraints be imposed only through user-defined parameters that can be modified without recompiling the program's source code.

3.5.8 *No split start codons*

A number of examples have been seen where the translation start site of a gene is interrupted in the DNA by an intron that is later spliced out to produce a valid start codon in the mature mRNA. To our knowledge, this situation is not yet addressed by any gene finders, nor is it known to what degree any specific solution to this problem will impact (positively or negatively) overall gene-finding accuracy.

3.5.9 *No split stop codons*

A number of early eukaryotic gene finders suffered from the shortcoming that they would occasionally predict genes having in-phase stop codons (other than the stop codon at the end of the CDS) which were interrupted by an intron. Such predictions are not valid coding segments (except under rare circumstances; see below) since translation into a protein product would be terminated before the end of the putative CDS. For this reason, most gene finders available today are explicitly prevented from predicting gene structures having in-phase stop codons interrupted by introns. Unfortunately, coding segments ending in such interrupted stop codons do (rarely) occur, though most gene finders cannot predict them. As with several previous items in this list, the magnitude of the benefits and/or costs of rectifying this problem have not, to our knowledge, been systematically studied.

3.5.10 *No alternative splicing*

As indicated previously, gene finders generally predict only the optimal parse for a given input sequence, whereas it is well known that many loci in the human genome can produce multiple protein products through the differential splicing of transcripts – a phenomenon referred to as *alternative splicing*. Much attention is now being directed at this problem (see section 12.1), though the incorporation of techniques for reliably predicting alternative splicing into state-of-the-art *ab initio* gene finders is likely a few years away, at the least (Guigó *et al.*, 2006).

3.5.11 *No selenocysteine codons*

As described in Chapter 1, there are 20 common amino acids. A relatively uncommon amino acid is one known as *selenocysteine*. This acid is currently known to occur only in a handful of species; unfortunately, it is encoded by the codon `TGA` (`UGA` in the mRNA), which also serves as a stop codon in most currently sequenced eukaryotic genomes. Because current gene finders forbid the occurrence of in-frame stop codons, the prediction of genes coding for selenocysteine-bearing proteins is precluded in most circumstances. Unfortunately, relaxing the restriction on in-frame stop codons would for most gene finders result in a significant degradation in gene prediction accuracy, and hence more research is required for the development of selenocysteine-aware gene finders.

3.5.12 *No ambiguity codes*

Ambiguity codes such as `R` (for purine) and `Y` (for pyrimidine) are often found in contigs resulting from low-coverage sequencing of genomes. Unfortunately, most gene finders either do not accept input containing such codes, or arbitrarily change them to a valid nucleotide upon reading their input. Another common ambiguity code is `N`, which some assemblers will place between fragments that are believed to lie close together on the genome, but for which the intervening gap has not been properly sequenced. Genomes sequenced to low coverage often produce large numbers of fragments and contigs which the assembler may then combine with arbitrarily long runs of `N`s between. In order to attain high sensitivity of exon predictions in such low-coverage genomes, it is generally necessary to enable partial gene prediction (see above) for these genomes; in these cases, runs of `N`s may be effectively interpreted by the gene finder as the end of the contig. To our knowledge, no systematic studies have been published describing the effects of various strategies for dealing with ambiguity codes.

3.5.13 *One haplotype only*

Recall from Chapter 1 that in diploid organisms such as humans, the genome consists of pairs of *haplotypes* – i.e., one set of chromosomes from each parent. Because a diploid organism's parents were also (in most cases) diploid, the process of *meiosis* (involved in the formation of *gametes* – i.e., *eggs* and *sperm*) in the parents was required to perform recombination of each parent's two haplotypes in order to form a single inherited haplotype for the offspring. In WGS-based sequencing (Chapter 1), because the DNA is sheared indiscriminately, we are unable upon reconstruction of the genome to differentiate between haplotypes, except where differences occur between the haplotypes (so-called *SNPs* – *single-nucleotide polymorphisms*). SNP density in humans is currently thought to be approximately 1 per 1000 bp. The issue of haplotypes and SNPs is currently ignored by gene finders, though it is conceivable that the processes of SNP detection and gene prediction may be linked in some way so as to mutually inform one another.

EXERCISES

3.1 Write a program to read in a FASTA file, reverse complement the sequence(s) in the file, and write the result into another well-formed FASTA file. Do not allow the lines of the output file to exceed 60 characters in length (except possibly for the defline).

3.2 Implement a procedure for building an ORF graph from a sequence.

3.3 Adapt the `highestScoringPath()` algorithm from Chapter 2 to perform gene finding on an ORF graph.

3.4 Modify the regular expression (3.8) to include phase constraints.

3.5 Modify the regular expression (3.8) so as to prohibit in-frame stop codons.

3.6 Devise an algorithm for identifying all signals in an ORF graph which are reachable from only one end of the graph or the other, and pruning them from the graph.

3.7 Download a sequence from Genbank (www.ncbi.nlm.nih.gov), print out the first 80 bases, and then draw the full ORF graph above the sequence. Report the number of valid parses.

3.8 What would happen if a donor site straddled an intron? Search the literature to determine whether this has ever been reported in a functional gene.

3.9 Suppose an intron was embedded within another intron. What effect would this have on the syntax of ORF graphs as described in this chapter? Can this situation be described using a regular expression? Elaborate on your answer.

3.10 Prove that $revcomp(S_{i..j}) = R_{L-1-j..L-1-i}$ for $R = revcomp(S)$, $L = |S|$, $0 \leq i < j < L$.

3.11 Augment the ORF graph formalism with new signal and edge types to represent promoters and polyadenylation signals.

3.12 Building on your answer to the previous question, augment the ORF graph formalism with new signal and edge types to represent cap sites and branch points.

3.13 Building on your answer to the previous question, augment the ORF graph formalism with new signal and edge types to represent exon splice enhancers and signal peptides. You may consult an appropriate molecular biology text for descriptions of these features.

3.14 Augment the ORF graph formalism to admit genes within the intron of another gene.

3.15 Augment the ORF graph formalism to admit genes overlapping on opposite strands.

3.16 Augment the phase requirements for ORF graphs and valid parses in section 3.4 to accommodate reverse-strand features.

3.17 Augment the five stipulations in section 3.4 to permit the prediction of *partial genes* (i.e, genes which may begin and/or end with an intron).

4

Gene finder evaluation

Before we begin our exploration of the various gene-finding strategies, we need to establish a set of methods for assessing and comparing the predictive accuracy of gene finders. The two main issues that must be addressed are: (1) the experimental protocol for obtaining a set of predictions for evaluation, including the collection, filtering, and preprocessing of a suitable set of test genes, and (2) the actual evaluation metrics and their computation. The importance of obtaining an objective assessment of gene finder accuracy should not be underestimated; the publication of flawed evaluations and comparative studies can effect a significant disservice to the field by reducing interest in what might otherwise be fruitful areas of inquiry or by fostering the additional expenditure of resources on methods having no real merit. For this reason, we consider this topic to be as important as that of the actual prediction methods employed.

4.1 Testing protocols

Although computational gene predictions can be utilized in a number of different ways by manual and automatic genome annotation efforts, it is arguably the case that the application of greatest practical significance for *ab initio* gene finders is the detection of functional genes which have not yet been discovered through other means. An obvious implication of this is that the greatest value of a gene finder comes from its ability to recognize even those genes that were not presented as examples to the gene finder during training. In the machine-learning and pattern-recognition literature, this capacity is referred to as *generalization* – i.e., the ability of the program to *generalize* the specific knowledge gleaned from a limited set of examples into a more general understanding of properties of the target class. In our case, the target class is the set of all genes in the genome of interest, and the limited set of examples is the training set. The opposite of generalization is *overtraining*: a model is overtrained to the extent that it is more representative

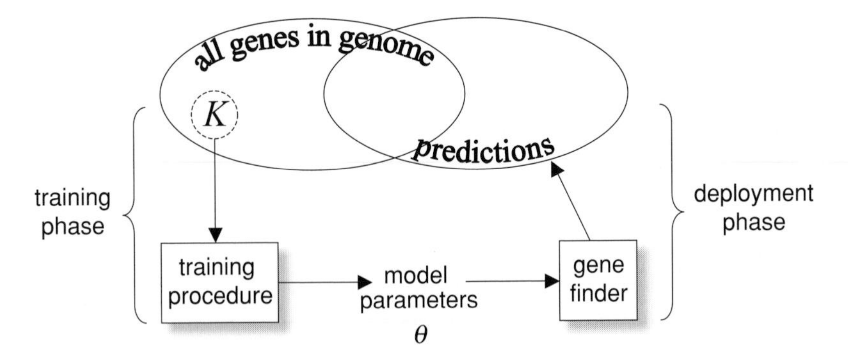

Figure 4.1 Training and deployment of a gene finder. The training set $T \subseteq K$ is selected from the set of all genes in the genome and provided to a training procedure which produces a set of model parameters, θ. The gene finder parameterized by θ produces a set of predictions for the entire genome. Estimating the expected degree of overlap between the set of predictions and the set of true genes is the goal of gene finder evaluation.

of the examples in the training set than of the unseen cases in the world at large.

Thus, the gene-finding practitioner is faced with two problems: (1) collection of a set of training genes that are sufficiently representative of the full complement of genes in the genome so as to adequately inform the gene finder during the training phase, and (2) the procurement and training of a gene finder that will be sufficiently capable of generalizing the knowledge of gene structure embodied in those training examples so as to be able to recognize novel genes when they are encountered in DNA during the deployment phase. These ideas are illustrated in Figure 4.1.

What concerns us in this chapter is how to predict reliably how well a gene finder will perform during its deployment phase, once it has been trained – that is, to assess its generalization ability. As suggested by Figure 4.1, this goal is roughly that of predicting the size of the overlap region between the set of correct gene structures and the set of predicted structures when the gene finder is applied to the entire genome. Since in the general case we will have possession of at most a small subset K of the correct gene structures prior to training, the problem is one of measuring the accuracy of the gene finder on (some subset of) that known set and then controlling for biases due to sampling error as we extrapolate those accuracy measurements into predictions of how well the gene finder is likely to perform on the whole genome.

Let us consider the composition of the set K of training examples. In practice, a training set for a gene finder may consist of ESTs, cDNAs, known genes from the genome G under study, and possibly genes from foreign genomes having significant BLAST homology to some segment in G. In the case of ESTs and BLAST hits, some form of mapping will typically be required to obtain a spliced alignment to the corresponding region in G so that the intron/exon structure for likely genes (whether whole or partial) in G can be obtained. A number of tools are available for performing spliced alignment of ESTs and cDNAs to a substrate sequence, including *BLAT* (Kent, 2002), *sim4* (Florea *et al.*, 1998), *gap2* (Huang *et al.*, 1997), *spidey* (Wheelan *et al.*, 2001), and *GeneSeqer* (Usuka *et al.*, 2000). Other programs now exist (e.g., *PASA* – Haas *et al.*, 2003) which can combine the alignments generated by these tools, to produce high-confidence annotations in an automated fashion, thereby aiding the annotation effort directly while also producing training data for a gene finder. In Chapter 9 we will consider the use of these various forms of evidence during the actual prediction process, rather than during training.

Regardless of how the training data have been obtained, it is always worthwhile to perform a number of automated checks to ensure data integrity. In particular, we recommend examining every gene to ensure that start codons, stop codons, and donor and acceptor splice sites occur exactly at the expected positions in the training sequences, given the feature coordinates to be used for training. Because coordinates of training features are often obtained from other parties that may use different conventions (such as 1-based rather than 0-based coordinates, or exon coordinates that do or do not include the final stop codon, or even different assemblies of the target genome), performing such "sanity checks" can help to catch training errors before they occur. Depending on the details of the training software, such errors, when they occur, might not be detected by the software and may only manifest themselves in poor gene-finder performance, making it difficult to determine the exact cause of the problem after the fact.

The process of data collection will sometimes produce a training set K that is exceptionally small. In such cases, there is, unfortunately, little choice but to train the gene finder on the full set K and to deploy the resulting program with little or no evidence regarding how well the gene finder is likely to perform during deployment. In such a case, objectively assessing the accuracy of the resulting gene finder can only be accomplished if one is prepared to expend considerable resources running the gene finder on anonymous DNA and then painstakingly examining the results – i.e., by performing a full manual annotation (or "curation") process on some (sufficiently large) random sample of the gene finder's output. Given the cost of this route, one might as well expend the same resources by instead enlarging the set K through similar curation activities, thereby furthering the twin goals of achieving a more accurate gene finder and at the same time procuring enough data to also allow for some form of evaluation of the gene finder.

Although it may be tempting in data-poor situations to simply evaluate the accuracy of the gene finder on the training set, it has been well documented in the machine-learning literature (e.g., Salzberg, 1999) that evaluating any predictive model on the training set generally produces artificially inflated accuracy

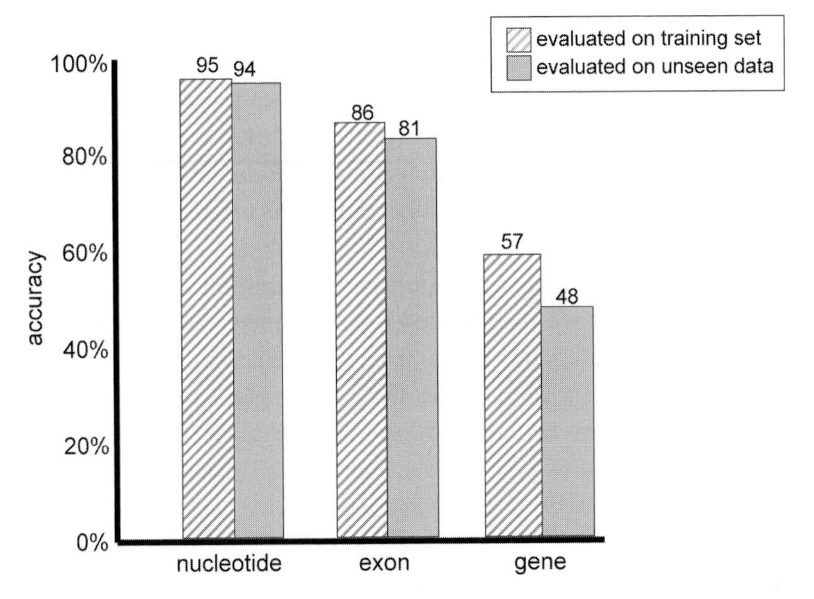

Figure 4.2 Results of evaluating a gene finder on the training set (hatched bars) vs. on unseen data (gray bars) for *Arabidopsis thaliana*. Training and test sets each contained 1000 genes. Metrics were, from left to right: nucleotide accuracy, exon *F*-measure, and gene sensitivity (see section 4.2 for definitions of *F*-score and sensitivity).

measurements which are not then reproduced when the model is later tested on a set of unseen test cases. Experiments have shown that gene finders are no exception to this rule (Burset and Guigó, 1996; Majoros and Salzberg, 2004). As a result, any evaluation results obtained on the training set should be considered as much a measure of the gene finder's susceptibility to overtraining as a measure of its likely accuracy on unseen data. Figure 4.2 illustrates this effect for a data set from the plant *Arabidopsis thaliana*.

There is one other option that is available when training data is very scarce or even nonexistent. By extracting long, continuous ORFs from anonymous DNA for the genome under study, one can obtain sequences that are (depending on their length and the overall {G,C} density of the genome) very likely to be real coding genes. The rationale for this is that under the absence of selective pressures to maintain open reading frames, long stretches of noncoding DNA become increasingly likely (as the length of the stretch increases) to have at least one in-frame stop codon in each frame. Thus, exceptionally long ORFs are, by this argument, overwhelmingly likely to be under positive selection for maintaining the ORF and therefore likely to be coding genes. It is for this reason that some bacterial gene finders (e.g., *Glimmer* – Delcher *et al.*, 1999a) obtain their training data almost exclusively by extracting long ORFs from the raw DNA sequences. Although not commonly practiced for eukaryotes, this same strategy is nonetheless available for those eukaryotic organisms for which few or no training data are otherwise available. Extracting such long ORFs is a simple matter, as it involves simply scanning through the successive trinucleotides in each of the three frames, keeping track of the position of the last stop codon seen in each frame, and emitting any ORF over a predetermined length. The resulting ORFs can then be used to train the coding model of the gene finder; other sources of data must obviously be procured to train the gene finder's splice site models and other components.

Yet another option in the absence of curated training data is the practice of *bootstrapping* (Korf, 2004; see also section 11.2). By employing a gene finder which has already been trained for a related organism, one can obtain a set of computational gene predictions for the genome under study. These predictions can then be taken as a very crude training set, with the obvious caveat that no guarantees can be given as to how many of these predictions are actually functional genes, nor how accurately the boundaries of those predicted functional genes have been identified. Nevertheless, the practice of bootstrapping, combined with the use of long ORFs, can sometimes provide enough training data to produce a gene finder with greater accuracy than one could obtain by simply applying a *foreign* gene finder (i.e., one trained for a different organism) to a novel genome (i.e., the practice of *parameter mismatching* – Korf, 2004).

Once we have obtained a reasonably large set K of example genes, we can entertain a number of possibilities for partitioning K into a *training set* T and a separate *hold-out set* H to be used for evaluating the gene finder. Ideally, T and H should be disjoint. It should be noted that the purpose of partitioning K in this way is solely for gene finder evaluation, and it need not be performed when training the final version of the program for deployment. Thus, we can train the gene finder on the subset T, evaluate its accuracy on H, and then re-train the program on the full set K for deployment. This protocol is illustrated in Figure 4.3.

The accuracy measurements A_H taken from the version of the gene finder trained only on T will of course be an imperfect estimate of the accuracy A_G of the final gene finder trained on the full set K and applied to the full genome G. The bias inherent in this estimation will depend on both the relation of K to the full complement of genes in the genome (i.e., how representative K is of all genes in G), and also on the manner in which we partition K into training and test sets. In practice we will typically have little control over the relation of K to the full genome, but we can at least consider how best to partition and otherwise preprocess K so as to obtain the best estimate of the final gene finder's accuracy.

The first task in preprocessing the set K of example genes is to eliminate redundancy. If a particular gene occurs multiple times in a training set, then the gene will exert a correspondingly larger influence on the training procedure than other genes that are less well represented in the set (since, as we will see in the chapters

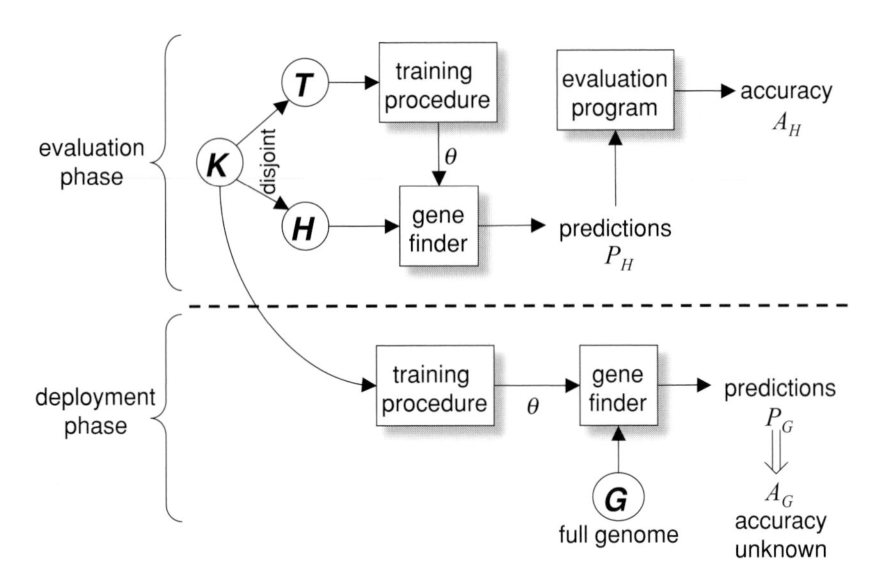

Figure 4.3 Estimating the accuracy of a re-trainable gene finder. Example genes K are partitioned into training set T and hold-out set H, but only for evaluation, producing accuracy estimate A_H. The deployed version of the program is trained on all of K and run on the full genome G. The unknown accuracy A_G on the genome can be estimated using A_H, though the estimate may be highly biased, depending on the contents of K and the manner of its partitioning into T and H.

ahead, training procedures for gene finders typically operate by observing frequencies of various features in the training set). This bias might be acceptable if a similar bias happened to be present in the genome as a whole, but more often it is the case that redundancy in the example set is merely an artifact of the data collection process (such as occurs, for example, when a number of genes are discovered and submitted to GenBank as a result of directed studies focusing on specific gene families or pathways). Eliminating redundancy in our set K of all known genes will also prevent the inadvertent duplication of identical or near-identical genes between sets T and H when we partition K; failure to eliminate such duplication can result in artificially inflated accuracy estimates by giving the gene finder an unfair advantage when it is tested on a gene in H which has directly influenced the gene finder's parameters via its simultaneous presence in T.

Redundancy can be eliminated from a set of genes K by clustering the elements of K by their pairwise alignment or BLAST scores. A common clustering strategy is to cluster together any two genes found to be more than 95% identical over at least 95% of their length. These percentages are typically chosen based on intuition rather than by any sort of principled deduction, and are sometimes chosen as low as 80% (i.e., 80% identical over 80% of length) or as high as 99%. The length in question is typically taken to be the alignment length of the spliced transcripts or their protein products. A very crude procedure for removing redundancy from a set of sequences is given in Algorithm 4.1.

Algorithm 4.1 A very crude procedure for eliminating redundant sequences from a set K, where two sequences are considered redundant if they are at least P_I% identical over at least P_L% of their aligned length.

```
1.  procedure elim_redundant(K,P_I,P_L)
2.    KK ←K × K;
3.    foreach (S_i,S_j) ∈ KK do
4.      if i≠j ∧ S_i∈K ∧ S_j∈K then
5.        a←alignment_score(S_i,S_j);
6.        L←alignment_length(S_i,S_j);
7.        if a≥P_I ∧ L≥P_L then K ← K−{S_j};
8.    return K;
```

In practice this algorithm can be optimized by utilizing BLAST to quickly find just those pairs having statistically significant similarity scores (see Korf *et al.*, 2003).

A more sophisticated approach to eliminating redundancy is to cluster the elements according to their alignment scores (where any two sequences not satisfying the aligned length criterion can be considered as having an effective alignment score of zero) and then keeping only a single element from each cluster. A hierarchical clustering algorithm such as *average link clustering* (Voorhees, 1986) can be used, with the tree-cutting criterion set at the desired maximum similarity; the UPGMA phylogeny reconstruction algorithm outlined in section 9.6.3 can be adapted for this purpose as well.

Once the set K of all available example genes has been rendered nonredundant, it can be partitioned into a training set T and a separate hold-out set H to be used for evaluation. In deciding on the relative sizes of these two sets, a number of different rules of thumb are in common practice, so that our recommendation is simply that neither T nor H should be so small as to make likely a high degree of bias due to sampling error, either in the training of the gene finder or its evaluation.

One very effective means of combating such bias is the use of *cross-validation* (e.g., Mitchell, 1997). We recommend at least a five-way cross-validation procedure, as follows. First, randomize the order of the example genes in K, and then partition K arbitrarily into five equal parts, designated K_i for $0 \leq i \leq 4$. Then perform five separate iterations of training and evaluation, with iteration i using:

$$T_i = \bigcup_{j=1}^{4} K_{(i+j) \bmod 5} \tag{4.1}$$

for training and using K_i for testing. The accuracy estimates A_i for the five iterations can then be averaged together to arrive at a single estimate for the accuracy of the gene finder.

When comparing gene finders it is important to minimize the difference in treatment of the programs and their input data. Using the language of experimental design, we say that the gene finders themselves should be the only *treatment* that differs among the trials of the experiment, with all else being held constant. Any violation of this protocol reduces the validity of the results of the comparison. This policy of equal treatment includes not only the set of known genes on which the programs are to be tested, but also the training data that are to be used to train the programs, and obviously the accuracy metrics that are to be used for the comparison. Unfortunately, many gene finders are not re-trainable by end-users, so that one does not always have control over the training data. A rather more severe consequence of this is that it may not be possible in many cases to ensure that the test set does not include genes that were present in the training set for one of the gene finders. Obviously, allowing one of the gene finders in the comparison to have exclusive access to a number of test genes during training can give an unfair advantage to that gene finder, thereby invalidating the results of the comparison.

In interpreting the results of any gene finder comparison, one should be very careful not to draw unfounded conclusions. In general, a comparison of N programs all trained on some set T and tested on some set H may indicate that some program P is superior to the others when trained on T and tested on H, but may not give a reliable indication of the relative accuracy of the respective programs under other training and testing scenarios. In the case of the large-scale competitions *GASP* (Reese *et al.*, 2000) and *EGASP* (Guigó *et al.*, 2006), the results may tell us very little about the merits of particular gene-finding algorithms, other than the obvious conclusion that comparative approaches (see Chapter 9) tend to outperform noncomparative methods, due to their use of more diverse sources of evidence. In practice, most gene finders comprise large, complex software systems; a count of noncomment, nonwhitespace lines of source code for four publicly available gene finders (*SNAP* – Korf, 2004; *GlimmerHMM* – Pertea, 2005; *GeneZilla* – Majoros *et al.*, 2004; *UNVEIL* – Majoros *et al.*, 2003) resulted in an average of 15 329 lines of C/C++ source code per system, with the smallest system comprising 9536 lines and the largest comprising 22 083 lines. Systems of this size unfortunately entail too many independent implementation details to allow us to reliably attribute differences in performance to particular algorithmic or modeling decisions as gleaned from the terse published descriptions of the respective systems. It is arguably the case that more reliable conclusions may be drawn from controlled experiments within the context of a single code base (possibly replicated over several independently developed

```
52  fumigatus  initial-exon  42155  42915  .  +  0  transgrp=10
52  fumigatus  internal-exon 42980  43210  .  +  2  transgrp=10
52  fumigatus  internal-exon 44004  44214  .  +  2  transgrp=10
52  fumigatus  internal-exon 44278  44525  .  +  0  transgrp=10
52  fumigatus  final-exon    44593  44758  .  +  2  transgrp=10
52  fumigatus  initial-exon  59987  60513  .  -  0  transgrp=16
52  fumigatus  final-exon    59549  59930  .  -  2  transgrp=16
52  fumigatus  single-exon   73702  74544  .  +  0  transgrp=24
```

Figure 4.4 An example GFF file. The fields are (left to right) substrate ID, source, feature type, begin, end, score, strand, phase, transcript ID.

code bases, and ideally replicated over a large number of genomes and training regimes), so that the differences in performance may be unambiguously attributed to individual algorithms, since all else would be held constant across treatments – i.e., the traditional scenario for rigorous scientific experiments. Unfortunately, the gene-finding literature is still very much lacking in large-scale, rigorous, controlled experiments of this sort, though the results of such experiments could be of enormous value in practical genome annotation efforts.

Finally, we consider the more mundane issue of file formats. Each gene finder's training program will have its own requirements regarding the input format for the training genes, but one format in particular is gaining momentum as a standard for specifying coordinates of features relative to the sequences in an associated FASTA file. *GFF*, or *General Feature Format*, specifies an easily parsable syntax for features and their coordinates. An excerpt from a sample GFF file is shown in Figure 4.4.[1]

Each line in a GFF file corresponds to a single feature (i.e., an exon, splice site, etc.), and consists of the following tab-separated fields:

substrate: typically the identifier of the contig on which the feature occurs
source: usually the species or the database from which the data was obtained
feature type: e.g., "exon" or "start-codon"
begin coordinate: the 1-based coordinate of the leftmost base of the feature
end coordinate: the 1-based coordinate of the rightmost base of the feature
score: a real number or dot (".") if not available
strand: a + or – to denote forward or reverse strand
phase: 0, 1, 2, or dot (".") if not applicable
extra fields: typically `transcript_id="X"` or `transgrp ="X"` for some identifier *X*.

Each GFF file is typically accompanied by one or more FASTA files which contain the contigs referenced in the first field of the GFF records. The contig identifiers

[1] The GFF standard is described in detail at www.sanger.ac.uk/Software/formats/GFF/GFF_Spec.shtml

are generally given as the first field on the defline (just after the >), so that it is simple to implement an automated procedure for extracting the features specified in the GFF file from the contigs in an associated FASTA file and to provide them to the gene finder training program. The exons of a gene are typically grouped together using a `transcript_id="X"` or `transgrp="X"` entry in the "extra fields" section of the GFF record. Note that characters such as quotes and semicolons are inconsistently used by various groups, so that a practical GFF parser must allow some flexibility in these extra fields. Score and phase are less often used for training and testing of the gene finder, though the score field can be used to specify weightings of different types of evidence, in the case where some of the example genes are less certain than others. GFF coordinates are always counted from the beginning of the substrate, regardless of strand, and they should always be interpreted as 1-based coordinates including both the first and last base of the feature.

4.2 Evaluation metrics

The accuracy of gene structure predictions can be assessed at three levels: that of individual nucleotides, that of individual exons, and that of complete genes. We refer to this distinction as the *level of granularity* of the evaluation. It is common in published evaluations to report measures at all three levels, since each measure may tell us something about the behavior of the gene finder that the other measures neglect.

A number of different evaluation metrics exist, each with a different mathematical formula, and each of these can in general be applied at any of the three levels of granularity. All of the metrics that we will present are based on the notion of *classification accuracy*. When classifying a number of objects according to their membership in some class C, we can distinguish between the following four events:

(1) **true positive**: an object that belongs in C has been correctly classified as belonging to C.
(2) **true negative**: an object that does not belong to C has been correctly classified as not belonging to C.
(3) **false positive**: an object that does not belong to C has been mistakenly classified as belonging to C.
(4) **false negative**: an object that belongs to C has been mistakenly classified as not belonging to C.

Given a set of classification attempts, we can denote by TP the number of true positives, TN the number of true negatives, FP the number of false positives, and FN the number of false negatives. Figure 4.5 illustrates this concept in the case of exons.

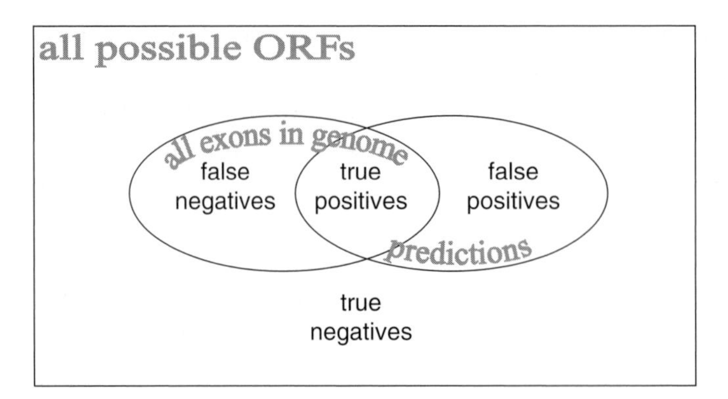

Figure 4.5 Classification events at the exon level: true positives are real exons that were correctly predicted, false positives are predicted exons that either do not exist or were not predicted perfectly, false negatives are real exons that were not predicted (correctly), and true negatives are all the possible ORFs that are not real exons and that were not predicted.

The simplest evaluation metric that we will consider is the *simple matching coefficient* (*SMC*) (Burset and Guigó, 1996), which is often referred to simply as "accuracy":

$$SMC = \frac{TP + TN}{TP + TN + FP + FN}. \tag{4.2}$$

SMC is popularly used to report the nucleotide-level accuracy of the gene finder, in which case the class *C* to which we referred earlier represents the set of coding nucleotides in the test set. That is, at the nucleotide level, a true positive is an individual nucleotide within a true coding segment that has been included in a predicted CDS. In this way, the numerator of Eq. (4.2) will evaluate to the number of nucleotides correctly classified by the gene finder as coding or noncoding, and the denominator is simply the total number of nucleotides classified (i.e., the total substrate length).

It is now widely recognized that nucleotide accuracy as measured by *SMC* is not an ideal measure, since in the case of a genome with low coding density (such as our own), a gene finder can produce a high *SMC* measure by simply not predicting any genes; in mammalian genomes one can often achieve >90% nucleotide accuracy with this (useless) strategy. One way to address this issue is to decompose the accuracy into *sensitivity* (*Sn*) and *specificity* (*Sp*):

$$Sn = \frac{TP}{TP + FN} \qquad Sp = \frac{TP}{TP + FP}. \tag{4.3}$$

Sensitivity and specificity are known in some other fields (such as information retrieval) as *recall* and *precision*, respectively. *Sensitivity* is the percentage of objects in class C that are correctly classified to C, and *specificity* is the percentage of those objects classified to C that actually belong in C.[2]

We have already seen how TP, TN, FP, and FN can be defined at the nucleotide level. At the level of individual exons, a *true positive* is an exon that was predicted exactly correctly – i.e., such that both the beginning and ending coordinates of the predicted exon were exactly correct. All other predicted exons are considered *false positives*, and any exon in the test set that is not predicted exactly correctly (or not predicted at all) is considered a *false negative*. Similarly, at the level of whole genes, only those genes in the test set which were predicted exactly correctly (i.e., with all exon coordinates for the gene being correctly identified) are considered true positives.

Thus, a reasonable set of metrics to report from a gene finder evaluation is:

SMC_{nuc} = nucleotide SMC
Sn_{nuc} = nucleotide sensitivity
Sp_{nuc} = nucleotide specificity
Sn_{exon} = exon sensitivity
Sp_{exon} = exon specificity
Sn_{gene} = gene sensitivity

Gene specificity has been omitted from this list because computing the specificity depends on knowing the number of false positives, which is difficult to assess in the case of genes (and to some degree in the case of exons as well), since the lack of evidence for a gene in any region of the test substrate does not strictly preclude the existence a true gene in that location. That is, a gene prediction in a place where no prediction was expected may indicate either that the prediction is faulty, or that our set of known genes in the test substrate is incomplete, and absent a reliable way to distinguish between these events we are unable to assess gene specificity.

For this reason, the test substrate is often taken to include only the immediate region of a test gene, so that false positives at the nucleotide and exon levels are less likely to be inflated due to unknown genes in the test substrate. For example, we might manufacture a test substrate by taking the known genes in the test set, extracting just the subsequences around the immediate vicinity of each gene (i.e., including the CDS of the gene itself, plus a margin of, say, 1000 or 50 000 bp on both sides of the known CDS), and then either providing these as inputs to separate runs of the gene finder, or concatenating these subsequences together into a single, synthetic substrate to be processed by a single run of the gene finder.

[2] Note that in other fields these terms are sometimes defined differently. The definitions we have given are the most commonly used in the gene-finding literature, and since we will consistently apply these same definitions throughout the book there should be little room for confusion.

Although this practice can be expected to reduce the bias in our measure of the false positive rate due to unknown genes in the test substrate, we will see that for many gene finders, especially those based on probabilistic models of gene structure, the distance between genes can strongly affect the gene finder's behavior, so that the accuracy measured on such a synthetic substrate may be higher or lower than the accuracy on a real substrate in which the spacing between genes has not been artificially manipulated. For this reason, we recommend using full-length test substrates and then simply ignoring gene predictions during the evaluation which do not overlap any known gene in the test set.

A number of other metrics have been championed in recent years to supplant the *SMC* metric, due to its tendency to be dominated by the true negative count in genomes with low coding density. One of these is the *F-measure*, which can be computed from the sensitivity and specificity:

$$F = \frac{2 \times Sn \times Sp}{Sn + Sp}. \tag{4.4}$$

Another is the *correlation coefficient*, *CC*:

$$CC = \frac{(TP \times TN) - (FN \times FP)}{\sqrt{(TP + FN) \times (TN + FP) \times (TP + FP) \times (TN + FN)}}. \tag{4.5}$$

Because the correlation coefficient can be undefined due to division by zero, an alternative has been proposed called the *average conditional probability*, or *ACP* (Burset and Guigó, 1996), which ranges between 0 and 1:

$$ACP = \frac{1}{n}\left(\frac{TP}{TP + FN} + \frac{TP}{TP + FP} + \frac{TN}{TN + FP} + \frac{TN}{TN + FN}\right), \tag{4.6}$$

where any ratio which would result in division by zero is itself replaced by 0 and n is the number of remaining ratios in the sum. From the *ACP* we can compute another metric which ranges from -1 to 1, called the *approximate correlation*, or *AC* (Burset and Guigó, 1996):

$$AC = 2(ACP - 0.5). \tag{4.7}$$

These latter measures (Eqs. (4.4–4.7)) are useful in that they replace the sensitivity and specificity scores with a single measurement, allowing for simpler comparison between the different treatments (i.e., gene finders, test sets, or other experimental conditions) of a gene-finding experiment, and they generally are less affected by coding density than the *SMC* (Burset and Guigó, 1996). Nevertheless, during gene-finder development or training it is often useful to have access to the sensitivity and specificity scores so that the cause of poor prediction accuracy can be investigated in terms of *under-prediction* (low sensitivity) and *over-prediction* (low specificity). Knowing that a particular gene finder has a tendency to under- or over-predict can in some cases guide us in the process of tuning the program's parameters (when they are

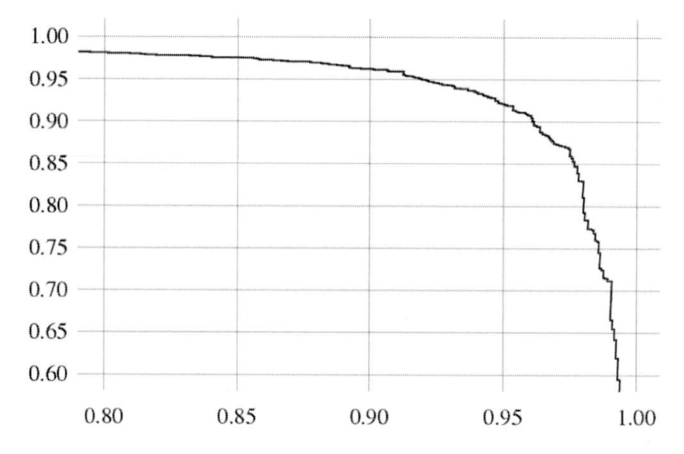

Figure 4.6 A sensitivity–specificity curve for the problem of distinguishing coding vs. noncoding DNA using a Markov chain trained on introns in *Aspergillus fumigatus*. Sensitivity is shown on the *y*-axis and specificity on the *x*-axis. Increasing either of these measures is generally made at the cost of decreasing the other measure.

available), or during gene finder development to identify program bugs or to aid in the selection of alternative algorithms.

Sensitivity and specificity are typically found to exhibit a trade-off in which high sensitivity is associated with low specificity, and vice versa. This can be seen by graphing the *receiver operating characteristic* (*ROC*) curve for a system, in which the *true positive rate* (sensitivity) is associated with the *y*-axis and the *false positive rate* (1-specificity) with the *x*-axis. A related plot, called the *sensitivity–specificity curve*, associates sensitivity with the *y*-axis and specificity (rather than 1-specificity as in the ROC curve) with the *x*-axis, as illustrated in Figure 4.6.

During algorithm development or gene finder training it is often useful to construct this curve by running the system at a set of different parameterizations, and then to choose a parameterization associated with an acceptable trade-off between sensitivity and specificity, as determined by the requirements of the application at hand. For genome annotation, sensitivity is often preferred over specificity, since *ab initio* gene finders are generally valued for their ability to find genes not found through other means, whereas for applications in synthetic biology and biomedicine one typically desires high specificity to reduce the cost of verifying predictions in the laboratory.

Finally, let us consider the problem of establishing a baseline performance for comparison. When no other gene finder is available for the purposes of comparison, we need some means for putting our accuracy measurements into perspective. One option is to compare the measured accuracy of our gene finder to the accuracy that we would expect if we were to emit predictions at random. A very simplistic way of modeling random predictions is to assume that a random gene predictor would guess, at each position in the sequence independently, whether the base at that position is coding or noncoding. Although this is an extremely primitive means of modeling random gene predictions, it at least provides an easily computable baseline value below which any accuracy score can be considered unacceptable for practical gene finding.

If we suppose that the random predictor predicts a coding base with probability p and that the test set has a *coding density* (ratio of coding bases to total sequence length) of d, then the expected sensitivity of the random predictor on the test set is:

$$Sn_{rand} = \frac{TP}{TP + FN} = \frac{dp}{dp + d(1-p)} = p, \tag{4.8}$$

and the expected specificity is:

$$Sp_{rand} = \frac{TP}{TP + FP} = \frac{dp}{dp + (1-d)p} = d, \tag{4.9}$$

so that the expected *F*-measure is:

$$F_{rand} = \frac{2 \times Sn \times Sp}{Sn + Sp} = \frac{2dp}{d+p}. \tag{4.10}$$

For a fixed coding density d, the maximal F is obtained by predicting all bases to be coding – i.e., $p = 1$ – so that our baseline F can be computed simply as:

$$F_{rand} = \frac{2d}{d+1}. \tag{4.11}$$

A baseline *SMC* value can be derived by choosing $p = round(d)$ – that is, $p = 0$ if $d < 0.5$, and $p = 1$ if $d \geq 0.5$:

$$SMC_{rand} = max(d, 1-d). \tag{4.12}$$

EXERCISES

4.1 Write a module or function in your favorite programming language to read from a file a set of gene structures in GFF format.

4.2 Write a program in your favorite programming language to read from a file a set of gene structures in GFF format and a FASTA file containing the substrate sequence for those genes, and to verify that all of the gene structures are well formed – i.e., that all start codons, stop codons, donor sites, and acceptor sites have the expected consensus sequences. Allow the user to specify the legal consensus sequences on the command-line.

4.3 Write a program to randomly sample a subset of the genes in a GFF file according to a partitioning percentage specified on the command line. Write out the selected genes into one file and the remaining genes into another file.

4.4 Implement Algorithm 4.1 in your favorite programming language.

4.5 Implement a hierarchical clustering algorithm for eliminating redundant sequences, using your favorite programming language. Refer to an appropriate reference (e.g., Voorhees, 1986; Durbin *et al.*, 1998) to obtain the details of such an algorithm.

4.6 Write a program to compute *Sn*, *Sp*, and *SMC*, given a GFF file of test genes and a GFF file of gene predictions.

4.7 Augment the program developed in the previous exercise to also compute the *F*-measure.

4.8 Augment the program developed in the previous exercise to also compute the *approximate correlation* (*AC*).

4.9 Prove that Eq. (4.12) follows from the $p = round(d)$ baseline strategy, for p the probability of predicting a coding base and d the actual coding density.

4.10 Prove that Eq. (4.10) is maximal at $p = 1$.

5

A toy exon finder

In this chapter we will construct a "toy" exon finder for a trivial genome, using mostly ad hoc techniques. While the level of computational sophistication and rigor will contrast starkly with that of later chapters, the purpose here will be to develop some intuition for the more salient aspects of the statistical structure of genes. Source code and sequence files for the examples in this chapter can be obtained through the book's website (see Appendix). The genome that we will be using is that of the fictional organism *Genomicus simplicans*. We begin by describing some of the features of this toy genome.

5.1 The toy genome and its toy genes

The *G. simplicans* genome has several distinguishing characteristics which make it suitable for illustrating the basic gene-finding problem:

- The genome is very dense with genes, with typically no more than 20 bp between successive genes on a chromosome.
- The exons within each gene tend to be very small, on the order of 20 bp.
- The introns within multi-exon genes tend to be very short, on the order of 20 bp.
- The exons incorporate fairly strong codon biases as well as a significant {G,C} bias, and the splice sites and start and stop codons are flanked by positions with fairly strong base composition biases.
- The codon usage statistics and base composition biases near splice sites and start and stop codons have been well characterized through extensive experimental work.
- Genes occur only on one strand (which we will arbitrarily call the *forward* strand).

An example *G. simplicans* contig is shown in Figure 5.1. The same contig is shown in Figure 5.2 with the genes superimposed to show the intron/exon structure. This sequence can be downloaded from the book's website (see Appendix). This particular

```
         10        20        30        40        50
 1 ACCAGGCTTTATAATGTGGGAGAGCCTCCGGGGGGGTTCATAATGCGATG
 2 AGATACGGTCATCGGTATTAATGGAGGAGGGGTGGGTCACTGCTCATCGT
 3 ATATAAGCCAGGGAGGCTGAAAGAGATCCGGTTACTATACTCTTCTTTAT
 4 AAGGGATGATGTCTCTCTGCGGTGATTCGGCGACTGGTTTAACCACAATA
 5 ATAATGACGCAAACGTACAAACGGATCCTCCGGTCACGTTAAGGCGAGAT
 6 AAAAGGGCACTGCTGGAGTACGTCACGTAGTTCCCGATAAGATTAAGCCA
 7 CAGTCCCTGGGCTGATAATGGTCATCGCATACCGGGGTCCAGATATTAGC
 8 ACGGCTGCTCAGGAGCAGGTGGGAGCCACTGCTGCCATGATTCGCAAAAA
 9 ATAACCTATGAACGGACTCCACTTCTAATGGCCCTGAGCATCTGGAGCCG
10 GAGCTAATGCGCAATAGTATGATAATGCGGTGGTCTACCCTAGAACTCGA
11 TGAGCGGTCAAGTCGCTTTCGAGGAAATGTGGCAAATTAGGGCTGGCTTG
12 GGAGCTGGGTTACTGCAGTCCCCATATAAGTCGAGCTGTGGTAATGCGCG
13 CTCATCGAGCAGGTTAGGAAGGAACGCAAGATGATGGGGCTATCTAGCAT
14 CAAATAAGGCGCTTCTGATCCCAACGCGTGGTGACCGTTAAGTAATAATG
15 AGGATCAAAAACAGCAAATGGTAGTGACCGAGCGTCGACCGAACATCGAC
16 TGATAATGCTGACGGAGGGGCGGTCGTCACATAAAAGTAGCGATGTATCT
17 TAAGGCGCGCCGAGGTTGATGATGGAGAGGTGGATCTGATGAGGCATTTG
18 ACTCCCTCGTGATGATGCTGATCTCTCAAGTTGCTTCATTGAATTATATA
19 AGAGCCTGGTCAGGTAGTGCGATACTAGGGACGTTCAATAAATGATAATG
20 AAAAGCACGTACTGGCAAGAGTCAAGTAGTACGTGGATAAAAGGAGTCGG
21 CGGCTGCTGGGTCCAGCTCTGCTGCCATGATTCGGCCGCGGCCACTAAAA
22 ATCCAAATAATGCCATTCGAGGTCAAAATCGTCAAGGACAGTTAAAAAAT
23 ATAAGAGGGCTACGATTGCCGTCACTTCGTTGCATACACCCCTTAAAAGT
24 TTGCAACGCTGTACCTGACGACGTCATCAAGGAGGTCTTAATGAGCATGC
25 AAACGCGACAGATACTCTGCTATGATTATGGTCCTGACGAAATACTGATG
26 AGCCGCTGTGATTGTCTGCTGTATTGCACTGACGATGCCATACTTATCAA
27 TTCGTCACTGATATAAGCACTCGCATCTAGGCGACGGTACACGGCAGGTT
28 AATGCGCAGTGTCATTAATAATGTGGGAGAAAGATTAGTGCGCTGACCTT
29 ATGATTTCTATGCAAAGTTCTCATGATGCGACATTACTGGAGGTAGGGCG
30 GGTACCCTAACAAG
```

Figure 5.1 A sample contig from the fictitious organism *Genomicus simplicans*.

sequence contains 29 exons comprising 17 genes. The exons range in length from 14 to 42 bp, with a mean of 24 ± 8 bp. The coding density is 47% coding nucleotides. The exon coordinates are given in Table 5.1. They can also be downloaded (in GFF format) from the book's website.

5.2 Random exon prediction as a baseline

As we develop our toy exon finder and as we incorporate successive enhancements intended to improve the predictive accuracy, it will be helpful if we have a baseline against which to measure our progress. For a baseline we will simply enumerate all ORFs, eliminate overlaps arbitrarily, and predict the set of all remaining ORFs as exons. When two or more ORFs overlap we will iteratively discard one at random until all overlaps have been eliminated.

Table 5.1 Exon coordinates for the contig shown in Figures 5.1 and 5.2; 1-based coordinates are shown

Exon	Begin	End	Exon	Begin	End	Exon	Begin	End
1	14	43	*11*	541	558	*21*	994	1030
2	71	85	*12*	581	594	*22*	1060	1080
3	108	130	*13*	634	657	*23*	1106	1120
4	154	175	*14*	698	724	*24*	1150	1191
5	204	232	*15*	756	775	*25*	1228	1248
6	256	280	*16*	788	803	*26*	1285	1303
7	318	336	*17*	822	839	*27*	1318	1352
8	350	390	*18*	865	879	*28*	1371	1388
9	428	457	*19*	903	917	*29*	1426	1446
10	475	511	*20*	948	970			

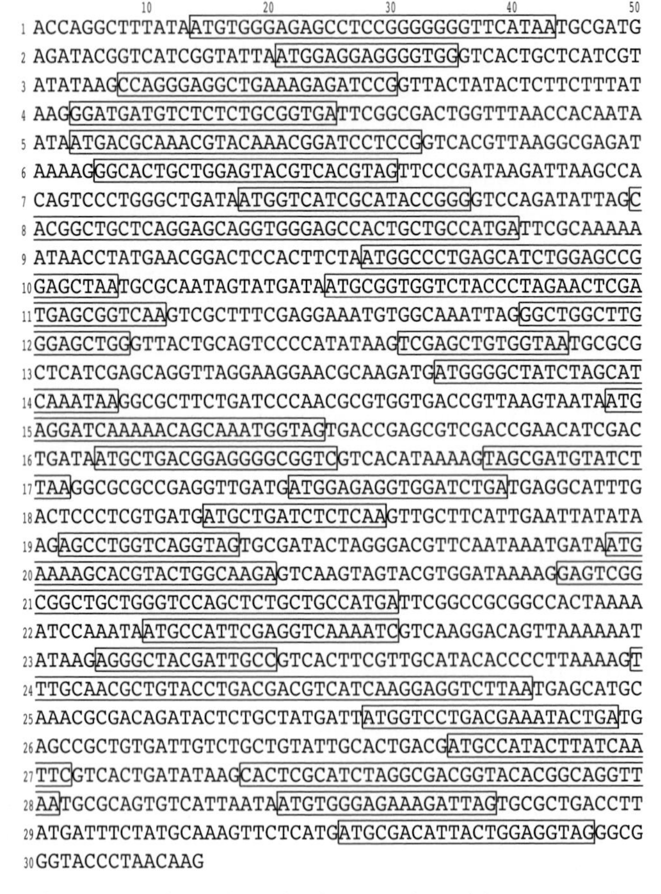

Figure 5.2 The same contig as shown in Figure 5.1, but with exons superimposed as rectangular boxes.

In this way we will obtain a set of nonoverlapping exon predictions selected without explicit reference to any compositional properties of the putative exons themselves. The only intrinsic property of an ORF which can influence its chances of selection under this procedure is the length, since longer ORFs will tend to take part in a larger number of comparisons and are thereby more likely to be discarded during overlap resolution. Indeed, while the 772 ORFs in this sequence have an average length of 64±77 bp, we will see that our random selection procedure will result in an average predicted length of 23±30 bp, which is significantly smaller. Fortunately, this is very close to the average length of the true exons, so this procedure should be suitable for use as a baseline.

Because this procedure is nondeterministic (i.e., it will generally produce a different set of predictions each time it is run), we will repeat the procedure several hundred times and report the mean nucleotide accuracy (± one standard deviation) as our baseline number. We can then compare this to the SMC_{rand} measure introduced in the previous chapter.

Because our toy exon finder will need to be able to enumerate ORFs as well, we will provide the pseudocode for this baseline procedure. Note that since we are only building a "toy" exon finder for a short contig, for the sake of simplicity the algorithms that we give will be highly inefficient (this will generally not be the case in the succeeding chapters). Nevertheless, it will be possible to implement any of these algorithms in an interpreted language such as Perl and to run them on our toy genome in a matter of seconds on a typical computer.

The first subroutine which we will need is `getAllSignals(S)`, which will return a list of all potential donors, acceptors, start codons, and stop codons occurring in sequence *S*. The procedure simply iterates through all positions of the sequence, taking note of any position where there is a GT or AG dinucleotide or an ATG, TAG, TGA, or TAA trinucleotide. The pseudocode is shown in Algorithm 5.1.

Algorithm 5.1 Obtaining a list ψ of all potential signals in a sequence *S*.

```
1.  procedure getAllSignals(S)
2.    L←|S|;
3.    for i←0 up to L-2 do
4.      s←S[i..i+1];
5.      if s∈{"GT","AG"} then listAppend(ψ,(s,i));
6.      if i <L−2 then
7.        s ← S[i..i+2];
8.        if s∈{"TAG", "TGA", "TAA"} then s←"TAG"
9.        if s∈{"ATG", "TAG"} then
10.          listAppend(ψ,(s,i));
11.   return ψ;
```

Given the list ψ of all potential signals in the sequence, we can construct the set of all forward-strand ORFs using Algorithm 5.2. This procedure finds all intervals in the sequence beginning with an ATG or AG and ending with a GT, TAG, TGA, or TAA.

Algorithm 5.2 Obtaining a list Ω of all forward-strand open reading frames in a sequence S.

```
1.  procedure getAllOrfs(S)
2.    ψ←getAllSignals(S);
3.    n←|ψ|;
4.    for i←0  up to n-2 do
5.      (s1,x1)← ψ[i];
6.      if s1∈{"ATG", "AG"} then
7.        if s1="AG" then x1← x1+2;
8.        for j←i+1 up to n-1 do
9.          (s2,x2)← ψ[j];
10.         if s2∈{"GT", "TAG"} then
11.           if s2="TAG" then
12.             x2←x2+3;
13.             if s1="ATG" and (x2-x1) mod 3 ≠ 0
14.               then next ;
15.           if x1<x2 then
16.             ω← (s1,x1,s2,x2-1);
17.             if hasOpenReadingFrame(ω,S) then
18.               Ω←ΩU{ω};
19.  return Ω;
```

Since the intervals considered by Algorithm 5.2 may contain stop codons in all phases, Algorithm 5.3 is called as a subroutine to filter out all those intervals which do not have at least one open reading frame.

The hasOpenReadingFrame() procedure works as follows. If a candidate interval begins with ATG, then we proceed along the sequence three bases at a time, ensuring at each trinucleotide that a stop codon does not occur in this position. If a candidate interval instead begins with an AG, we walk along the sequence in each of the three frames to see whether any of them are open (i.e., whether they lack an in-frame stop codon). The exception to this is the case of terminal exons, which begin with an AG and end with a stop codon. For these cases we use *mod 3* arithmetic (see Chapter 2) to determine which is the only acceptable frame at the beginning of the interval (line 7), and then we consider only that frame as we look for in-frame stop codons.

Algorithm 5.3 Determining whether an interval ω in sequence S has at least one open reading frame.

```
1.  procedure hasOpenReadingFrame(ω,S)
2.     (s1,b,s2,e) ← ω;
3.     Δ ← {0,1,2};
4.     if s1="ATG" then Δ ←{0};
5.     if s2="TAG" then
6.       e←e-3;
7.       Δ ←{(e-b+1) mod 3};
8.     foreach δ ∈ Δ do
9.         stop←false;
10.        for x←b+δ up to e-2 by 3 do
11.          codon←S[x..x+2];
12.          if codon ∈ {"TAG", "TGA", "TAA"} then
13.            stop←true;
14.        if stop=false then return true;
15.    return false;
```

Determining whether two ORFs overlap is simple; `orfsOverlap()` in Algorithm 5.4 provides this functionality.

Algorithm 5.4 Determining whether two ORFs ω_1 and ω_2 overlap.

```
1.  procedure orfsOverlap(ω1,ω2)
2.     (sb1,b1,se1,e1) ← ω1;
3.     (sb2,b2,se2,e2) ← ω2;
4.     return b1≤e2 and b2≤e1;
```

Given these utility functions, it is now easy to implement our baseline procedure, `RANDORF()`, as shown in Algorithm 5.5.

Upon running this algorithm several hundred times on the *G. simplicans* contig we find that its nucleotide accuracy is on average 53±3%, which matches perfectly the SMC_{rand} (Chapter 4) value of $max(d, 1 - d) = 53\%$ for coding density $d = 47\%$. According to the *empirical rule* from Chapter 2, if the observed distribution of accuracy scores for this algorithm is approximately mound-shaped (which it is – see Exercise 5.3), then roughly 95% of all observations under this distribution should be between 47% and 59% accuracy. Thus, our goal will now be to see if we can devise a simple exon finder for this genome with greater than 59% nucleotide accuracy – that is, if we can do better than expected by chance.

Algorithm 5.5 Predicting random ORFs as exons in sequence S.

```
1.  procedure RANDORF(S)
2.    Ω ←getAllOrfs(S);
3.    foreach (ω1,ω2)∈ Ω × Ω do
4.      if orfsOverlap(ω1,ω2) then
5.        if random (0..1) < 0.5 then Ω ← Ω-{ω1};
6.        else Ω ← Ω-{ω2};
7.    return Ω;
```

5.3 Predicting exons based on {G,C} bias

Like a number of real organisms, *G. simplicans* exhibits a {G,C} bias in the coding portions of its exons. What this means is that on average, the proportion of G and C bases in coding exons is higher than in the genome as a whole. This does not mean that *every* exon has an elevated {G,C} density, but it does mean that there is a detectable bias across the entire set of exons.

Fortunately, the existence of this bias is well known to the *G. simplicans* research community, who inform us that the average {G,C} density in this organism's exons (not including start and stop codons) is approximately 58%, whereas the overall {G,C} density is only 48%. This suggests a very crude method of identifying at least some of the ORFs likely to be real exons. We need only score the ORFs by their {G,C} density and then predict those ORFs with the highest scores as being probable exons. This will be the approach employed by the first version of our toy exon finder.

We begin with a simple procedure to measure the {G,C} density of an ORF, as shown in Algorithm 5.6.

Algorithm 5.6 Measuring {G,C} density of an ORF ω in sequence S.

```
1.  procedure orfGcDensity (ω, S)
2.    (s1,b,s2,e) ← ω;
3.    gc←0;
4.    acgt←0;
5.    for i←b up to e do
6.      if S[i]∈{"G", "C"} then gc←gc+1;
7.      if S[i]∈ {"A", "C", "G", "T"} then
8.        acgt← acgt +1;
9.    return gc/acgt
```

It is worth noting that in Algorithm 5.6 we do not simply divide gc by the length of the sequence, but instead go through the trouble of actually counting all {A,C,G,T} bases and dividing by that count. As mentioned in Chapter 3, it is not uncommon in practice to encounter sequences with symbols other than {A,C,G,T}, either when a sequencer was unable to resolve the exact identity of a base (in which case an *ambiguity code* may be substituted for the unknown nucleotide) or when the assembler inserts a number of copies of the symbol N to denote a gap between contigs.

Now we can formulate our first toy exon finder, TOYSCAN_1(), as shown in Algorithm 5.7. This procedure retains only those ORFs having a {G,C} density greater than the specified threshold, t. Overlaps are resolved by simply discarding the ORF with the lower {G,C} density.

Algorithm 5.7 A toy exon finder based on {G,C} content. Inputs are a sequence S and a {G, C} density threshold, t.

```
1.  procedure TOYSCAN_1(S,t)
2.    Ω ←getAllOrfs(S);
3.    foreach ω ∈ Ω do
4.      if orfGcDensity(ω,S)<t then Ω ← Ω-{ω};
5.    foreach (ω1,ω2)∈ Ω × Ω do
6.      if orfsOverlap(ω1,ω2) then
7.        if orfGcDensity(ω1,S)<orfGcDensity(ω2,S)
8.        then Ω ← Ω-{ω1};
9.        else Ω ← Ω-{ω2};
10.   return Ω;
```

Applying this procedure to the *G. simplicans* contig shown in Figure 5.1, we find that the nucleotide accuracy of TOYSCAN_1() varies depending on the {G,C} threshold value. If we run the exon finder iteratively over a range of threshold values, the optimal parameter turns out to be approximately 0.68, giving a nucleotide accuracy of 63%. This is more than three standard deviations greater than the mean accuracy of 53% for the random predictor RANDORF(), demonstrating that the {G,C} bias present in the coding portions of this genome can be used to improve prediction accuracy significantly over that expected from random guessing. However, this still leaves much room for improvement; the problem is that {G,C} density is not a perfect predictor of the *coding potential* of ORFs in this organism. In the next section we will consider a more reliable statistic than {G,C} density for identifying probable exons.

5.4 Predicting exons based on codon bias

It turns out that our toy genome also exhibits a fairly strong bias in the codon usage within coding exons, and so it is reasonable for us to consider whether the

Table 5.2 Codon usage statistics for *G. simplicans*; values are relative frequencies of codons observed in coding exons

Codon	P	Codon	P	Codon	P	Codon	P	Codon	P
AAA	0.0446	AAT	0.0014	AAC	0.0245	AAG	0.0004	ATA	0.0004
ATT	0.0135	ATC	0.0446	ATG	0.0008	ACA	0.0014	ACT	0.0008
ACC	0.0014	ACG	0.0446	AGA	0.0008	AGT	0.0008	AGC	0.0446
AGG	0.0446	TAT	0.0014	TAC	0.0446	TTA	0.0025	TTT	0.0008
TTC	0.0025	TTG	0.0014	TCA	0.0245	TCT	0.0446	TCC	0.0014
TCG	0.0008	TGT	0.0004	TGC	0.0008	TGG	0.0446	CAA	0.0446
CAT	0.0008	CAC	0.0004	CAG	0.0025	CTA	0.0014	CTT	0.0014
CTC	0.0245	CTG	0.0812	CCA	0.0446	CCT	0.0008	CCC	0.0008
CCG	0.0008	CGA	0.0446	CGT	0.0004	CGC	0.0014	CGG	0.0812
GAA	0.0014	GAT	0.0245	GAC	0.0014	GAG	0.0446	GTA	0.0025
GTT	0.0004	GTC	0.0245	GTG	0.0025	GCA	0.0245	GCT	0.0245
GCC	0.0245	GCG	0.0025	GGA	0.0008	GGT	0.0245	GGC	0.0014
GGG	0.0245	TAG	0.0000	TGA	0.0000	TAA	0.0000		

exploitation of this bias may give us greater discriminatory power than simple {G,C} statistics. What we mean by *codon bias* is the preferential use of some of the 64 possible codons over others within coding exons. If we had reliable statistics on the frequencies of all the codons as they appear in the already-annotated portions of the *G. simplicans* genome, then we could use this much as we did the {G,C} density information to identify those ORFs with codon usage most similar to that observed in known *G. simplicans* exons. Fortunately, we do have such information, in the form of Table 5.2

Note that the frequencies of the stop codons are given as 0.0000, since this table applies only to the coding portions of exons up to but not including the termination signal.

The scoring of a putative exon can be accomplished via the `codonBias()` procedure shown in Algorithm 5.8. Note that this procedure evaluates an ORF in all three frames, regardless of whether the ORF is actually open in all the frames. If the ORF contains an in-frame stop codon, the logarithm on line 12 will evaluate to $-\infty$, ensuring that the frame is not selected (since we know the ORF has at least one open frame that will not evaluate to $-\infty$). In the case of initial and terminal exons, in which there is only one valid frame, to consider the other frames is simply invalid, and we do not recommend doing so in practice. However, for the purpose of our toy exon finder, we expect the impact on the accuracy of our predictions to be negligible.

Algorithm 5.8 Scoring the codon bias of an ORF ω in all frames on sequence S, given the codon bias model θ. The codon bias of the most biased frame is returned.

```
1.  procedure codonBias(ω,S,θ)
2.     (s1,b,s2,e)← ω;
3.     if s2="TAG" then e←e-1;
4.     Hbest←-∞;
5.     for f←0 up to 3 do
6.       b←b+(3-f) mod 3;
7.       n← ⌊(e-b+1)/3⌋;
8.       λ ←0;
9.       for i←0 up to n-1 do
10.          codon←S[b+3*i..b+3*i+2];
11.          q=θ[codon];
12.          λ ← λ+log(q);
13.      H← λ/n;
14.      if H>Hbest then Hbest←H;
15.    return Hbest;
```

We can now modify our toy exon finder so that it uses this new scoring function. The result is shown in Algorithm 5.9.

Algorithm 5.9 A toy exon finder based on codon bias. Inputs are a sequence S, a codon bias model θ, and a codon bias threshold, t.

```
1.  procedure TOYSCAN_2 (S,θ,t)
2.     Ω ←getAllOrfs (S);
3.     foreach ω ∈ Ω do
4.       if codonBias(ω,S,θ)<t then Ω ← Ω-{ω};
5.     foreach (ω1,ω2)∈ Ω × Ω do
6.       if orfsOverlap(ω1,ω2) then
7.         if codonBias (ω1,S,θ)<codonBias (ω2,S,θ)
8.         then Ω ← Ω-{ω1};
9.         else Ω ← Ω-{ω2};
10.    return Ω;
```

TOYSCAN_2() represents a considerable improvement over TOYSCAN_1(), as its accuracy on the sample contig is 87% – a 24% improvement. We have not explained, however, why we formulated our codon bias scoring function in the way that we did, and so it may not be obvious why our strategy paid off. We will explain that now.

The loop comprising lines 9 through 12 of Algorithm 5.8 iterates through all codons in the current reading frame. On lines 11 and 12 we accumulate the sum of codon log frequencies, $\log q_i$, according to the codon bias model θ, and then on line 13 we divide by n, the number of codons considered:

$$\frac{1}{n}\sum_{i=1}^{n}\log q_i .$$

If we group the $\log q_i$ terms by the 64 codon types, we can rearrange algebraically as follows:

$$\frac{1}{n}\sum_{i=1}^{n}\log q_i = \frac{1}{n}\sum_{i=1}^{64} n_i \log q_i = \sum_{i=1}^{64}\frac{n_i}{n}\log q_i = \sum_{i=1}^{64} p_i \log q_i ,$$

where $\sum_i n_i = n$, for n_i the number of times codon i appears in the ORF. The term on the far right, $\sum_i p_i \log q_i$, is known as the negative of the *cross entropy*, a well-known measure of the similarity of distributions and the basis for the *relative entropy* measure described in section 2.10. This term achieves its maximum when $\forall_i(p_i = q_i)$; i.e., when the observed distribution $\{p_i\}$ of codons within an ORF is identical to what our model θ tells us is the "correct" distribution $\{q_i\}$ of codons in an exon.

Thus, our scoring function ranks ORFs according to the similarity of their codon bias to the known codon bias of real exons. For this example we found through trial and error that the optimal threshold value for the negative cross entropy was -4.05. It should be noted that the practice of identifying the optimal parameter settings of any gene finder via trial and error on the test set is essentially equivalent to evaluating the gene finder on the *training set*, which as we noted in Chapter 4 can produce artificially inflated accuracy measurements and is therefore strongly discouraged in real gene-finding studies. For the purposes of this chapter, however, the relative merits of our alternate ORF-scoring functions will nevertheless be amply demonstrated; since this is our overriding objective at present, in the interest of simplicity we will not be overly concerned with rigorous evaluations in this chapter.

5.5 Predicting exons based on codon bias and WMM score

The codon bias measure which we introduced above captures much of the information present within exons. One of its weaknesses, however, is in determining the precise boundaries of exons – i.e., deciding, once it has found a region of decidedly biased codon usage, where exactly that biased region begins and ends. A more suitable model for evaluating putative exon boundaries is the *weight matrix*, which we describe next (see section 7.3.1 for a more thorough description of weight matrices).

Table 5.3 Donor site weight matrix for *G. simplicans*; columns refer to downstream positions from the GT consensus

	Position				
	+1	+2	+3	+4	+5
A	0.0064	0.8013	0.1603	0.0064	0.0321
C	0.8013	0.0321	0.8013	0.0321	0.0064
G	0.0321	0.1603	0.0064	0.1603	0.1603
T	0.1603	0.0064	0.0321	0.8013	0.8013

Table 5.4 Acceptor site weight matrix for *G. simplicans*; columns refer to upstream positions from the AG consensus

	Position				
	−5	−4	−3	−2	−1
A	0.8013	0.1603	0.8013	0.1603	0.8013
C	0.0064	0.0064	0.0321	0.0321	0.0064
G	0.0321	0.0321	0.1603	0.0064	0.1603
T	0.1603	0.8013	0.0064	0.8013	0.0321

Table 5.5 Start codon weight matrix for *G. simplicans*; columns refer to upstream positions from the *ATG* consensus

	Position				
	−5	−4	−3	−2	−1
A	0.1603	0.1603	0.8013	0.0321	0.8013
C	0.0064	0.0064	0.1603	0.1603	0.0064
G	0.0321	0.8013	0.0321	0.0064	0.1603
T	0.8013	0.0321	0.0064	0.8013	0.0321

Tables 5.3–5.6 give the weight matrices (WMMs) for the donors, acceptors, start codons, and stop codons of *G. simplicans* coding exons. Consider the donor WMM shown in Table 5.3. This matrix specifies, for the individual positions within a fixed window downstream from the donor site, the distribution of the four bases at that position in all known donor sites for this organism. Thus, for example, the "+1" column of this matrix tells us that in roughly 80% of actual donor sites for this organism, the base immediately following the GT dinucleotide is a C; 16%

Table 5.6 Stop codon weight matrix for *G. simplicans*; columns refer to downstream positions from the TAG/TGA/TAA consensus

	Position				
	+1	+2	+3	+4	+5
A	0.0064	0.0321	0.01603	0.0064	0.0321
C	0.0321	0.0064	0.8013	0.1603	0.8013
G	0.1603	0.8013	0.0064	0.8013	0.1603
T	0.8013	0.1603	0.0321	0.0321	0.0064

of cases have a T in this position, 3% have a G, and fewer than 1% have an A. The "+2" column gives us statistics about the next position after that one, and so on.

We can use these distributions in the same way we used the codon distributions – namely, via negative cross entropy: $\sum p_i \log q_i$. The q_i values are the nucleotide probabilities at a given position according to the WMM, and the p_i values reflect the distribution of bases actually seen at a given position near a putative donor site. Since only one base can occupy a given position at a time, these p_i values are always either 0 or 1, denoting whether a given base is absent or present, respectively. In this way, the negative cross entropy for a single position within the WMM window simplifies to $\log q_i$, for $q_i = P(\text{base } b_i \text{ occurs at this position in the window})$.

The negative cross entropy value for each position within each WMM at either end of an ORF can then be added to the negative cross entropy value for the codon usage of the ORF to produce a single score for the entire ORF. Under such a scoring scheme, the "best" ORF would be one having a codon usage exactly matching the distribution given in Table 5.2 and having signals at either end flanked by the most probable base in each position of the respective weight matrices for those signal types. For example, in the case of an internal exon, the best flanking bases for the acceptor and donor sites at either end are:

ATATAAG . . . . GTCACTT

where the acceptor consensus AG and the donor consensus GT have been underlined. For a single-exon gene the best flanking bases are:

TGATAATG . . . . TAGTGCGC

The other WMMs are given by Tables 5.4–5.6.

The scoring of a subsequence using a weight matrix is implemented by Algorithm 5.10. Matrix M is indexed as M[position][base] to produce the frequency in which the given base is expected at the given position of the current window, relative to the leftmost base in the window.

Algorithm 5.10 Scoring a weight matrix M of length L at location x in sequence S.

```
1. procedure scoreWMM(M,x,S,L)
2.    r←0;
3.    for i←0 up to L-1 do
4.       r←r+ log (M[i][S[x+i]]);
5.    return r;
```

Now we can formulate the `scoreOrf()` procedure (Algorithm 5.11) which will score an ORF based on both its codon bias and the WMM scores of its flanking signals.

Algorithm 5.11 Scoring a forward-strand ORF ω based on codon bias model θ and weight matrices M_{donor}, $M_{acceptor}$, M_{start}, and M_{stop}.

```
1.  procedure scoreOrf(ω,S,θ,Mdonor,Macceptor,Mstart,Mstop)
2.     (s1,b,s2,e) ← ω;
3.     if s1= "ATG" then M1←Mstart;
4.     else
5.        M1←Macceptor;
6.        b←b-2;
7.     if s2="TAG" then M2←Mstop;
8.     else
9.        M2←Mdonor;
10.       e←e+2;
11.    L1←windowLength(M1);
12.    L2←windowLength(M2);
13.    return codonBias(ω,S,θ) +scoreWMM(M1,b-L1,S,L1)
14.       +scoreWMM(M2,e+1,S,L2);
```

Finally, we can formulate our newest version of the toy exon finder, `TOYSCAN_3()`, as shown in Algorithm 5.12.

Running this procedure on the test contig gives a nucleotide accuracy of 97%, far better than the 53% expected by random guessing, and approaching the upper limits of accuracy for this organism. The optimal threshold was found via trial and error to be -15.8 for this data set.

Algorithm 5.12 A toy exon finder based on codon bias and WMM score. Inputs are a sequence S, a codon bias model θ, weight matrices M_{donor}, $M_{acceptor}$, M_{start}, and M_{stop}, and a scoring threshold, t.

```
1.  procedure TOYSCAN_3(S,θ,M_donor,M_acceptor,M_start,
2.                         M_stop,t)
3.     Ω ←getAllOrfs(S);
4.     foreach ω ∈ Ω do
5.       if scoreOrf(ω,S,θ,M_donor,M_acceptor,M_start,
6.                   M_stop)<t then
7.           Ω ← Ω-{ω};
8.     foreach (ω1,ω2)∈ Ω × Ω do
9.       if orfsOverlap(ω1,ω2) then
10.          if scoreOrf(ω1,S,θ,M_donor,M_acceptor,M_start,M_stop)
11.              <
12.              scoreOrf(ω2,S,θ,M_donor,M_acceptor,M_start,M_stop)
13.          then
14.            Ω ← Ω-{ω1};
15.          else Ω ← Ω-{ω2};
16.    return Ω;
```

Table 5.7 Accuracy results for different versions of the toy exon finder

Exon finder	Scoring function	Score threshold	Nucleotide accuracy
RANDORF()	Random	n/a	53±3%
TOYSCAN_1()	{G,C} percent	0.68	63%
TOYSCAN_2()	Codon bias	−4.05	87%
TOYSCAN_3()	Codon bias + WMM score	−15.8	97%

5.6 Summary

Accuracy scores for the three successive versions of the toy exon finder are shown in Table 5.7.

In this chapter we considered the use of {G,C} bias, codon bias, and weight matrix scores to evaluate ORFs as potential exons, and it was seen that the latter two techniques in combination allowed for fairly accurate exon prediction in our toy genome. It is important to note, however, that the levels of accuracy achieved here cannot be expected to be reproduced by these techniques on all genomes. The biases present in the sequences of individual genomes can be vastly different, which is why any gene finder should ideally be trained for the specific organism on which it is to be applied, whenever possible. Furthermore, because we have not followed rigorous evaluation protocols in assessing predictive accuracy, the actual accuracy measurements given in this chapter are at best a rough indication of the relative merits of these ad hoc ORF-finding algorithms for finding exons in this synthetic genome. Nevertheless, our treatment has hopefully afforded the reader some intuition for the statistical nature of both the problem and the solutions commonly employed in gene prediction.

In the next chapter we will leave these ad hoc approaches behind and begin our exploration of more rigorous probabilistic models for gene prediction. In the process we will consider a lengthy example illustrating the construction of a real gene finder, which we will apply to real genomic DNA.

EXERCISES

5.1 Write a program to count the number of forward-strand ORFs in the sequence shown in Figure 5.1. Download the sequence from the book's website (see Appendix) and run the program to verify that the count is 772.

5.2 Write a program to evaluate the nucleotide accuracy of a set of exon predictions.

5.3 Implement the `RANDORF()` baseline algorithm to predict random ORFs as exons. Run your program several times on the test contig, evaluating the accuracy each time, and compute the mean and standard deviation of the accuracy values obtained, to show that the expected accuracy is near 53%. Also verify that the distribution is roughly mound-shaped.

5.4 Implement the `TOYSCAN_1()` algorithm and evaluate its accuracy on the test contig.

5.5 Implement the `TOYSCAN_2()` algorithm and evaluate its accuracy on the test contig.

5.6 Implement the `TOYSCAN_3()` algorithm and evaluate its accuracy on the test contig.

5.7 See if you can improve the accuracy of `TOYSCAN_3()` on the example contig by manipulating not the algorithm itself, but the various model parameters (i.e., components of the weight matrices and the codon usage table).

6

Hidden Markov models

Hidden Markov models (HMMs) form the basis for the majority of gene finders in use today, though as we will see in later chapters, a number of extensions to the HMM formalism exist which have been found invaluable in achieving the accuracy and flexibility required of a practical, state-of-the-art gene finder. In this chapter we will limit our attention to the use of "pure" HMMs. We will show that even this basic formulation permits surprisingly accurate prediction of gene structures in real genomes. In the succeeding chapters we will explore the extensions to this framework that are now in common use, as well as those that are now at the forefront of current research.

6.1 Introduction to HMMs

In section 2.15 we introduced the notion of a *hidden Markov model* as a stochastic machine denoted by a 6-tuple:

$$M = (Q, \alpha, P_t, q^0, q^f, P_e) \tag{6.1}$$

for state set Q, alphabet α, transition distribution $P_t : Q \times Q \mapsto \mathbb{R}$, initial state q^0, final state q^f, and emission distribution $P_e : Q \times \alpha \mapsto \mathbb{R}$. It was explained that a machine M operates by starting in state q^0, transitioning stochastically from state to state according to $P_t(y_i|y_{i-1})$, for $\{y_i, y_{i-1}\} \subseteq Q$, and terminating in state q^f. Upon entering a state q, the machine emits a symbol s according to $P_e(s|q)$. There are no transitions into q^0, and none out of q^f, and neither state emits any symbols. This is the formulation of HMMs that we will use in this book. An alternative formulation (e.g., Jelinek, 1997) associates emissions with the transitions rather than the states, and different authors differ on whether to explicitly model the initial and final states as silent states (e.g., Durbin *et al.*, 1998), as we do here. Such differences in formulation are largely inconsequential in that the representational power of the resulting models are in most cases equivalent, and adoption of one or the other set of conventions is in most cases based on convenience and notational preference.

As a general rule, the operation of an HMM is considered *opaque*, in the sense that as the machine is running, only its outputs are visible to us; the actions performed internally by the machine (i.e., its transitions and in particular the sequence of states visited by the machine) occur within a "black box," and cannot be directly observed. This is the sense in which hidden Markov models are *hidden*. We especially draw attention to the fact that such models are by nature *generative*, in that they are formulated as stochastic processes which generate linear streams of output data. An HMM is not by itself a gene-prediction or parsing algorithm, but is instead merely a descriptive model of the statistical and syntactical properties of sequences (genomes in our case). However, through the use of an appropriate *decoding* algorithm (section 6.2), an HMM can be used to drive a heuristic parser for finding the most promising gene prediction. In section 6.9 we will consider the optimality (in terms of expected accuracy) of parsers derived in this way.

Let us establish some notational conventions. We will reserve the symbol q for particular states in the model: $Q = \{q_0, \ldots, q_{m-1}\}$, for $m = |Q|$. Often we will wish to form a list or sequence of states, where each particular state q_i may occur any number of times in the list (since an HMM is free, in principle, to visit any state multiple times during a given run). In this case we may denote the elements of the list using some generic variable, such as y – i.e., $\phi = (y_0, y_1, \ldots, y_{n-1})$ for $n = |\phi|$. It should be clear that we may have $y_i = y_j$ for $i \neq j$, whereas $q_i = q_j$ will always imply that $i = j$, since q_i is taken to be the unique name for the ith state in Q. Since $q^0 \in Q$ and $q^f \in Q$, we must have $q^0 = q_i$ for some i, and likewise $q^f = q_j$ for some j. For convenience, we will always assume $q^f = q^0 = q_0$ – that is, the 0th state in Q will always serve the function of initial and final state for the HMM, so that every run of the machine begins and ends in state q_0 (which is therefore a silent state); we will nevertheless often refer to this state explicitly as q^0, as a reminder of the special status of this state. Thus, we can now denote an HMM more compactly as:

$$M = (Q, \alpha, P_t, P_e). \tag{6.2}$$

In a similar way, we will reserve the letter s for the elements of the alphabet $\alpha = \{s_0, \ldots, s_{k-1}\}$ for $k = |\alpha|$. When dealing with an input sequence S we will use a generic variable such as x to denote the individual symbols in the sequence: $S = x_0, \ldots, x_{L-1}$, for $L = |S|$. Since any particular symbol s_i may occur in a sequence S zero or more times, we may have $x_i = x_j$ for $i \neq j$, whereas $s_i = s_j$ will always imply that $i = j$, since s_i is taken to be the unique name for the ith symbol in α. Thus, for s_i we take i to be an index into the alphabet α, whereas for x_j we take j to be an index into a sequence.

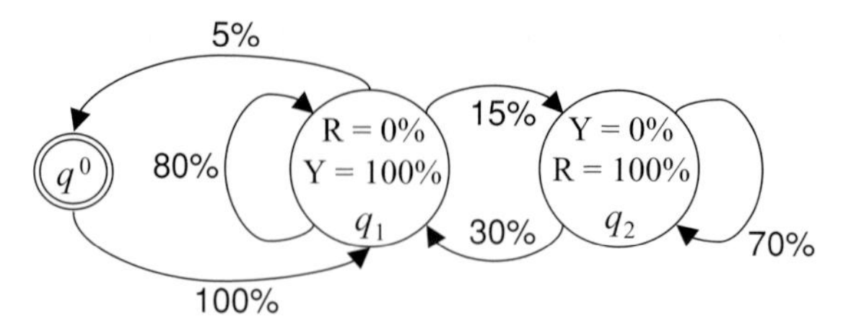

Figure 6.1 An example HMM. Emission probabilities are shown inside each state. Each edge is shown with its transition probability.

6.1.1 *An illustrative example*

Let us consider a simple example. Define $M_1 = (\{q_0, q_1, q_2\}, \{\mathtt{Y}, \mathtt{R}\}, P_t, P_e)$, for

$$P_t = \{(q_0, q_1, 1), (q_1, q_1, 0.8), (q_1, q_2, 0.15), \\ (q_1, q_0, 0.05), (q_2, q_2, 0.7), (q_2, q_1, 0.3)\} \tag{6.3}$$

(where we have omitted the zero-probability transitions), indicating that, e.g., M_1 can transition from state q_1 to state q_2 with probability 15%; and

$$P_e = \{(q_1, \mathtt{Y}, 1), (q_1, \mathtt{R}, 0), (q_2, \mathtt{Y}, 0), (q_2, \mathtt{R}, 1)\}, \tag{6.4}$$

so that state q_1 emits only the symbol Y, whereas q_2 emits only R. A graphical illustration of this HMM is shown in Figure 6.1.

A single run of M_1 might produce the sequence:

$$\mathtt{YRYRY}. \tag{6.5}$$

On another run of the HMM we might observe:

$$\mathtt{YRYYRYRRY}. \tag{6.6}$$

Although both of these sequences can be emitted by the model, they are not equally likely. Because each nonsilent state in this HMM can emit only one of the two symbols in the alphabet, we can compute the probability that any given run of M_1 results in a given sequence by multiplying together the transition and emission probabilities of the unique path corresponding to the generated sequence. In the case of sequence (6.5), we have:

$$P(\mathtt{YRYRY}|M_1) = \\ a_{0\to1} \times b_{1,\mathrm{Y}} \times a_{1\to2} \times b_{2,\mathrm{R}} \times a_{2\to1} \times b_{1,\mathrm{Y}} \times a_{1\to2} \times b_{2,\mathrm{R}} \times a_{2\to1} \times b_{1,\mathrm{Y}} \times a_{1\to0} = \\ 1 \times 1 \times 0.15 \times 1 \times 0.3 \times 1 \times 0.15 \times 1 \times 0.3 \times 1 \times 0.05 = \\ 0.000\,101\,25, \tag{6.7}$$

where $a_{i\to j}$ denotes $P_t(q_j|q_i)$ and $b_{i,s}$ denotes $P_e(s|q_i)$. Note that when the model M_1 is implicitly known we will typically abbreviate $P(S|M_1)$ as $P(S)$. For the longer sequence (6.6), multiplying the non-unity factors together produces:

$$P(\texttt{YRYYRYRRY}|M_1) =$$
$$0.15 \times 0.3 \times 0.8 \times 0.15 \times 0.3 \times 0.15 \times 0.7 \times 0.3 \times 0.05 =$$
$$0.000\,002\,551\,5, \tag{6.8}$$

which it may be observed is smaller than the probability for sequence (6.5). As a general rule, longer sequences will tend to be less probable than shorter sequences, since they involve the multiplication of larger numbers of emission and transition probabilities. We will see that this has two important ramifications, namely (1) that we will sometimes wish to condition the probability of a sequence S on its length – i.e., $P(S|L)$ for $L = |S|$ – in order to control for differences in length when comparing the probabilities of sequences, and (2) that as we multiply larger and larger numbers of probability terms together we risk causing *numerical underflow* (section 2.1) in the computer by requiring ever more bits for the accurate representation of the product. We will address these issues as they arise.

6.1.2 Representing HMMs

An HMM can be represented very simply in software by utilizing two matrices, one for the emission probabilities and one for the transition probabilities. For a state set $Q = \{q_0, q_1, \ldots, q_{n-1}\}$ and alphabet $\alpha = \{s_0, s_1, \ldots, s_{m-1}\}$, we can utilize an $n \times m$ *emission matrix*, **E**, by establishing $\mathrm{E}_{ij} = P_e(s_j|q_i)$. Similarly, we can designate an $n \times n$ *transition matrix*, **T**, such that $\mathrm{T}_{ij} = P_t(q_j|q_i)$. For the purpose of computational gene finding, upon reading the input sequence into memory it is useful to translate the DNA sequence into an array of symbol identifiers according to a mapping such as $\nabla = \{0 \leftrightarrow \texttt{A}, 1 \leftrightarrow \texttt{C}, 2 \leftrightarrow \texttt{G}, 3 \leftrightarrow \texttt{T}\}$. Thus, ATTAAA may be represented in the computer as the string "033000". The superiority of such a representation over the native ASCII codes should be apparent when considering the need to index into compact arrays using individual symbols as indices.

For large HMMs (i.e., those having many states), there are several enhancements we can make to our representation to improve the efficiency of the algorithms which we will be presenting later in this chapter. The first is to associate with each state q_i an explicit list λ_{trans} of those states q_j from which the HMM can transition into q_i:

$$\lambda_{trans}[i] = \{q_j | P_t(q_i|q_j) > 0\}. \tag{6.9}$$

Without λ_{trans}, determining the set of possible transitions into a state q_i would involve searching through an entire column of the transition matrix. For large HMMs having sparse transition matrices (i.e., with relatively few nonzero entries), the use of a λ_{trans} array can greatly accelerate any algorithm which needs to consider all possible paths through the model. Similarly, we can construct an explicit list

λ_{emit} of states capable of emitting any given symbol s_k:

$$\lambda_{emit}[k] = \{q_i | P_e(s_k|q_i) > 0\}. \tag{6.10}$$

The use of λ_{emit} arrays is profitable only when the HMM contains specialized states which can emit only a subset of the full alphabet α of the HMM. Later in this chapter we will be developing a simple gene finder based on such an HMM.

6.2 Decoding and similar problems

The example HMM shown in Figure 6.1 had the convenient property that for any given sequence S, there was at most one series of states, or *path*, (denoted ϕ) by which the model could generate that particular sequence. We say that such an HMM is *inherently unambiguous*. Unfortunately, the HMMs that are used in practice tend to be highly ambiguous, making it impossible (due to the model's opacity) in many cases for us to determine with certainty which path the machine followed as it was generating a particular output sequence. In such circumstances one generally settles for finding the most *probable* path for a particular sequence, given the parameters (i.e., the emission and transition probabilities) of the model. Finding that most probable path is the *decoding problem*, which we will address next. A slightly different formulation of the decoding problem, called *posterior decoding*, will be considered in section 6.10.

6.2.1 *Finding the most probable path*

Decoding with an HMM can be performed using a dynamic programming procedure called the *Viterbi algorithm* (Viterbi, 1967). Given a model $M = (Q, \alpha, P_t, P_e)$ with m states and a nonempty sequence $S = x_0 x_1 \ldots x_{L-1}$, the algorithm operates by progressively computing, for each state q_i $(0 \le i < m)$ and each position k $(0 \le k < L)$ in the sequence, the most probable path

$$\phi_{i,k} = y_0 \ldots y_{k+1} \; (\forall_{0 \le j \le k+1} y_j \in Q; y_0 = q^0, y_{k+1} = q_i)$$

ending in state q_i at position k whereby the model M could have generated the subsequence $x_0 \ldots x_k$ – i.e., such that the model was in state q_i when it emitted symbol x_k:

$$\phi_{i,k} = \begin{cases} \underset{\phi_{j,k-1}+q_i}{argmax} \left[\begin{array}{c} P(\phi_{j,k-1}, x_0 \ldots x_{k-1}) \cdot \\ P_t(q_i|q_j) P_e(x_k|q_i) \end{array} \right] & \text{if } k > 0 \\ q^0 q_i & \text{if } k = 0 \end{cases} \tag{6.11}$$

where $\phi + q$ denotes the parse resulting from the concatenation of state q onto the end of parse ϕ, and:

$$P(\phi_{i,k}, x_0 \ldots x_k) = \begin{cases} \underset{j}{max} \left[\begin{array}{c} P(\phi_{j,k-1}, x_0 \ldots x_{k-1}) \cdot \\ P_t(q_i|q_j) P_e(x_k|q_i) \end{array} \right] & \text{if } k > 0 \\ P_t(q_i|q^0) P_e(x_0|q_i) & \text{if } k = 0. \end{cases} \tag{6.12}$$

Once we have computed $\phi_{i,k}$ for all states q_i and all positions k in the sequence, it is then a simple matter to select the most probable path for the full sequence S by comparatively enumerating all paths ending at the last symbol, x_{L-1}, with the additional consideration that the last act of the machine after emitting x_{L-1} must have been to transition into state q^0:

$$\phi' = \underset{\phi_{i,L-1}}{argmax}\, P(\phi_{i,L-1}, S)P_t(q^0|q_i)$$

$$\phi^* = \phi' + q^0 = \underset{\phi}{argmax}\, P(\phi, S), \tag{6.13}$$

that is, the optimal path ϕ^* is the one ending in the state q_i which maximizes the product $P(\phi_{i,L-1}, S)P_t(q^0|q_i)$, with q^0 appended after q_i. This latter probability is precisely $P(\phi, S)$ for parse ϕ beginning and ending in state q^0. Following section 2.8, this maximization is equivalent to $argmax_{\phi} P(\phi|S)$, and ϕ^* may be referred to as the *maximum a posteriori* (*MAP*) parse. To see why the *maximum likelihood* (*ML*) parse, $argmax_{\phi} P(S|\phi)$, would be less desirable than the MAP parse, one need only note that the MAP parse is influenced by vital information which the ML parse ignores, namely, the transition probabilities $P_t(q_i|q_j)$ which encode prior probabilities on the raw gene structures.

The Viterbi algorithm utilizes the following dynamic programming recurrence to efficiently compute the probabilities $P(\phi_{i,k}, S_{0..k})$ of the prospective paths:

$$V(i,k) = \begin{cases} \max\limits_{j} V(j, k-1)P_t(q_i|q_j)P_e(x_k|q_i) & \text{if } k > 0, \\ P_t(q_i|q^0)P_e(x_0|q_i) & \text{if } k = 0. \end{cases} \tag{6.14}$$

$V(i,k)$ represents the probability $P(\phi_{i,k}, S_{i..k})$ of the most probable path $\phi_{i,k}$ which ends at state q_i and emits the subsequence $x_0 \ldots x_k$. While computing this probability we also keep track of the optimal *predecessor link* $T(i,k)$ which gave rise (via Eq. 6.14) to $V(i,k)$:

$$T(i,k) = \begin{cases} \underset{j}{argmax}\, V(j, k-1)P_t(q_i|q_j)P_e(x_k|q_i) & \text{if } k > 0, \\ 0 & \text{if } k = 0. \end{cases} \tag{6.15}$$

Each element $T(i,k)$ is thus a state index j for the optimal predecessor q_j of q_i at position k, in the sense that the optimal path $\phi_{i,k}$ must have as its last transition $q_j \rightarrow q_i$. Furthermore, in the event that the optimal global path ϕ^* puts the machine in state q_i when emitting the kth symbol x_k of the sequence, the first $k+1$ states of the optimal path ϕ^* can then be identified (in reverse) by iteratively following the $T(\bullet, \bullet)$ pointers, starting at $T(i,k)$ and proceeding recursively through $T(\ldots T(T(T(i,k), k-1), k-2)\ldots)$ until $T(\bullet, \bullet) = 0$ is reached, much like the *traceback* procedure in a dynamic programming alignment algorithm (see Durbin *et al.* (1998) for a thorough overview of alignment methods).

The full procedure is given in Algorithm 6.1. Figure 6.2 depicts the operation of this algorithm. A dynamic programming matrix of size $|Q| \times |S|$ is computed from left to right along the sequence, with each cell being defined in terms of the best cell found in the previous column (weighted by the appropriate transition probability). V and T values are stored in each cell. The V values are computed using logarithms, to avoid numerical underflow in the computer.

Algorithm 6.1 The *Viterbi algorithm* for finding the most probable path through model $M = (Q, \alpha, P_t, P_e)$, given an output sequence $S \neq \varepsilon$. The λ_{trans} and λ_{emit} arrays are assumed to be defined as described in section 6.1.2. Returns a stack ϕ of states comprising the path, with the beginning of the path on top of the stack.

```
procedure viterbi(Q,α,Pt,Pe,S,λtrans,λemit)
1.    for k ← 0 up to |S|−1 do
2.       for i ← 0 up to |Q|−1 do
3.          V[i][k] ← −∞;
4.          T[i][k] ← NIL;
5.    for i ← 1 up to |Q|−1 do
6.       V[i][0] ← log(Pt(qi|q0))+log(Pe(S[0]|qi));
7.       if V[i][0] > −∞ then T[i][0] ← 0;
8.    for k ← 1 up to |S|−1 do
9.       foreach qi ∈ λemit[S[k]] do
10.         foreach qj ∈ λtrans[qi] do
11.            v ← V[j][k − 1]+log(Pt(qi|qj))+
12.                     log(Pe(S[k]|qi));
13.            if v > V[i][k] then
14.               V[i][k] ← v;
15.               T[i][k] ← j;
16.   y ← 1;
17.   push φ, 0;
18.   for i ← 2 up to |Q|−1 do
19.      if V[i][|S|−1]+log(Pt(q0|qi)) >
20.            V[y][|S|−1]+log(Pt(q0|qy)) then y ← i;
21.   for k ← |S|−1 down to 0 do
22.      push φ, y;
23.      y ← T[y][k];
24.   push φ, 0;
25.   return φ;
```

This procedure has a running time of $O(|Q| \times |S|)$ for sparse HMMs (i.e., those for which each state has some small, fixed maximal number of predecessors), due to the use of the λ_{trans} array.

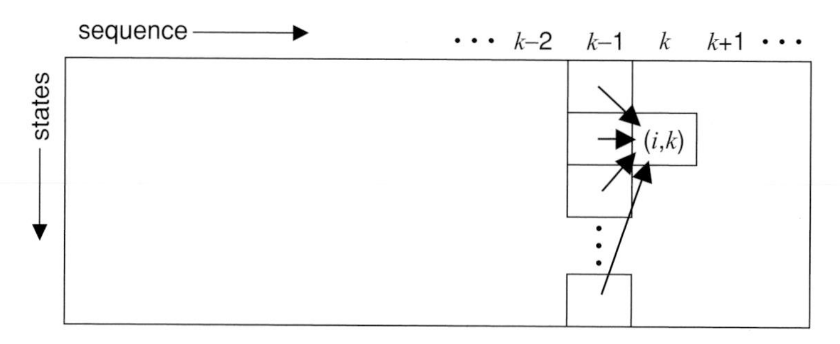

Figure 6.2 Computing the cells of the dynamic programming matrix in the Viterbi and Forward (section 6.2.2) algorithms. Each cell is computed based on the maximum (Viterbi) or sum (Forward) of the cells in the previous column, weighted by the appropriate transition probabilities.

The details of the procedure are as follows. Lines 1–4 initialize the matrix to zero probability, which corresponds to $-\infty$ in log space. Lines 5–7 initialize the first column and enforce the requirement that any valid path must start in state q_0. Lines 8–15 constitute the main part of the algorithm; they compute the cells for the rest of the matrix by finding the optimal predecessor at each cell (i, k), based on the *inductive score* $V(j, k-1)$ at each prospective predecessor and the transition probability $P_t(q_i|q_j)$ from the predecessor to the current state. The score of the best predecessor times the emission probability $P_e(x_k|q_i)$ for the current cell is stored as the inductive score for the new cell. Lines 16–20 select the optimal ending cell in the final column of the matrix while requiring that the machine end in the state q_0, and lines 21–23 perform a standard *traceback* procedure starting at this optimal ending cell and following the T links backward through the matrix until the start state q_0 is reached (line 24). Line 25 returns the optimal path ϕ.

We will see in section 6.4 that the most probable path, as computed by the Viterbi algorithm for an appropriately structured HMM, can be interpreted as a gene prediction. A relatively simple modification of this algorithm can be used to find the n best parses rather than only the single best (see Exercise 6.38), though it should be noted that the problem of finding the n best *labelings* (rather than parses), in the case of ambiguous HMMs, is rather more difficult (we refer the interested reader to Schwartz and Chow, 1990; Krogh, 1997).

Note that we have assumed above that the sequence S is nonempty and also that the HMM is capable of emitting S; modifications to the algorithm to relax these assumptions are trivial and are left as an exercise for the reader (Exercise 6.40).

6.2.2 Computing the probability of a sequence

A procedure very similar to the Viterbi algorithm can be used to find the probability that a given model M emits (nonempty) sequence S during any given run of the machine. Because M may potentially emit S via any number of paths through the states of the model, to compute the full probability of the sequence we need to sum over all possible paths emitting S. This is accomplished trivially by replacing the maximization step in the Viterbi algorithm with a summation step, resulting in the *Forward algorithm*:

$$F(i,k) = \begin{cases} 1 & \text{for } k = 0, i = 0 \\ 0 & \text{for } k > 0, i = 0 \\ 0 & \text{for } k = 0, i > 0 \\ \sum_{j=0}^{|Q|-1} F(j, k-1)P_t(q_i|q_j)P_e(x_{k-1}|q_i) & \text{for } 1 \le k \le |S|, \\ & 1 \le i < |Q| \end{cases} \tag{6.16}$$

where we have used a slightly different indexing scheme from Eq. (6.14), to simplify the pseudocode; we now employ a slightly larger matrix of size $|Q| \times (|S|+1)$. $F(i,k)$ represents the probability $P(S_{0...k-1}, q_i)$ that the machine emits the subsequence $x_0 \ldots x_{k-1}$ by any path ending in state q_i – i.e., so that symbol x_{k-1} is emitted by state q_i. Given this definition, we can compute the *posterior probability* of a sequence given a model as:

$$P(S|M) = \sum_{i=0}^{|Q|-1} F(i, |S|)P_t(q_0|q_i) \tag{6.17}$$

where the F and P_t terms are implicitly conditioned on the parameters of M. Because the Forward algorithm deals with sums of probabilities (rather than products), working with logarithms is no longer convenient. The procedure given in Algorithm 6.2 shows the Forward algorithm computed with raw probabilities, which for long sequences can cause numerical underflow. We will present a solution to this latter complication in section 6.6.3.2.

Algorithm 6.2 The (unscaled) *Forward algorithm* for computing the probability that a model $M = (Q, \alpha, P_t, P_e)$ emits a nonempty sequence S. The λ_{trans} and λ_{emit} arrays are assumed to be defined as described in section 6.1.2. The first procedure computes (and returns) the dynamic programming matrix F; the second procedure returns the probability of the sequence.

```
procedure unscaledForwardAlg(Q, Pt, Pe, S, λtrans, λemit)
1.    ∀i∈[0,|Q|−1] ∀k∈[0,|S|] F[i][k] ← 0;
2.    F[0][0] ← 1;
3.    for k ← 1 up to |S| do
4.        s ← S[k − 1];
5.        foreach qi ∈ λemit[s] do
6.            sum ← 0;
7.            foreach qj ∈ λtrans[i] do
8.                sum ← sum + Pt(qi|qj)*F[j][k − 1];
9.            F[i][k] ← sum*Pe(s|qi);
10.   return F;
```

```
procedure unscaledForwardProb(Q, Pt, Pe, S,λtrans,λemit)
1.     F ← unscaledForwardAlg(Q, Pt, Pe, S,λtrans,λemit);
2.     sum ← 0;
3.     len ← |S|;
4.     s ← S[len − 1];
5.     foreach qi ∈ λemit[s] do
6.         sum ← sum + F[i][len]*Pt(q0|qi);
7.     return sum;
```

Figure 6.2 again serves to provide a basic conceptualization of how this algorithm operates. Lines 1–2 of `unscaledForwardAlg()` initialize all the cells of the matrix to probability 0, except for cell (0,0), which is initialized to probability 1 to reflect the fact that at the beginning of the HMM's operation it will be in state q_0 and will not have emitted any symbols yet. Lines 3–9 compute the rest of the matrix inductively, left-to-right, by summing the probabilities of all paths emitting $x_0 \ldots x_{k-1}$ which end at q_i. This block consists of a pair of nested loops which consider those states $\lambda_{\texttt{emit}}$[s] capable of emitting the symbol at the current position in the sequence (line 5), and for each of these consider all possible predecessor states $\lambda_{\texttt{trans}}$[i] (line 7). Line 8 sums the products of predecessors' inductive scores (F[j][k − 1]) and the transition probabilities from those predecessors into the current cell ($P_t(q_i|q_j)$). Line 9 multiplies this sum by the probability of emitting the current symbol (s) from the current state (q_i). Finally, `unscaledForwardProb()` forms the sum F[i][len]*$P_t(q_0|q_i)$ over all states q_i which can emit the last symbol in the sequence and transition into the final state q_0, as dictated by Eq. (6.17).

6.3 Training with labeled sequences

Given the parameters of an HMM, we now know how to compute the probability that the model emits any given sequence S, as well as how to identify the most probable path ϕ through the model whereby S could have been generated. We turn now to the problem of constructing an HMM and setting its parameters so as to most accurately model the class of sequences in which we are interested. This is the problem of *parameter estimation*, or of *training* the HMM.

For now we will assume that the individual symbols comprising the training sequences have been explicitly labeled with the states in the HMM corresponding to the only possible path through the model for that sequence; we are thus limiting our attention for the moment to inherently unambiguous HMMs. In a later section we will explore several alternative training procedures that are applicable when the HMM is ambiguous, or when only unlabeled training sequences are available. In the next section we will be working through a rather lengthy example of a real

Table 6.1 Transition frequencies computed from sequence (6.18)

		To state		
		0	1	2
From state	0	0 (0%)	1 (100%)	0 (0%)
	1	1 (4%)	21 (84%)	3 (12%)
	2	0 (0%)	3 (20%)	12 (80%)

Table 6.2 Emission frequencies computed from sequence (6.18)

		Symbol			
		A	C	G	T
State	1	6 (24%)	7 (28%)	5 (20%)	7 (28%)
	2	3 (20%)	3 (20%)	2 (13%)	7 (47%)

HMM for gene finding, and there we will show how one can go about labeling a set of sequences for use in training an appropriately structured HMM.

Let us suppose we wish to train a 3-state HMM from the following labeled sequence:

```
 CGATATTCGATTCTACGCGCGTATACTAGCTTATCTGATC
0111111122222221111112222111111222211110
```
(6.18)

where we have given the state labels below the corresponding symbols in the sequence; for models with more than 10 states, an alternative format must obviously be devised to allow room for multi-digit state labels. The zeros at the beginning and end of the bottom row represent the fact that the model must begin and end in the silent state q_0, as usual, and can be omitted from the labeling since they are implicitly assumed to be present. Although this is an extremely small amount of data for training an HMM, it will serve to illustrate the idea.

To compute the transition probabilities, we merely observe the number of times state label i is followed by label j and then convert this to a relative frequency by dividing by the total number of observations involving state q_i. Similarly, for the emission probabilities we merely count the number of times each symbol s_i was observed in association with state q_j and then normalize this into a probability by dividing by the total sample size for state q_j. Tables 6.1 and 6.2 give the resulting probabilities for this example. These values are the *maximum likelihood estimates* (Chapter 2) for the true parameters of the HMM assumed to have generated the training data. In this particular case, because the training set was so small, these estimates are likely to deviate quite a bit from the true parameters, due to sampling error.

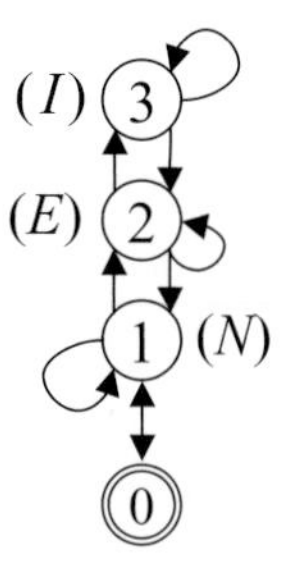

Figure 6.3 Model H_3, a 4-state HMM for gene finding.

More formally, we can derive MLE parameters for an HMM from a set of labeled training examples S by tabulating the number $A_{i,j}$ of times that state q_j is observed to immediately follow state q_i in the training set, and also the number $E_{i,k}$ of times that state q_i is observed to emit symbol s_k, for $\alpha = \{s_k | 0 \leq k < 4\}$. Given these counts we can then compute:

$$a_{i,j} = \frac{A_{i,j}}{\sum_{h=0}^{|Q|-1} A_{i,h}} \qquad e_{i,k} = \frac{E_{i,k}}{\sum_{h=0}^{|\alpha|-1} E_{i,h}} \tag{6.19}$$

where $a_{i,j}$ provides an estimate of $P_t(q_j | q_i)$ and $e_{i,k}$ provides an estimate of $P_e(s_k | q_i)$.

6.4 Example: Building an HMM for gene finding

We now possess all the necessary tools for designing, training, and deploying a rudimentary HMM-based gene finder. We will begin with the simple 4-state HMM shown in Figure 6.3.

We refer to this example HMM as H_3, since the model contains only three nonsilent states. The state identifiers (i.e., indices into Q) are shown inside the states, and the emissions are characterized abstractly by letters within parentheses next to each state. In this way, state q_1 represents intergenic (N) sequence, state q_2 represents exons (E), and state q_3 represents introns (I). All three states will be capable of emitting letters from the DNA alphabet $\alpha = \{\texttt{A,C,G,T}\}$.

This HMM always begins by generating intergenic sequences, since state q_1 is the only state to which H_3 can transition out of the start state. Once in the intergenic state, the model can stay in that state arbitrarily long by selecting the *self-transition* back to q_1. Eventually the model must leave q_1 by either transitioning into the exon state q_2 or by terminating the run by transitioning back into q_0. The exon state also has a self-transition, so that each exon symbol generated can be followed by an arbitrary number of additional exon bases. From the exon state the HMM can transition either into the intron state or back into the intergenic state. The intron state also has a self-transition, as well as a transition back into the exon state. Thus, H_3 models the basic exon/intron/exon pattern of eukaryotic genes, albeit in a very crude way; the explicit modeling of start and stop codons and splice sites will be considered later as we progressively extend the model.

Let us evaluate how well such a simple HMM is able to accurately predict gene structures in a real genome. We will use a set of 500 *Arabidopsis thaliana* genes as a training set, and a separate set of 500 *A. thaliana* genes as a test set, with each gene in both sets having a margin of up to 1000 bp on either side of the CDS. The training and test sets are provided in GFF and FASTA formats (see Appendix to obtain these files). The coordinates of the exons within each gene are specified in the GFF file; these coordinates refer to specific positions within the sequences given in the associated FASTA file (see section 4.1). Note that throughout this example we will assume for simplicity that all genes occur on the forward strand. The generalization to both strands is trivial, and is left as an exercise to the enthusiastic reader.

Before we can train the HMM we must label the training sequences with the appropriate model states corresponding to the features given in the GFF file. This can be done very simply as shown in Algorithm 6.3.

Algorithm 6.3 Labeling a sequence S for a 4-state HMM according to the exon coordinates given in an array E. S and E are assumed to have been loaded from corresponding FASTA and GFF files, respectively. Elements of E are assumed to be pairs of 0-based (*begin*, *end*) coordinates for exons, where *begin* is the coordinate of the first base in the exon and *end* is the coordinate of the last base of the exon. All exons are assumed to lie on the forward strand; generalization to both strands is left as an exercise (Exercise 6.34).

```
procedure labelExons(S, E)
1.    E ← sort(E);
2.    for i ← 0  up to |S|−1 do
3.       label[i] ← 1;
4.    foreach (b, e) ∈ E do
5.       for i ← b up to e do
6.          label[i] ← 2;
7.    for i ← 0 up to |E|−2 do
8.      if same_gene (i, i + 1) then
9.          (b1, e1) ← E[i];
10.         (b2, e2) ← E[i + 1];
11.         for i ← e1+1 up to b2−1 do
12.            label[i] ← 3;
13.   return label;
```

Line 1 sorts the exons from left to right along the sequence. Lines 2–3 initialize the label array to all intergenic (state q_1). Lines 4–6 iterate through all the exons and initialize the corresponding regions in the label array to the label 2

Table 6.3 Accuracy results for model H_3, as compared to the baseline. All values are percentages, except the number of correct genes

	Nucleotides			Splice sites		Start/stop codons		Exons			Genes	
	Sn	Sp	*F*	Sn	Sp	Sn	Sp	Sn	Sp	*F*	Sn	Number
Baseline	100	28	44	0	0	0	0	0	0	0	0	0
H_3	53	88	66	0	0	0	0	0	0	0	0	0

Sn, sensitivity; Sp, specificity; F = *F*-measure (see section 4.2 for definitions of these terms)

(exons – state q_2). Lines 7–12 label all regions falling between two exons of the same gene as introns (label 3 – state q_3). Finally, line 13 returns the array of state labels, not including the initial and final states. An auxiliary procedure would then be required to print out the sequence and the labels in some convenient format (e.g., each symbol and its state on a separate line).

One final task before evaluating the accuracy of the HMM on the test set is to assess a baseline accuracy for comparison. Following section 4.2 we first assess the mean coding density of the genes in the test set (including up to 1000 bp margins on either side of each gene) to arrive at 28% (the percentage of coding bases in each test "chunk"). Equations (4.8–4.11) then allow us to assess the baseline nucleotide sensitivity, specificity, and *F*-measure as 100%, 28%, and 44%, respectively.

Given the labeled training sequences produced with Algorithm 6.3, we can now train the HMM using the formulas given in Eq. (6.19). We can then use the trained HMM to predict genes on the test sequences by applying the Viterbi algorithm to obtain the most probable path through the model and predicting all contiguous bases associated by that path with state q_2 to be exons. Evaluating the resulting predictions according to the methods described in section 4.2 produces nucleotide sensitivity, specificity, and *F*-measure values of 53%, 88%, and 66%.

Table 6.3 summarizes these results. Note that the baseline sensitivity of 100% results from the strategy of setting $p = 1$ in the random baseline predictor in order to maximize the F_{rand} score (see section 4.2). This example illustrates well the difficulty in comparing predictive accuracies via sensitivity and specificity, and underscores the value of a single, unified statistic such as the *F*-measure.

As shown in the table, the *F*-score for H_3 is substantially better than that of the baseline, showing that even a trivial 4-state HMM can perform better than expected from random guessing (where the baseline scores effectively provide an upper bound on the accuracy of a random guesser, since we set $p = 1$ to maximize F_{rand} – see section 4.2). Nevertheless, there is clearly much room for improvement. In particular, we can see that none of the splice sites, start or stop codons, exons, or whole genes was predicted correctly.

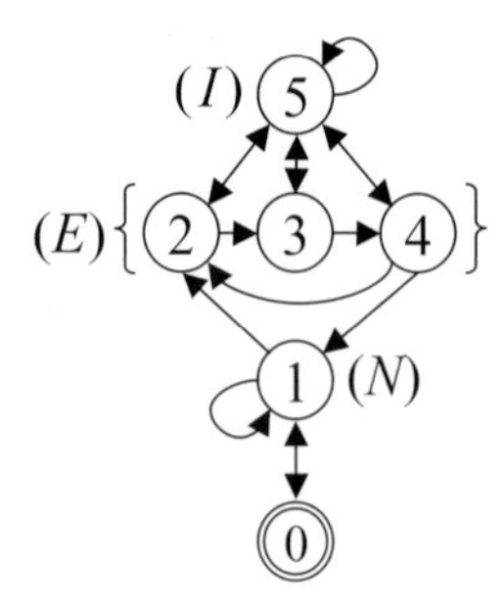

Figure 6.4 Model H_5, a 6-state HMM for gene finding.

There are a number of enhancements that we can make to the model in hopes of improving prediction accuracy. The first enhancement we will consider is to expand the single exon state into a set of three states representing the three *phases*, or codon positions. Figure 6.4 depicts the resulting model, which we will call H_5.

Now each exon is emitted as a series of codons, represented by states $\{q_2, q_3, q_4\}$. The intron state has been relabeled q_5. Introns can occur in any phase; hence the transitions from states $\{q_2, q_3, q_4\}$ into q_5. Producing labeled training sequences for this HMM can be accomplished by making a set of trivial modifications to Algorithm 6.3 so as to label exonic bases using a cyclic series of states (i.e., $\ldots 2, 3, 4, 2, 3, 4, 2, 3, 4, \ldots$). Further modifications to the labeling algorithm will be necessary to handle the additional enhancements that we will be introducing below; these changes to the algorithm are trivial and are left as an exercise for the reader (Exercises 6.14–6.18).

Note that there is no mechanism in H_5 for "remembering" into which phase an intron spliced when the model transitions from the intron state back into an exon state. For real eukaryotic genes, an intron which interrupts a codon in phase ω will also splice out in the same phase, so that the very next coding base will occur in the next available codon position as if the intron were not even present. H_5 does not enforce this discipline, however, since the model is free to transition, for example, from state q_2 (the first codon position) into state q_5 (intron) and then back into state q_4 (the third codon position). We will address this problem later.

For now, we can see that the explicit modeling of codon positions in the exon states provides an improvement of $\Delta F = 10\%$ at the nucleotide level, as shown in Table 6.4. H_5 was also able to identify a small number of splice sites and start and stop codons correctly, though it is still not predicting any exons exactly correctly.

For our next set of enhancements we will incorporate a number of additional states for explicitly modeling splice sites and start and stop codons, in hopes that by improving the identification of these important elements of gene structure we will thereby improve overall prediction accuracy at all levels, especially at the whole exon level. It is worthwhile to note that the improved modeling of any specific

Table 6.4 Accuracy results for model H_5, as compared to H_3; all values are percentages, except the number of correct genes

	Nucleotides			Splice sites		Start/stop codons		Exons			Genes	
	Sn	Sp	*F*	Sn	Sp	Sn	Sp	Sn	Sp	*F*	Sn	Number
H_3	53	88	66	0	0	0	0	0	0	0	0	0
H_5	65	91	76	1	3	3	3	0	0	0	0	0

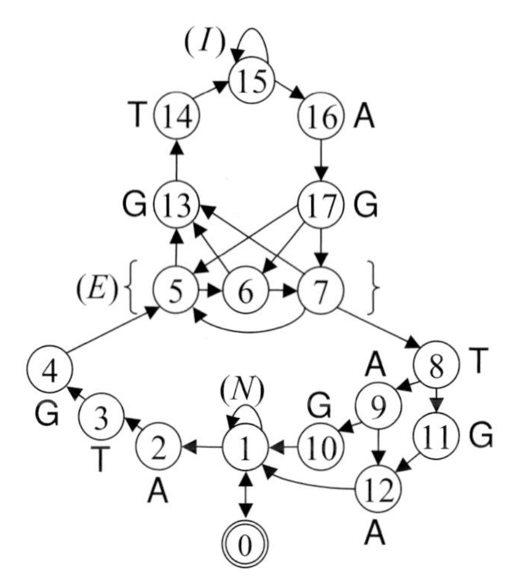

Figure 6.5 Model H_{17}, an 18-state HMM for gene finding.

feature of gene structure may conceivably improve predictive accuracy at all levels, due to the strong dependencies among syntactic elements that are induced by the very nature of the parsing problem – i.e., by the obvious fact that sequence intervals comprising a parse are by definition mutually exclusive and jointly account for the entire input sequence.

Our newest enhancements are as follows. We will add a pair of states $\{q_{13}, q_{14}\}$ to model the GT dinucleotide of donor sites, another pair of states $\{q_{16}, q_{17}\}$ for the AG dinucleotide of acceptor sites, a triple of states $\{q_2, q_3, q_4\}$ to model the common ATG start codon, and a set of five states $\{q_8, q_9, q_{10}, q_{11}, q_{12}\}$ to model the three common stop codons, TAG, TGA, and TAA. The rest of the model will remain unchanged, except that the intron state will be relabeled as q_{15}, the exon states will be relabeled $\{q_5, q_6, q_7\}$, and transitions into and out of the intron state from the exon states will be replaced with appropriate transitions into and out of the donor and acceptor splice site states; the transitions between the intergenic and exon states will also be modified accordingly to accommodate the start and stop codon states. The resulting model is depicted in Figure 6.5.

Table 6.5 Accuracy results for model H_{17}, as compared to H_5; all values are percentages, except the number of correct genes

	Nucleotides			Splice sites		Start/stop codons		Exons			Genes	
	Sn	Sp	*F*	Sn	Sp	Sn	Sp	Sn	Sp	*F*	Sn	Number
H_5	65	91	76	1	3	3	3	0	0	0	0	0
H_{17}	81	93	87	34	48	43	37	19	24	21	7	35

The new model, H_{17}, has 18 states (including the initial/final state q_0). Although the model has grown significantly in size, it should be clear that the overall structure remains very much the same, in the sense that the paths through the model that were present in H_5 are largely retained unchanged in H_{17} except for the insertion of specific series of states at key points along those paths to correspond to the generation of splice sites and start and stop codons. The reader should verify that the set of states $\{q_8, q_9, q_{10}, q_{11}, q_{12}\}$ matches precisely the set {TAG, TGA, TAA} of common eukaryotic stop codons. The results of applying H_{17} to the test set are given in Table 6.5.

It can be seen from these results that the addition of states for modeling splice sites and start and stop codons has resulted in a significant increase in prediction accuracy. Not only has the nucleotide *F*-score increased by 11%, but the sensitivity and specificity of splice site and start/stop codon identification have increased dramatically, accounting for our ability to now predict roughly 21% of the exons in the test set exactly correctly, as well as 7% of the genes – a large improvement indeed.

The next enhancement that we would like to consider is to track phase across introns. Referring to model H_{17}, let us define the set $I = \{q_{13}, q_{14}, q_{15}, q_{16}, q_{17}\}$, consisting of the donor and acceptor states and the intron state; we will refer to this set of states within H_{17} as the *intron submodel* of H_{17}, since it generates whole introns. We can then enforce phase constraints on introns (i.e., that each intron splices out in the same phase in which it splices in) by replicating the intron submodel three times to produce sets $I_0 = \{q_{13}, q_{14}, q_{15}, q_{16}, q_{17}\}$, $I_1 = \{q_{18}, q_{19}, q_{20}, q_{21}, q_{22}\}$, and $I_2 = \{q_{23}, q_{24}, q_{25}, q_{26}, q_{27}\}$, where the index ω for submodel I_ω denotes the phase which will be maintained across introns generated by submodel I_ω. The actual enforcement of this discipline is then easily maintained by manipulating the transition patterns in the composite model in such a way that introns occurring immediately after the first codon position are only generated by I_1, those following the second codon position are generated by I_2, and those following the third codon position are generated by I_0. In this way, the model can no longer enter an intron in one phase and then return to the exon submodel in a different phase, since each intron is phase-specific. The resulting model, H_{27}, is shown in Figure 6.6.

Table 6.6 Accuracy results for model H_{27}, as compared to H_{17}; all values are percentages, except the number of correct genes

	Nucleotides			Splice sites		Start/stop codons		Exons			Genes	
	Sn	Sp	F	Sn	Sp	Sn	Sp	Sn	Sp	F	Sn	Number
H_{17}	81	93	87	34	48	43	37	19	24	21	7	35
H_{27}	83	93	88	40	49	41	36	23	27	25	8	38

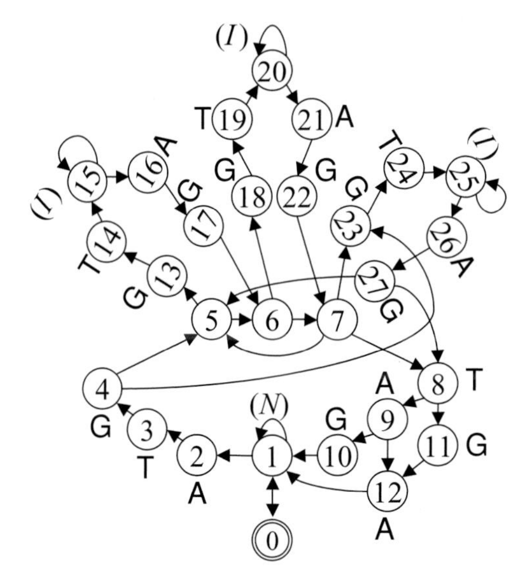

Figure 6.6 Model H_{27}, a 28-state HMM for gene finding.

The results given in Table 6.6 show that the improvement in accuracy deriving from our modification to enforce phase constraints was very modest: 1% at the nucleotide level, 1% at the whole gene level, and 4% at the exon level. Examination of the other columns shows that the improvement appears to have derived mainly from improved splice site identification, which is not unexpected, given the nature of the modification.

Unfortunately, our ability to accurately identify start and stop codons has suffered slightly as a result of the change; this is possibly due to the fact that we have now effectively reduced our sample size for each intron state by classifying each intron into one of the three phase categories and estimating the emission probabilities for the intron states from the smaller, phase-specific training examples. It should be kept in mind that due to the dependencies among different gene elements (which are induced by the nature of the parsing problem), inadequacies in the modeling of any given feature may negatively impact the prediction accuracy for other features in the model. We will attempt to rectify the sample size issue in the next run by pooling the introns from the three phases when training the intron submodels. This practice is referred to as *parameter tying*; though the intron submodel of H_{17} has been duplicated into three submodels in H_{27}, in our next iteration of the model we will treat these submodels as effectively identical for the purposes of parameter estimation, distinguishing between these three intron submodels only during Viterbi decoding.

It should be noted that the above mechanism for tracking phase across introns does not solve the problem of in-frame stop codons straddling an intron:

x**GT**...**AG**yz

xy**GT**...**AG**z

where xyz ∈ {TAG, TGA, TAA} is a stop codon and GT...AG is a putative intron. In the first case, the stop codon will be in-frame if the intron occurs in phase 1, and in the second case the stop codon will be in-frame if the intron occurs in phase 2. In both cases the *decoder* (i.e., decoding algorithm) must explicitly check, when computing the dynamic programming score for exon states immediately following an acceptor site, whether an in-frame stop codon would be formed, and must assign a probability of 0 (or $-\infty$ in log-space) for a dynamic programming predecessor which would cause a parse to contain such an in-frame stop codon. A number of early HMM-based gene finders did not attend to this situation, and were thus found occasionally to emit illegal gene parses due to in-frame stop codons straddling an intron. Modifications to a gene finder's decoding procedure to handle these cases properly are conceptually simple but somewhat tedious and implementation-specific, and are left as an exercise for the reader (Exercise 6.38).

One other enhancement that we will make to our model is to add several linear series of states to capture statistical patterns at fixed positions relative to splice sites and start and stop codons. We will see in Chapters 7 and 8 that such patterns can be handled very simply in generalized HMMs using a type of signal sensor called a *WMM* (*weight matrix method*), or *PSM* (*position-specific matrix*); such models were utilized by the "toy" exon finder developed in Chapter 5. The new model topology is shown in Figure 6.7. The wavy boxes represent linear HMM modules as shown in Figure 6.8. Note that these linear submodels have been inserted only in noncoding regions of the model.

Results of evaluating H_{77} on the test set are shown in Table 6.7. We can see from the table that the modeling of base frequencies at fixed positions around splice sites and start and stop codons has resulted in a significant improvement: 4% at the nucleotide level, 21% at the exon level, and 5% at the whole gene level.

As a final enhancement we will consider modeling base frequencies at fixed positions in the coding sequence immediately flanking splice sites. Recall that the linear submodels (Figure 6.8) inserted into H_{77} occurred only in regions of the HMM corresponding to noncoding sequence. We will now augment H_{77} by inserting a

Table 6.7 Accuracy results for model H_{77}, as compared to H_{27}; all values are percentages, except the number of correct genes

	Nucleotides			Splice sites		Start/stop codons		Exons			Genes	
	Sn	Sp	*F*	Sn	Sp	Sn	Sp	Sn	Sp	*F*	Sn	Number
H_{27}	83	93	88	40	49	41	36	23	27	25	8	38
H_{77}	88	96	92	66	67	51	46	47	46	46	13	65

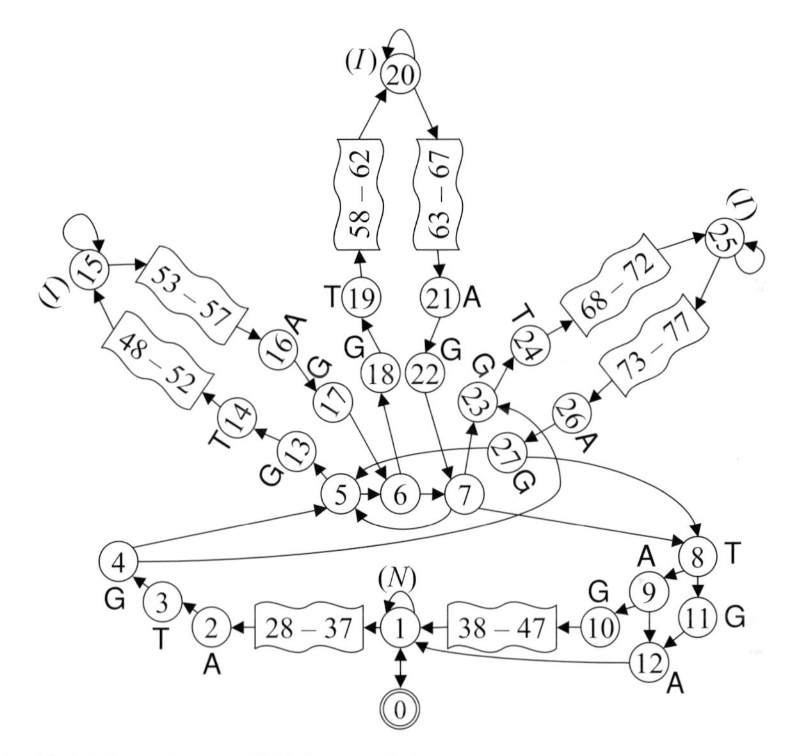

Figure 6.7 Model H_{77}, a 78-state HMM for gene finding.

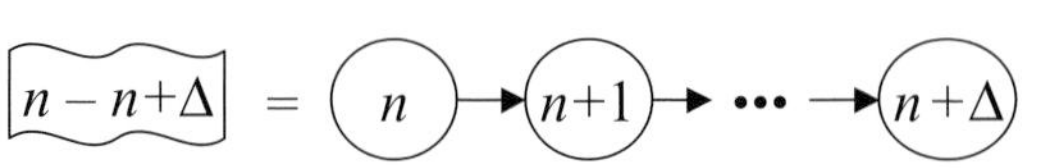

$$\boxed{n - n{+}\Delta} \;=\; (n) \rightarrow (n{+}1) \rightarrow \cdots \rightarrow (n{+}\Delta)$$

Figure 6.8 A linear HMM module effecting a weight matrix.

linear sequence of 3 states before each donor site and after each acceptor site, resulting in:

$$(3 \text{ states}) \times (2 \text{ locations per intron submodel}) \times (3 \text{ intron submodels}) = 18$$

additional states. Using a multiple of 3 states in each location preserves the phase constraints encoded into our present topology, though in principle any number of states could have been added, with appropriate modification to the transition

Table 6.8 Accuracy results for model H_{95}, as compared to H_{77}; all values are percentages, except the number of correct genes

	Nucleotides			Splice sites		Start/stop codons		Exons			Genes	
	Sn	Sp	*F*	Sn	Sp	Sn	Sp	Sn	Sp	*F*	Sn	Number
H_{77}	88	96	92	66	67	51	46	47	46	46	13	65
H_{95}	92	97	94	79	76	57	53	62	59	60	19	93

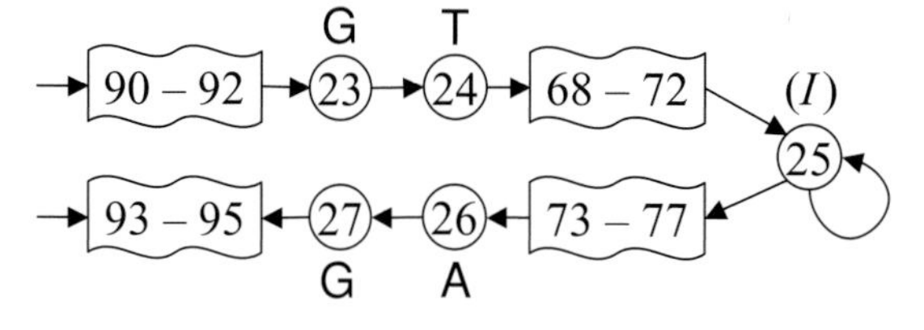

Figure 6.9 An improved intron submodel for H_{95}. Linear series of states before and after each donor and acceptor site model the base composition biases at fixed positions around the splice sites.

patterns into and out of the intron submodels. In the interests of space, the resulting 96-state model, H_{95}, is not depicted it its entirety; Figure 6.9 illustrates just the new intron model, which will of course be duplicated three times, just as before. The resulting model is very similar to that shown in Figure 6.7, but with the new intron submodels substituted in for the old.

Once again, a set of simple modifications must be made to the labeling procedure in order to produce the labeled sequences needed for training the augmented model, and these are again left as an easy exercise for the reader (Exercise 6.18). Table 6.8 shows that the results of this modification again represent a significant improvement over the previous model. Model H_{95} is 2% more accurate (by *F*-measure) at the nucleotide level, 6% more accurate at the whole gene level, and 14% more accurate at the whole exon level than H_{77} on this particular test set.

Figure 6.10 summarizes these experiments by showing how the nucleotide accuracy (by *F*-measure) has increased as the number of states in the HMM has increased. Table 6.9 combines the foregoing results into a single table. We will see in a later section of this chapter that significant improvement is still possible by making fundamental changes to the HMM without increasing the number of states in the model.

In the case study that follows we will examine the eukaryotic gene finders VEIL and UNVEIL, which both utilize an HMM structured much like H_{95}, but with many more states.

Table 6.9 Accuracy results for models H_3 through H_{95} on a set of 500 *A. thaliana* genes; all values are percentages, except the number of correct genes

	Nucleotides			Splice sites		Start/stop codons		Exons			Genes	
	Sn	Sp	*F*	Sn	Sp	Sn	Sp	Sn	Sp	*F*	Sn	Number
baseline	100	28	44	0	0	0	0	0	0	0	0	0
H_3	53	88	66	0	0	0	0	0	0	0	0	0
H_5	65	91	76	1	3	3	3	0	0	0	0	0
H_{17}	81	93	87	34	48	43	37	19	24	21	7	35
H_{27}	83	93	88	40	49	41	36	23	27	25	8	38
H_{77}	88	96	92	66	67	51	46	47	46	46	13	65
H_{95}	92	97	94	79	76	57	53	62	59	60	19	93

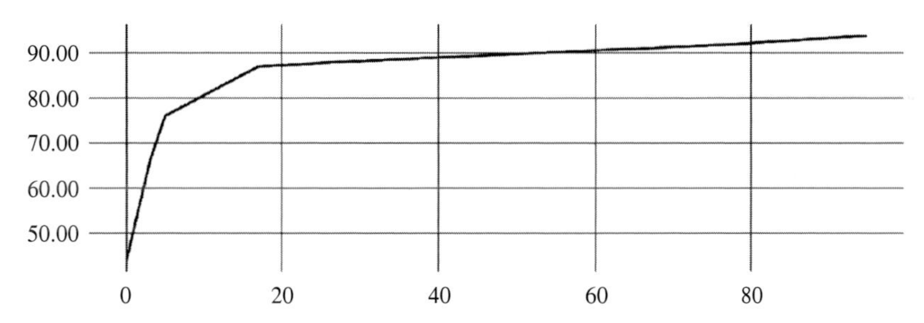

Figure 6.10 Nucleotide *F*-score (*y*-axis) on a test set of 500 *A. thaliana* genes, as a function of number of states (*x*-axis) in an HMM for a simple gene finder.

6.5 Case study: VEIL and UNVEIL

There are now a number of publicly available HMM-based gene finders, including *GeneMark.hmm* (Lukashin and Borodovsky, 1998), *HMMgene* (Krogh, 1994, 1997, 1998), *VEIL* (Henderson *et al.*, 1997), and *UNVEIL* (Majoros *et al.*, 2003). We will consider the latter two.

VEIL and UNVEIL are both open-source programs which can be downloaded over the internet (see Appendix). Since they utilize the same model topology, we will base our discussion on the more recent of the two implementations. UNVEIL is based on a 283-state HMM, and is capable of predicting genes on either strand. The HMM contains 141 states dedicated to forward-strand genes, with a mirror image of this 141-state topology being used to model reverse-strand genes; the two are joined in a single intergenic state by which the model can switch between generation of forward- and reverse-strand genes.

Although structured much like the H_{95} model which we presented in section 6.4, there are a number of significant differences. The primary difference, and the

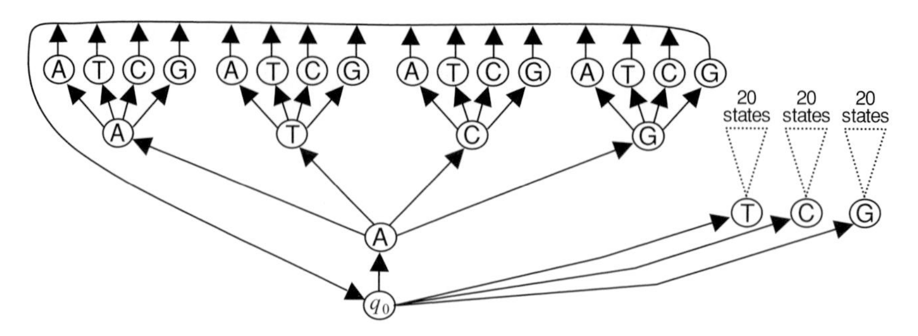

Figure 6.11 The 85-state hierarchical exon submodel of VEIL/UNVEIL.

one that contributes most to the difference in number of states between UNVEIL and H_{95}, is the exon model, which is depicted in Figure 6.11.

As can be seen from the figure, this 85-state submodel is formed in a recursive tree structure, with a subtree rooted at each of the four DNA bases. The symbol labeling each state is the only symbol which that state can emit; all other symbols in that state have probability zero. In the central part of the figure is shown the main subtree rooted at the A nucleotide; those for C, G, and T are depicted in iconic format on the right, but are in fact structured identically to the tree rooted at the A. Each of the subtrees consists of three levels, corresponding to the three codon positions.

Thus, whereas H_{95} contained one state for each codon position, UNVEIL models every possible codon using a different triple of states. In particular, due to the tree structure of the exon submodel, the statistics for the second codon position are partitioned according to the nucleotide occurring in the first position; similarly, the statistics for the third codon position are partitioned according to the previous two nucleotides. In this way, we can say that the probabilities of the four nucleotides in each of the three codon positions are conditioned on zero, one, or two nucleotides occurring before the present one. This is a very powerful technique; as can be seen from Table 6.10, the accuracy of UNVEIL on our 500 *A. thaliana* test genes of section 6.4 is considerably higher, especially at the level of whole genes, than that of H_{95}, owing largely to the expanded exon model.

Whereas the UNVEIL exon submodel effectively conditions its transition probabilities on a small number of previous states, a very similar effect can be achieved by explicitly conditioning the emission probabilities of all states in the HMM on some number of previously emitted symbols. This technique, referred to as the use of *higher-order emissions*, is much more flexible and powerful than the explicit tree-based method used in UNVEIL, in that similar accuracy benefits can be achieved using far fewer states, as we will soon see. Methods for training and using such higher-order HMMs are described in section 6.7.

Table 6.10 Accuracy results for UNVEIL on the 500 *A. thaliana* test genes, as compared to model H_{95} from section 6.4; all values are percentages, except the number of correct genes

	Nucleotides			Splice sites		Start/stop codons		Exons			Genes	
	Sn	Sp	F	Sn	Sp	Sn	Sp	Sn	Sp	F	Sn	Number
H_{95}	92	97	94	79	76	57	53	62	59	60	19	93
UNVEIL	94	98	96	79	83	75	71	68	71	69	37	183

6.6 Using ambiguous models

The model topologies which we utilized in the running example of section 6.4 had the useful property that any gene structure could be unambiguously assigned a unique path through the model such that the features of the gene structure (e.g., exons and introns) would correspond exactly with those same features in the HMM. Such unambiguous models are attractive because they permit the use of the simple labeled sequence training procedure described earlier. There are many possible HMM topologies, however, which do not admit such an unambiguous labeling of training sequences, and in those cases we must resort to more sophisticated parameter estimation techniques. We will consider two such methods: *Viterbi training*, and the *Baum–Welch* algorithm. It must be noted that without additional modifications these procedures are useful only for training individual *submodels* – i.e., sets of states which collectively generate a sequence feature of a particular type, such as an exon or intron. We will show in section 6.6.2 how these submodels, once they have been individually trained on their corresponding training sequences, can be merged back into a complete model of gene structure.

6.6.1 *Viterbi training*

We have already seen that the Viterbi decoding algorithm can be used to determine the most probable path ϕ, for any particular sequence, through a given model M. We can use this procedure to train the individual *submodels* (sets of states) of an HMM as follows.

Suppose an HMM has been partitioned into a set of submodels M^{Γ} corresponding to individual feature types Γ – i.e., M^{exon}, M^{intron}, etc. Now consider the set of transitions $\tau^{in} = \{q_i \to q_j | q_j \in M^{\Gamma} \wedge q_i \notin M^{\Gamma}\}$ leading into a submodel M^{Γ} and the set of transitions $\tau^{out} = \{q_i \to q_j | q_i \in M^{\Gamma} \wedge q_j \notin M^{\Gamma}\}$ leaving M^{Γ}. It should be clear that we can easily treat M^{Γ} as a complete HMM H^{Γ} having some transition set E^{Γ}, by taking the set of states in M^{Γ} and augmenting them with an artificial start/stop state $q^{0(\Gamma)}$ having transitions to and from the other states of M^{Γ} as defined by τ^{in} and τ^{out}: i.e., $q^{0(\Gamma)} \to q_j \in E^{\Gamma}$ iff $\exists_i q_i \to q_j \in \tau^{in}$, and similarly $q_i \to q^{0(\Gamma)} \in E^{\Gamma}$ iff

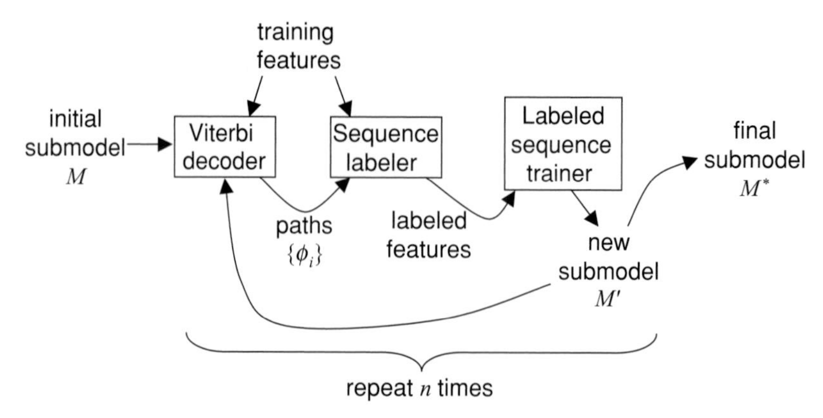

Figure 6.12 Iterative Viterbi training of an inherently ambiguous HMM submodel.

$\exists_i q_i \rightarrow q_j \in \tau^{out}$. Thus, we may subject M^{Γ} to any of the available algorithms on complete HMMs.

Now suppose we have initialized the parameters of the submodel M^{Γ} with random emission and transition probabilities (possibly assigning some particular transitions a probability of zero in order to prevent their use). Then for each feature f of the appropriate type Γ in the training set, we can use the Viterbi algorithm to determine the path ϕ by which the submodel M^{Γ} is most likely to emit f. We can then use the series of states comprising ϕ as the labeling for f, and proceed to re-estimate the parameters of M^{Γ} using this newly acquired set of labeled sequences.

This procedure can be repeated iteratively, as depicted in Figure 6.12, so that the submodel $M^{\Gamma\prime}$ resulting from the previous training run is used as an input to the Viterbi algorithm once again to relabel the training features, which are then used to re-estimate the submodel parameters yet again. The full algorithm is given in Algorithm 6.4. We have simplified the call to `viterbi()` by omitting the expansion of the model into its components as well as the λ_{trans} and λ_{emit} arrays defined by Eqs. (6.9, 6.10); the function `labeledSequenceTraining()` can be inferred easily from Eq. (6.19).

Algorithm 6.4 Iterative Viterbi training. The initial submodel M (which is passed by reference) is subjected to n iterations of parameter re-estimation using training features $T = \{f_i | 0 \leq i < |T|\}$.

```
procedure viterbiTraining( ref M,T,n)
1.  for i ← 1 up to n do
2.     T_labeled ← {};
3.     foreach f ∈ T do
4.        φ ← viterbi(M,f);
5.        push T_labeled, (f,φ);
6.     M ← labeledSequenceTraining(T_labeled, M);
```

Since we initialized the states of the submodel M^{Γ} with random emission and transition probabilities before passing M^{Γ} to `viterbiTraining()`, it may be advisable to repeat the entire process several times, each time re-randomizing the model parameters and then subjecting them to Viterbi training, and then to keep the model which performs most accurately on a held-out test set.

6.6.2 *Merging submodels*

It is important to note that Viterbi training as we have presented it is applicable only to the individual submodels of an HMM rather than to the entire gene finder, and that by *unlabeled sequences* we do not mean sequences for which the correct gene structure is unknown, but rather training features (such as individual exons or introns) for which we do not know, a priori, the precise series of states which the corresponding submodel could be expected to have followed when emitting those features. Thus, the prescribed procedure for training an HMM-based gene finder having ambiguous submodels is to segment the training sequences into individual features, to split the HMM into individual submodels corresponding to those feature types, and then to use Viterbi training (or the Baum–Welch algorithm described in section 6.6.3) to estimate the parameters for the individual submodels. The final HMM can then be obtained by merging these independently trained submodels back into a composite model (with additional transition probabilities between submodels; see below). We will describe a systematic method for doing this.

Let us first define the notion of a *metamodel*. Let $M^{meta} = (Q^{meta}, \alpha, P_t^{meta}, P_e^{meta})$ be an HMM in which the individual states of Q^{meta} correspond to entire submodels M^X – i.e, so that for each state $q_X \in Q^{meta}$ there is associated an entire HMM $M^X = (Q^X, \alpha, P_t^X, P_e^X)$ having initial/final state $q_0^X \in Q^X$. Then the merged HMM, denoted $M^* = (Q^*, \alpha, P_t^*, P_e^*)$, can be constructed as follows.

First, we define the set of states Q^* for the composite model by taking the union of the states from the submodels, replacing the individual start/stop states q_0^Y of the submodels M^Y with a single start/stop state q_0^* for the merged model:

$$Q^* = \{q_0^*\} \cup \bigcup_{q_Y \in Q^{meta}} Q^Y - \{q_0^Y\}. \tag{6.20}$$

Next, we define transitions from any state a in one submodel to any state b in any other submodel (or within the same submodel, if a self-transition on the submodel is present in the metamodel topology):

$$\underset{\substack{q_A \in Q^{meta} \\ q_B \in Q^{meta}}}{\forall} \; \underset{\substack{a \in Q^A \\ b \in Q^B}}{\forall} \; P_t^*(b|a) = P_t^A\left(q_0^A|a\right) P_t^{meta}(q_B|q_A)\, P_t^B\left(b|q_0^B\right) \tag{6.21}$$

where $P_t^A(q_0^A|a)$ can be thought of as taking us from state a within submodel M^A to the associated metastate q_A in M^*; $P_t^{meta}(q_B|q_A)$ takes us from metastate q_A to metastate q_B; and finally, $P_t^B(b|q_0^B)$ descends from metastate q_B into its associated submodel M^B to arrive at state b.

Transitions within a submodel are unchanged by the merging process:

$$\underset{q_Y \in Q^{meta}}{\forall} \quad \underset{\substack{(a,b)\in \\ Q^Y - \{q_0^Y\}}}{\forall} \quad P_t^*(b|a) = P_t^Y(b|a). \tag{6.22}$$

Transitions out of the composite model's start state, q_0^*, can be handled in a similar manner to transitions between metastates (i.e., similar to Eq. (6.21)):

$$\underset{q_Y \in Q^{meta}}{\forall} \; \underset{y \in Q^Y}{\forall} \; P_t^*(y|q_0^*) = P_t^{meta}\left(q_Y|q_0^{meta}\right) P_t^Y\left(y|q_0^Y\right) \tag{6.23}$$

and likewise for transitions into the composite model's final state (q_0^*):

$$\underset{q^Y \in Q^{meta}}{\forall} \; \underset{y \in Q^Y}{\forall} \; P_t^*(q_0^*|y) = P_t^Y\left(q_0^Y|y\right) P_t^{meta}\left(q_0^{meta}|q^Y\right). \tag{6.24}$$

Finally, the emission probabilities can be imported directly from the submodels into the composite model:

$$\underset{q^Y \in Q^{meta}}{\forall} \; \underset{\substack{y \in Q^Y \\ s \in \alpha}}{\forall} \; P_e^*(s|y) = P_e^Y(s|y). \tag{6.25}$$

Note that the emission probabilities P_e^{meta} of M^{meta} are never used and are therefore not needed. The transition probabilities P_t^{meta} can be estimated from labeled sequences in the usual way, using the MLE counting method given in Eq. (6.19), where the counts $A_{i,j}$ are derived by simply counting the number of times a feature of type Γ_i is followed by a feature of type Γ_j in the gene structures of the training set, where Γ_X denotes the feature type represented by submodel M^X.

Using a metamodel/submodel architecture in this way not only facilitates the use of ambiguous submodels, but also eases the task of making changes to individual submodels, such as re-training them or modifying their internal topologies. It will be seen in Chapter 8, however, that these same ideals can be served by abandoning the simple HMM formalism in favor of the more flexible *GHMM*, or *generalized hidden Markov model*.

6.6.3 *Baum–Welch training*

A rather more sophisticated procedure than Viterbi training for training ambiguous submodels is the *Baum–Welch* algorithm, which is an instance of the well-known *EM (expectation maximization)* family of algorithms (Baum *et al.*, 1970; Baum, 1972). Like Viterbi training, the Baum–Welch algorithm is an iterative procedure which successively re-estimates model parameters based on the current values of the parameters and their relation to the features in the training set. Unfortunately, implementing Baum–Welch is complicated by the fact that it tends to cause numerical underflow in the computer when applied to long sequences, and unlike the Viterbi algorithm, it is not convenient to use the log transformation since the algorithm involves sums (rather than exclusively products) of probabilities. Thus, we will first present the "naive" version of the algorithm in order to briefly establish the general workings of the procedure, and then we will give the detailed "scaling" version which prevents numerical underflow and can therefore be implemented directly in software.

6.6.3.1 Naive Baum-Welch algorithm

The Baum–Welch algorithm utilizes the *Forward algorithm* which we described in section 6.2.2 as well as a related procedure called the *Backward algorithm*, which we will present shortly. Our description is adapted from the one provided by Durbin *et al.* (1998). The top-level Baum–Welch procedure is shown in Algorithm 6.5.

Algorithm 6.5 The Baum–Welch algorithm. The initial submodel M is subjected to n iterations of parameter re-estimation using training features $T = \{S_i | 0 \leq i < N_T\}$, $N_T = |T|$.

```
procedure baumWelch (ref M,T,n)
1.   (Q,α,Pt,Pe) ← M;
2.   for h ← 1 up to n do
3.      ∀i∈[0,|Q|−1] ∀j∈[0,|Q|−1] A[i][j] ← 0;
4.      ∀i∈[0,|Q|−1] ∀k∈[0,|α|−1] E[i][k] ← 0;
5.      foreach S ∈ T do
6.         F ← forwardAlgorithm(M,S);
7.         B ← backwardAlgorithm(M,S);
8.         updateCounts(A,E,F,B,Q,S);
9.      updateModel(M,A,E);
```

The outer loop causes the algorithm to perform n iterations of re-estimation. Lines 3–4 initialize two arrays, A and E, which will be used to count the expected uses of each transition and emission in the model, respectively, for the features in the training set. The inner loop starting at line 5 iterates through the training features, running the Forward and Backward algorithms for each. The dynamic programming matrices F and B produced by the Forward and Backward algorithms, respectively, are used by `updateCounts()` to accumulate expected transition and emission counts in A and E, as we describe below. Finally, at the end of each iteration `updateModel()` updates the model parameters by normalizing the counts in A and E into transition and emission probabilities; this can be done exactly as described by Eq. (6.19), where $a_{i,j}$ provides an estimate of $P_t(q_j|q_i)$ and $e_{i,k}$ provides an estimate of $P_e(s_k|q_i)$. As the procedure `updateModel()` may be trivially inferred from Eq. (6.19), we omit it from the text.

Recall that the Forward algorithm computes $F(i,k)$, defined as the probability that a machine M will emit the subsequence $x_0 \ldots x_{k-1}$ with the final symbol of the subsequence being emitted by state q_i. In like fashion, the Backward algorithm uses a dynamic programming strategy to efficiently compute $B(i,k)$, defined as the probability that the machine M will emit the subsequence $x_k \dots x_{L-1}$ and then terminate, given that M is currently in state q_i (which has already emitted x_{k-1}). The alert reader will no doubt have noticed that the product $F(i,k)B(i,k)$ can be interpreted as the probability of machine M emitting the complete sequence $x_0 \dots x_{k-1} \dots x_{L-1} = S$ and being in state q_i at the time x_{k-1} is emitted; this will be useful when we re-estimate $P_e(x_{k-1}|q_i)$.

Equation (6.26) gives the formal definition of $B(i,k)$.

$$B(i,k) = \begin{cases} \sum_{j=1}^{|Q|-1} P_t(q_j|q_i)P_e(x_k|q_j)B(j,k+1) & \text{if } k < L\,, \\ P_t(q_0|q_i) & \text{if } k = L\,. \end{cases} \tag{6.26}$$

A dynamic programming algorithm for efficiently computing this relation is given in Algorithm 6.6.

Algorithm 6.6 The (unscaled) *Backward algorithm*. Parameters are as defined for the Forward algorithm. Returns the dynamic programming array B.

```
procedure unscaledBackwardAlg (Q, Pt, Pe, S, λtrans, λemit)
1.  len ← |S|;
2.  ∀i∈[0,|Q|−1] B[i][len] ← Pt(q0|qi);
3.  for k ← len − 1 down to 0 do
4.    for i ← 0 up to |Q|−1 do
5.      sum ← 0;
6.      for j ← 1 up to |Q|−1 do
7.        sum ← sum+Pt(qj|qi) ∗ Pe(S[k]|qj) ∗ B[j][k+1];
8.      B[i][k] ← sum;
9.  return B;
```

What remains is to specify the implementation of `updateCounts()`, which we do in Algorithm 6.7. The `updateCounts()` algorithm operates as follows. On line 1 we obtain the probability of the sequence S from the (0,0) cell of the B matrix. Then for each state q_i and each symbol $s = S_{k-1}$ in the sequence, we update `E[i][s]` by the product of the forward and backward values for this state at this position in the sequence (as we alluded to earlier), but weighted by the reciprocal of the probability P of the training sequence so as to produce a conditional probability $P(q_i, x_{k-1}|S)$. The sum of these conditional probabilities over the training set gives the expectation of the number of times each emission event will occur during generation of the training set.

On lines 6–7 we update `A[j][i]` for states q_i and q_j with the product:

$$F(j,k)P_t(q_i|q_j)P_e(S_k|q_i)B(i,k+1)/P\,. \tag{6.27}$$

Reading the terms from left to right, $F(j,k)$ gives us the probability of emitting the sequence $x_0 \ldots x_{k-1}$ with state q_j providing the emission of x_{k-1}; we then require that the machine transition from q_j to q_i and that q_i then emit symbol x_k; multiplying by the backward variable $B(i,k+1)$ takes us from state q_i at position k through to the end of the sequence by any (and all) possible paths ending in the final state q^0. Finally, we again weight the product by the reciprocal of the full probability P of the training sequence S to produce a conditional probability $P(q_j, q_i, x_{k-1}, x_k|S)$; summing these over the training set again produces the desired expectation – in this case the expected number of times the given transition would occur if the model were to generate the training set.

Algorithm 6.7 Updating the A and E arrays for the Baum–Welch algorithm. The dynamic programming arrays F and B from the Forward and Backward algorithms, respectively, are taken as inputs. A and E are passed by reference.

```
procedure updateCounts (ref A, ref E, F, B, Q, S)
1.  P ← B[0][0];
2.  for i ← 1 up to |Q|−1 do
3.    for k ← 0 up to |S|−1 do
4.      E[i][S[k]] ← E[i][S[k]]+F[i][k]*B[i][k]/P;
5.      for j ← 0 up to |Q|-1  do
6.        A[j][i] ← A[j][i]+F[j][k]*Pt(qi|qj)*
7.                            Pe(S[k]|qi)*B[i][k+1]/P;
8.    A[i][0] ← F[i][|S|]*Pt(q0|qi)/P;
```

What may not be so obvious from the foregoing description is that the update equations described above are designed in such a way as to guarantee monotonic convergence of the Baum–Welch algorithm to a local maximum for the likelihood of the training set – i.e., that this algorithm results in a *maximium likelihood* solution; proof of this assertion can be found in Durbin *et al.* (1998) and Baum (1972).

6.6.3.2 Baum–Welch with scaling

In order to avoid numerical underflow in the computer when dealing with long sequences, it is necessary to re-scale the probabilities used in the Baum–Welch algorithm so as to keep them within a reasonable range. This can be accomplished by multiplying the Forward and Backward variables by a scaling factor computed at each position in the sequence based on the current inductive score. The following treatment is adapted from Durbin *et al.* (1998).

As a scaling factor for the forward algorithm we can use:

$$s_k^{fw} = \sum_{j=1}^{|Q|-1} \left(P_e(x_{k-1}|q_j) \sum_{i=0}^{|Q|-1} F'(i, k-1) P(q_j|q_i) \right) \quad (6.28)$$

where $F'(i, k-1)$ denotes the scaled version of $F(i, k-1)$, as we describe below. Initialization of the Forward variable is unaffected by scaling:

$$\underset{1\le i<|Q|}{\forall} F'(i,0)=0, \qquad \underset{1\le k\le L}{\forall} F'(0,k)=0, \qquad F'(0,0)=1. \tag{6.29}$$

The recurrence part of the Forward algorithm must incorporate the scaling term in the denominator:

$$\underset{1\le k\le L}{\forall}\ \underset{1\le i<|Q|}{\forall} F'(i,k)=\frac{P_e(x_{k-1}|q_i)}{s_k^{fw}}\sum_{j=0}^{|Q|-1} F'(j,k-1)P_t(q_i|q_j). \tag{6.30}$$

The scaled probability P'_j of sequence S_j is now given by:

$$P'_j=\sum_{i=0}^{|Q|-1} F'(i,L)P_t(q_0|q_i) \tag{6.31}$$

for L the length of the sequence. For the Backward algorithm we recommend using a separate set of scaling terms, as we have found that using the identical terms from the Forward algorithm (as per Durbin *et al.*, 1998) can sometimes fail to protect against numerical underflow in the computer:

$$s_k^{bw}=\sum_{i=0}^{|Q|-1}\sum_{j=1}^{|Q|-1} P_t(q_j|q_i)P_e(x_{k-1}|q_j)B'(j,k) \tag{6.32}$$

where $B'(i,k)$ denotes the scaled version of $B(i,k)$, which is given below. The scaling terms for the Backward algorithm must again occur in the denominator of the recursion:

$$\underset{0\le k<L}{\forall}\ \underset{0\le i<|Q|}{\forall} B'(i,k)=\frac{1}{s_{k+1}^{bw}}\sum_{j=1}^{|Q|-1} P_t(q_j|q_i)P_e(x_k|q_j)B'(j,k+1). \tag{6.33}$$

As in the Forward algorithm, no changes are necessary for the initialization of the Backward variables:

$$\underset{0\le i<|Q|}{\forall} B'(i,L)=P_t(q_0|q_i). \tag{6.34}$$

Because the above modifications invalidate the invariants required for convergence of the Baum–Welch algorithm, the scaling factors that we have introduced into the Forward and Backward algorithms must be explicitly cancelled out when we re-estimate the model parameters. The following ratio will prove useful for this purpose:

$$\underset{1\le k\le L}{\forall} r_k=\log\left(\frac{1}{s_L^{fw}}\prod_{h=k}^{L-1}\frac{s_{h+1}^{bw}}{s_h^{fw}}\right). \tag{6.35}$$

Initialization of the transition counts can be performed as follows:

$$A_{0,0}=0, \qquad \underset{1\le i<|Q|}{\forall} A_{i,0}=\sum_{\substack{\text{sequences}\\ j}}\frac{e^{r_L}s_L^{fw}F'(i,L)P_t(q_0|q_i)}{P'_j}. \tag{6.36}$$

During the recursion step we force cancellation of the additional scaling terms as follows:

$$\underset{0\le i<|Q|}{\forall}\ \underset{1\le j<|Q|}{\forall}\ A_{i,j}=\sum_{\substack{\text{sequences}\\ h}}\sum_{k=0}^{L-1}\frac{e^{r_{k+1}}F'(i,k)P_t(q_j|q_i)P_e(x_k|q_j)B'(j,k+1)}{P'_h}. \tag{6.37}$$

Similarly, cancellation of scaling terms for the emission counts must be performed as well:

$$\underset{1\le i<|Q|}{\forall}\ \underset{s\in\alpha}{\forall}\ E_{i,s}=\sum_{\substack{\text{sequences}\\ j}}\sum_{\substack{1\le k\le L,\\ x_{k-1}=s}}\frac{e^{r_k}F'(i,k)B'(i,k)s_k^{fw}}{P'_j}. \tag{6.38}$$

Given the foregoing modifications, re-estimation of the transition and emission probabilities may proceed exactly as before:

$$\underset{0\le i<|Q|}{\forall}\ \underset{0\le j<|Q|}{\forall}\ P_t(q_j|q_i)=\frac{A_{i,j}}{\sum_{h=0}^{|Q|-1}A_{i,h}},\qquad \underset{1\le i<|Q|}{\forall}\ \underset{s\in\alpha}{\forall}\ P_e(s|q_i)=\frac{E_{i,s}}{\sum_{t\in\alpha}E_{i,t}}. \tag{6.39}$$

When implementing this algorithm in software, it is useful to monitor the successive changes to the log-likelihood of the model over the course of the training process. When properly implemented, the log-likelihood should be nondecreasing except for very small fluctuations due to rounding errors in the computer. Any but the smallest of decreases in the likelihood are indications of an error in the implementation – hence the utility of monitoring this quantity during software development.

The log-likelihood may be computed as follows:

$$\lambda=\sum_{\substack{\text{sequences}\\ j}}\log(P'_j)+\sum_{k=1}^{L}\log\left(s_k^{fw}\right). \tag{6.40}$$

A sample log-likelihood curve graphed as a function of time during Baum–Welch training is shown in Figure 6.13, where it can be seen that the log-likelihood increases monotonically as the algorithm progresses through successive iterations.

As noted earlier, it is essential that all scaling terms which were applied during the Forward and Backward algorithms be properly cancelled before updating the expected counts in the A and E arrays. To see that the above formulas achieve this cancellation, we can convert the recurrences to closed-form expressions and substitute these back into the original equations to see that all of the scaling terms cancel. The closed-form expressions for P'_j, F', and B' are given below. Verifying from these that the proper cancellation occurs as necessary is left as an exercise

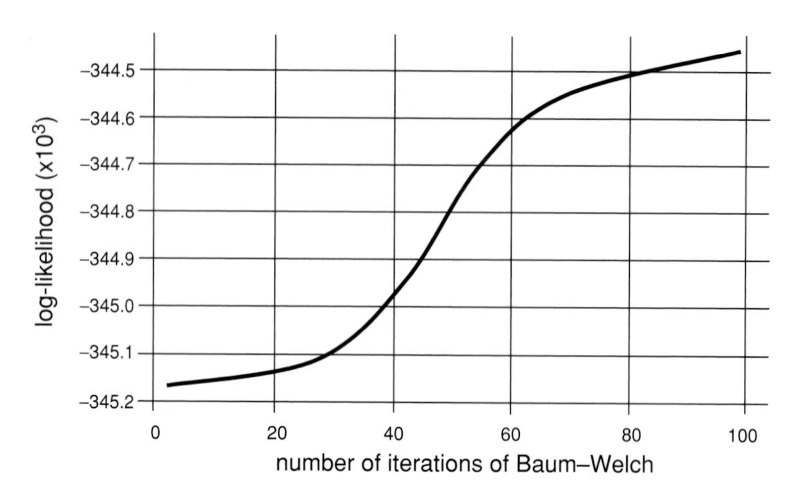

Figure 6.13 Log-likelihood (y-axis, $\times 10^3$) of a model as a function of iteration number (x-axis) of the Baum–Welch algorithm, for a particular HMM.

for the reader (Exercise 6.27).

$$P'_j = \frac{1}{\prod_{k=1}^{L} s_k^{fw}} \sum_{i=1}^{|Q|-1} F(i, L) P_t(q_0|q_i) \tag{6.41}$$

$$\forall_{1 \le j < |Q|} \forall_{1 \le k \le L} F'(j,k) = \frac{F(j,k)}{\prod_{h=1}^{k} s_h^{fw}}, \qquad \forall_{0 \le i < |Q|} \forall_{0 \le k < L} B'(i,k) = \frac{B(i,k)}{\prod_{h=k}^{L-1} s_{h+1}^{bw}} \tag{6.42}$$

A working implementation of this algorithm can be downloaded from the book's website (see Appendix). Note that we have avoided the use of *pseudocounts* (e.g., Durbin *et al.*, 1998) in our implementation of Baum–Welch, as their presence requires additional modifications to the algorithm to ensure numerical stability (Exercise 6.35). An alternative to the use of explicit pseudocounts is to simply augment the training set with an additional sequence specifically constructed so as to exhibit each type of event for which we do not wish the observed count to be zero (see Exercise 6.41).

A simpler alternative to the scaled Baum–Welch algorithm involves the use of the relation:

$$\log\left(\sum_{i=0}^{n-1} p_i\right) = \log p_0 + \log\left(1 + \sum_{i=1}^{n-1} e^{\log p_i - \log p_0}\right), \tag{6.43}$$

where we assume the p_i have been sorted so that the largest is denoted p_0. This relation may be used in the Forward and Backward algorithms to compute the sums of probabilities in log-space. In the log-space version of these algorithms, one merely replaces the raw probabilities p_i with their logarithmic counterparts, $\log p_i$, and applies Eq. (6.43) whenever the probabilities are to be summed. Evaluation of the $e^{\log p_i - \log p_0}$ term should generally not result in numerical underflow in practice, since this term evaluates to p_i/p_0, which for probabilities of similar events should not deviate too far from unity. Trade-offs between time efficiency and numerical precision for this and the scaling method described earlier in this section await rigorous characterization for the specific task of gene prediction.

6.7 Higher-order HMMs

Higher-order emissions (i.e., those conditioned on previous emissions) can be incorporated into HMMs with surprising ease. The essential insight is to recode the emission alphabet α by grouping short, overlapping substrings and atomizing them into discrete symbols from a much larger alphabet. To illustrate, consider the sequence:

$$\texttt{ACGATTATCTACG} \tag{6.44}$$

We can recode this sequence over the 16-symbol alphabet α^2 by taking overlapping pairs of bases:

$$\begin{array}{l} \texttt{AC AT AT TA} \\ \quad \texttt{CG TT TC AC} \\ \quad\quad \texttt{GA TA CT CG} \end{array} \tag{6.45}$$

Thus, the first pair is AC, the second pair is CG, the third pair is GA, the fourth is AT, and so on, until we arrive at the final pair CG. Each pair is considered a discrete symbol, so that, e.g., the pair AC becomes the single symbol which we might denote s_{AC}. Since there are four symbols in the DNA alphabet α, taking pairs of these symbols results in 16 combinations, and hence a 16-letter, 1st-order alphabet α^2.

In general, an nth-order alphabet α^n is one that is formed by taking substrings of length $n+1$ over the DNA alphabet α, giving rise to 4^{n+1} discrete symbols of nth-order. We will see that training and decoding with higher-order alphabets can be accomplished with relatively simple modifications to the corresponding algorithms for zeroth-order models.

6.7.1 *Labeled sequence training for higher-order HMMs*

Given a sequence $S = x_0 \ldots x_{L-1}$ and a series $\phi = y_0, \ldots, y_{L-1}$ of corresponding labels, we can easily modify the procedure outlined in section 6.3 to handle nth-order emissions. Rather than counting the number of times each symbol s_i is observed in association with state q_j in the training set, we instead observe the number of times each $(n+1)$-gram $G = g_0 \ldots g_n$ ends with a symbol that has been labeled with state q_j, and use this count to estimate $P_e(g_n|g_0 \ldots g_{n-1}, q_j)$, the emission probability of 0th-order symbol g_n in state q_j conditional on having just emitted the n-gram $g_0 \ldots g_{n-1}$.

Formally, let $C(g_0 \ldots g_n, q_j)$ denote the number of times the $(n+1)$-gram $G = g_0 \ldots g_n$ was seen with final element g_n labeled q_j. Then we can estimate the nth-order $(n > 0)$ emission probability $P_e(g_n|g_0 \ldots g_{n-1}, q_j)$ of state q_j emitting symbol

g_n conditional on the prior emissions $g_0 \ldots g_{n-1}$ as:

$$P_e(g_n | g_0 \ldots g_{n-1}, q_j) \approx \frac{C(g_0 \ldots g_n, q_j)}{\sum_{s \in \alpha} C(g_0 \ldots g_{n-1} s, q_j)} \tag{6.46}$$

so long as $\sum_{s \in \alpha} C(g_0 \ldots g_{n-1} s, q_j) > 0$. It may be observed, however, that in cases where the denominator is zero, the numerator must be zero also, so that we may quite reasonably assign a probability of zero in these cases – though we will see in section 6.8 that the use of variable-order models provides a more elegant solution.

The only remaining difficulty pertains to the handling of the first n symbols in a training sequence, since these do not fall at the end of an $(n + 1)$-gram, and would therefore go unused by the training procedure despite their having been assigned state labels. The obvious solution is to estimate lower-order emission probabilities from these initial symbols, with the lower-order probabilities being used only when assessing emission probabilities for the corresponding initial positions in an input sequence. Unfortunately, sample sizes for such initial positions will generally be much smaller than those at other positions, since they are limited by the number of training sequences, irrespective of the sequence lengths. Thus, it is preferable to instead estimate (using complete training sequences) models of all orders up to the desired order n, and to use the lower-order emission probabilities only when evaluating the initial positions in an input sequence.

6.7.2 *Decoding with higher-order HMMs*

Once we have trained an HMM with higher-order emissions, it is a simple matter to perform decoding of input sequences using a slightly modified Viterbi algorithm in which all instances of the term $P_e(x_k | q_i)$ are replaced with the appropriate higher-order emission probability function $P_e(x_k | x_{k-n} \ldots x_{k-1}, q_i)$. For the sake of efficiency one may wish to preprocess the input sequence by extracting all the n-grams, mapping them to unique, higher-order symbol identifiers, and then storing the resulting series of higher-order symbols in an array for repeated use during decoding. The potential for improved efficiency will be more apparent when we consider specific types of content sensors for generalized hidden Markov models (section 7.2.3).

Mapping n-grams to unique identifiers can be accomplished by employing the base-4 number system. First, define a mapping $\nabla = \{\mathrm{A} \leftrightarrow 0, \mathrm{C} \leftrightarrow 1, \mathrm{G} \leftrightarrow 2, \mathrm{T} \leftrightarrow 3\}$ between individual nucleotides s_i and their indices i in the alphabet. To account for lower-order states we need to define:

$$\Lambda(L) = \sum_{i=0}^{L-1} 4^i \tag{6.47}$$

which gives the total number of n-grams of length less than L over the DNA alphabet. Converting an n-gram $S = x_0 \ldots x_{L-1}$ from base-4 to base-10 can be accomplished as follows:

$$\beta^{(4)}(S) = \sum_{i=0}^{L-1} 4^i \nabla(x_{L-i-1}). \tag{6.48}$$

Finally, an n-gram S can be mapped to a higher-order symbol $s_{\psi(S)}$ using:

$$\psi(S) = \Lambda(|S|) + \beta^{(4)}(S). \tag{6.49}$$

6.8 Variable-order HMMs

A potentially very detrimental aspect of higher-order HMMs is that the sample sizes available for estimating emission probabilities at higher orders are necessarily smaller than those for lower-order models. This can be demonstrated by the following simple argument. If we suppose that some subsequence $S_1 = x_i \ldots x_{i+m-1}$ of length m occurs c times in a given training sequence S_T, then by removing the first symbol x_i from S_1 we obtain a shorter sequence $S_2 = x_{i+1} \ldots x_{i+m-1}$ which must occur *at least* c times in S_T, since S_2 is a subsequence of S_1; it may occur exactly c times, or possibly more than c times, but certainly not fewer. Thus, the sample sizes for longer n-grams are always less than or equal to those for the shorter n-grams contained within them, so that as n increases, we can expect (in general) to reduce the fidelity of our model by increasing the incidence and severity of sampling error in our estimation of the model parameters.

An obvious means of reducing the effects of sampling error is to simply avoid using higher-order models altogether. A more useful solution is to adaptively tailor the order of model used for each conditional probability, based on the available sample size for the corresponding n-gram used for conditioning. A number of techniques have been devised for doing precisely this, and we will consider two that are in common use: *backing-off*, and *interpolation*.

6.8.1 *Back-off models*

A simple method for reducing sampling error in estimations involving n-gram frequencies is to enforce a *minimum sample size* rule. Such a rule operates by replacing any probability estimate based on fewer than K data points, for some K, with an estimate based on shorter n-grams:

$$P_e^{backoff}(g_n|g_0 \ldots g_{n-1}, q_j) = \begin{cases} \dfrac{C(g_0 \ldots g_n, q_j)}{\sum\limits_{s\in\alpha} C(g_0 \ldots g_{n-1}s, q_j)} & \text{if } C(g_0 \ldots g_{n-1}, q_j) \geq K \text{ or } n = 0 \\ P_e^{backoff}(g_n|g_1 \ldots g_{n-1}, q_j) & \text{otherwise} \end{cases} \tag{6.50}$$

Table 6.11 Accuracy results for higher-order versions of model H_{95} (96 states), as compared to the original zeroth-order model (top row) and to the 141-state UNVEIL model (bottom row)

Model	Order	Nucleotides			Splice sites		Start/stop codons		Exons			Genes	
		Sn	Sp	F	Sn	Sp	Sn	Sp	Sn	Sp	F	Sn	Number
H_{95}	0	92	97	94	79	76	57	53	62	59	61	19	93
H_{95}	1	95	98	97	87	81	64	61	72	68	70	25	127
H_{95}	2	98	98	98	91	82	65	62	76	69	72	27	136
H_{95}	3	98	98	98	91	82	67	63	76	69	72	28	140
H_{95}	4	98	97	98	90	81	69	64	76	68	72	29	143
H_{95}	5	98	97	98	90	81	66	62	74	67	70	27	137
UNVEIL		94	98	96	79	83	75	71	68	71	70	37	183

As argued above, the count for any n-gram G_n occurring within another n-gram G_{n+1} is always greater than or equal to that for G_{n+1}, so that reducing the size of the n-gram by one or more nucleotides often produces an acceptably large sample size, for reasonably chosen K. Minimum sample sizes of between 150 and 300 are common for gene-finding applications (e.g., Burge, 1998). Assuming that every nucleotide occurs at least once in the training set, we can be sure that a lower-order estimate will be available for any n-gram, since in the extreme case we can always fall back on 0th-order probabilities. In the event that even this requirement fails, we can install *pseudocounts* (i.e., a small number of artificial observations of each type of event, inserted into the training set) to ensure that the 0th-order probabilities remain nonzero. A special case of pseudocounts is the use of the *Laplace prior*, in which we add a single spurious observation of each point in a sample space to the training set before estimating a distribution (Durbin *et al.*, 1998).

The practice of imposing a minimum sample size for probability estimates based on n-grams is known as a simple form of *backing-off*, as proposed in a slightly more complicated form by Katz (1987). Elaborations on the method can be found in Jelinek (1997) and Manning and Schütze (1999).

6.8.2 *Example: Incorporating variable-order emissions*

To illustrate the power of using higher-order emission probabilities in HMMs, we have adapted our model H_{95} from section 6.4 to use variable-order emissions via a back-off strategy, and retested the model on the same 500 *A. thaliana* test genes. We initially imposed a minimum sample size of 80 to train H_{95} at a maximum order of 1, 2, 3, 4, and 5. The variable-order models of order 4 and 5 were then found to perform better with minimum sample sizes of 200. These results are shown in Table 6.11. As can be seen from the table, the use of higher-order models has significantly improved the accuracy of the gene finder over that of the 0th-order model.

Although the accuracy of the higher-order models is still below that of UNVEIL at the whole gene level, the accuracy of the higher-order models (particularly 3rd-order) at the level of exons and nucleotides is clearly competitive with that of the 141-state UNVEIL model. It can also be seen that the 4th- and 5th-order models did not significantly outperform the 3rd-order model, and in some cases even trailed the 3rd-order model, suggesting that a moderate degree of sampling error is still present despite the use of the minimum sample size constraints.

6.8.3 Interpolated Markov models

Rather than simply discarding higher-order probability estimates which were based on insufficient sample sizes, one can instead opt to use a weighted average of estimates of different orders weighted by some function of their sample sizes. This is the strategy utilized by *interpolated Markov models (IMMs)* (Salzberg *et al.*, 1998a,b), which are very similar to the *deleted interpolation* models used in speech recognition (e.g., Jelinek and Mercer, 1980; Bahl *et al.*, 1986; Potamianos and Jelinek, 1998). The goal of using IMMs is to benefit from the information present in higher-order probability estimates while reducing sampling error through the simultaneous use of lower-order probabilities. This is a form of *smoothing* (section 2.7), since we are modifying our estimates for a probability distribution in hopes of reducing sampling error.

Formally, define:

$$P_e^{IMM}(s\,|\,g_0 \dots g_{k-1}) = \begin{cases} \lambda_k^G P_e(s\,|\,g_0 \dots g_{k-1}) + (1-\lambda_k^G) P_e^{IMM}(s\,|\,g_1 \dots g_{k-1}) & \text{if } k > 0 \\ P_e(s) & \text{if } k = 0 \end{cases} \tag{6.51}$$

where the weight λ_k^G is specific to n-gram $G = g_0 \dots g_{k-1}$, and P_e is the fixed-order emission distribution. We say that $P_e^{IMM}(s\,|\,g_0 \dots g_{k-1})$ is the *kth-order interpolated estimate* of $P_e(s\,|\,g_0 \dots g_{k-1})$; it should be clear that these probabilities are also conditional on a state q, which we omit for simplicity. As can be seen from Eq. (6.51), the 0th-order interpolated model simply evaluates to the relative frequency of symbol s. For all higher-order models, we interpolate between orders k and $k{-}1$ using weights λ_k^G and $(1-\lambda_k^G)$, respectively. Note that while the kth-order term is a function of the raw Markov probabilities $P_e(s\,|\,g_0 \dots g_{k-1})$, the lower-order terms are incorporated in a recursive fashion via $P_e^{IMM}(s\,|\,g_1 \dots g_{k-1})$, the interpolated $(k{-}1)$th-order probability. Thus, if we were to "unwind" the recurrence defined by Eq. (6.51) for any given n-gram, we would find that the corresponding IMM probability would be defined in terms of a weighted average over all orders between zero and some maximum order k selected at our discretion (typically 5 or 8 – see Figure 2.7 in Chapter 2).

The weights λ_k^G used in the weighted average can be defined in various ways. In the microbial gene finder *Glimmer* (Salzberg *et al.*, 1998a) they are defined as:

$$\lambda_k^G = \begin{cases} 1 & \text{if } m \geq 400 \\ 0 & \text{if } m < 400 \text{ and } c < 0.5 \\ \frac{c}{400} \sum_{x \in \alpha} C(g_0 \ldots g_{k-1}x) & \text{otherwise} \end{cases} \tag{6.52}$$

where $m = \sum_{x \in \alpha} C(g_0 \ldots g_{k-1}x)$ is the sample size for the k-gram $g_0 \ldots g_{k-1}$ providing a kth-order context for symbol s, and c is a measure of our confidence that the kth-order probability estimate $P_e(s|g_0 \ldots g_{k-1})$ does not significantly differ from the IMM estimate for the next lower order. This "confidence" value is defined as $c = 1 - P(\chi^2)$, where χ^2 is the test statistic of the χ^2 *test of homogeneity*, which is essentially the same as the χ^2 *goodness-of-fit test* described in section 2.9, and $P(\chi^2)$ is the associated *P*-value resulting from the test. Thus, to obtain the test statistic χ^2, we simply apply Eq. (2.53), using $C(g_0 \ldots g_{k-1}x)$ as our observed count c_i for symbol x, and $m \cdot P_e^{IMM}(x|g_1 \ldots g_{k-1})$ as the expected count c_i' (under the null hypothesis that the higher-order distribution does not differ significantly from the interpolated, lower-order distribution).

Note that the value of 400 used in Eq. (6.52) was chosen somewhat arbitrarily, and can be replaced with any other reasonable value. Other interpolation schemes have been investigated both within the field of gene finding (e.g., Azad and Borodovsky, 2004) and outside of it (e.g., Jelinek and Mercer, 1980; Ristad and Thomas, 1997).

It should also be noted that the above construction is applied only during training of the IMM, so that once the model has been trained, it can be represented in memory exactly as any other HMM, and all the standard HMM algorithms can be applied directly to the IMM with no modifications. Thus, interpolation constitutes a novel training method for standard HMMs, and IMMs are therefore simply HMMs for which interpolation has been employed during training.

6.9 Discriminative training of HMMs

The training of HMMs for gene finding has traditionally been carried out using some form of *maximum likelihood estimation* (MLE), either via the Baum–Welch algorithm or using a simpler procedure based on labeled sequences such as the one presented in section 6.3. However, a number of alternatives do exist, and some of these are arguably more suitable for the task of gene finding than MLE.

The effect of training an HMM using MLE at the submodel level (followed by an appropriate merging procedure as detailed in section 6.6.2) is to maximize the joint probability of the training sequences and their parses:

$$\theta^* = \underset{\theta}{argmax} \left(\prod_{(S,\phi) \in T} P(S, \phi|\theta) \right) \tag{6.53}$$

where the summation is over all (*sequence*, *parse*) pairs $(S, \phi) \in T$ in the training set T. What renders this approach attractive is that it is easily and efficiently computable (especially for unambiguous models, for which we can use labeled sequence training) while often producing reasonably well-trained HMMs. It is not optimal, however, in the sense that it does not attend to the global optimality of the HMM as a predictor of maximally accurate gene structures when used in conjunction with Viterbi decoding (Jelinek, 1997, p53; see also section 12.4). That is, MLE does not guarantee *maximum discrimination* between the genomic elements we wish to delineate via parsing. Any training regime having maximum discrimination as its goal is known as *discriminative training*.

Ideally, one would like to maximize the expected accuracy of the gene finder on unseen data. If we accept the training set as a surrogate for this "unseen data," then we may consider maximizing the combined probability of all the prescribed gene parses in the training set:

$$\theta^* = \underset{\theta}{argmax} \left(\prod_{(S,\phi)\in T} P(\phi|S,\theta) \right) = \underset{\theta}{argmax} \left(\prod_{(S,\phi)\in T} \frac{P(S,\phi|\theta)}{P(S|\theta)} \right) \tag{6.54}$$

where the collection of model parameters making up the HMM is denoted θ. This *argmax* gives us the parameterization θ^* under which the full gene *parses* in the training set will be maximally likely (on average), given the training sequences, and is known as the *conditional maximum likelihood* (*CML*) parameterization. Considering the fact that the Viterbi algorithm during gene prediction also maximizes $P(\phi|S,\theta)$, this would seem to be a more appropriate training goal than that prescribed by MLE. Unfortunately, this goal presents a rather difficult computational problem, as there are no known methods for directly solving Eq. (6.54) (Rabiner, 1989).

Several methods have been developed, however, which at least approach the ideal of discriminative training. These developments are largely due to efforts in the fields of speech recognition (e.g., Bahl *et al.*, 1986; Rabiner, 1989; Krogh, 1994; Reichl and Ruske, 1995; Johansen, 1996; Normandin, 1996; Markov *et al.*, 2001; Schlüter *et al.*, 2001) and natural-language parsing (e.g., Toutanova *et al.*, 2003), though some few efforts have been undertaken to bring discriminative training to bear on the problem of training HMM-based gene finders (e.g., Krogh, 1997) as well as for training HMMs for other bioinformatics applications (e.g., Eddy *et al.*, 1995; Durbin *et al.*, 1998). We will see in section 8.6.2 that the need for discriminative training in generalized HMMs (GHMMs) has recently begun to receive some attention as well (e.g., Majoros and Salzberg, 2004).

The major variants of discriminative training are *conditional maximum likelihood* (*CML*), *maximum mutual information* (*MMI*), and *minimum classification error* (*MCE*), though other variants have been proposed (e.g., Markov *et al.*, 2001). Each variant V dictates a slightly different objective function $f_V(\theta)$ to be optimized. In the case of CML, the objective function $f_{CML}(\theta)$ is derived directly from Eq. (6.54) by segmenting the training examples into individual features S_i and their "correct" feature types Γ_i:

$$f_{CML}(\theta) = \sum_{(S_i,\Gamma_i)\in F} \log P(\Gamma_i|S_i,\theta) \tag{6.55}$$

where F is the set of all (S_i, Γ_i) pairs across all training examples, and $P(\Gamma_i|S_i,\theta)$ denotes the probability of classifying sequence S_i as being a feature of type Γ_i. Optimization proceeds by maximizing this objective function; i.e., $\theta^* = argmax_\theta$ $f_{CML}(\theta)$. The MMI criterion $f_{MMI}(\theta)$ is instead based on the pointwise *mutual information* (see section 2.10) between the feature sequences S_i and their feature types Γ_i:

$$MI(S_i, \Gamma_i) = \log\left(\frac{P(\Gamma_i, S_i|\theta)}{P(\Gamma_i|\theta)P(S_i|\theta)}\right) = \log\left(\frac{P(S_i|\Gamma_i,\theta)}{P(S_i|\theta)}\right), \tag{6.56}$$

so that the MMI objective function becomes:

$$f_{MMI}(\theta) = \sum_{(S_i,\Gamma_i)\in F} \log\left(\frac{P(S_i|\Gamma_i,\theta)}{\sum_{\Gamma} P(S_i,\Gamma|\theta)}\right). \tag{6.57}$$

MCE is very similar to MMI except that in the denominator we sum only over the "wrong" feature types rather than all feature types as in MMI:

$$f_{MCE}(\theta) = \sum_{(S_i,\Gamma_i)\in F} \log\left(\frac{P(S_i|\Gamma_i,\theta)}{\sum_{\Gamma\neq\Gamma_i} P(S_i,\Gamma|\theta)}\right) \tag{6.58}$$

with the intention being to maximize the log difference between the probability assigned by the correct model M^i of type Γ_i and that assigned by all other models; in practice, Eq. (6.58) is generally evaluated after transformation by a sigmoidal function (Reichl and Ruske, 1995).

All of these criteria can in principle be evaluated by a *gradient ascent* procedure (section 2.5; see also section 10.15) to iteratively explore incremental modifications to the model parameters, keeping only those modifications that are observed to improve the value of the objective function. In practice, this can be a very slow process, as the number of parameters in the model may be very large. In an attempt to speed up this optimization, methods have been devised based on modifications to the Baum–Welch procedure, though it seems that all of these have met with difficulty in that the resulting update equations are either *unstable* (i.e., not guaranteed to always move in the direction of increasing gradient) or are stable only for certain algorithm configurations, where the configuration is controlled by yet another parameter, which must be determined through trial and error. Additional research will thus be necessary to bring these methods into mainstream use for HMM-based gene finding.

6.10 Posterior decoding of HMMs

The decoding problem for HMM's has been addressed by the Viterbi algorithm, which finds the most probable parse ϕ under the scoring function $P(\phi|S)$, as described in section 6.2.1. The so-called *posterior decoding* problem is somewhat different in nature, in that it does not seek an optimal parse of the input sequence, but merely the assignment of a probability value to each potential exon (or other feature) in the sequence. That is, for a given exon $E = [i, j]$ expressed as an interval on the input sequence S, we wish to compute $P(E \text{ is an exon}|S)$ – i.e., the probability that E is an exon, conditional only on S (and the model), and not conditional on any particular parse. This probability can be expressed as a summation over all possible (syntactically valid) parses of the input sequence which contain exon E:

$$P(E\,\mathit{is\,an\,exon}|S) = \sum_{\substack{\mathit{all}\,\phi\\ \mathit{containing}\,E}} P(\phi|S)$$

and may be computed using the Forward and Backward variables, F and B:

$$\begin{aligned} &P(E = [i, j]\,\mathit{is\,an\,exon}|S) \\ &= \sum_{\substack{y \neq q_{exon},\\ z \neq q_{exon}}} F(y, i) P_t(q_{exon}|q_y) \left(\prod_{k=i}^{j} P_e(x_k|q_{exon}) \right) \\ &\times P_t(q_{exon}|q_{exon})^{j-i} P_t(z|q_{exon}) P_e(x_{j+1}|z) B(z, j+2). \end{aligned} \tag{6.59}$$

Recall that $F(y, i)$ represents the probability that the machine emits the subsequence $x_0 \ldots x_{i-1}$ by any path ending in state q_y (with symbol x_{i-1} being emitted by state q_y), and that $B(z, k)$ is the conditional probability that the machine will emit $x_k \ldots x_{L-1}$ and then terminate, given that the machine is currently in state q_z (which has already emitted x_{k-1}), for a sequence of length L. Thus, Eq. (6.59) merely stipulates that the machine must emit the subsequence $x_i \ldots x_j$ from the exon state q_{exon} (which may be replaced with a set of states in the case where exons are modeled using multiple states – see Exercise 6.39), and that the machine must transition into q_{exon} from some other state y and transition out of q_{exon} into some other state z (where y and z would presumably be intron or intergenic/UTR states) at the precise boundaries of the exons. Use of the Forward and Backward variables attends to the summing over all possible partial parses of the sequences to the left and right of the exon. Thus, we fix the states which emit the exon and allow everything else about the parse to vary, as crudely illustrated in Figure 6.14.

Posterior decoding, as we have formulated it, provides a probability estimate for a single exon (or any other feature, if we replace q_{exon} with some other state), and may be interpreted as our confidence that the interval in question truly is an exon. The program GENSCAN uses this interpretation when computing exon scores (Burge and Karlin, 1997). After performing Viterbi decoding in the usual way for a

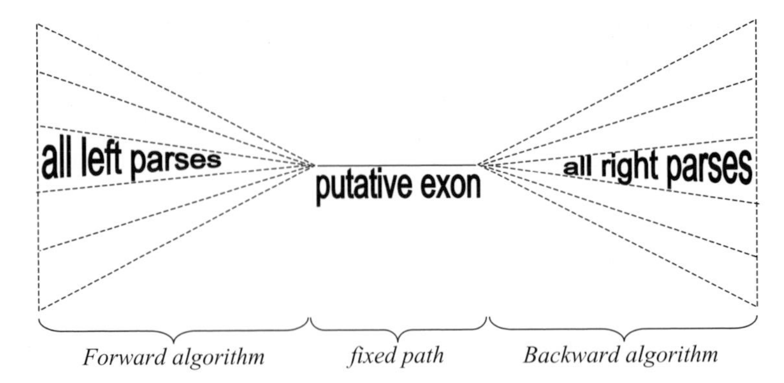

Figure 6.14 Crude representation of dynamic programming matrix for posterior decoding. A putative exon is modeled using a (more or less) fixed path, while the left and right subsequences outside the exon may be generated by any legal path through the model, as enumerated by the Forward and Backward algorithms, respectively.

GHMM (see section 8.3), the program applies posterior decoding to each predicted exon in order to compute a confidence score for the exon, which is then merely emitted for the user to see. Thus, GENSCAN does not use the posterior decoding scores in choosing its parse of the input sequence.

Fariselli *et al.* (2005) introduced a posterior decoding algorithm, called *posterior Viterbi (PV)*, which does in fact produce a global parse of the full input sequence based on posterior probabilities. This algorithm computes the posterior of the label τ_i for the nucleotide at position k in the input sequence S:

$$P(\tau_i|S,k) = \sum_{\substack{q_j \ni \\ label(q_j)=\tau_i}} F(j,k+1)B(j,k+1)$$

and then substitutes these posteriors in place of products of transition and emission probabilities in the Viterbi algorithm to obtain a modified Viterbi decoder (hence the name *posterior Viterbi*):

$$PV(i,k) = \begin{cases} \max_j PV(j,k-1)\delta_{trans}(q_j,q_i)P(\tau_i|x_k) & \text{if } k > 0, \\ \delta_{trans}(q^0,q_i)P(\tau_i|x_0) & \text{if } k = 0, \end{cases} \tag{6.60}$$

where $\delta_{trans}(q_j,q_i)$ evaluates to 1 if $P(q_i|q_j) > 0$ and evaluates to 0 otherwise. A related algorithm is the so-called *optimal accuracy decoder (OAD)* of Käll *et al.* (2005), which aims to maximize the expected nucleotide-level decoding accuracy; the only algorithmic difference between PV and OAD is that the latter sums the posterior probabilities along a prospective parse, whereas the former multiplies them.

Table 6.12 compares the use of these alternate decoders to standard Viterbi. Unfortunately, these results suggest little or no advantage to the use of these alternate decoders for the task of gene prediction, with any gain in nucleotide-level

Table 6.12 Results of applying an HMM-based gene finder (UNVEIL) to 50 *Oryza sativa* genes, 200 *Arabidopsis thaliana* genes, and 50 *Aspergillus fumigatus* genes, using Viterbi decoding, posterior Viterbi (PV), and the optimal accuracy decoder (OAD); Nucleotide *F*-score, exon *F*-score, and gene sensitivity are shown

	O. sativa			*A. thaliana*			*A. fumigatus*		
Algorithm	Nucleotide	Exon	Gene	Nucleotide	Exon	Gene	Nucleotide	Exon	Gene
Viterbi	85%	42%	24%	96%	68%	36%	94%	50%	30%
PV	89%	41%	26%	97%	64%	33%	96%	39%	22%
OAD	89%	41%	26%	97%	61%	30%	96%	39%	22%

accuracy coming largely at the expense of exon and/or gene-level accuracy. In Chapter 8 we will consider the use of modified versions of these decoders for generalized HMMs.

EXERCISES

6.1 Given the example HMM specified by Eqs. (6.3, 6.4), compute the probabilities of the following sequences:
(a) YYYYRRRRYYYYY
(b) YRRRRRRRRRRRY
(c) YRYRYRYRYRYRY
(d) YRYYRYYRYRYYR
(e) YYYYYYYYYYYYY

6.2 Compute the probability that the HMM in problem 1 emits *all* the sequences (a)–(e) in that problem, in order, during successive runs of the HMM. Compute the same probability for emitting the strings in the reverse order – i.e., (e)–(a). Are these two probabilities the same? Explain your answer.

6.3 Given the example HMM specified by Eqs. (6.3, 6.4), compute the probabilities of the following sequences:
(a) Y
(b) YY
(c) YYY
(d) YYYY
(e) YYYYY

6.4 For the sequences in the previous exercise, compute the ratio of probabilities of each sequence to that of its preceding sequence – e.g., the ratio of (b) to (a), (c) to (b), etc. Do you notice a pattern? Explain your findings.

6.5 Continuing in the same vein as the previous exercise, compute the probability $P(\mathtt{Y}^n)$ for $1 \leq n \leq 50$, for the HMM shown in Figure 6.1. Plot a curve of the resulting probabilities as a function of n.

6.6 Develop a module in your favorite programming language for reading an HMM into memory and representing it using a set of efficient data structures.

6.7 Given the HMM depicted in Figure 6.15, determine the most probable path for the following sequences:

(a) TAGCTGATCGT

(b) ATCGTA

(c) CGATTCGC

(d) GCATCGGATC

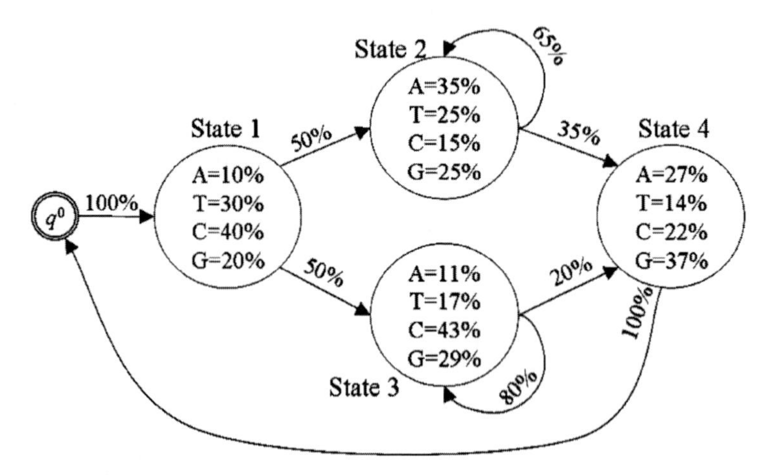

Figure 6.15

6.8 Implement the Viterbi algorithm using your favorite programming language. Use the program to solve Exercise 6.7.

6.9 Draw the dynamic programming matrix that would be computed by the Viterbi algorithm for each of the sequences in Exercise 6.7, given the same HMM. Use arrows to show the traceback links, and write the probability inside each cell of the matrix (i.e., use one matrix, not two).

6.10 For each of the sequences in Exercise 6.7, quantify the amount of computation that is saved by utilizing the λ_{emit} and λ_{trans} arrays in representing the HMM. Express your answer in terms of the total number of arithmetic operations and memory accesses, combined into a single number.

6.11 Compute the probability for each of the sequences in Exercise 6.7, given the same HMM used in that exercise.

6.12 Implement the Forward algorithm in your favorite programming language. Use the program to answer Exercise 6.11.

6.13 Assuming a 3-state HMM with states $Q = (q_0, q_1, q_2)$, compute the maximum likelihood estimates for the emission and transition probabilities of a model trained (separately) with each of the following labeled sequences. Assume an implicit zero label at the beginning and end of each sequence to denote the initial/final state q_0.

(a) AATCGGGATTAGCTACGTAGCTACTTATCGATCG
    1111112222222111111122222211111111

(b) CGATGCGCGCGGCCCATTCTCTACTGTACCTTAT
    1111112222222111111122222211111111

(c) GCTATCTCTTTTCTAGATCGACTCTATCATATTC
    1111112222222111111122222211111111

(d) CGCGCTATATCGGCTATCGTATCTCTCTTTCATG
    1111112222222111111122222211111111

(e) ATATATCGCGCTATGCTCCTCTCTAGTCTCATCT
    1111112222222111111122222211111111

6.14 Modify the procedure `labelExons()` in Algorithm 6.3 to perform labeling for the 6-state model H_5.

6.15 Modify the procedure `labelExons()` in Algorithm 6.3 to perform labeling for the 18-state model H_{17}.

6.16 Modify the procedure `labelExons()` in Algorithm 6.3 to perform labeling for the 28-state model H_{27}.

6.17 Modify the procedure `labelExons()` in Algorithm 6.3 to perform labeling for the 78-state model H_{77}.

6.18 Modify the procedure `labelExons()` in Algorithm 6.3 to perform labeling for the 96-state model H_{95}.

6.19 Modify the topology of model H_5 to keep track of phase across introns. You may add at most two states to the model.

6.20 Write a module in your favorite programming language to take as input a state path for model H_{95}, as produced by the Viterbi algorithm, and emit a set of zero or more gene predictions in GFF format (see section 4.1).

6.21 Give the state paths through model H_{27} for the following labeled sequences, where N denotes intergenic sequence, E denotes exon sequence (including start and stop codons), and I denotes intron sequence (including splice sites).

(a) AATCATGGGGATTAGCTACGTATAGACTTATCGA
    NNNNEEEEEEEEEEEEEEEEEEEEENNNNNNNNN

(b) CATGGCGTGCAGCCCATTAGTTACTGTACCTAGT
    NEEEEEIIIIIIIIIIIIIIEEEEEEEEEEEEEN

(c) GCATGGTCTTAGGTGTATCGACAGTGTAGTAATC
    NNEEEIIIIIIIEEIIIIIIIIIIEIIIIEEENN

(d) CATGCTATATGTGCAGTCGTGTCTCAGTTTGATG
    NEEEEEEEEEIIIIIIEEEEIIIIIIIEEEEENN

(e) ATATATCATGCTATGCTCCTCTTGTGAGAGTGAT
    NNNNNNNEEEEEEEEEEEEEEEEIIIIIEEEEEN

6.22 Examine sequence (e) in Exercise (6.21). What is wrong with this gene structure? Can this structure be predicted by model H_{27}?

6.23 In Figure 6.7 we presented a model in which some number of bases (represented by the wavy box) at fixed distances from splice sites and start and stop codons are modeled by linear series of states. Consider that increasing the length of these linear submodels can preclude the gene finder from predicting features (such as introns) which are shorter than some minimum length. Devise a means of allowing these modules to predict arbitrarily short features. Do not add any additional states.

6.24 Download the HMM-based gene finder UNVEIL (see Appendix), compile the program for your system if necessary, and then train the gene finder using a training set obtained from the same web site. Evaluate the accuracy of the gene finder on a separate hold-out set, either from the same species or from a different species.

6.25 Implement an iterative Viterbi training program in your favorite programming language. Demonstrate that it works by training an HMM on some sample data and examining the resulting model.

6.26 Download the HMM-based gene finder UNVEIL (see Appendix). Use the included "baum-welch" program to train an ambiguous HMM from an arbitrary set of training sequences. Specify 100 iterations on the command line. Did any messages appear regarding reductions in log-likelihood? If so, report the largest reduction observed.

6.27 Using the closed-form expressions given in Eqs. (6.41, 6.42), verify that all scaling terms cancel, including both the Forward and Backward scaling terms and also the scaling ratio r. Show your work.

6.28 Using Eqs. (6.47–6.49), compute the higher-order state labels for the following n-grams:

(a) `ATC`

(b) `CGAT`

(c) `TCAGT`

(d) `TCAGA`

6.29 Referring to Eqs. (6.47–6.49), devise an algorithm for converting a higher-order symbol h to an n-gram over the DNA alphabet.

6.30 Referring to Eqs. (6.47–6.49), convert the following higher-order symbols to 7-mers over the DNA alphabet:

(a) 12

(b) 116

(c) 234

(d) 1026

6.31 Give a formal proof of the second equality in Eq. (6.54).

6.32 Give a formal proof of the second equality in Eq. (6.56).

6.33 Modify model H_{27} to model genes on both strands. Draw the transition diagram for the resulting model. Be sure to label consensus states with the symbol which they emit.

6.34 Continuing with the previous exercise, generalize the `labelExons(S,E)` procedure to handle genes on both strands.

6.35 Modify the Baum–Welch algorithm to permit the use of pseudocounts. Implement the algorithm in software and demonstrate that the algorithm is numerically stable when pseudocounts are used.

6.36 Modify the Viterbi algorithm to return the n highest-scoring *parses* (not labelings) rather than only the single best. Hint: consider duplicating the dynamic programming matrices into n sets of matrices and allowing any cell in any matrix to link back to any cell (in the preceding column) in any other matrix.

6.37 Implement the H_{95} model in software. Apply it to a set of sequences with known gene structures and report the accuracy of your program.

6.38 Continuing with the previous exercise, modify your Viterbi decoding procedure to prohibit the prediction of in-frame stop codons straddling an intron. Test the modified program on the same test set as in the previous exercise. Report the difference in prediction accuracy as a result of this modification.

6.39 Generalize Eq. (6.59) to permit multiple exon states in place of q_{exon}.

6.40 Modify Algorithm 6.1 to handle empty sequences. Also, check whether the input sequence can be emitted by the HMM and return NIL if it cannot.

6.41 Explain how an artificial (labeled) training sequence can be constructed for an (inherently unambiguous) HMM which will ensure that no event will have an observed count of zero during training, as suggested in section 6.6.3.2.

7

Signal and content sensors

In the previous chapter we considered hidden Markov models and their use for *ab initio* gene prediction. Before advancing to the more popular successor to the HMM, namely, the *generalized hidden Markov model* (GHMM), it will be useful to develop some familiarity with the several types of submodel employed by modern gene finders for the detection of discrete *signals* (such as putative splice sites and start/stop codons) and the scoring of the variable-length regions of DNA stretching between those putative signals. The former class of submodels are termed *signal sensors* and the latter *content sensors*; together, these are referred to as *feature sensors*. In this chapter we will consider the submodel types most commonly employed for feature sensing by modern gene finders. Several of the less commonly used model types such as neural networks and other machine learning algorithms are considered in Chapter 10. In the next chapter we will show how these submodels may be integrated into efficient parsing systems.

7.1 Overview of feature sensing

Within the probabilistic framework which we will develop in the next chapter, entire gene features (e.g., exons, introns) are to be modeled as single-event emissions by states in a stochastic machine (a notion to which we hinted with our introduction of the *metamodel* concept in section 6.6.2). Thus, at the level of individual submodels, what will be desired is an assessment of the probability that the given submodel would, during any given emission event, emit a particular subsequence, or *feature*, S:

$$P(S|\theta_q) = P_e(S|q), \tag{7.1}$$

where θ_q denotes the model parameters for the sensor inhabiting state q of the metamodel. It is worth pointing out that in GHMM-based gene finding we are generally not interested (except in the case of *posterior decoding* – section 8.5.4) in directly evaluating $P(q|S)$ – the probability that the sequence S constitutes a feature of type q. Instead, we will for the most part be interested solely in evaluating the likelihood given by Eq. (7.1). The reasons for this will become clearer when we present the GHMM framework in the next chapter.

7.2 Content sensors

One advantage of the GHMM framework is that it permits the explicit modeling of feature length, as we will later see. Thus, we will refine the formulation of Eq. (7.1) by conditioning our emission probabilities on the feature length:

$$P(S|\theta_q, d) = \frac{P(S|\theta_q)}{\sum_{S' \in \alpha^d} P(S'|\theta_q)}, \tag{7.2}$$

for the DNA alphabet α, and where $d = |S|$ will be referred to interchangeably as the *length* of the feature S, and as the *duration* of the emission event responsible for generating S. Since signals are by definition fixed-length features, for a signal state q we have $P(S|\theta_q, d) = P(S|\theta_q)$ whenever d matches the length prescribed by the signal sensor, and $P(S|\theta_q, d) = 0$ otherwise. In the case of content sensors, the novice reader will find it a worthwhile exercise to verify, for each of the model types that we describe, that the probabilities which we construct are indeed conditional on duration – i.e., that:

$$\underset{d \geq 0}{\forall} \left[\sum_{S \in \alpha^d} P(S|\theta_q, d) = 1 \right]. \tag{7.3}$$

7.2.1 *Markov chains*

The operation of Markov chains will seem intuitive enough, given the material presented in Chapter 6. We define an nth-order Markov chain as a 2-state HMM with nth-order emissions:

$$M = (Q, \alpha, P_t, P_e),$$

where:

$$\begin{aligned} Q &= \{q_0, q_1\}, \\ P_t &= \{(q_0, q_1, 1), (q_1, q_1, p), (q_1, q_0, 1-p)\}. \end{aligned} \tag{7.4}$$

Note that the transition distribution is effectively defined by only one parameter, p, the probability of self-transition in state q_1. Taking advantage of this fact while also treating the state set Q as implicit allows us to simplify our notation to:

$$M = (\alpha, p, P_e), \tag{7.5}$$

where the conditioning argument to P_e is now implicitly taken to be state q_1 – i.e., $P_e(s) = P_e(s|q_1)$. Figure 7.1 depicts a Markov chain as a 2-state HMM.

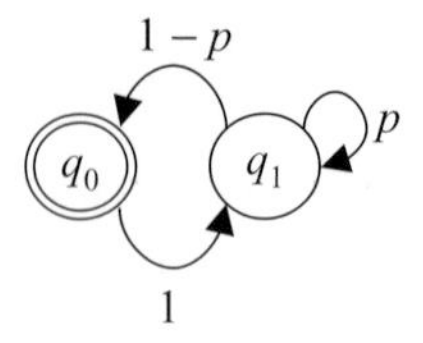

Figure 7.1 Defining a Markov chain as a 2-state HMM. The transition distribution is determined entirely by parameter p, the probability of self-transition in state q_1. State q_0 is the silent initial/final state.

Given this definition of a Markov chain M, we can compute the conditional probability that M emits a string S as:

$$P(S|\theta, d) = \frac{\left(\prod_{i=0}^{d-1} P_e(x_i)\right) p^{d-1}(1-p)}{\sum_{S' \in \alpha^d} \left(\prod_{i=0}^{d-1} P_e(x'_i)\right) p^{d-1}(1-p)}, \tag{7.6}$$

where θ encapsulates the model parameters of M, $S = x_0x_1 \ldots x_{d-1}$ is the feature sequence, and $S' = x'_0x'_1 \ldots x'_{d-1}$ iterates over all strings of length d. $P_e(x_i)$ would obviously be replaced by $P_e(x_i|x_{i-n} \ldots x_{i-1})$ in the case of higher-order chains. The factor $p^{d-1}(1-p)$ can be cancelled from the numerator and denominator, and we leave it as an exercise for the reader (Exercise 7.1) to show that:

$$\sum_{S' \in \alpha^d} \left(\prod_{i=0}^{d-1} P_e(x'_i)\right) = 1, \tag{7.7}$$

so that Eq. (7.6) can be simplified to:

$$P(S|\theta, d) = \prod_{i=0}^{d-1} P_e(x_i). \tag{7.8}$$

From Eq. (7.8) we observe that the parameter p does not affect $P(S|\theta, d)$, so that making the alphabet $\alpha = \{\mathrm{A, C, G, T}\}$ implicit allows us to further simplify our notation for a Markov chain to:

$$M = (P_e). \tag{7.9}$$

Thus, an nth-order Markov chain M for GHMM-based feature sensing in DNA is fully specified by a collection of nth-order emission probabilities.

Given the foregoing, it can be seen that the training and decoding problems for Markov chains can be trivially solved. For training, we need only count the $(n+1)$-gram occurrences and convert these to conditional probabilities via the simple method described in section 6.7.1:

$$P_e(g_n|g_0 \ldots g_{n-1}) \approx \frac{C(g_0 \ldots g_n)}{\sum_{s \in \alpha} C(g_0 \ldots g_{n-1}s)}, \tag{7.10}$$

where $C(g_0 \ldots g_n)$ is the number of times the $(n+1)$-gram $G = g_0 \ldots g_n$ was seen in the set of training features for this state.

The decoding problem is easier still, since there is only one possible path through the chain for any given sequence, namely, $\phi = q_0, q_1, q_1, \ldots, q_1, q_0$ for $|\phi| = |S| + 2$.

Finally, as shown in Eq. (7.8), the conditional probability of a sequence can be computed as a simple product of (possibly higher-order) emission probabilities for each symbol in the sequence. Because this probability rapidly approaches zero as the length of the sequence increases, it is advisable to instead compute the log probability by summing the logs of the individual emission probabilities:

$$\log P(S|\theta, d) = \sum_{i=0}^{d-1} \log P_e(x_i) \tag{7.11}$$

for a 0th-order chain. We will see in the next chapter that the log transformation is especially convenient within the GHMM framework; all of the feature sensors presented in this chapter will thus be formulated in log space. For nth-order chains $(n > 0)$ we have:

$$\log P(S|\theta, d) = \sum_{i=0}^{n-1} \log P_e(x_i|x_0 \ldots x_{i-1}) + \sum_{i=n}^{d-1} \log P_e(x_i|x_{i-n} \ldots x_{i-1}), \tag{7.12}$$

where we have conditioned positions 0 through $n-1$ on fewer preceding bases, as per the available left margin of the base x_i. It should be noted that the decomposition prescribed by Eq. (7.12) is possible only because we make the *Markov assumption* – that is, we assume that the sequence S was generated by a Markov process with limited memory, so that any given base is dependent on at most n previous bases, and is thus conditionally independent of all other preceding bases in the sequence (as well as some – but not all – of the bases which will follow; see Exercise 7.23). We will see later in this chapter that other conditional independence assumptions may be applied in modeling the dependence structure of a sequence.

To briefly illustrate the application of the above formulae, consider the putative feature $S =$ TCGAT. The probability for this sequence under a 3rd-order model would be $P(\text{T}) \cdot P(\text{C}|\text{T}) \cdot P(\text{G}|\text{TC}) \cdot P(\text{A}|\text{TCG}) \cdot P(\text{T}|\text{CGA})$, or $\log P(\text{T}) + \log P(\text{C}|\text{T}) + \log P(\text{G}|\text{TC}) + \log P(\text{A}|\text{TCG}) + \log P(\text{T}|\text{CGA})$ in log space.

The graph in Figure 7.2 shows a series of log-likelihood ratios from a 2nd-order Markov chain trained on human exons versus a chain trained on noncoding sequences – i.e., $\log P(S|\theta_{exon})/P(S|\theta_{noncoding})$ – applied to a sequence containing three exons (gray bars). The ratios were computed along a 2230 bp sequence and then smoothed using 15 iterations of smoothing via Algorithm 2.4, using a window size of 13. As can be seen from the figure, the scores are mostly postive in the exonic regions and mostly negative in the noncoding regions, though the boundaries of the exons are not perfectly delineated, and there are several short regions where the smoothed line is positive in a noncoding region. Nevertheless, it is clear that

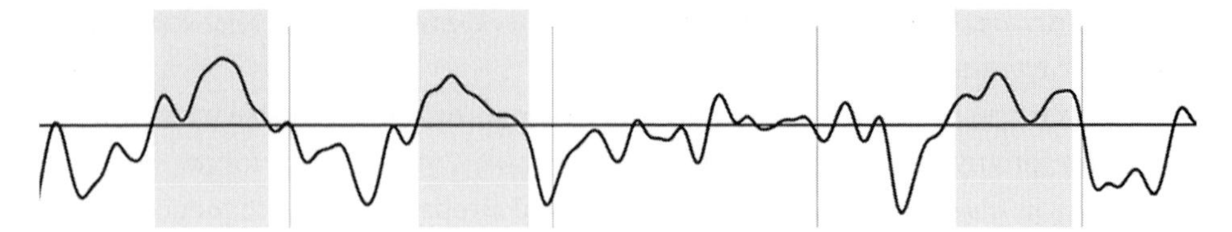

Figure 7.2 Log-likelihood ratio scores (*y*-axis) smoothed along a ~2000 bp sequence. Gray bars show the locations of exons. Ratios were computed using two 2nd-order Markov chains, one trained on exons and the other on noncoding DNA. The horizontal line shows where the curve passes through zero, and serves as a reasonable decision boundary for this problem.

for this particular sequence the combined model is able to distinguish coding from noncoding DNA with considerable accuracy.

Unfortunately, not all sequences admit this level of parsing accuracy using such simple models. In later sections of this chapter we will consider more sophisticated content sensors, and in the next chapter we show how to integrate these with signal sensors and duration and transition probabilities within the GHMM framework for more accurate sequence parsing.

7.2.2 *Markov chain implementation*

Markov chains can be implemented most conveniently using *hash tables*. A hash table is a data structure which can be used to efficiently implement a function of a single variable, $h{:}D \mapsto R$, where D in this particular case would be the set of strings α^d of length d, and $R = \mathbb{R}$. Since a Markov chain is effectively defined by its emission probability function P_e, we can represent any nth-order chain M using a hash table h which maps strings from α^{n+1} to their normalized log probabilities:

$$h(x_0 \ldots x_n) = \log P_e(x_n | x_0 \ldots x_{n-1}). \tag{7.13}$$

That is, we store the log of the conditional probability $P_e(x_n|x_0 \ldots x_{n-1})$, which conditions x_n on the n-gram $x_0 \ldots x_{n-1}$ preceding it, at the location associated with the $(n{+}1)$-gram $x_0 \ldots x_n$ in the hash table. For most practical hash functions, the cost of each look-up (including computation of the hash key) will be $O(n)$, so that the cost of evaluating an entire feature of length d will be $O(nd)$. See Cormen *et al.* (1992) for a detailed description of hash tables.

7.2.3 *Improved Markov chain implementation*

Although the hash table representation for Markov chains is a reasonably efficient one, a significant speed-up can be obtained by using an alternative representation based on the notion of a *finite state transducer* (*FST*), which is an extension of the *deterministic finite automaton* (*DFA*) introduced in section 2.15.

Recall that a DFA is defined as a 5-tuple $(Q, \alpha, \delta, q^0, F)$ for state set Q, alphabet α, deterministic transition function δ, initial state $q^0 \in Q$, and final state set

$F \subseteq Q$. A transducer extends the notion of a DFA by emitting a symbol upon entering each state (in the case of a *Moore machine* – Moore, 1956) or upon following a transition (in the case of a *Mealy machine* – Mealy, 1955). We will adopt the former convention. Formally, a *Moore machine*, or *transducer*, is a 7-tuple $(Q, \alpha, \delta, q^0, F, \xi, \Omega)$ in which Q, α, δ, q^0, and F are as defined for a DFA, and $\xi : Q \mapsto \Omega$ is a function associating with each state $q \in Q - \{q^0\}$ a single output value $\xi(q)$ from the finite output alphabet Ω. For the purposes of implementing Markov chains, we can take $\Omega = \mathbb{R}$ (noting that $\mathbb{R}$ is technically a finite alphabet when implemented using standard floating-point numbers in a particular computer system), so that the machine can output a log probability for each nucleotide read from the input. In particular, for an nth-order Markov chain $M = (P_e)$ we can define a Moore machine $D = (Q, \alpha, \delta, q_0, F, \xi, \mathbb{R})$ in which $Q = \{q_i | 0 \le i \le \Lambda(n+2)\}$, $F = \{q^0\}$, δ is defined by:

$$\delta(\psi(x_0 \ldots x_{k-1}), s) = \begin{cases} q_{\psi(x_0 \ldots x_{k-1} s)} & \text{if } k \le n \\ q_{\psi(x_1 \ldots x_{k-1} s)} & \text{if } k = n+1 \end{cases} \tag{7.14}$$

and ξ is defined as:

$$\xi(\psi(x_0 \ldots x_n)) = \log P_e(x_n | x_0 \ldots x_{n-1}), \tag{7.15}$$

where we have made use of the Λ and ψ functions from section 6.7.2:

$$\Lambda(L) = \sum_{i=0}^{L-1} 4^i,$$
$$\psi(S) = \Lambda(|S|) + \beta^{(4)}(S),$$
$$\beta^{(4)}(S) = \sum_{i=0}^{L-1} 4^i \nabla(x_{L-i-1}),$$

for $\nabla = \{\text{A} \leftrightarrow 0, \text{C} \leftrightarrow 1, \text{G} \leftrightarrow 2, \text{T} \leftrightarrow 3\}$, $L = |S|$. Figure 7.3 illustrates the transition function for part of a 5th-order Markov chain implemented as a transducer. Note that this same figure served as an illustration in section 2.15 of a 6th(not 5th!)-order Markov chain (rather than a transducer) – see Exercise 7.22.

What we have done is to create a transducer in which each state is labeled with a unique $(k+1)$-gram for values of k between -1 and n, inclusively. We have defined the transition function of this transducer so that, e.g., upon seeing input symbol s while in state $x_0 \ldots x_n$, the machine transitions deterministically into state $x_1 \ldots x_n s$, where we have appended s to the end of the previous state's label and shortened the label from the left as necessary to avoid exceeding the state label length limit of n+1. Finally, we have defined the output function ξ in such a way that upon entering state $x_0 \ldots x_n$ the machine will emit the log-probability $\log P_e(x_n | x_0 \ldots x_{n-1})$. By running the transducer over an input sequence S we will therefore obtain as output a series of emission log-probabilities for the successive

Table 7.1 Time and memory consumption measured for a gene finder applied to a 1.8 Mb sequence. The gene finder's content sensors (all Markov chains) were configured to use a hash table implementation for the first run; for the second run they used a transducer implementation. The transducers were over two times faster and required less memory than the hash tables, even though the hash tables were implemented in a highly efficient manner using native C character arrays

	Time (min:sec)	Memory
Hash table	1:15	53 Mb
Transducer	0:34	44 Mb

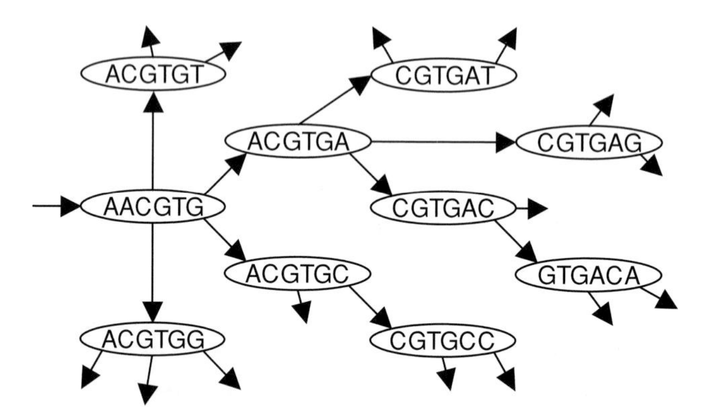

Figure 7.3 A transducer representing a 5th-order Markov chain. The state labels represent the higher-order input symbols associated with the transition into a state; the output of each state – the log probability of the associated Markov emission – is omitted from the figure for clarity.

elements of S; summing these will give us our (logarithmic) conditional probability, $\log P(S|\theta, d)$.

Table 7.1 shows the results of running a gene finder on a 1.8 Mb sequence, using Markov chain content sensors implemented either as hash tables or as transducers. As can be seen from the table, a speed-up of approximately $2\times$ was achieved, while also reducing memory requirements, by employing transducers rather than hash tables.

7.2.4 Three-periodic Markov chains

In section 6.4 it was seen that modeling the emission probabilities for the three codon positions separately within an HMM can improve the accuracy of the resulting gene finder by detecting codon usage biases in coding regions. A similar technique can be applied when using Markov chains. By utilizing three separate chains in periodic fashion we can model the three codon positions, resulting in a *three-periodic Markov chain*, a specific type of *inhomogeneous Markov chain* (Borodovsky and McIninch, 1993) or *nonstationary Markov chain* (section 7.2.6).

In order to train a three-periodic Markov chain (3PMC) for coding regions, we must first know the phase of the training exons. Suppose we are given a forward-strand training exon $E = x_0x_1x_2\ldots x_{m-1}$ and are told that exon E begins in phase ω. Then we can train a set of three Markov chains M_0, M_1, and M_2 on E by modifying the normal training procedure to observe the rule that only n-grams ending at a position $y = 3x + (i - \omega) \bmod 3$ relative to the beginning of the exon can be used in the training of chain M_i, for all x such that $0 \leq 3x + (i-\omega) \bmod 3 < m$. The collection $M = (M_0, M_1, M_2)$ constitutes a three-periodic Markov chain.

Applying a 3PMC M to compute the conditional probability of a sequence S in phase ω on the forward strand can be achieved via:

$$P(S|M, d, \omega) = \prod_{i=0}^{d-1} P_e(x_i|M_{(\omega+i) \bmod 3}) \tag{7.16}$$

for $d = |S|$; modification of Eq. (7.16) for use on the reverse strand is relatively straightforward (see Exercise 7.8). The use of feature probabilities conditional on phase will be amply illustrated in the next chapter.

7.2.5 Interpolated Markov chains

The interpolation techniques described in section 6.8.3 for smoothing higher-order emission probabilities in HMMs are directly applicable to Markov chains, with the result variously referred to as an *interpolated Markov model* (IMM) or an *interpolated Markov chain* (IMC). Thus, an IMC is a model $M = (P_e^{IMM})$, for interpolated emission distribution P_e^{IMM} defined as in section 6.8.3.

Interpolation is equally applicable to inhomogeneous Markov chains. Performing interpolation on each of the chains in a three-periodic Markov chain results in a *three-periodic interpolated Markov chain* (3PIMC).

Figure 7.4 illustrates the effect of interpolation on classification accuracy for a particular task involving the discrimination of human exons from an equal number of noncoding regions of similar size, with the latter extracted from known human introns and intergenic regions. For this example we employed log-likelihood ratios of the form:

$$score(S) = \log \frac{P(S|M_{exon})}{P(S|M_{noncoding})}$$

where the model M_{exon} was trained on several hundred human exons and $M_{noncoding}$ was trained on a several hundred noncoding (i.e., intergenic and intronic) sequences

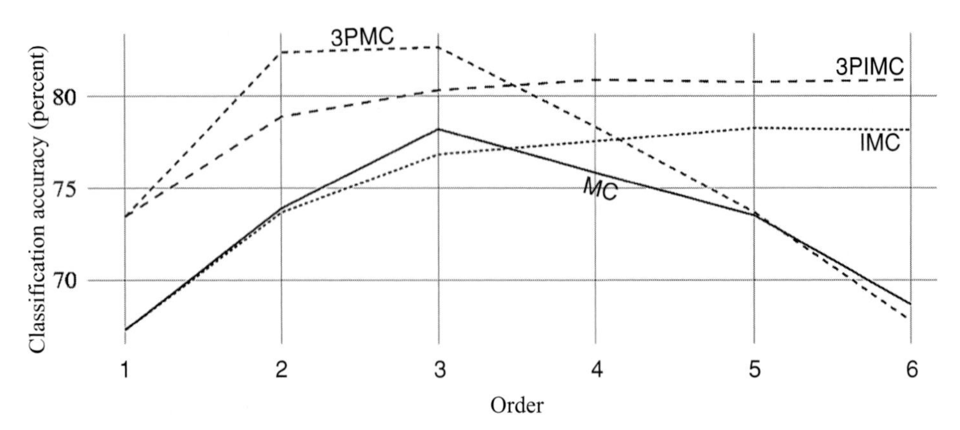

Figure 7.4 Relative accuracy of Markov chains (MC), IMCs, three-periodic MCs (3PMC), and three-periodic IMCs (3PIMC) on a particular task involving classification of human DNA sequences as coding versus noncoding.

from the human genome; classification was performed via the rule:

$$class(S) = \begin{cases} exon & \text{if score} \geq \gamma \\ noncoding & \text{otherwise} \end{cases}$$

for the empirically optimal threshold γ. From the figure it can be seen that the use of periodicity confers significant advantage at most orders, while the use of interpolation appears mainly to counteract the effects of sampling error at higher orders (in which sample sizes are obviously reduced due to longer *n-gram* lengths – see Exercise 7.21).

7.2.6 Nonstationary Markov chains

Most of the models described so far have the property that their probability distributions are *stationary* – that is, it is assumed that the model parameters are constant, and in particular, that they do not change over the course of a single run of the model. One exception is the three-periodic Markov chain, which, when viewed as a single model rather than a collection of three models, effectively changes its parameters at every position along the sequence in periodic fashion in order to track the progression of codon phases. As another example we might consider the operation of a complete HMM for gene finding, as the alternation between coding and noncoding regions corresponding to distinct submodels in the HMM gives rise to nonstationary nucleotide composition frequencies when viewed as the parameters of a single-state model.

It is reasonable to suppose that there might be other instances of nonstationarity which could be usefully harnessed by a gene finder. For example, it is known that local {G,C} density varies significantly as one traverses segments of the human genome, with continuous regions of relatively uniform {G,C} density being referred to as distinct *isochores* (Bernardi *et al.*, 1985). The gene finder GENSCAN exploits this nonstationarity by measuring the {G,C} density of the input sequence and using this to choose a set of precomputed isochore-specific parameters to apply during decoding (Burge, 1997); in the case of GENSCAN, the nonstationarity is implicitly partitioned as the user segments the target genome (during a preprocessing phase) into "bite-sized" chunks of a size suitable for GENSCAN's consumption (see section 8.4.1 for an alternative approach to isochore modeling during GHMM decoding).

More generally, we might consider modeling nonstationarity by partitioning individual features such as exons or introns into a small number of contiguous regions each modeled by a different set of parameters. Two challenges to such an approach are to efficiently compute the probability of an entire feature, given all possible partitionings into stationary subsequences (since the actual partitioning is *unobservable*), and also to reliably estimate model parameters; additional research is needed to explore these possibilities.

7.3 Signal sensors

Signal sensors are used to identify putative signals such as splice sites or start/stop codons. These fixed-length features provide the "anchor" points delimiting the ends of the variable-length features evaluated by the content sensors.

The types of signals sought by signal sensors can usually be characterized by a fairly consistent *consensus sequence*. Some canonical signal consensus sequences in eukaryotes are:

- ATG for start codons
- TAG, TGA, or TAA for stop codons
- GT for donor splice sites
- AG for acceptor splice sites
- AATAAA or ATTAAA for polyadenylation signals
- TATA for the TATA-box portion of a human promoter signal.

We stress that these are merely the common consensus sequences for these signals in many eukaryotes, and that any given organism may have a number of other consensus sequences – called *noncanonical consensus sequences* – which occur in genuine signals recognized by the cell (e.g., the ~10 polyadenylation signals identified by Beaudoing *et al.*, 2000). Still other organisms may have entirely different *canonical* signal consensuses; prokaryotic start codons are a prime example (see section 7.5).

The usual method for applying signal sensors is to evaluate each sensor at every position along the sequence, from left to right, using a "sliding window" approach as illustrated in Figure 7.5. During each evaluation the sensor has access to a fixed-length window around the *consensus region* – i.e., the fixed positions within the sliding window where the consensus bases are expected to occur. More commonly,

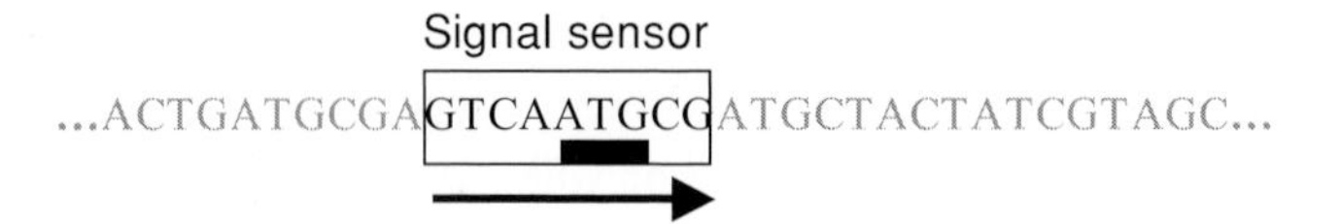

Figure 7.5 A signal sensor, conceptualized as a "sliding window" that is superimposed at successive positions along the input sequence. The sensor is evaluated at each position using the bases captured in the fixed-length window. The thick underbar shows the consensus region, where the signal's consensus sequence is expected.

each sensor will be evaluated only at those positions in the input sequence where the proper signal consensus occurs, since most signal sensors will evaluate to 0 (or $-\infty$ in log space) at sites having an invalid consensus sequence.

Because the consensus sequences for most signal types are very short (e.g., GT for donor sites), they can be expected to occur many times at random in any lengthy DNA sequence of roughly uniform nucleotide composition, so that only a small fraction of putative signals are actually recognized by all the relevant cellular machinery on the pathway to a functional protein. We will refer to those putative signals which are not involved in the production of functional proteins as *pseudosignals*. The goal of signal sensing is therefore to assign a score to each putative signal in a way that maximally separates the true signals from the pseudosignals – that is, so that true signals are assigned scores much higher than the pseudosignals. In this way, we will lessen the chances that a pseudosignal will be included in the optimal GHMM parse (GHMM-based parsing is described in section 8.3), and therefore that a pseudosignal will give rise to an erroneous gene prediction.

In practice, the training of signal sensors tends to be a tedious task in which many different window sizes and different positionings of the consensus region within the window, as well as different model types, are evaluated and the best configuration selected for deployment in the gene finder. Given a set ψ^+ of positive examples (true signals) and a set ψ^- of negative examples (pseudosignals) of a given type, we can evaluate the *classification accuracy* of a newly trained signal sensor M by scoring each of the positive and negative examples using the signal sensor and finding the threshold t^* which maximizes the *separability* of the positive and negative examples:

$$t^* = \underset{t}{argmax} \frac{N^+(t) + N^-(t)}{N}$$

where:

$$N^+(t) = |\{c \in \psi^+ | P(c|M) \geq t\}|$$

is the number of positive examples c satisfying the threshold,

$$N^-(t) = |\{c \in \psi^- | P(c|M) < t\}|$$

A = 31%	A = 18%	A	T	G	A = 19%	A = 24%
T = 28%	T = 32%				T = 20%	T = 18%
C = 21%	C = 24%				C = 29%	C = 26%
G = 20%	G = 26%	100%	100%	100%	G = 32%	G = 32%

consensus region

Figure 7.6 A hypothetical weight matrix for a eukaryotic start codon. Fixed positions in the vicinity of the ATG consensus sequence are modeled. Base frequencies at the modeled positions are shown as percentages.

is the number of negative examples failing the threshold, and N is the total number of test cases.

An alternative approach is to train two versions of the sensor: M^+, which is trained on positive examples, and M^-, which is trained on negative examples. Classification of a test case c may then be performed by using the likelihood ratio $P(c|M^+)/P(c|M^-)$, with the estimation of classification accuracy proceeding as above, using the likelihood ratio instead of the probability $P(c|M)$.

7.3.1 *Weight matrices*

Perhaps the simplest signal sensor is the *weight matrix*, also known as a *WMM* (*weight matrix method*) or *PSM* (*position-specific matrix*), which models nucleotide biases at independent positions in a fixed-length window (Staden, 1984). Weight matrices were briefly introduced in section 5.5.

The example matrix in Figure 7.6 has a *length* of 7, with two positions on either side of the consensus region being included in the model. Although the consensus positions are shown in this example to have fixed emissions (100%) corresponding to the consensus bases ATG, nonobligatory consensus bases can be accommodated easily enough by making appropriate changes to the emission probabilities at those positions. Weight matrices can have any length, and need not be centered on the consensus region.

A common visual representation for a weight matrix is the *pictogram* (Burge *et al.*, 1999) or *DNA-gram*, in which residues are sized in proportion to their frequency of occurrence at particular positions within the WMM window, as shown in Figure 7.7. From the figure it can be seen, for example, that a bias toward Gs exists at positions 3 and 8 (using zero-based indices relative to the left end of the window) of human donor sites.

Computing the conditional probability of a sequence with a WMM can be accomplished by superimposing the matrix onto the region of interest and multiplying together the appropriate probabilities, as illustrated in Figure 7.8.

Formally, we define a weight matrix M as a triple $M = (L_M, P_M, \alpha)$ for length $L_M \in \mathbb{N}$, alphabet α, and emission function $P_M : \mathbb{N} \times \alpha \mapsto \mathbb{R}$ mapping position i $(0 \leq i < L_M)$ to a conditional probability distribution $P(s|i)$ over symbols $s \in \alpha$.

Figure 7.7 A DNA-gram of human donor sites. The GT consensus is shown at positions 4 and 5 (using zero-based indices) from the left. The height of each letter indicates its relative frequency in the corresponding position of the matrix.

A = 31% T = 28% C = 21% G = 20%	A = 18% T = 32% C = 24% G = 26%	**A** 100%	**T** 100%	**G** 100%	A = 19% T = 20% C = 29% G = 32%	A = 24% T = 18% C = 26% G = 32%
C	T	**A**	**T**	**G**	A	C
0.21	0.32	**1.0**	**1.0**	**1.0**	0.19	0.26

$$P(\text{CT}\underline{\textbf{ATG}}\text{AC}|q_{\,start\_codon})$$
$$= 0.21 \times 0.32 \times 1.0 \times 1.0 \times 1.0 \times 0.19 \times 0.26$$
$$= 0.00331968$$

Figure 7.8 Computing the score of a putative signal CTATGAC using a WMM. Bases in the consensus region are shown in bold.

Given this definition we can then compute the probability of a putative signal $f = x_0 \ldots x_{m-1}$ for $m = L_M$ as:

$$P(f|M, L_M) = \prod_{i=0}^{L_M-1} P_M(i, x_i). \tag{7.17}$$

More generally, given an input sequence $S = x_0x_1 \ldots x_{|S|-1}$ to a gene finder, we can evaluate the conditional probability for a putative signal of class *type*(*M*) occupying positions $[i, i + L_M - 1]$ via:

$$P(x_i \ldots x_{i+L_M-1}|M, L_M) = \prod_{j=0}^{L_M-1} P_M(j, x_{i+j}) \tag{7.18}$$

or, in log space:

$$\log P(x_i \ldots x_{i+L_M-1}|M, L_M) = \sum_{j=0}^{L_M-1} \log P_M(j, x_{i+j}). \tag{7.19}$$

Because the weight matrix is a fixed-length model, we will generally omit the duration parameter L_M from the conditional probability term $P(f|M, L_M)$, with the understanding that $P(f|M, m) = 0$ for any feature f having length $m \neq L_M$.

7.3.2 Weight array matrices

The alert reader has no doubt noticed that the emission probabilities used in the weight matrices described above are of 0th order – i.e., they are assumed to be independent of the bases inhabiting other positions in and around the matrix. In order to take into account any dependences that may exist, we can instead condition each of the emission probabilities in the matrix on some number of previous bases. The resulting model is called a *weight array matrix*, or *WAM* (Zhang and Marr, 1993).

Formally, we define an nth-order weight array matrix M as a 4-tuple $M = (L_M, n, P_M, \alpha)$ for length $L_M \in \mathbb{N}$, order $n \in \mathbb{N}$, alphabet α, and emission function $P_M : \mathbb{N} \times \alpha \times \alpha^n \mapsto \mathbb{R}$ mapping position $j\,(0 \leq j < L_M)$ in the matrix window to a probability distribution $P(s\,|\,j, x_0 \ldots x_{n-1})$ over symbols s conditional on the preceding n bases $x_0 \ldots x_{n-1}$ as well as on the position j in the matrix.

The log-space computation of conditional probability for a feature $f = x_i \ldots x_{i+m-1}$ with $m = L_M$ is then given by:

$$\log P(x_i \ldots x_{i+L_M-1}|M, L_M) = \sum_{j=0}^{L_M-1} \log P_M(j, x_{i+j}, x_{i+j-n} \ldots x_{i+j-1}) \tag{7.20}$$

where the function $P_M(j, x_{i+j}, x_{i+j-n} \ldots x_{i+j-1})$ evaluates to $P(x_{i+j}|j, x_{i+j-n} \ldots x_{i+j-1})$, the probability of base x_{i+j} occurring at position j in the matrix, given the preceding n bases, $x_{i+j-n} \ldots x_{i+j-1}$.

Table 7.2 provides some accuracy results on a set of classification experiments comparing the ability of WMMs and WAMs of orders 1 through 3 to distinguish various signals (start/stop codons, donor sites, acceptor sites) from decoy sites having the same consensus sequence. Classification was performed using an empirically determined threshold based on likelihood ratios of models trained on positive and negative examples, as in the earlier example for content sensors. As can be seen from the table, the WAM models provide a slight advantage over the WMM, though for some of the signal types we see evidence of overtraining in the higher-order models.

Note that the conditional probabilities for positions near the beginning of the matrix are sometimes conditioned only on as many preceding bases as can fit within the matrix; thus, for positions $m < n$ we would condition on only m previous bases rather than n. This is generally only done as a matter of convenience during training when the training features have been given as short, fixed-length sequences with no surrounding context. Although restricting the model in this way might be expected to negatively impact prediction accuracy, for sensors with an adequate

Table 7.2 Classification accuracy of WMM and WAM models of orders 1–3 on the task of discriminating true signals from decoy signals

Model	Order	Stop codon	Start codon	Donor splice site	Acceptor splice site
WMM	0	0.758 495	0.675 649	0.927 073	0.805 862
WAM	1	0.767 597	0.691 617	0.936 064	0.815 633
WAM	2	0.756 675	0.698 603	0.935 065	0.819 881
WAM	3	0.746 966	0.694 611	0.934 565	0.819 031

number of explicitly modeled positions 5′ of the consensus region, the effect is likely negligible.

A convenient way of implementing WAMs in software is simply to use arrays of Markov chains, with each position in the matrix having a Markov chain associated with it for the purpose of computing the conditional probabilities at that position. Given an existing software module for training and evaluating Markov chains (or perhaps IMCs – section 7.2.5), and a module for training and evaluating WMMs, it is an easy task to combine the two to produce an efficient WAM implementation; we leave the details as an exercise for the reader (Exercise 7.20).

7.3.3 *Windowed weight array matrices*

A variant of the WAM has been devised which pools the n-gram frequencies from nearby locations when estimating the parameters at each position in the matrix:

$$P_e(g_n | j, g_0 \ldots g_{n-1}) = \frac{\sum_{k=-2}^{2} C(j+k, g_0 \ldots g_n)}{\sum_{k=-2}^{2} \sum_{s \in \alpha,} C(j+k, g_0 \ldots g_{n-1}s)} \tag{7.21}$$

where $C(i, g_0 \ldots g_n)$ denotes the number of times the $(n+1)$-gram $g_0 \ldots g_n$ was seen to occur ending at position i in the training features. The summation term $\sum_k$ effectively pools these counts over nearby positions. Such pooling is useful in cases where training data is very scarce, and will generally have the effect of increasing sample sizes, though with some loss of positional specificity. This type of model has been called a *windowed WAM*, or *WWAM* (Burge and Karlin, 1997). The indices $k \in [-2, 2]$ can obviously be taken over an arbitrary interval, though to our knowledge, the effect of using different interval sizes has not been systematically studied. Another possible generalization to this method would utilize nonuniform weightings for the pooled counts, and has also not (to our knowledge) been investigated.

7.3.4 *Local optimality criterion*

When identifying putative signals in DNA, we may choose to completely ignore low-scoring candidates in the vicinity of higher-scoring candidates. That is, when

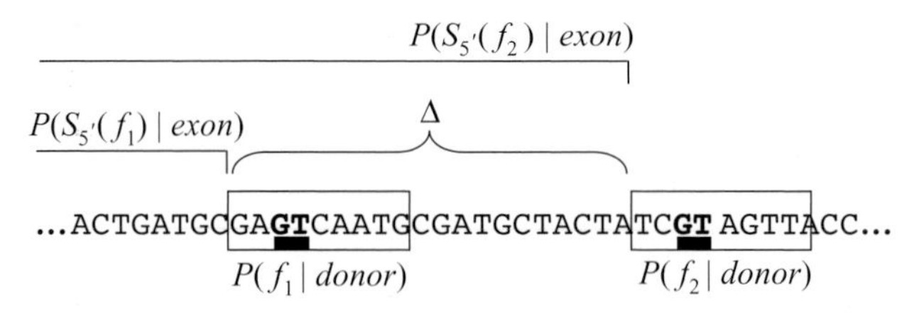

Figure 7.9 The rationale for imposing a local optimality criterion for signal sensing. The probability $P(S_{5'}(f_2)|exon)$ can be seen to correspond to a region only slightly larger than that of $P(S_{5'}(f_1)|exon)$. If the contribution Δ to the difference in these two terms varies, due to statistical fluctuation, more than the difference between $P(f_1|donor)$ and $P(f_2|donor)$, then the lower-scoring donor site may be inadvertently favored during gene prediction.

$P(g_i \ldots g_{i+m}|M) < P(g_j \ldots g_{j+m}|M)$ for two putative signals occurring at positions i and j in the input sequence, with $|i - j| < \varepsilon$ for some small ε defining the size of our locality, we may wish to set $P(g_i \ldots g_{i+m}|M) = 0$ in order to remove $g_i \ldots g_{i+m}$ from consideration as a putative signal. The rationale for such a policy is that in the context of the full gene finder, the difference between $P(g_i \ldots g_{i+m}|M)$ and $P(g_j \ldots g_{j+m}|M)$ may not be great enough to overcome any statistical fluctuations in the content scores of the (short) region lying between these two putative signals. Figure 7.9 illustrates the situation for the forward strand.

In the figure, two putative features f_1 and f_2 are found in close proximity by the donor site signal sensor, which assigns them scores of $P(f_1|donor)$ and $P(f_2|donor)$, respectively. If we consider these scores within the context of a complete gene finder, we may assume that in the highest-scoring parse including feature f_i, $i \in \{1, 2\}$, there is a putative exon ending at f_i and having score $P(S_{5'}(f_i)|exon)$ – where we denote the corresponding portion of the input sequence 5′ of feature f_i as $S_{5'}(f_i)$. Because f_1 and f_2 are in such close proximity, the difference between $P(S_{5'}(f_1)|exon)$ and $P(S_{5'}(f_2)|exon)$ will be determined by the coding potential of a very short region, which we have identified with Δ in the figure. Our fear is that since this region is very short, it may – purely by chance – appear to have a higher-than-average coding potential according to our content sensor, even if the region is not exonic; in some cases, the effect of this "misinformation" from our content sensor may be large enough to dominate the information coming from our signal sensor, resulting in a prediction error that could conceivably have been avoided if the signal sensor scores had been weighted more highly than the content sensor score.

The purpose of the local optimality criterion is to apply such a weighting in cases where two putative signals are very close together, with the chosen weight being 0 for the lower-scoring signal and 1 for the higher-scoring one. Such a rule is reasonable if we suspect that our signal sensors possess greater discriminative power at the granularity of short sequences than our content sensors. Variations on this approach could conceivably apply other weighting schemes besides this crude 0/1 rule. The local optimality criterion has been utilized by Brendel and

Kleffe (1998) and in the popular program *GeneSplicer* (Pertea *et al.*, 2001), which we describe in section 7.3.6. Nonbinary weightings of signals bordering exons have been utilized in the form of so-called "exon optimism" parameters (e.g., Majoros and Salzberg, 2004) which are incurred when a prospective parse passes through an exon. Other nonprobabilistic "fudge factors" have been employed in a number of probabilistic gene finders (e.g., "conservation score coefficient" in Gross and Brent, 2005; "coding bias" in Siepel and Haussler, 2004b), illustrating what some believe to be a fundamental shortcoming in current probabilistic models for gene finding (see section 12.4).

7.3.5 Coding–noncoding boundaries

A key observation regarding splice sites and start and stop codons is that all of these signals delimit the boundaries between coding and noncoding regions within genes (although the situation becomes more complex in the case of *alternative splicing*; see section 12.1). One might therefore consider weighting a signal score by some function of the scores produced by the coding and noncoding content sensors applied to the regions immediately 5′ and 3′ of the putative signal. As we saw in the discussion of the local optimality criterion, the use of weighting schemes for signal scores is not without merit, even if such weights are not easily justified on purely probabilistic grounds, so long as they are found to improve prediction accuracy in practice.

Such a weighting scheme is employed in the program *GeneSplicer* (Pertea *et al.*, 2001), which we describe in the next section. In this scheme, the score of a putative donor site is weighted by a pair of likelihood ratios:

$$\frac{P(S_{5'}(f)|coding)}{P(S_{5'}(f)|noncoding)}P(f|donor)\frac{P(S_{3'}(f)|noncoding)}{P(S_{3'}(f)|coding)} \tag{7.22}$$

whereas the score of a putative acceptor site is instead weighted via:

$$\frac{P(S_{5'}(f)|noncoding)}{P(S_{5'}(f)|coding)}P(f|acceptor)\frac{P(S_{3'}(f)|coding)}{P(S_{3'}(f)|noncoding)}. \tag{7.23}$$

In this way, the score of a putative donor site will be elevated in proportion to the extent that the region immediately 5′ of the putative site appears (to our content sensors) to be coding sequence and the region 3′ of it appears to be noncoding; for acceptor sites this pattern is obviously reversed. The same technique can be applied to start and stop codons.

One difficulty with this method is that the lengths of the upstream and downstream regions (or *margins*) $S_{5'}(f)$ and $S_{3'}(f)$ must be chosen carefully, since longer regions, while less susceptible to sampling error, may in practice extend beyond the coding or noncoding features that they are intended to model. For example, if the likelihood ratios for a putative donor site are measured over a 500 bp margin upstream and downstream of every putative signal, then if we encounter a true donor site at the end of a 100 bp exon and/or at the beginning of a 100 bp intron, the weighting prescribed by Eq. (7.22) is likely to do more harm than good. For this reason, one should take into consideration the typical lengths of coding and noncoding features in the organism of interest when applying such a weighting scheme. In the case of GeneSplicer, which was originally designed for *Arabidopsis thaliana* and *Homo sapiens*, the margin size was set to 80 bp.

7.3.6 *Case study: GeneSplicer*

Both the local optimality criterion and the coding–noncoding boundary detection methods just discussed are used in the program GeneSplicer (Pertea *et al.*, 2001), which provides predictions of splice sites in raw DNA. Although the program is not a complete gene finder, its predictions have been found to be quite useful both in manual annotation efforts and as inputs to a *combiner* program (see section 9.2).

GeneSplicer also utilizes the MDD procedure, introduced by Burge (1997) and described in the next section, to evaluate $P(f|donor)$ and $P(f|acceptor)$ for each putative donor and acceptor site, respectively. At the leaves of the MDD tree, GeneSplicer uses 1st-order weight array matrices (WAMs) to evaluate a 16 bp window surrounding each putative donor site, or a 29 bp window around each acceptor site. For the evaluation of the 5′ and 3′ margins, a coding content sensor and a noncoding content sensor were constructed by training two 2nd-order Markov chains on corresponding 80 bp regions on either side of known donor and acceptor sites in a training set. Finally, a local optimality criterion is applied to any pair of putative signals (of the same type – i.e., donor/acceptor) that are within N bp of each other (for some user-defined N) and on the same strand.

7.3.7 *Maximal dependence decomposition*

Given a sufficiently large set of training signals, it is reasonable to consider whether we might partition the training set according to some important characteristic of the training signals and to train separate sensors for each subset. During gene finding, we can then partition the set of putative signals of each type according to the same characteristics and evaluate those signals using the specific signal sensor trained on the corresponding partition of the training set. This is the idea behind the *maximal dependence decomposition*, or *MDD*, procedure introduced by Burge (1997) and employed in the popular gene finder GENSCAN (Burge and Karlin, 1997).

The MDD method is based on the notion of a *decision tree*, which we will consider in more detail in section 10.6. A decision tree is a special kind of tree in which each internal node denotes a decision to be made, and each leaf node denotes a solution. An MDD decision tree for human donor sites is shown in Figure 7.10(a); this is the tree used in the gene finder GENSCAN. The diamonds represent internal nodes labeled with boolean predicates; at the leaves of the tree are submodels M_i,

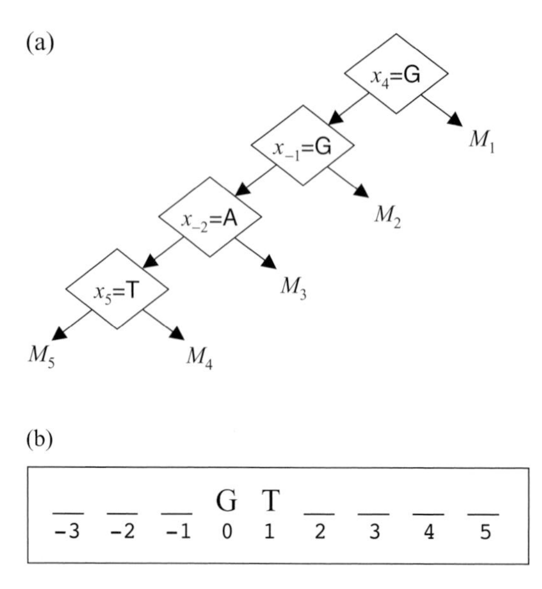

Figure 7.10 (a) The MDD decision tree used by GENSCAN to model human donor sites. A submodel M_i is selected by starting at the top node and descending through the tree according to the predicates at each node, going left when the predicate is true and right when the predicate is false. x_j denotes the base at position j relative to the consensus region of the putative donor site. (b) Indexing relative to the consensus region of a putative donor site in a 9 bp window.

which in GENSCAN are simple weight matrices. The reader may be interested to compare this tree to the DNA-gram in Figure 7.7.

The MDD tree operates as follows. Upon encountering a possible donor site $x_{-3}x_{-2}\ldots x_5$ (with $x_0x_1 = \text{GT}$), we begin at the topmost node in the tree and descend recursively to lower nodes according to the dictates of the predicates labeling each node. These predicates take the form $x_i = N$ for nucleotide N. When a predicate evaluates to *true* we descend to the left child of the current node and incur a probability score associated with the traversed branch; otherwise we descend to the right and incur the probability associated with the right branch. These branch probabilities are assessed during training by observing how many of the training sequences reaching the parent node then flow to the left versus to the right under the given predicate. Continuing with our donor site example, upon reaching a leaf node labeled with a submodel M_i, we then apply signal sensor M_i to the putative donor site; the score assigned by the MDD tree to this putative signal is then the score produced by M_i, weighted by the product of branch probabilities incurred on the path from the root of the tree to the selected leaf node.

For example, given the sequence ACG**GT**ACGC with the consensus bases GT located at the fourth and fifth positions, we can renumber the bases as shown in Figure 7.10(b) so that the consensus region begins at position 0; this is done merely for convenience of discussion, and need not be performed in software

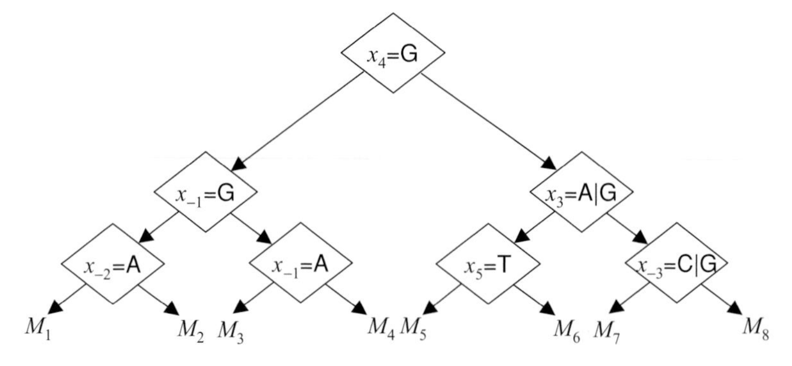

Figure 7.11 An MDD decision tree constructed by TIGRscan to model donor sites in *Arabidopsis thaliana*. A number of the predicates match those in the human MDD tree shown in Figure 7.9. Predicates of the form $x_i = Y|Z$ indicate logical disjunction (see text).

implementations. Now starting at the top of the tree, we test whether a G occurs at position 4 (i.e. $x_4 = \text{G}$; second position from the right in the sensor window). It does, and so we proceed to the left child, testing whether a G occurs at position -1, just 5′ of the consensus bases. Since it does, we proceed to the left again, testing this time whether an A occurs at position -2. It does not – a C occurs there instead – and so we proceed to the right child, which is a leaf node labeled M_3. At this point we can use the weight matrix M_3 to evaluate the putative signal, terminating the MDD search procedure. The score assigned to the putative donor site will be the score returned by M_3, weighted by the product of the scores associated with the three branches traversed on the way to the M_3 node. The next time we encounter a putative donor site we will start again from the top of the tree, potentially descending through a different path to a different leaf node.

MDD trees can be constructed for any signal type and for any organism for which we have adequate training data. An MDD tree for donor sites in *Arabidopsis thaliana* is shown in Figure 7.11. Comparing this tree to the previous one, we can see that all of the predicates occurring in the human tree also occur in the *A. thaliana* tree, presumably reflecting similarities in the biology of splicing in the two organisms.

Let us consider how to construct an MDD tree from a set of training signals. Let $T = \{S_i | 0 \leq i < |T|\}$ denote the training sequences, which we will assume are of the proper length L for the signal sensor (and we of course assume that their consensus regions all align properly with the consensus bases of the sensor window). For each position j in the sensor window we define $c_j \in \{\text{A, C, G, T}\}$ to be the most frequently occurring nucleotide at position j in T (breaking ties arbitrarily). Then for each sequence $S_i = x_{i,0} \ldots x_{i,L-1}$ in T and each position j in the sequence, $0 \leq j < L$, we can define a binary indicator variable $C_{ij} \in \{0, 1\}$ such that $C_{ij} = 1$ iff $x_{i,j} = c_j$. That is, $C_{ij} = 1$ if the base at position j in sequence S_i matches the consensus base for this position; otherwise $C_{ij} = 0$. Next, we can recode each sequence as

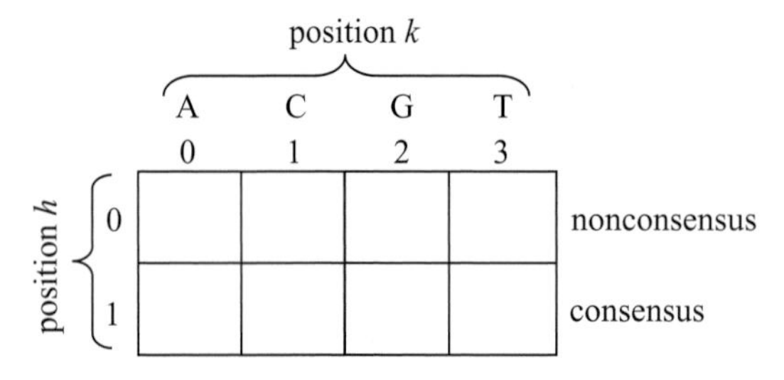

Figure 7.12 A contingency table for MDD. Rows correspond to whether position h matches the consensus for that position, and columns record the distribution of bases at position k. A χ^2 test applied to the table thus detects dependencies between the base at position k and the consensus/nonconsensus status of position h. A different table is constructed for every pair of positions in the signal sensor.

$S_i = X_{i,0} \ldots X_{i,L-1}$, where $X_{i,j} = \nabla(x_{i,j})$ for our usual DNA↔index mapping: $\nabla = \{\mathrm{A} \leftrightarrow 0, \mathrm{C} \leftrightarrow 1, \mathrm{G} \leftrightarrow 2, \mathrm{T} \leftrightarrow 3\}$.

Now for every pair of positions $h \neq k$ in the sensor window, we can form a 2×4 contingency table $R_{hk} = [o_{zb}]$ in which o_{zb} denotes the number of sequences S_i in the training data for which $C_{ih} = z$ and $X_{i,k} = b$, for all $z \in \{0, 1\}$ and all $b \in \{0, 1, 2, 3\}$, as illustrated in Figure 7.12.

Applying the χ^2 *test of independence* (section 2.9), we obtain a test statistic $\chi^2_{h,k}$ for each such pair of positions. Summing over all second positions k gives us $\delta_h = \sum_{0 \leq k < L, k \neq h} \chi^2_{h,k}$, which we can interpret as a measure of the dependence of all other positions k on the status (i.e., whether matching the consensus base or not) of position h.

We can build an MDD tree τ based on these dependence values δ_h, as follows. Starting with the set of all training signals $\gamma_I = T$, we identify the position h for which δ_h is maximal (i.e., $h = argmax_h\, \delta_h$), and we construct an internal node I and label it with the predicate $x_h = c_h$. Now we can partition γ_I into two sets, $\gamma_{left} = \{S_i \in \gamma_I | x_{i,h} = c_h\}$ and $\gamma_{right} = \{S_i \in \gamma_I | x_{i,h} \neq c_h\}$, and repeat this procedure recursively for each of these partitions, with the root of the subtree built from γ_{left} becoming the left child of node I and the root of the subtree built from γ_{right} becoming its right child. The branch probability $p_{I \to r}$ for the branch $I \to r$ connecting parent node I to child r is taken to be $p_{I \to r} = |\gamma_r|/|\gamma_I|$ – i.e., the proportion of training signals which traverse branch $I \to r$ (conditional on having reached I in the first place). The recursion continues until either all positions h have been tested by a predicate in the tree, none of the $\chi^2_{h,k}$ test statistics proves significant at the 0.05 level, or the set γ_r dwindles to a size deemed too small for further partitioning. At each point l in the incipient tree where the recursion terminates, we instantiate a leaf node and associate with it a weight matrix trained from the remaining training sequences in γ_l; it should be clear that these are precisely the subset of sequences in the original training set T that flow down to the current leaf node from the root of the tree as dictated by the predicates encountered in the internal nodes along the way. Thus, the weight matrix attached to a given leaf node constitutes a model specifically focused on the class of signals which all share the set of characteristics defining the path from the root to the leaf node.

Several things should be noted about this procedure. As originally defined in Burge and Karlin (1997), once we have instantiated a node with a predicate testing some position h in the sensor window, the procedure should not test that position again, since at all nodes beneath the original test of that position, the result of the test is already predetermined and need not be tested again. However, if we generalize the tests to be of the form $x_i \in \rho_k$ for some $\rho_k \subset \{\mathrm{A, C, G, T}\}$, $0 < |\rho_k| < 4$, then it makes sense to allow multiple tests for a given position, since the tests can utilize different subsets ρ_k.

Secondly, a number of options are available for terminating the recursion. In GENSCAN the minimum sample size of 175 is applied so that if all of the partitionings suggested by the alternative predicates at a given point in the recursion would result in a γ_{left} or γ_{right} subset with fewer than 175 elements, the recursion is terminated. Obviously, the minimum sample size of 175 may be modified in *de novo* implementations of MDD, if desired. Also, one might consider imposing a maximum tree depth, so that the recursion terminates as soon as the tree reaches a certain pre-specified depth.

Finally, we will see in the next section that an MDD tree can be interpreted as a form of *Bayesian network*, and that other methods for inferring Bayesian networks for signals in DNA have been applied in the context of eukaryotic gene finding. Bayesian networks are described more fully in section 10.4.

7.3.8 *Probabilistic tree models*

The MDD procedure described above allows us to detect and exploit dependencies between nonadjacent positions in a sensor window. Another technique which exploits nonadjacent dependencies is the *probabilistic tree model*, or *PTM* (Delcher *et al.*, 1999b; Cai *et al.*, 2000).

Let us consider first the dependency structure which is implicitly assumed when we use a 1st-order Markov chain M to evaluate the probability $P(S|M)$ of a sequence $S = x_0 \ldots x_{L-1}$. Recall that the *Markov assumption* (section 7.2.1) allows us to evaluate $P(S|M)$ via:

$$P(x_0 \ldots x_{L-1}|M) = P(x_0|M) \prod_{0<i<L} P(x_i|x_{i-1}, M). \tag{7.24}$$

In this way, we have conditioned the probability of each base in the sequence (other than the very first base) on the identity of the preceding base: $P(x_i|x_{i-1}, M)$. Writing out the terms of this product explicitly (and omitting the conditioning on the model, which is implicit) gives us $P(S) = P(x_0)P(x_1|x_0)P(x_2|x_1) \ldots P(x_{L-1}|x_{L-2})$.

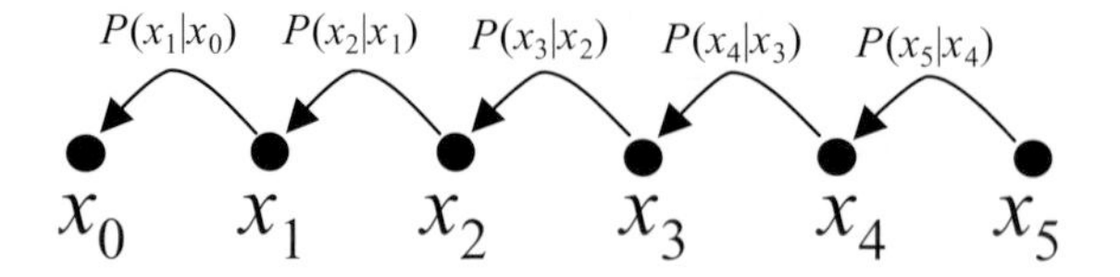

Figure 7.13 A dependency graph for a 1st-order Markov chain.

The first two terms can be combined using the conditional probability rule to give $P(x_0, x_1)$ so that the full expression becomes $P(x_0, x_1)P(x_2|x_1) \ldots P(x_{L-1}|x_{L-2})$. Now if we assume that x_2 is conditionally independent of x_0 – that is, that $P(x_2|x_1) = P(x_2|x_1, x_0)$ – then we can collapse the first two terms in this new expression via $P(x_0, x_1)P(x_2|x_1, x_0) = P(x_0, x_1, x_2)$ to give the new form of the full expression: $P(x_0, x_1, x_2)P(x_3|x_2) \ldots P(x_{L-1}|x_{L-2})$. Continuing in this way, we can collapse the entire expression to the joint probability $P(x_0, x_1, \ldots, x_{L-1})$, so long as we are willing to accept the *nth-order Markov assumption* for (in this case) $n = 1$: that each position in the sequence is conditionally independent of all earlier positions except the preceding n.

Applying the above argument in reverse provides a recipe for decomposing the joint probability $P(x_0x_1 \ldots x_{L-1})$ for any sequence into a product of conditional probabilities according to Eq. (7.24). These conditional probabilities $P(x_i|x_{i-1})$ can in turn be interpreted as edges in a dependency graph, such as the one illustrated in Figure 7.13.

A *dependency graph* is a directed acyclic graph in which each vertex represents a variable (such as x_i, the variable denoting the symbol at position i in the sequence) and each edge denotes a 1st-order dependency. The dependency structure of such a graph is valid insofar as we accept the conditional independence assumptions implied by the underlying undirected graph – i.e., that any variable x_i having neighbors Ω in the graph is conditionally independent of any other variable $x_k \notin \Omega$, given the values of all the members of $\Omega : P(x_i|\Omega, x_k) = P(x_i|\Omega)$.

We can thus consider other dependency graphs in which the strictly right-to-left Markov chain dependency structure is replaced with an arbitrary acyclic structure. Such a graph can be converted into a probabilistic statement relating the joint and conditional probabilities, merely by multiplying together the conditional probabilities implied by the graph:

$$P(x_0x_1 \ldots x_{L-1}) = \left(\prod_{\substack{\{i|\neg\exists \\ (i\to j)\in E\}}} P(x_i) \right) \left(\prod_{\substack{\{i|\exists \\ (i\to j)\in E\}}} P(x_i|succ(x_i)) \right) \tag{7.25}$$

for dependency graph $G=(V, E)$ with vertices $V = \{x_i|0 \leq i < L\}$ and edges $E \subseteq V \times V$, where $succ(x_i) = \{x_j|i \to j \in E\}$. If the dependency graph is a tree, Eq. (7.25) simplifies to:

$$P(x_0 x_1 \ldots x_{L-1}) = \left(\prod_{\substack{\{i \mid \neg\exists \\ (i \to j) \in E\}}} P(x_i) \right) \left(\prod_{(i \to j) \in E} P(x_i | x_j) \right). \quad (7.26)$$

Whether a given dependency graph accurately reflects the true and complete set of dependencies that exist between the variables is another question, but as we saw in the case of the MDD method, we can test for such dependencies in sample data by applying a statistical test, such as the χ^2 test utilized by MDD.

In the PTM approach to dependence decomposition, dependencies in training data are modeled using *mutual information* (section 2.10). We begin by constructing an undirected, complete graph $G = (V, E)$ in which $V = \{x_i | 0 \leq i < L\}$ for L the signal sensor window length and E the set of all possible edges between different vertices: $E = \{(i, j) | i \neq j\}$. We then weight each edge (i, j) in E with the negative of the mutual information $M(x_i, x_j)$ between variables x_i and x_j, where x_i denotes the nucleotide residing in position i for a given sequence. Using the MST() procedure from section 2.14 we can then extract a minimal spanning tree $\tau = (V, E_\tau)$ from G, where $E_\tau \subseteq E$. This undirected tree τ is then converted to a directed tree τ^* by arbitrarily designating x_0 as the root of the tree and orienting each edge to point toward the root (see Exercise 7.19 for the details of this algorithm). The resulting tree τ^* can now be interpreted as a 1st-order dependency graph, and applying Eq. (7.26) gives us a decomposition of the joint probability that can be used to evaluate $P(S|M)$ for the signal sensor.

The rationale of the PTM method is based on the assumption that by using the mutual information to measure dependence between sensor window positions and by choosing the spanning tree with the maximal mutual information, we will obtain a dependency graph which is a reasonably accurate model for the actual dependency structure of the bases surrounding the modeled signal. To the extent that this assumption is valid, the resulting conditional probability decomposition should provide a reasonably accurate estimate of the joint probability $P(S|M)$, which is precisely the goal of our signal sensor.

Dependency graphs are also referred to as *Bayesian networks* (see section 10.4) and have been applied to a number of other problems in machine learning (see, e.g., Chow and Liu, 1968; Pearl, 1991).

7.3.9 Case study: Signal sensing in GENSCAN

The popular eukaryotic gene finder *GENSCAN* (Burge and Karlin, 1997) has a total of six signal sensors, corresponding to donor sites, acceptor sites, start codons, stop codons, promoters, and polyadenylation signals. Although various other elements of GENSCAN were designed to be sensitive to the specific *isochore* (i.e., {G,C} density) from which the input sequence was taken, the signal sensors are not isochore-specific.

The modeling of donor sites using the MDD procedure has already been described. In GENSCAN, the models used at the leaves of the tree are weight matrices (WMMs) of length 9 bp, with the GT splice consensus located at offset 3 from the beginning of the matrix. MDD is not used for any of the other sensors in GENSCAN. Acceptor sites are modeled using a 23 bp WAM with the AG splice consensus located at offset 20 within the matrix. The 9 bp immediately 5′ of this matrix are modeled using a WWAM; this region is known as the *branch point* (see sections 1.1 and 11.3). Start and stop codons are modeled using simple weight matrices, the start codon being modeled by a 12 bp WMM with the consensus region at offset 6, and the stop codon being modeled by a 6 bp WMM with the consensus bases occurring at offset 0. Polyadenylation signals are modeled using a 6 bp WMM which includes the entire AATAAA consensus sequence. Splice site modeling in GENSCAN is described in some detail in Burge (1998).

Promoters are modeled in GENSCAN using a combination of two fixed-length models separated by a short region that is allowed to vary over a very limited range, so that the signal sensor, while technically not of fixed length, is very nearly so. The two main promoter features modeled by GENSCAN are the TATA-box and the CAP site (sections 1.1 and 11.3). The TATA-box is modeled by a 15 bp WMM, and the CAP site is modeled by an 8 bp WMM. The region between these sub-sensors is constrained to have a length between 14 and 20 bp, and the sequence in that space is modeled by a 0th-order Markov chain trained from noncoding sequence.

Exons of all types (initial, internal, final, and single) are modeled in GENSCAN by one of two 5th-order three-periodic Markov chains (3PMCs) – one trained for regions with 0–43% {G,C} density, and the other trained for regions with 43–100% {G,C} content. Note that these two models were trained from *all four* exon types; the states for all exon types in fact share the *same* Markov chain. All noncoding regions, including introns, 5′ and 3′ UTRs, and intergenic regions, are likewise modeled by one of two (homogeneous) 5th-order Markov chains, one for each of the two {G,C} densities (0–43%, 43–100%) as with the exon models.

7.4 Other methods of feature sensing

A number of other modeling techniques have been investigated for use in signal and content sensors. One which seems particularly promising at present is the use of predicted RNA *secondary structure* (e.g., Patterson *et al.*, 2002; Marashi *et al.*, 2006), which effectively models the two-dimensional structure of the folded mRNA molecule, flattened into the two-dimensional plane (Durbin *et al.*, 1998). Because the splicing process is a phenomenon involving single-stranded mRNA molecules, the structures that such molecules assume after being transcribed may certainly influence the efficiency of splicing and even whether a given splice site is recognized by the eukaryotic spliceosome in vivo (Buratti and Baralle, 2004). Although we will postpone a description of the RNA secondary structure prediction problem to section 12.2, we may remark that the efficient integration of RNA structure prediction into prevailing gene-parsing frameworks for the purpose of modifying the scoring functions of gene parsers is a problem that yet awaits thorough treatment in the literature. DNA (rather than RNA) secondary structure is also potentially of use by gene finders, since the local structure of the double-stranded DNA molecule may influence the potential for the transcriptional machinery to assemble at the appropriate place on the DNA substrate and to begin transcription of a gene. Beyond this we may also consider the state of the chromatin in which the DNA is packaged in its native state, and the so-called *histone code*, of which relatively little is yet known (Jenuwein and Allis, 2001). The successful incorporation of these higher-order phenomena into the prediction of one-dimensional gene structure awaits further research.

Several other mathematical techniques with which various gene researchers have dabbled include the use of *Fourier analysis* (Saeys, 2004), *maximum entropy* modeling (Yeo and Burge, 2004), *wavelet transforms* and *local Hölder exponents* (Kulkarni *et al.*, 2005), *estimation of distribution algorithms* (*EDA* – Saeys *et al.*, 2004), and *support vector machines* (*SVMs* – Sonnenburg, 2002; Zien *et al.*, 2000; see also section 10.14). These methods also await further investigation, and as of yet have evaded widespread adoption in the field, though SVMs in particular continue to gain momentum for various bioinformatic applications.

7.5 Case study: Bacterial gene finding

A number of the methods for feature sensing described in this chapter were originally introduced within the context of *prokaryotic* or *bacterial gene finding*. In contrast to the eukaryotic case (our primary focus in this book), the prokaryotic gene-finding problem is considerably simplified by the fact that prokaryotic genes do not contain introns, so that the problem effectively reduces to the accurate identification of start and stop codons for single-exon genes. Because these genomes also tend to be more densely packed with genes than in eukaryotes, the problem of false positives can be less severe, due to the existence of fewer pseudogenes. The problem is far from trivial, however, due to the occurrence of overlapping genes in these genomes (Lukashin and Borodovsky, 1998), and to the greater diversity of start codons, which in bacteria can have the consensuses ATG, GTG, and TTG (and very rarely CTG – Besemer *et al.*, 2001). The prokaryotic gene-finding problem is made slightly more complicated in the case of circular genomes, in which a gene may span the arbitrary "cutting point" of the circular molecule.

Early attempts at finding genes in prokaryotic DNA focused on the use of primitive content sensing methods such as *n*-mer statistics and simple Markov chains (e.g., Borodovsky *et al.*, 1994), though the use of IMMs were later introduced (Salzberg *et al.*, 1998a; Delcher *et al.*, 1999b). A common strategy for early prokaryotic gene finding was to use a *sliding window* approach, in which a fixed-length window was scored at each position along the sequence for *n*-mer statistics (with *n* typically ~5), and to then predict any ORF having an average window score greater than some predetermined threshold as a gene. This is somewhat reminiscent of the approach taken by our exon finder TOYSCAN_2 in Chapter 5, which scored ORFs by codon frequencies and applied an empirically determined threshold to select those ORFs to be predicted as exons.

The bacterial gene finder Glimmer (Salzberg *et al.*, 1998a) introduced the use of IMMs for content scoring, but still applied a length-normalized probability threshold to each ORF in deciding whether to predict the ORF as a gene. These early programs mostly predicted only nonoverlapping genes, and thus employed various overlap-resolution rules when two or more ORFs scoring above the threshold overlapped. As it became clearer that the phenomenon of overlapping genes was in fact relatively common in many prokaryotes, these overlap-resolution rules were generalized so as to allow gene predictions to overlap when the overlapping genes were all of sufficiently high confidence, according to the content sensor scores (e.g., Delcher *et al.*, 1999b).

The use of Viterbi decoding of HMMs for extracting a global parse of a DNA sequence was introduced by the programs *ECOPARSE* (Krogh *et al.*, 1994) and *GeneMark.hmm* (Lukashin and Borodovsky, 1998). While the former program did include special states for modeling overlapping genes and the latter program has been generalized to allow multiple parses in different reading frames (Shmatkov *et al.*, 1999), the use of HMM-based Viterbi parsing for prokaryotic gene finding has not been as universally embraced as in eukaryotic gene finding, presumably due to the phenomenon of overlapping genes. Instead, the ORF-based approach as exemplified by the earlier GeneMark and Glimmer systems seems to remain fairly dominant for prokaryotic gene-finding systems (though it should be noted that a recently released version of Glimmer does in fact utilize a Viterbi-like parsing framework, and it is conceivable that other systems may eventually follow suit). In section 12.1 we will see that in fact a similar situation arises in the case of *alternative splicing* in eukaryotic genes – a phenomenon now known to be much more common, at least in mammalian genomes, than was previously guessed.

Signal sensing via WMMs has also been utilized in prokaryotic systems, with the primary signals of interest being the start and stop codons and the *ribosomal binding site* (*RBS*; also called the *Shine–Delgarno sequence*). Whereas in eukaryotes the signals most commonly modeled upstream from the start codon are the core promoter elements such as *transcription factor binding sites* (see sections 11.3 and 12.3), in prokaryotes it is possible to model the actual location where the ribosome binds to the DNA before initiating transcription of a gene (e.g., Suzek *et al.*, 2001). The occurrence of one of these signals within roughly 20 bp upstream of a putative start codon can be used to strengthen our confidence in the putative start codon, and indeed much of the effort in improving prokaryotic gene finders has been directed at determining more accurately the correct start codon once a gene has been located. Whereas early approaches simply maximized the length of a predicted gene by choosing the most 5′ start codon in a reading frame, it has been shown that this is not an overly reliable heuristic, and that more sophisticated efforts at identifying the correct start codon (such as via the modeling of the RBS) can significantly improve accuracy (e.g., Lukashin and Borodovsky, 1998).

Other methods that were largely pioneered within the context of prokaryotic gene finding include the use of *posterior decoding* (e.g., Borodovsky *et al.*, 1994), the use of long ORFs as training genes (Salzberg *et al.*, 1998a; see also section 11.2), the use of bootstrapping (Besemer *et al.*, 2001; see section 11.2), and the use of homology information (Borodovsky *et al.*, 1994; see Chapter 9).

EXERCISES

7.1 Prove Eq. (7.7).

7.2 For each of the sequences given below, estimate the 0th-, 1st-, and 2nd-order Markov chain emission probabilities using Eq. (7.10).
  (a) AATCGGGATTAGCTACGTAGCTACTTATCGATCG
  (b) CGATGCGCGCGGCCCATTCTCTACTGTACCTTAT
  (c) GCTATCTCTTTTCTAGATCGACTCTATCATATTC
  (d) CGCGCTATATCGGCTATCGTATCTCTCTTTCATG
  (e) ATATATCGCGCTATGCTCCTCTCTAGTCTCATCT

7.3 Continuing with the previous exercise, compute the likelihood of each sequence (a)–(e) based on the 0th-, 1st-, and 2nd-order emission probabilities computed above.

7.4 Using your favorite programming language, implement a Markov chain module using a hash table (see Cormen *et al.*, 1992). You may reuse a hash table implementation obtained from elsewhere.

7.5 Using your favorite programming language, implement a Markov chain module using a transducer.

7.6 Implement two Markov chain modules, one using a hash table and the other using a transducer. Train and test them on the same data sets and measure the difference in execution time.

7.7 Draw a state transition diagram for a transducer having the 2nd-order DNA alphabet $\{A,C,G,T\}^2$ as its state labels, and with a transition function defined by Eq. (7.14).

7.8 Using the definitions given in section 3.3 regarding phase and frame, rewrite Eq. (7.16) to give the probability of a feature on the reverse strand. You may assume that ω is the phase of the leftmost base in the putative feature.

7.9 Use the WMM shown in Figure 7.6 to evaluate the probabilities of the following putative start codons:

(a) CCATGAC

(b) TGATGTA

(c) AAATGGC

(d) ACATGCT

(e) GATGCTT

7.10 Give an algorithm for evaluating a sequence using a weight matrix; use precise pseudocode as defined in section 2.4.

7.11 Implement a WMM evaluation module in your favorite programming language. Use the log transformation.

7.12 Compute the emission probabilities of a 1st-order WAM model for the sequences in Exercise 7.9. Use 0th-order probabilities for the first position.

7.13 Repeat the previous exercise using a 1st-order windowed WAM model (WWAM), averaging over a 3 bp window rather than a 5 bp window as prescribed in Eq. (7.21).

7.14 Download the GeneSplicer program over the internet, compile the program if necessary, and run it on a sequence of moderate length. Examine the results. How many signals of each type did it predict per kilobase? Repeat the experiment for a sequence from another organism and compare the density of signals (per kb). Report which organism appears to have a higher density of putative splice sites.

7.15 Obtain a set of known human gene sequences from GenBank. For each donor site in each gene, apply the decision tree shown in Figure 7.10(a) to classify each donor site into the set $\{M_1, M_2, M_3, M_4, M_5\}$. Report how often each model M_i was selected.

7.16 Flip a pair of pennies simultaneously, one in each hand, and record the results. Repeat this 10 times, tabulating the results in a 2×2 contingency table. Compute a χ^2 statistic to determine whether the results support the hypothesis of independence.

7.17 Explain why a dependency graph must be acyclic. Your answer should make use of the definition of joint probability.

7.18 For each of the following expressions, draw a dependency graph representing the conditional dependence structure implied by the expression. Assume there are six variables labeled x_0, x_1, x_2, x_3, x_4, x_5.

(a) $P(x_0)P(x_1|x_0)P(x_2)P(x_3|x_2)P(x_4)P(x_5|x_4)$

(b) $P(x_0|x_1)P(x_1|x_2)P(x_2|x_3)P(x_3|x_4)P(x_5)P(x_4|x_5)$

(c) $P(x_0|x_4)P(x_4|x_1)P(x_5|x_1)P(x_3|x_0)P(x_2|x_0)P(x_1)$

(d) $P(x_0|x_5)P(x_5|x_4)P(x_2|x_1)P(x_4|x_3)P(x_3)P(x_1|x_3)$

(e) $P(x_5|x_4)P(x_4|x_2)P(x_2|x_0)P(x_0|x_1)P(x_1|x_3)P(x_3)$

7.19 In section 7.3.8 it was stated that a minimum spanning tree τ could be converted to a directed tree τ^* by designating a root and orienting each edge to point toward the root. Design a formal algorithm for accomplishing this. Hint: consider performing a *depth-first search* (see Cormen *et al.*, 1992) starting at the root of the tree, and replacing each undirected edge (i, j) with a directed edge $i \to j$ after it is traversed.

7.20 Give formal algorithms for the training and evaluation of a WAM.

7.21 Explain why higher-order Markov chains suffer from sampling error more than lower-order chains. Is the effect a linear or a super-linear one? Explain your answer.

7.22 Explain why Figure 7.3 can be interpreted as part of a 6th-order HMM or as part of a transducer (i.e., a Moore machine) for a 5th(not 6th)-order Markov chain. Hint: consider the difference between a *Moore machine* and a *Mealy machine* as described in section 7.2.3.

7.23 Prove that $P(Z|Y, X) = P(Z|Y)$ implies $P(X|Y, Z) = P(X|Y)$. Interpret this result in the context of a 1st-order Markov chain.

8

Generalized hidden Markov models

In Chapter 6 we presented the hidden Markov model (HMM) and showed how this simple and elegant formalism can be used to model the statistical properties of genes. In the present chapter we will show how this formalism can be extended so as to admit more generalized gene finding approaches within one unifying framework. We will see that a number of state-of-the-art gene finders currently in wide use are based on this *generalized hidden Markov model* (GHMM) approach, and in succeeding chapters we will explore various ways of incorporating homology and phylogeny information into this framework and in the process delve into several of the most active areas of gene-finding research. In this way, it should be seen that the GHMM provides a very solid foundation for a broad array of computational gene-finding techniques.

8.1 Generalization and its advantages

We saw in Chapter 6 that an HMM could be defined (in abbreviated form) as a 4-tuple:

$$M = (Q, \alpha, P_t, P_e), \tag{8.1}$$

for states $Q = \{q_i | 0 \leq i < |Q|\}$, alphabet α, transition distribution P_t, and emission distribution P_e. The denoted machine operated by transitioning stochastically from state to state, emitting a single symbol upon entering each state. The key change that we wish to make in generalizing this class of models is to allow the machine to emit an arbitrary string of symbols upon entering each state, rather than only a single symbol. That is, P_e will denote a function $P_e : Q \times \alpha^* \mapsto \mathbb{R}$ mapping each state in Q to a probability distribution over strings of arbitrary length. To account for any bias that the machine may have toward strings of particular lengths, we will also introduce a state-specific *duration distribution* $P_d : Q \times \mathbb{N} \mapsto \mathbb{R}$, so that $P_d(d|q)$ will denote the probability that the next emission from state $q \in Q$ will be of length

d. The complete model, which we call a *generalized hidden Markov model* or *GHMM* (after Kulp *et al.*, 1996), is therefore specified by a 5-tuple:

$$M = (Q, \alpha, P_t, P_e, P_d), \tag{8.2}$$

where we have again opted to leave the silent start/stop state q^0 implicit.

Decoding with such a model involves the same type of optimization problem as with the simpler HMM – namely, finding the most probable path (or *parse*) ϕ^* through the states of the model, given the input sequence S:

$$\begin{aligned}\phi^* &= \underset{\phi}{argmax}\, P(\phi|S) \\ &= \underset{\phi}{argmax}\, \frac{P(\phi, S)}{P(S)} \\ &= \underset{\phi}{argmax}\, P(\phi, S) \\ &= \underset{\phi}{argmax}\, P(S|\phi)P(\phi).\end{aligned} \tag{8.3}$$

Note that on the third line of this derivation we have made use of the fact that over the $argmax_\phi$ the prior probability $P(S)$ is unchanging, since $P(S)$ is not conditional on ϕ, thereby allowing us to drop this term from the *argmax*. Next, we can decompose ϕ into a series of state $\times$ duration pairs $((q^0, 0), (y_1, d_1), \ldots, (y_n, d_n), (q^0, 0))$ for $y_i \in Q$, where a pair (y_i, d_i) denotes that the ith emitting state visited during the operation of the GHMM emitted a string of length d_i (i.e., the state experienced a *duration* d_i). The duration of the silent start/stop state q^0 is obviously constrained to be 0.

Given this decomposition of ϕ, we can then express $P(\phi)$ as:

$$P(\phi) = P_t(q^0|y_n) \prod_{i=1}^{n} P_t(y_i|y_{i-1}) P_d(d_i|y_i) \tag{8.4}$$

and we can express $P(S|\phi)$ as:

$$P(S|\phi) = \prod_{i=1}^{n} P_e(S_i|y_i, d_i) \tag{8.5}$$

where S_i is the subsequence of length d_i presumed to have been emitted by the ith state in ϕ, so that we have the concatenation $S = S_0 S_1 \ldots S_{|\phi|-1}$. Each S_i constitutes a putative feature, such as an exon or a splice site (except S_0 and $S_{|\phi|-1}$, which are both empty; i.e., $S_0 = S_{|\phi|-1} = \varepsilon$).

The full decoding problem now becomes:

$$\underset{\phi}{argmax}\, P_t(q^0|y_n) \prod_{i=1}^{n} P_t(y_i|y_{i-1}) P_d(d_i|y_i) P_e(S_i|y_i, d_i). \tag{8.6}$$

The term $P_t(y_i|y_{i-1})$ is handled in precisely the same way as in the pure HMM. The P_d and P_e terms (duration and emission) deserve some elaboration, since they encapsulate much of the flexibility provided by this framework.

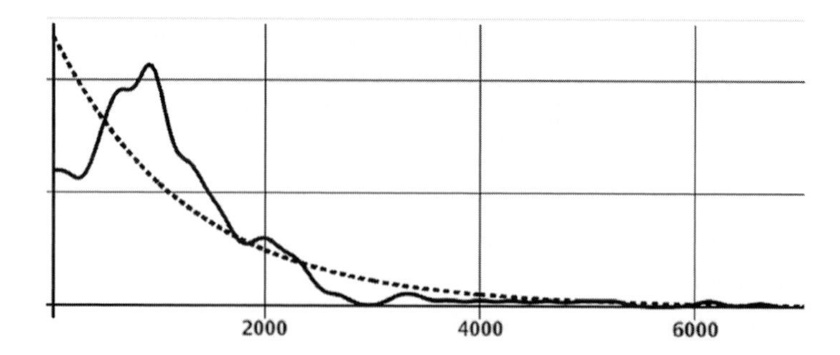

Figure 8.1 Exon length (bp) distribution for single-exon genes in the fungus *Aspergillus fumigatus* (solid line) and a geometric distribution (dashed line) with parameter 0.00079428. Whereas HMMs generally enforce a geometric distribution on feature lengths, with a GHMM we are free to use a histogram of observed feature lengths, for more accurate modeling.

To see how the use of an explicit duration term can be advantageous, consider the effect of modeling a feature such as an exon using a single state in an HMM, as we did in our model H_3 in section 6.4. Under such a model, the probability of a putative exon $x_0 \ldots x_{d-1}$ given the model parameters θ can be expressed as:

$$P(x_0 \ldots x_{d-1}|\theta) = \left(\prod_{i=0}^{d-1} P_e(x_i|\theta)\right) p^{d-1}(1-p), \tag{8.7}$$

where d is the length of the putative exon and p is the exon state's self-transition probability, $P(q_{exon}|q_{exon})$. The term $1-p$ denotes the probability of any transition other than the self-transition – that is, it is the probability that the exon ends immediately after the current base. We have thus factored the probability of the putative exon into a product of emission probabilities together with the transition probabilities $p^{d-1}(1-p)$, so that if this single-state exon submodel were to be utilized by a GHMM we would be compelled to treat $p^{d-1}(1-p)$ as the duration function P_d for the state.

The astute reader will recognize the term $p^{d-1}(1-p)$ from section 2.7 as describing a *geometric distribution* with probability of success $1-p$. Thus, regardless of the true characteristics of the exon length distribution for a given organism, a single-state HMM exon submodel will always model exon lengths as being geometrically distributed, and it can be shown that a similar statement can be made for typical multi-state HMM submodels for variable-length features as well (see Exercises 8.3 and 8.4). As illustrated in Figure 8.1, however, not all genomic features have geometrically distributed lengths.

Fortunately, we saw in the previous chapter that there is a simple way to implement Markov chains and other related content sensor models so as to condition the feature probabilities on the feature length, by simply omitting the transition probabilities from the computation. This provides us with $P_e(S_i|y_i, d_i)$, which is precisely what is called for in Eq. (8.6).

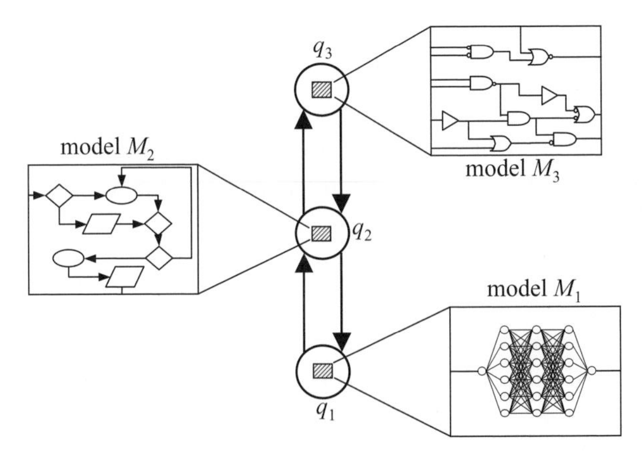

Figure 8.2 Submodel abstraction in a GHMM. Each state q_i in the GHMM has associated with it a submodel M_i which implements the state-specific P_e function for the state. Submodels may be implemented as HMMs, neural networks, or any of a large number of other algorithms, so long as they compute a value which can be interpreted as a probability.

In this way we are able to separate the duration probability from the rest of the submodel's computation and are then free to implement P_d as we see fit – in particular, we are free to use a P_d function that more accurately models the length distribution of the feature being modeled. Duration functions are usually evaluated in practice using an efficient histogram implementation or by interpolating from a curve which has been fit to the observed distribution (see section 2.7). We will see that for noncoding features, a geometric distribution is often a reasonable approximation to the observed length distribution, and indeed, we will make critical use of this fact in deriving efficient decoding algorithms for GHMMs.

Perhaps the greatest source of flexibility in GHMM modeling, however, is provided by the emission term $P_e(S_i|y_i, d_i)$ included in Eq. (8.6). Because we are free to implement this term using an arbitrary state-specific function $f_q(S, d)$, we are afforded the flexibility of modeling each feature type using any conceivable modeling technique, including, but not limited to: Markov models, *n*-gram frequency models, neural networks, decision trees, or any of the various machine-learning algorithms (see Chapter 10) which happen to be popular at the moment. In this way, the GHMM framework provides a form of *submodel abstraction* (Figure 8.2), with the different classes of submodel being treated generically as interchangeable components within a *metamodel* – very much like the metamodel construction presented in section 6.6.2 for pure HMMs.

In theory, the only requirement for a GHMM submodel is that it return values which can be interpreted as probabilities of the form $P(S_i|y_i, d_i)$, though we will see later that any function which can be interpreted as a likelihood ratio is acceptable as well. In practice, we would also like the function $f_q(S, d)$ to be efficiently computable, and in particular, we will favor those functions that can be factored into a product of d terms (i.e., with one term per nucleotide). All of the content sensors described in Chapter 7 satisfy this latter criterion, and so the use of any of these within a GHMM will satisfy our efficiency requirements. Models that violate this criterion may be used, but will result in a slower gene finder.

Submodel abstraction affords a number of significant advantages for gene finder development. Given a set of submodel types satisfying the above criteria, one can readily explore the effects on gene-finder accuracy of employing different combinations of submodel types for the states in the GHMM. Because the quality and quantity of the features within a training set can vary, one is free to apply different criteria and estimation parameters (such as a model's emission order) when training the submodels for different states of the GHMM. The interchangeable nature of submodels also brings with it all of the advantages of software reuse and modularity during the development of actual gene-finder software – e.g., division of labor, ease of debugging, reduction of system complexity, etc.

The utility of generalizing HMMs to allow variable state duration has been recognized for quite some time (e.g., Rabiner, 1989), with the result being variously referred to as *semi-Markov models*, *explicit duration Markov models*, or *generalized hidden Markov models*. The GHMM formalism was first applied to the problem of gene finding in *GENIE* (Kulp *et al.*, 1996), followed soon after by *GENSCAN* (Burge and Karlin, 1997). Since then, a fair number of GHMM-based gene finders have appeared, including *GENSCAN++* (Korf *et al.*, 2001), *Phat* (Cawley *et al.*, 2001), *AUGUSTUS* (Stanke and Waack, 2003), *Exonomy* (Majoros *et al.*, 2003), *SNAP* (Korf, 2004), *TIGRscan/GeneZilla* (Majoros *et al.*, 2004), and *GlimmerHMM* (Majoros *et al.*, 2004).

8.2 Typical model topologies

One advantage of the GHMM framework is that in practice GHMM-based systems tend to have fewer states than an equivalent HMM. Figure 8.3(a) shows a GHMM model topology which is functionally identical to the one used by the (original version of the) gene finder GENIE (Kulp *et al.*, 1996). In order to facilitate comparison with other model topologies, we have rendered the model slightly differently from the way it is depicted in (Kulp *et al.*, 1996). In particular, we have explicitly shown both the fixed-length states (the diamonds) and the variable-length states (the ovals).

As can be seen from Figure 8.3(a), there are separate states for each type of exon (initial, internal, final, and single). The model always begins in the 5′ UTR state and always ends in the 3′ UTR state, so that each execution of the model generates exactly one gene. As with the HMMs illustrated in Chapter 6, we have explicitly shown only the transitions having nonzero probabilities. The $\times 3$ label on the transition entering the donor state denotes the modeling of phase-specific introns. We have expanded upon this feature in Figure 8.3(b), in which it can be seen that there are three intron states and three internal exon states. In practice, one rarely

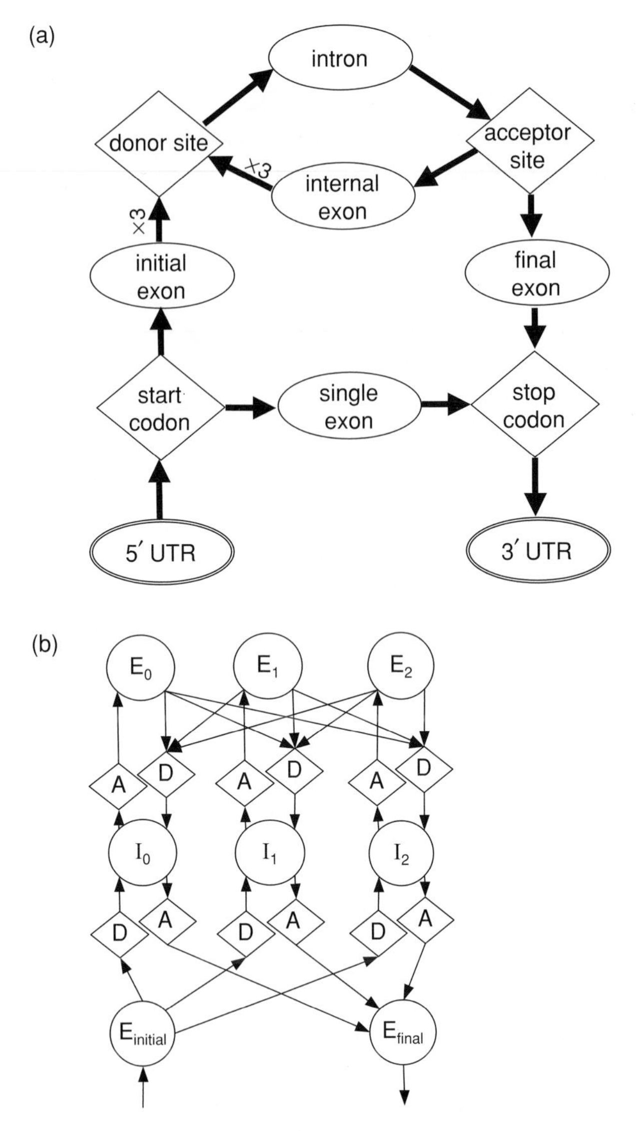

Figure 8.3 (a) The state topology used by the gene finder GENIE (Kulp *et al.*, 1996). Fixed-length states are shown as diamonds, variable-length states are shown as ovals, valid transitions are shown as arrows, and states having a transition to or from the silent initial/final state (not shown) have a double border. (b) Phase-specific introns, which were indicated by the ×3 transition label in (a). D, donor; A, acceptor; E, exon; and I, intron. Subscripts indicate the phase of the 5′ end of a feature.

implements these as separate submodels, however, so that the number of actual submodels in a typical GHMM tends to be quite small. The purpose of showing the intron states in triplicate is to denote the fact that the phase is tracked across introns, and that the transition probabilities entering the introns of different phases can be different – a practice that does seem to mirror the reality of intron phase distributions in at least some eukaryotic genomes.

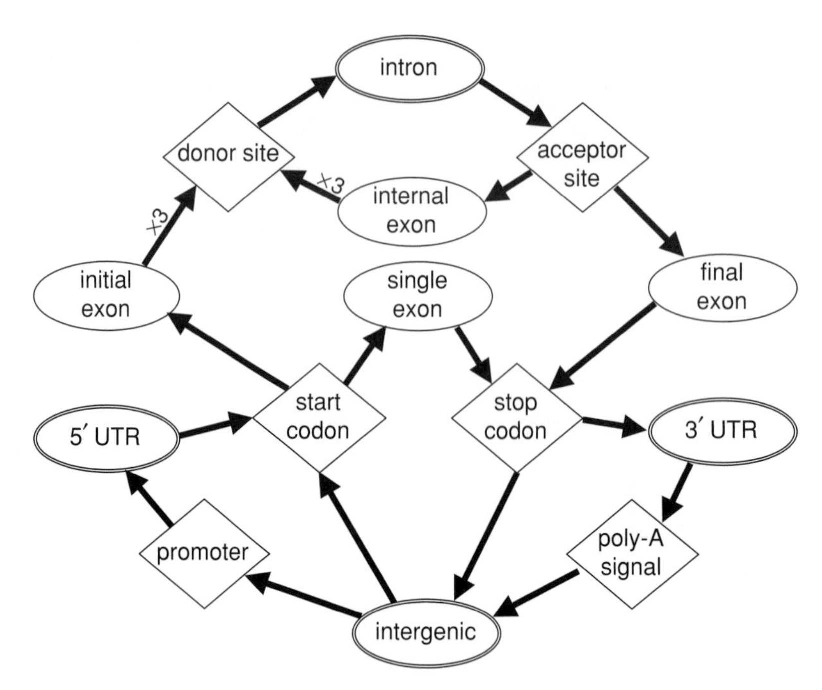

Figure 8.4 The state topology used by the gene finder GENSCAN (Burge and Karlin, 1997). The reverse-strand states are omitted for clarity. Fixed-length states are shown as diamonds, variable-length states are shown as ovals, transitions are shown as arrows, and states having a transition to or from the silent initial/final state (not shown) have a double border. Phase-specific introns are indicated by the ×3 label on the transitions entering the donor state.

Regarding the tracking of phase across introns, the reader will recall that we did in fact utilize three separate intron submodels for this very purpose in our pure HMM model H_{27} in section 6.4. This was necessary in the case of an HMM, because there is no other way in the traditional HMM formalism to condition specific transition probabilities on events occurring arbitrarily far back into the past. As we will see, in the GHMM framework we may adopt a less theoretically "pure" but more practical approach in which higher-level processing at the whole-system level is available for keeping track of events and quantities such as phase. Another example of such higher-level processing is the detection and avoidance of in-frame stop codons straddling an intron, which as we remarked in Chapter 6 is somewhat inconvenient to implement for a pure HMM; this problem is simpler to handle in the GHMM setting (see section 8.3.1; Exercise 8.22).

In Figure 8.4 we show the model topology of the popular gene finder GENSCAN (Burge and Karlin, 1997); this model is slightly more complicated than the one shown in Figure 8.3. The key differences are: (1) the GENSCAN model can generate any number of genes, since it includes an intergenic state which can be re-entered arbitrarily many times, (2) the model can generate partial genes by starting and/or ending in an intron state (as indicated by the states with a double border in the state diagram), and (3) the model can generate UTR delimited by a promoter signal in the case of 5′ UTR or a polyadenylation signal in the case of 3′ UTR.

In both of the previous two figures it should be noted that only the forward-strand models are depicted. In practice, one can create a mirror image of the forward-strand model to obtain a suitable reverse-strand model (see section 8.2.2). The forward- and reverse-strand models can then be merged through a common intergenic state to obtain a single model capable of generating genes on both strands.

In section 11.3 we will consider the utility of incorporating states for novel features such as *CpG islands*, *branch points*, and *signal peptides*.

8.2.1 One exon model or four?

It was remarked previously that duplicated states in a GHMM topology need not be implemented by distinct submodels in the actual gene finder. Indeed, it is a (poorly known) fact that even though many published accounts of gene finders include diagrams showing four exon states (corresponding to initial, internal, final, and single exons), most of these gene finders actually utilize only one exon model – that is, they populate four exon states with only one set of emission parameters. We alluded to this practice in Chapter 6 when we duplicated our intron model into three identical models corresponding to the three phases at which those introns could splice into a CDS. In that case we *tied* the corresponding parameters of the duplicate models by *pooling* the training data for the three intron classes into a single training set for all three models.

This practice is generally motivated either by a lack of training data (such that partitioning the data into smaller subsets would only exacerbate the problem of sampling error) or by a belief that the statistical properties of the three feature variants (i.e., the three phases of introns) are not significantly different. When training data are plentiful, however, it is reasonable to consider whether any gain in prediction accuracy might be obtained by using distinct sets of parameters for duplicate states – that is, by *not* pooling the training data.

Table 8.1 shows the results of using four distinct sets of parameters to model the four exon types (initial, internal, final, single) in the fungus *Aspergillus fumigatus*. It can be seen that the pooled model performed somewhat better than the model with four distinct exon submodels, even though the training set consisted of over 9000 genes – many more than are typically available for training a gene finder in practice. Thus, while it is conceivable that the practice of using distinct exon submodels may improve prediction accuracy on some organism, we have yet to see any quantifiable advantage in any of the cases which we have investigated.

Table 8.1 The advantages of pooling data to improve sample size. The "pooled" model utilized only one exon submodel, trained on all exon types (initial, internal, final, single) and used in all four exon states. The "separate" model utilized distinct submodels for the four exon types. Training data consisted of 9368 *Aspergillus fumigatus* genes; the test set comprised 360 held-out genes

	Nucleotides			Splice sites		Start/stop codons		Exons			Genes
	Sn	Sp	F	Sn	Sp	Sn	Sp	Sn	Sp	F	Sn
Pooled	93	94	94	51	58	49	54	39	44	41	24
Separate	92	94	93	50	56	47	52	38	42	40	20

8.2.2 One strand or two?

Although the models depicted in the foregoing sections have explicitly attended only to forward-strand features, one would clearly like to find genes on whichever strand they inhabit. We remarked earlier that a double-stranded model can be obtained from a single-stranded model by forming the mirror image of the forward-strand model and merging the resulting reverse-strand model with the forward-strand model through a common intergenic state. In this way, the model can transition at will between genes on either strand by passing through the intergenic state.

Actually obtaining the reverse-strand model $M_{reverse}$ from the forward-strand model $M_{forward}$ can be done in the following way. For each state q_i^f in $M_{forward}$ we obtain a reverse-complementary state q_i^r for $M_{reverse}$ by training the submodel for q_i^r from the reverse-complements of the sequences used to train q_i^f. Similarly, the transition probabilities for the reverse-strand model can be estimated by reversing the order of all signals in each training gene and estimating the transition frequencies from the corresponding state sequences, as in a standard HMM. State duration distributions are the same for the two models. Finally, the intergenic state q_N^r in $M_{reverse}$ is removed and all transitions into and out of q_N^r are replaced with equivalent transitions into and out of q_N^f, the intergenic state present in the forward-strand model, with the resulting set of outgoing transitions from q_N^f being renormalized so their probabilities sum to 1.

Two-stranded models obviously have many more states than single-stranded models, and in the case of *PSA decoding* (section 8.3.1), they can require significantly more memory during gene finding. Two-stranded models also require additional software development (especially debugging) and training effort, thereby making them somewhat less convenient to use in practice. Thus, it is reasonable to consider employing a single-stranded model for two-stranded gene prediction by running a forward-strand model over both the input sequence S and its reverse-complement S', and then applying some post-processing routine to disambiguate between any conflicting predictions – i.e., cases in which a predicted gene on one strand overlaps a predicted gene on the other strand. The advantage of using a two-stranded model is that such disambiguation is achieved automatically by the GHMM formalism, since the model cannot assume multiple states simultaneously. However, this "advantage" can work as a disadvantage in cases of true overlapping genes.

For this reason, several gene finders including *SNAP* (Korf, 2004) and *AUGUSTUS* (Stanke and Waack, 2003) provide the option to perform independent single-stranded prediction on both strands, so that overlapping genes may be predicted.

8.3 Decoding with a GHMM

It was seen earlier that the decoding problem for GHMMs can be formulated as the following MAP optimization problem:

$$\begin{aligned}\phi^* &= \underset{\phi}{argmax}\ P(\phi|S)\\ &= \underset{\phi}{argmax}\ P_t(q^0|y_n)\prod_{i=1}^{n} P_t(y_i|y_{i-1})P_d(d_i|y_i)P_e(S_i|y_i, d_i).\end{aligned} \qquad (8.8)$$

From the first equation, we can see that this is analogous to the computation which is carried out by the Viterbi algorithm for standard HMMs (section 6.2.1), and indeed, the solution to this problem for GHMMs has often been described as a "generalized Viterbi algorithm" (e.g., Choo *et al.*, 2004).

The main difficulty in formulating a Viterbi-like solution to Eq. (8.8) is in dealing efficiently with the variable-length nature of the state emissions. Recall from section 6.2.1 that in the standard Viterbi algorithm for HMMs, the updating of an individual cell in the dynamic programming matrix can be performed by selecting the best predecessor cell from among those in the preceding column. Thus, for a sequence S and an HMM with state set Q, each cell requires $O(|Q|)$ operations, so that for the entire matrix, which is of size $|Q| \times |S|$, the run time becomes $O(|Q|^2 \times |S|)$. For a GHMM, since each variable-length state can emit a string of arbitrary length, it is not sufficient to consult only the cells in the preceding column when computing the value for a new cell; we are instead forced to consider linking arbitrarily far back in the matrix as we consider emissions of arbitrary length (see Exercise 8.7), so that the run-time increases to $O(|Q|^2 \times |S|^3)$ – cubic in the sequence length for a fixed HMM.

Fortunately, by making several very reasonable assumptions and observations, we can reduce the time complexity for GHMMs with *sparse* state topologies (i.e., with all in-degrees bounded by a small, fixed constant) to $O(|Q| \times |S|)$. The three key assumptions are that:

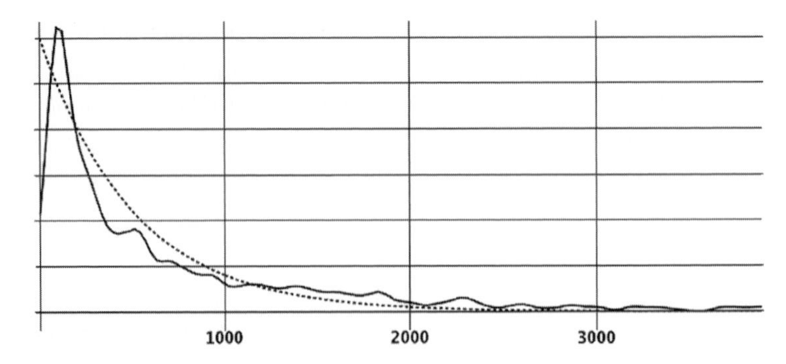

Figure 8.5 Intron length (bp) distribution for *Homo sapiens* (solid line) and a geometric distribution (dashed line) with parameter $p = 0.002$.

(1) The emission functions P_e for variable-length features are *factorable* in the sense that they can be expressed as a product of terms evaluated at each position in a putative feature – i.e.,

$$factorable(P_e) \Leftrightarrow \exists_f P_e(S) = \prod_i f(S, i)$$

for $0 \leq i < |S|$ over any putative feature S.

(2) The lengths of noncoding features in genomes are geometrically distributed.

(3) The model contains no transitions between variable-length states.

The last assumption simplifies the decoding problem considerably by reducing the granularity of the dynamic programming problem to the level of putative signals (i.e., splice sites, etc.) rather than individual nucleotides, as we describe more fully below.

Regarding the first assumption, we have already remarked (in section 8.1) that this assumption is valid for all of the standard content sensors described in Chapter 7, including Markov chains and their most popular variants. For content sensors not satisfying this criterion, the run-time can be expected to grow to a miserable $O(|Q| \times |S|^2)$ for sparse GHMM topologies.

Regarding the second assumption, that noncoding feature lengths are geometrically distributed, we offer Figure 8.5 as an anecdotal observation that for at least some noncoding features in some organisms, the assumption is not a terribly unreasonable one. In section 8.4.2 we will explore methods for relaxing this assumption.

A key observation in bounding the time complexity of GHMM decoding relates to the expected number of eligible predecessor cells which one needs to consider when linking back from a cell in the dynamic programming matrix – i.e., when establishing *traceback* pointers (section 6.2.1) for later reconstruction of the optimal parse. We will see, in the descriptions of the two decoding algorithms that follow, that the granularity of the linking operation is reduced from the level of individual nucleotides in the case of a pure HMM to the level of putative splice sites and other signals in the case of the GHMM.

Recall from Chapter 3 that the gene prediction problem can be formulated as one of choosing the optimal path through a directed acyclic graph (an *ORF graph*) in which each vertex is a putative signal and each edge is a variable-length feature such as an exon, intron, or intergenic region. It will be seen shortly that in both of the algorithms that follow, the computation can be conceptualized as proceeding from left to right along the vertices of the underlying ORF graph, where at each vertex we choose the optimal predecessor from among those preceding signals having an edge in common with the current signal. We will now argue that for typical eukaryotic genomes, the underlying ORF graph (or at least that portion of it which needs to be examined during standard decoding) tends to be locally sparse, so that each signal tends to have at most some fixed number N_{pred} of valid predecessors, where N_{pred} is determined by the average nucleotide frequencies in the genome, and is insensitive to the length of the input sequence.

Let us first classify all putative signals into the following sets:

$$\begin{aligned}\Gamma_{bc} &= \{\text{signals that may begin a coding feature}\}\\ \Gamma_{ec} &= \{\text{signals that may end a coding feature}\}\\ \Gamma_{bn} &= \{\text{signals that may begin a noncoding feature}\}\\ \Gamma_{en} &= \{\text{signals that may end a noncoding feature}\}.\end{aligned}$$

For example, putative donor sites would be members of both Γ_{ec} and Γ_{bn}, start codons would be members of Γ_{en} and Γ_{bc}, and promoters would be members of Γ_{en} and Γ_{bn}.

Consider first any signal in Γ_{ec}. Although it might at first seem that a predecessor to such a signal could occur arbitrarily far back in the sequence, in practice the accumulation of randomly occurring stop codons in the intervening sequence increasingly favors the termination of all three reading frames as the number of such randomly occurring stop codons increases (except of course for true exons, of which there are generally a much smaller number than the number of potential coding edges, so that their impact on the average time complexity is negligible). Thus, the further back we look for valid *coding predecessors* (i.e., signals in Γ_{bc}), the less likely we are to find any as all three reading frames become increasingly likely to be terminated by randomly occurring stop codons.

One could formalize this argument by showing that the expected number of open reading frames steadily decreases as a function of sequence length, so that we could say that with 95% confidence, no ORF longer than some fixed L_{max} should be expected to occur in a particular stretch of random DNA (as an aside, this argument can be used to identify likely exons for training purposes, and indeed, is used by the bacterial gene finder Glimmer in identifying likely training genes;

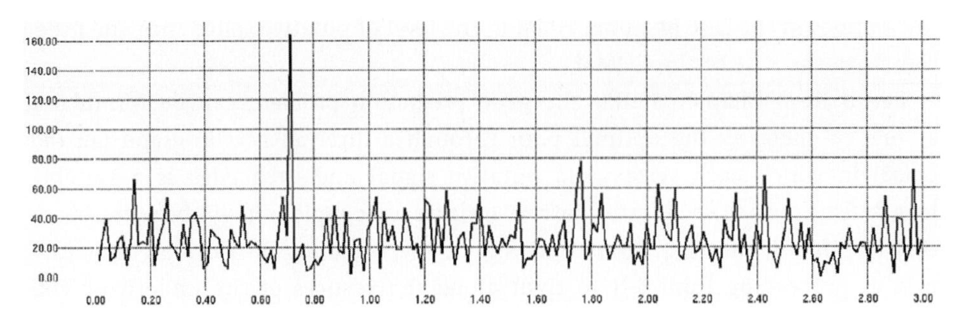

Figure 8.6 Number of potential coding predecessors (Γ_{bc}) as a function of sequence length (in megabases). Data are from a 3 Mb human DNA sequence. Linear regression resulted in a slope of −0.000 001, demonstrating that the number of coding predecessors does not increase without bound; the intercept indicated 27.3 predecessors on average.

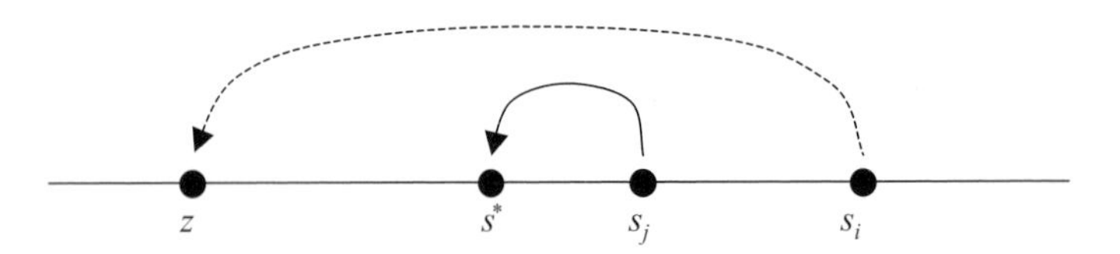

Figure 8.7 Once an optimal noncoding predecessor s^* is identified for signal s_j, no signal z preceding s^* can be selected as a predecessor of a later signal s_i, assuming a geometric length distribution and a factorable content scoring function.

see Exercises 8.20 and 8.21; also, sections 7.5 and 11.2). Rather than laboring the point, we will trust that the foregoing suffices to convince the reader that for genomes not especially lacking in random stop codons, the expected number of potential coding predecessors for any given signal in Γ_{ec} can be effectively limited to some constant N_{pred} as dictated by the statistical properties of the particular genome under consideration. Note that we are *not* advocating that N_{pred} actually be determined and then artificially imposed on the predecessor selection mechanism; rather, the above argument merely states that the *expected* number of predecessors which *will* be examined does not increase asymptotically with the sequence length, thereby limiting the time complexity of the decoding algorithm. Figure 8.6 provides empirical support for this claim.

The case of noncoding predecessors (i.e., signals in Γ_{bn}) can be handled using a different argument, once we have accepted assumptions (1) and (2) given previously. In particular, we would like to argue that at most one member of Γ_{bn} of each type (and phase, in the case of introns) need be considered as potential predecessors during the dynamic programming computation for any given signal in Γ_{en}. We formalize this in the following theorem, attributable to Burge (1997). The reader may wish to refer to Figure 8.7 for clarification.

Theorem 8.1 *(Burge's noncoding predecessor theorem) Suppose that for some signal $s_j \in \Gamma_{en}$, the optimal noncoding predecessor was found to be some $s^* \in \Gamma_{bn}$. If the next signal encountered after s_j is some $s_i \in \Gamma_{en}$, then under assumptions (1) and (2) above, s^* is also the optimal predecessor (of its type and phase) for s_i.*

Proof: From Eq. (8.8) we can infer that the dynamic programming score for a signal s_i should be given by:

$$V(s_i; s^*) = P_t(s_i|s^*)P_d(d_i|s_i)P_e(S_i|s_i, d_i)V(s^*; pred(s^*))$$

for an optimal predecessor s^*, where S_i is the subsequence stretching from s^* to s_i, d_i is the length of S_i, and *pred*(s^*) is the optimal predecessor of s^* in the appropriate phase. Let us suppose that for some signal $s_j \in \Gamma_{en}$, the optimal noncoding predecessor was found to be some $s^* \in \Gamma_{bn}$. Then if the next signal encountered after s_j is $s_i \in \Gamma_{en}$, then given assumptions (1) and (2) above, s^* is also the optimal predecessor (of its type and phase) for s_i. Assume by way of contradiction that this is not so – i.e., that there is some signal z for which $V(s_i; z) > V(s_i; s^*)$. Because s_i was assumed to be the next putative signal following s_j, z must occur before s_j, so that z was available as a potential predecessor of s_j. Because s^* was chosen rather than z as s_j's predecessor, we know that $V(s_j; s^*) > V(s_j; z)$. But we know that $V(s_i, z) = V(s_j; z)p^{\Delta}P_e(S_{j:i}|s_i, \Delta)$ where p is the parameter to the geometric length distribution for the noncoding feature ending at s_i, $S_{j:i}$ is the subsequence accumulated between signals s_j and s_i, and $\Delta = |S_{j:i}|$ is the length accumulated since s_j (including the bases comprising s_j but not s_i). Likewise, $V(s_i; s^*) = V(s_j; s^*)p^{\Delta}P_e(S_{j:i}|s_i, \Delta)$. Thus, we have $V(s_j; s^*) > V(s_j; z) \rightarrow V(s_i; s^*) > V(s_i; z)$, which contradicts our supposition $V(s_i; z) > V(s_i; s^*)$, showing that s^* must be the optimal predecessor (of its type and phase) for s_i as well.

Restating the above informally, upon reaching s_i, the attractiveness of z as a predecessor cannot have improved in relation to s^* since s_j, because the dynamic programming scores of partial parses passing through z and s^* have changed at the same rate – our assumption of a geometric distribution for the length of noncoding features ensures that the P_d terms change at the same rate (p^{Δ}), and our assumption that the emission functions are factorable ensures that the P_e terms change at the same rate also, so long as the P_e terms are evaluated by the same content sensors, which they are, since z and s^* were assumed to be of the same type. ❑

Thus, for each type of noncoding predecessor, only the single optimal predecessor in each phase need be remembered as we perform a decoding pass over the sequence. Newly encountered signals of the same type may displace the current optimal predecessor, but no other predecessor z occurring earlier in the sequence than

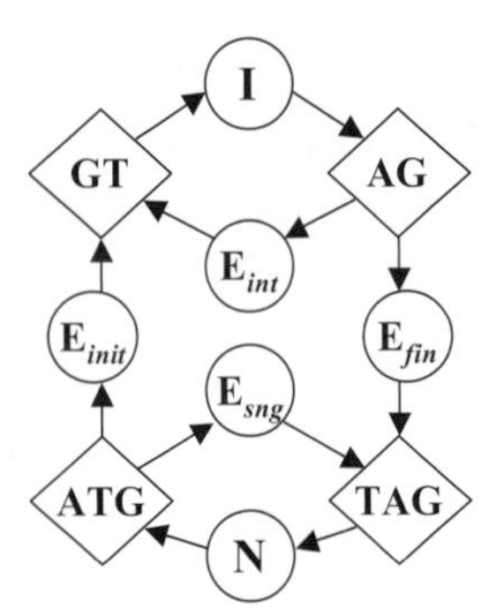

Figure 8.8 A simple GHMM topology to facilitate discussion of decoding. I, intron; N, intergenic; GT, donor site; AG, acceptor site; ATG, start codon; TAG, any stop codon; E_{int}, internal exon; E_{sng}, single exon; E_{init}, initial exon; E_{fin}, final exon.

the current optimal predecessor s^* may do so, and therefore, such z need not be remembered during decoding.

8.3.1 *PSA decoding*

We now present the details of the most commonly used decoding algorithm for GHMM-based gene finders. Although the time complexity of this algorithm is quite satisfactory, as characterized above, the space complexity is less than ideal, as the algorithm requires the storage of $O(|Q| \times |S|)$ scores in memory, which for long sequences can become quite large. In the we will describe an algorithm with the same time complexity but greatly reduced memory requirements on the order of $O(|Q| + |S|)$, where the $|S|$ term has a very small constant factor.

To facilitate the current discussion, let us for the time being assume the simple GHMM topology shown in Figure 8.8.

The strategy which is commonly employed for decoding with a GHMM is to allocate several arrays, one per variable-length feature state, and to evaluate the arrays left-to-right along the length of the input sequence, according to a dynamic programming algorithm, which we will detail below. We refer to this approach as the *prefix sum arrays (PSA)* approach, since the values in the aforementioned arrays represent cumulative scores for prefixes of the sequence. We will see that using such a data structure allows us to very quickly evaluate an arbitrary interval of the input sequence by performing a subtraction between the array elements at either end of the interval.

The first step of the PSA algorithm is to compute a *prefix sum array* α for each content sensor. For noncoding states such as introns, UTRs, and intergenic regions, this initialization can be performed as shown below:

$$\underset{0 \le i < |S|}{\forall} \alpha[i] = \begin{cases} \alpha[i-1] + \log M(S, i) & \text{if } i > 0 \\ \log M(S, i) & \text{if } i = 0 \end{cases} \tag{8.9}$$

where $M(S, i)$ denotes the probability returned by the model M associated with the noncoding state, for the ith nucleotide in sequence S.

In the case of exon states, it is important to distinguish between the three different reading frames, since a priori we do not know, for any putative exon, in which phase the exon begins. Thus, we allocate three separate prefix sum arrays α_ω, for $\omega \in \{0, 1, 2\}$, to represent the three reading frames. Given a factorable content sensor M for coding features, we can imagine separating the sensor into three distinct models, M_0, M_1, and M_2, corresponding to the three exon phases, where we denote by $M_\omega(S, i)$ the probability, according to model M_ω, of the ith nucleotide in sequence S, given that the nucleotide occurs in phase ω. Initialization of the prefix sum arrays for a forward-strand coding state with associated content sensor model M can thus be achieved via:

$$\underset{\substack{0 \le i < |S|, \\ 0 \le \omega < 3}}{\forall} \alpha_\omega[i] = \begin{cases} \alpha_\omega[i-1] + \log M_{(\omega+i) \bmod 3}(S, i) & \text{if } i > 0 \\ \log M_\omega(S, i) & \text{if } i = 0\,. \end{cases} \tag{8.10}$$

It should be apparent from this definition that the three arrays α_ω will be phase-shifted from one another, with each element in an array storing the cumulative score of the prefix up to and including the current nucleotide. The first nucleotide of array α_ω is taken to be in phase ω.

Initializing the arrays for reverse-strand states can be achieved by simply reverse-complementing the DNA sequence, applying Eq. (8.9) or (8.10) as appropriate, and then reversing the order of the resulting arrays (keeping in mind later that the reverse-strand arrays tabulate their sums from the right, rather than the left, and that ω is the phase of the last array entry rather than the first).

Once the prefix sum arrays have been initialized for all variable-duration states, we make another left-to-right pass over the input sequence to look for all possible matches to the fixed-length states, via the signal sensors. Recall from section 7.3 that a typical signal sensor θ models the statistical biases of nucleotides at fixed positions surrounding a signal of a given type, such as a start codon. Whenever an appropriate consensus is encountered (such as ATG for the start codon sensor), the signal sensor's fixed-length window is superimposed around the putative signal (i.e., with a margin of zero or more nucleotides on either side of the signal consensus) and evaluated to produce a logarithmic signal score $R_S = \log P(x_h \ldots x_{h+n-1}|\theta)$, where h is the position of the beginning of the window, n is the window length, and x_i is the ith base in S. If signal *thresholding* is desired, R_S can be compared to a pre-specified threshold and those locations scoring below the threshold can be eliminated from consideration as putative signals.

The remaining candidates for signals of each type are then inserted into a type-specific *signal queue* for consideration later as possible predecessors of subsequent signals in a prospective parse. As each new signal is encountered, the optimal predecessors for the signal are selected from among the current contents of the signal queues, using a scoring function described below. For the purposes of our discussion, we will assume the following minimal set of transition patterns between signals, corresponding to the simple GHMM in Figure 8.8:

$$\begin{aligned}
\text{ATG} &\rightarrow \text{TAG} \\
\text{ATG} &\rightarrow \text{GT} \\
\text{GT} &\rightarrow \text{AG} \\
\text{AG} &\rightarrow \text{GT} \\
\text{AG} &\rightarrow \text{TAG} \\
\text{TAG} &\rightarrow \text{ATG}.
\end{aligned} \tag{8.11}$$

Associated with each of these patterns is a transition probability, $P_t(y_i|y_{i-1})$, which is included in the scoring of a prospective predecessor; this probability can be assessed in constant time by indexing into a two-dimensional array. The logarithmic transition score will be denoted $R_T(y_{i-1}, y_i) = \log P_t(y_i|y_{i-1})$. Note that we have collapsed the transitions between signal states and content states, since they are generally unambiguous given the pair of signals at either end of a variable-length feature. For example, the transition pattern ATG → $\text{EXON}_{\text{SINGLE}}$ → TAG has been collapsed into the single transition ATG → TAG, with the latter taking the probability of ATG → $\text{EXON}_{\text{SINGLE}}$, since the transition $\text{EXON}_{\text{SINGLE}}$ → TAG can be assumed to have a probability of 1.

The distance (i.e., length of intervening sequence) from a prospective predecessor to the current signal is also included in the evaluation in the form of $P_d(d_i|q_{i:j})$ for distance (= duration) d_i and content type (= state) $q_{i:j}$. This probability can usually be obtained relatively quickly, depending on the representation of the duration distributions. If the distributions have been fitted to a curve with a simple algebraic formula, then evaluation of the formula may be a constant-time operation. If a histogram is instead maintained, then a *binary search* is typically required to find the histogram bin containing the given distance (recall from section 2.13 that the binary search is a logarithmic-time operation). We denote the logarithmic duration score $R_D(q_i, q_j) = \log P_d(d_i|q_{i:j})$ where d_i is the length of the content region delimited by signals s_i and s_j, and $q_{i:j}$ is the variable-length state corresponding to that content region (with q_i and q_j being the fixed-length states delimiting the region).

Following Eq. (8.8), the final component of the scoring function is the emission probability $P_e(S_i|q_i, d_i)$. For a fixed-length state, this is simply the score produced by the signal sensor for that state. For a variable-length state q_i, P_e can be evaluated very quickly (in log space) by indexing into the prefix sum array $\alpha_{i,\gamma}$ for state q_i and phase γ at the appropriate indices for the two signals and simply performing subtraction:

$$R_C(s_{pred}, s_{cur}, \omega) \leftarrow \alpha_{i,\gamma}[wpos(s_{cur}) - 1] - \alpha_{i,\gamma}[wpos(s_{pred}) + wlen(s_{pred}) - 1], \tag{8.12}$$

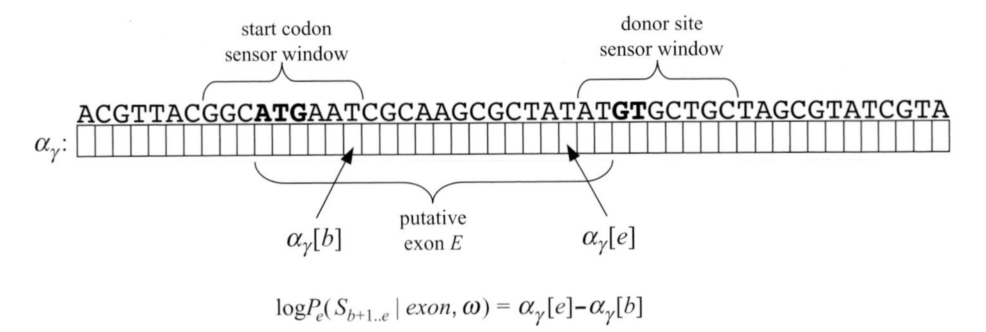

Figure 8.9 Using a prefix sum array to compute the emission probability for a putative exon. Subtraction is performed between array elements at either end of the [*b,e*] interval, which covers only the portion of the exon not covered by either signal sensor.

where *wpos*(*s*) is the 0-based position (within the full input sequence) of the first nucleotide in the context window for signal *s*, *wlen*(*s*) is the length of the context window for signal *s*, and s_{pred} and s_{cur} are the predecessor and current signals, respectively. In the case of coding features, γ is the phase of the array and $\omega = (\gamma + pos(s_{cur})) \bmod 3$ is the phase of s_{cur}, for $pos(s_{cur})$ the position of the leftmost consensus base of s_{cur} in the input sequence *S*. For reverse-strand features, since the prefix sum arrays tabulate their sums from the right instead of the left, the subtraction must be reversed:

$$R_C(s_{pred}, s_{cur}, \omega) \leftarrow \alpha_{i,\gamma}[wpos(s_{pred}) + wlen(s_{pred})] - \alpha_{i,\gamma}[wpos(s_{cur})], \tag{8.13}$$

and $\omega = (\gamma + L - pos(s_{cur}) - 1) \bmod 3$, for $L = |S|$ the sequence length. For noncoding features, the phases ω and γ can be ignored when computing R_C, since there is only one array per noncoding state. The above scheme is illustrated in Figure 8.9.

The reader will note that in formulating the above equations, we have taken care to ensure that the interval over which a putative content region is evaluated is strictly nonoverlapping with the regions falling within the signal sensor windows corresponding to the signals at either end of the putative content region. This is necessary in order to avoid "double-counting" nucleotides within any given gene parse, which would create an unintended bias toward smaller numbers of exons and genes in the final prediction. This idea is illustrated in Figure 8.10.

The resulting optimization function is:

$$\underset{s_i}{argmax}\ R_I(s_i, \gamma_i) + R_T(s_i, s_j) + R_D(s_i, s_j) + R_C(s_i, s_j, \gamma_j), \tag{8.14}$$

for current signal s_j and predecessor signal s_i; $R_I(s_i, \gamma_i)$ denotes the logarithmic inductive score for signal s_i in phase γ_i. For forward-strand coding features, the phases γ_i and γ_j are related by:

$$\gamma_i = (\gamma_j - \Delta) \bmod 3, \tag{8.15}$$

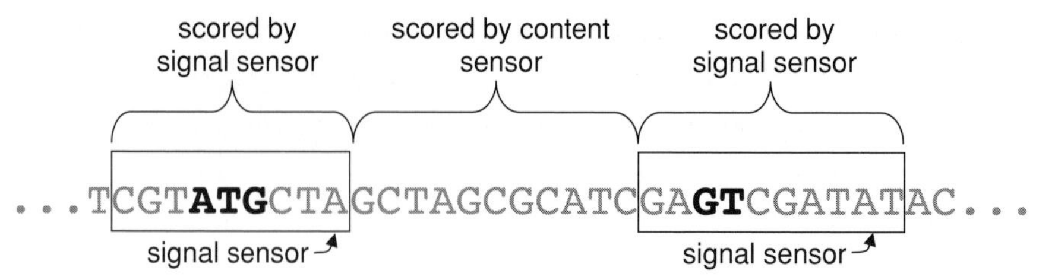

Figure 8.10 Nonoverlapping sensor windows in a putative gene parse. Ensuring that sensor windows do not overlap within a single gene parse is necessary to avoid an unintended bias toward smaller numbers of features in the final gene prediction.

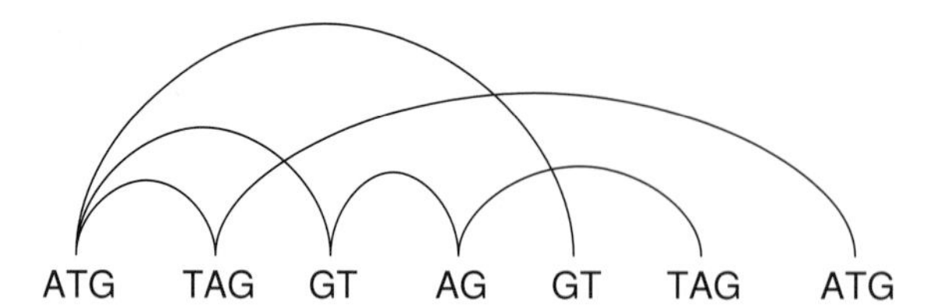

Figure 8.11 Hypothetical portion of a trellis. Each vertex has at most three trellis links (one per phase) to predecessors in the trellis. Arrows are omitted for clarity; edges implicitly point leftward.

for Δ the putative exon length, or, equivalently,

$$\gamma_j = (\gamma_i + \Delta) \bmod 3. \tag{8.16}$$

These relations (Eqs. 8.15–8.16) can be modified for application on the reverse strand by swapping + and −. For introns, $\gamma_i = \gamma_j$, since an intron must splice out in the same phase in which it spliced in. For intergenic features, the phase of s_i or s_j will always be 0 for a forward strand signal and 2 for a reverse strand signal (since on the reverse strand the leftmost base of a start or stop codon would be in phase 2).

The result of Eq. (8.14) is the optimal predecessor for signal s_j. This scoring function is evaluated for all appropriate predecessor signals, which are readily available in one or more queues, as described above. A pointer called a *trellis link* is then created, pointing from the current signal to its optimal predecessor. In the case of those signals that can terminate an exon or an intron, three optimal predecessors must be retained, one for each phase. The inductive score $R_I(s_j, \gamma_j)$ of the new signal s_j is then initialized from the selected predecessor s_i as follows:

$$R_I(s_j, \gamma_j) \leftarrow R_I(s_i, \gamma_i) + R_T(s_i, s_j) + R_D(s_i, s_j) + R_C(s_i, s_j, \gamma_j) + R_S(s_j), \tag{8.17}$$

where $R_S(s_j)$ is the logarithmic score produced by the signal sensor for signal s_j – i.e., $\log P_e(s_j|state(s_j))$.

Creation of trellis links in this way gives rise to a directed graph called a *trellis*. A section of a sample trellis is depicted in Figure 8.11. Although the trellis is reminiscent of the *ORF graph* which we introduced in section 3.4 (and to which we will return later in this chapter), the trellis is generally much sparser than the full ORF graph, since each vertex in the trellis has at most three trellis links (one per phase) pointing to signals left of it.

A final step to be performed at each position along the input sequence is to drop from each queue any signal that has been *eclipsed* from all subsequent positions due to intervening stop codons. Except for the final stop codon of a CDS, *in-phase* (i.e., in phase 0) stop codons are generally not permitted in coding exons; for this reason, any potential stop codon (regardless of its signal score) will eclipse any preceding start codon or acceptor site (or, on the reverse strand, any stop codon or donor site) in the corresponding phase, where by *eclipse* we mean that subsequently encountered signals along the sequence from left-to-right may not link back directly (i.e., via a single trellis link) to any eclipsed signal.

The procedure shown in Algorithm 8.1 addresses this issue by dropping any fully eclipsed signal (i.e., eclipsed in all three phases) from its queue. Procedure `drop(s,G)` removes `s` from queue `G`. For the reverse strand, line 2 of `eclipse()` should be changed to:

$$\omega \leftarrow (\texttt{p-pos(s)-len(s)-1}) \ \textbf{mod} \ 3;$$

where `len(s)` is the length of the consensus sequence for signal `s` (e.g., 3 for `ATG`). Note that by `x` **mod** 3 we mean the *positive* remainder after division of `x` by 3; in some programming languages (such as C/C++), a negative remainder may be returned, in which case 3 should be added to the result (see section 2.1). It should be obvious that the reverse-strand version of Algorithm 8.1 should be invoked only for reverse-strand coding queues (i.e., those containing reverse-strand stop codons and donor sites) *and* reverse-strand stop codons; likewise, forward-strand stop codons may eclipse only those signals in forward-strand coding queues (i.e., those containing putative start codons and acceptor sites).

Algorithm 8.1 Eclipsing signals in coding queue `G` when a stop codon has been encountered at position `p`. `pos(s)` is the position of the first base of the signal's consensus sequence (e.g., the `A` in `ATG`). `len(s)` is the length of the signal's consensus sequence (e.g., 3 for `ATG`).

```
procedure eclipse(ref G, p)
1.    foreach s ∈ G do
2.      ω←(pos(s) + len(s) - p) mod 3;
3.      eclipsed_s[ω] ← true;
4.      if eclipsed_s[(ω + 1) mod3] and
5.         eclipsed_s[(ω + 2) mod3]
6.        then drop(s,G);
```

A special case of eclipsing which is not handled by `eclipse()` is that which occurs when a stop codon straddles an intron; this can be handled fairly simply by checking for such when considering each donor signal as a prospective predecessor for an acceptor signal (or vice versa on the reverse strand). As each predecessor is evaluated, the bases immediately before the donor and immediately following the acceptor are examined, and if a stop codon is formed, the predecessor is no longer considered eligible for selection in the corresponding phase (see Exercise 8.22).

As shown in Algorithm 8.1, when a signal has been eclipsed in all three phases it can be removed from its queue. In this way, as a signal falls further and further behind the current position in the sequence, the signal becomes more and more likely to be eclipsed in all three phases as randomly formed stop codons are encountered in the sequence, as argued above, so that coding queues (e.g., those holding forward-strand start codons and acceptors, or reverse-strand donors and stop codons) tend not to grow without bound, but to be limited on average to some maximal load determined by the nucleotide composition statistics of the sequence. Because of this effect, the expected number of signals which must be considered during predecessor evaluation can be considered effectively constant in practice, as we have argued previously.

In the case of noncoding queues (e.g., those holding forward-strand donors or stop codons, etc.), the assumption that noncoding features follow a geometric distribution allows us to limit these queues to a single element (per phase), because once a noncoding predecessor has been selected in a given phase, no other noncoding predecessor which has already been compared to the selected predecessor can ever become more attractive, as established by Theorem 8.1.

Because the coding and noncoding queues are effectively limited to a constant load, the expected processing time at each nucleotide is $O(|Q|)$ for sparse GHMM topologies (i.e., with all in-degrees bounded by a small, fixed constant – see Exercise 8.7) and therefore the entire algorithm up to this point requires time $O(|S| \times |Q|)$. It will be seen that the traceback procedure described below requires time $O(|S|)$, so that the entire PSA algorithm for typical eukaryotic genomes requires time $O(|S| \times |Q| + |S|) = O(|S| \times |Q|)$, which is equivalent to the optimized Viterbi for (sparse) HMMs utilizing a λ_{trans} array (sections 6.1.2 and 6.2.1).

Once the end of the sequence is reached, the optimal parse ϕ^* can be reconstructed by tracing back through the trellis links. In order for this to be done, a set of virtual *anchor signals* (one of each type) must be instantiated at either terminus of the sequence (each having signal score $R_S = 0$). Those at the left terminus will have been entered into the appropriate queues at the very start of the algorithm as prospective targets for the first trellis links (and having inductive scores $R_I = 0$), and those at the right terminus are the last signals to be evaluated and linked into the trellis. The highest scoring of these right-terminal anchor signals is selected (in its highest-scoring phase) as the starting point for the traceback procedure. Traceback consists merely of following the trellis links backward while adjusting for phase changes across exons, as shown in Algorithm 8.2.

Algorithm 8.2 Reconstruction of the optimal parse by tracing back through trellis links. Parameters are the selected right-terminus signal s and its chosen phase ω. Returns a stack of signals constituting the optimal parse, with the top signal at the beginning of the parse and the bottom signal at the end. `exon_length(p,s)` denotes the number of coding nucleotides between signals p and s.

```
procedure traceback(s,ω)
1.    stack K;
2.    push K,s;
3.    while ¬ left_terminus(s) do
4.      p←pred(s,ω);
5.      push K,p;
6.      if type(p)∈{ATG,TAG} then ω←0;
7.      elsif type(p)=AG then
8.        ω←(ω-exon_length(p,s))mod3;
9.      s←p;
10.   return K;
```

Modifications to Algorithm 8.2 for features on the reverse strand include changing the AG on line 7 to GT, changing the subtraction on line 8 to addition, and changing the 0 on line 6 to 2.

The PSA algorithm is summarized in Algorithm 8.3.

Algorithm 8.3 Overview of the PSA decoding algorithm. See text for details.

```
procedure PSA(S,θ)
1.    Initialize arrays via Equations 8.9 & 8.10
2.    At each position along the sequence do:
3.      Perform eclipsing via Algorithm 8.1
4.      Apply signal sensors at current location
5.      If a putative signal si is detected then:
6.        Link si back to optimal predecessors
7.                                      via Eq. 8.14
8.        Append si to appropriate signal queues
9.    Form the optimal parse φ* via Algorithm 8.2
10.   Convert φ* to a set of gene predictions
```

It should be clear from the foregoing that the space requirements of the PSA decoding algorithm are $O(|S| \times |Q|)$. If, for example, array elements are 8-byte double-precision floating-point numbers, then a typical GHMM would require at least 14 prefix sum arrays (4 exon states $\times$ 3 phases + 1 intergenic state + 1 intron state), resulting in a memory requirement of at least 112 bytes per nucleotide. For a GHMM which can model both DNA strands simultaneously, this would increase to 216 bytes per nucleotide, so that processing of a 1 Mb sequence would require at least 216 Mb of RAM just for the arrays. Adding states for 5′ and 3′ UTRs would increase this to 248 Mb of RAM for a 1 Mb sequence, or over 1 Gb of RAM for a 5 Mb sequence. For the purposes of comparative gene finding (Chapter 9) on multiple organisms with large genes, these requirements seem less than ideal, especially when one considers the possibility of adding yet other states.

The memory requirements can be reduced in several ways. First, Markov chains can be shared by similar states – a form of *parameter tying*, as mentioned earlier. For example, the intron and intergenic states can share a single Markov chain trained on pooled noncoding DNA, and all the exon states can use the same three-periodic Markov chain trained on pooled coding DNA.

Second, the models for exons can be modified so as to utilize likelihood ratios instead of probabilities. If the models for exons are re-parameterized to compute:

$$\frac{P(S|coding)}{P(S|noncoding)}, \tag{8.18}$$

and the noncoding models are modified to compute:

$$\frac{P(S|noncoding)}{P(S|noncoding)}, \tag{8.19}$$

then the latter can be seen to be unnecessary, since it will always evaluate to 1. Such a modification is valid and will have no effect on the mathematical structure of the optimization problem given in Eq. (8.6) as long as the denominator is evaluated using a factorable model, since the effect of the denominator on inductive scores will then be constant across all possible predecessors for any given signal (see Exercise 8.23). Using such ratios allows us to skip the evaluation of all emission terms for noncoding states, so that the number of prefix sum arrays required for a double-stranded version of our example GHMM in Figure 8.8 would be only 6 (assuming the previous optimization is applied as well), corresponding to the three exon phases on two strands. Furthermore, to the extent that these likelihood ratios are expected to have a relatively limited numerical range, lower-precision floating-point numbers can be used, or the ratios could instead be multiplied by an appropriate scaling factor and then stored as 2-byte integers (Burge, 1997). This is a significant reduction, though asymptotically the space complexity is still $O(|S| \times |Q|)$. An additional consideration is that the log-likelihood strategy makes unavailable (or at least inseparable) the raw coding and noncoding scores, which might be desired later for some other application, such as posterior decoding (see section 8.5.4).

A third method of reducing the memory requirements is to eliminate the prefix sum arrays altogether, resulting in what we call the *dynamic score propagation (DSP)* algorithm.

8.3.2 DSP decoding

Informally, the DSP algorithm is similar to the PSA algorithm, except that rather than computing feature scores via subtraction between elements of precomputed arrays, we instead keep track of all putative variable-length features which have not yet been terminated during a left-to-right pass over the sequence, updating the inductive score for each active feature at each newly evaluated nucleotide, according to the content sensor for the corresponding feature type.

Associated with each signal is a *propagator* variable which represents the log probability (in each phase) of the highest-scoring *partial parse* up to and including this signal. As processing proceeds left-to-right along the sequence, these propagators are updated so as to extend these partial parses up to the current position. In this way, the inductive score of each signal is incrementally propagated up to each potential successor signal that is encountered during processing; when a signal is eclipsed in all phases by stop codons (i.e., removed from its respective queue), propagation of that signal's inductive score halts, since further updates would be useless beyond that point. Because no prefix sum arrays are allocated, and because the signal queues are effectively limited in size (as argued previously), the expected memory requirements of DSP will be seen to be $O(|S| + |Q|)$, where the constant factor associated with the $|S|$ term is small, reflecting only the memory required to store the sequence itself.

Let us introduce some notation. We define a *propagator* π to be a 3-element array, indexed using the notation $\pi[i]$ for $0 \leq i \leq 2$; when dealing with multiple propagators, $\pi_j[i]$ will denote element i of the jth propagator.

Each signal s_i will now have associated with it a propagator, denoted π_i. For signals which can be members of multiple queues (such as start codons, which can be members of both the *initial exon* queue and the *single exon* queue), the signal will have one propagator per queue, but it will be clear from the context to which propagator we refer. Each queue will also have a propagator associated with it, though to avoid confusion we will refer to these as *accumulators* and represent them with the symbol α. The purpose of the accumulators is to reduce the number of updates to individual signal propagators; otherwise, every signal propagator in every queue would need to be updated at every position in the input sequence. The accumulator for a given queue will accumulate additions to be made to the propagators of the signals currently in the queue. The updating of signal propagators from their queue's accumulator is delayed as long as possible, as described below. Accumulator scores are initialized to zero, as are the propagator scores for the left terminus anchor signals; the general case of propagator initialization will be described shortly.

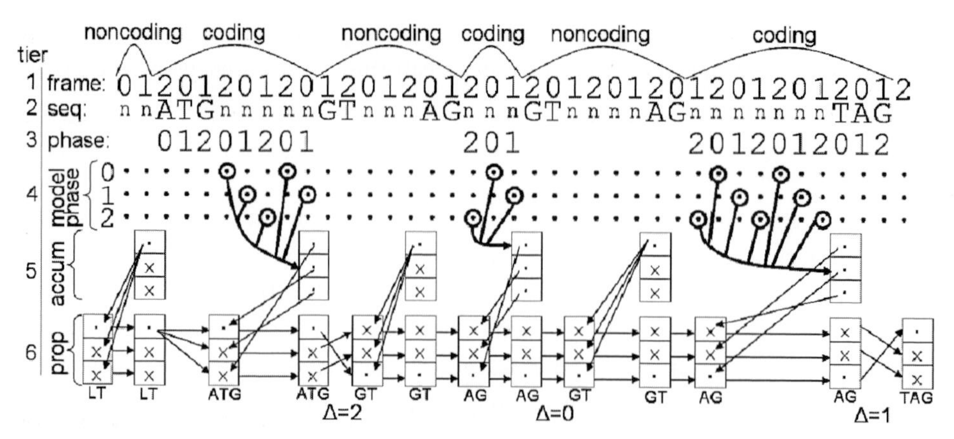

Figure 8.12 Operation of the dynamic score propagation (DSP) algorithm. Each signal has an associated propagator (sixth tier) which is propagated along the sequence by periodic updates from a queue-specific accumulator (fifth tier). Accumulators are updated at each position of the sequence in a phase-specific manner from the model scores (shown as dots in fourth tier), which are computed from the raw sequence (second tier; nonsignal nucleotides are shown as n's for clarity). Negative infinity scores are indicated with the symbol $\times$. See section 8.3.4 for a detailed description.

The operation of the dynamic score propagation (DSP) algorithm is very crudely illustrated in Figure 8.12.

Updating of a propagator π from an accumulator α is simple in the case of a noncoding queue:

$$\forall_{0\leq\omega\leq 2}\, \pi[\omega] \leftarrow \pi[\omega] + \alpha[0]. \tag{8.20}$$

For coding queues, the update must take into account the location of the signal s associated with the propagator π, in order to synchronize the periodic association between phase and array index:

$$\forall_{0\leq\omega\leq 2}\, \pi[\omega] \leftarrow \pi[\omega] + \alpha[(\omega - pos(s) - len(s)) \bmod 3], \tag{8.21}$$

or, on the reverse strand:

$$\forall_{0\leq\omega\leq 2}\, \pi[\omega] \leftarrow \pi[\omega] + \alpha[(\omega + pos(s) + len(s)) \bmod 3]. \tag{8.22}$$

Given a content sensor M, a coding accumulator can be updated according to the rule:

$$\forall_{0\leq\omega\leq 2}\, \alpha[\omega] \leftarrow \alpha[\omega] + \log P_{M[(\omega+f) \bmod 3]}(x_f), \tag{8.23}$$

or, on the reverse strand:

$$\forall_{0\leq\omega\leq 2}\, \alpha[\omega] \leftarrow \alpha[\omega] + \log P_{W[(\omega-f) \bmod 3]}(x_f), \tag{8.24}$$

where f is the position of the current nucleotide x_f, $P_{M[\omega]}(x_f)$ is the probability assigned to x_f by the content sensor M in phase ω, and W is the reverse-complementary model to M which computes the probability of sequences on the opposite strand and takes contexts (in the case of higher-order Markov chains – see section 7.2) from the right rather than from the left. This update occurs once at each position along the input sequence. Use of f provides an absolute frame of reference when updating the accumulator. This is necessary because the accumulator for a queue has no intrinsic notion of phase: unlike an individual signal, a queue is not rooted at any particular location relative to the sequence.

For noncoding queues, only the 0th element of the accumulator must be updated:

$$\alpha[0] \leftarrow \alpha[0] + \log P_M(x_f). \tag{8.25}$$

All that remains is to specify the rule for selecting an optimal predecessor and using it to initialize a new signal's propagator. We first consider new signals which terminate a putative exon. Let s_i denote the predecessor under consideration and s_j the new signal. Denote by Δ the length of the putative exon. Then on the forward strand, we can compare predecessors with respect to phase ω via the scoring function $R_{CI} + R_D + R_T$, where R_D and R_T are the duration and transition scores described earlier and R_{CI} includes the content score and the inductive score from the previous signal:

$$\forall_{0\leq\omega\leq 2}\, R_{CI}(s_i, \omega) \leftarrow \pi_i[(\omega - \Delta) \bmod 3]. \tag{8.26}$$

On the reverse strand we have:

$$\forall_{0\leq\omega\leq 2}\, R_{CI}(s_i, \omega) \leftarrow \pi_i[(\omega + \Delta) \bmod 3]. \tag{8.27}$$

For introns it is still necessary to separate the three phase-specific scores to avoid greedy behavior (see Exercise 8.24), though the phase does not change across an intron, so no Δ term is necessary:

$$\forall_{0\leq\omega\leq 2}\, R_{CI}(s_i, \omega) \leftarrow \pi_i[\omega]. \tag{8.28}$$

When the preceding feature is intergenic we need only refer to phase zero of the preceding stop codon:

$$R_{CI}(s_i, \omega) \leftarrow \pi_i[0], \tag{8.29}$$

or, on the reverse strand, phase 2 of the preceding start codon (since the leftmost base of the reverse-strand start codon will reside in phase 2).

Once an optimal predecessor with score $R_{CI} + R_D + R_T$ is selected with respect to a given phase ω, the appropriate element of the new signal's propagator can be initialized directly:

$$\pi_j[\omega] \leftarrow R_{CI}(s_i, \omega) + R_D(s_i, s_j) + R_T(s_i, s_j) + R_S(s_j), \tag{8.30}$$

where $R_S(s_j) = P(context(s_j)|\theta_j)$ is the score assigned to the context window of the new signal s_j by the appropriate signal sensor θ_j. An exception to Eq. (8.30) occurs when ω is not a valid phase for signal s_j (e.g., phase 1 for a start codon), in which case we instead set $\pi_j[\omega]$ to $-\infty$.

One final complication arises from the fact that the algorithm, as we have presented it, does not permit adjacent signals in a prospective parse to have overlapping signal sensor windows; to allow such would be to permit "double-counting" of nucleotide probabilities, thereby biasing the probabilistic scoring function, as we have previously noted. It is a simple matter to reformulate the algorithm so that signal sensors score only the two or three consensus nucleotides of the signals under consideration; this would allow adjacent signals in a prospective parse to be as close as possible without actually overlapping (i.e., a single exon consisting of the sequence ATGTAG would be permitted, even if the start codon and stop codon context windows overlapped). However, doing so may be expected to decrease gene finder accuracy, for two reasons: (1) statistical biases occurring at fixed positions relative to signals of a given type can in general be better exploited by a signal sensor specifically trained on such positions than by a content sensor trained on data pooled from many positions at variable distances from the signal, and (2) in the case of Markov chains and interpolated Markov models (section 7.2), probability estimates for nucleotides immediately following a signal can be inadvertently conditioned on the few trailing nucleotides of the preceding feature (assuming the chain has a sufficiently high order), even though the models are typically not trained accordingly. For these reasons, we recommend the use of signal sensors that impose a moderate margin around their respective signals, both to detect any biologically relevant biases which might exist within those margins, and to ensure that content sensors condition their probabilities only on nucleotides within the same feature.

Given the foregoing, it is necessary to utilize a separate *holding queue* for signals which have recently been detected by their signal sensors but which have context windows still overlapping the current position in the DSP algorithm. The reason for this is that propagator updates via Eqs. (8.20–8.22) must not be applied to signals having context windows overlapping any nucleotides already accounted for in the accumulator scores, since to do so would be to double-count probabilities. It is therefore necessary to observe the following discipline.

Associated with each signal queue G_i there must be a separate *holding queue*, H_i. When a signal is instantiated by a signal sensor it is added to the appropriate H_i rather than to G_i. As the algorithm advances along the sequence, at each new position we must examine the contents of each holding queue H_i to identify any signal having a context window which has now passed completely to the left of the current position. If some nonempty set ζ of such signals are identified, then we first update the propagators of all the signals in the main queue G_i using Eqs. (8.20–8.22), then zero-out the values of the accumulator α_i for that queue, and then allow the recently passed signals ζ to graduate from H_i to G_i. Observe that at this point all the signals in G_i have in their propagators scores which have effectively been

Table 8.2 Comparison of memory and time requirements of the PSA vs. DSP decoding algorithms on a sample 922 kb sequence; DSP requires far less memory, while achieving the same speed as PSA

	RAM/state(Mb)	Seconds/state
DSP	0.95	2.8
PSA	14	2.8

Source: Adapted from Majoros *et al.* (2005a).

propagated up to the same point in the sequence, and that point is immediately left of the current position; this invariant is necessary for the proper operation of the algorithm. All content sensors are then evaluated at the current position and their resulting single-nucleotide scores are used to update the accumulators for their respective queues. Finally, whenever it becomes necessary to evaluate the signals in some queue G_i as possible predecessors of a new signal, we must first update the propagators of all the elements of G_i as described above, so that the comparison will be based on fully propagated scores.

The DSP algorithm is summarized in Algorithm 8.4.

Algorithm 8.4 Overview of the DSP decoding algorithm. See text for details.

```
procedure DSP(S,θ)
1.    At each position along the sequence do:
2.      Evaluate all content sensors
3.      Update accumulators with content scores
4.      Perform eclipsing via Algorithm 8.1
5.      Allow mature signals to graduate from
6.                                their holding queues
7.      Apply signal sensors at current location
8.      If a putative signal s_i is detected then:
9.        Link s_i back to optimal predecessors
10.       Propagate appropriate queue elements
11.       Append s_i to appropriate holding queues
12.   Form the optimal parse φ* via Algorithm 8.2
13.   Convert φ* to a set of gene predictions
```

A comparison of memory and time usage by the DSP and PSA algorithms is given in Table 8.2, from which it can be seen that DSP is significantly more space-efficient while achieving the same speed as PSA.

8.3.3 Equivalence of DSP and PSA

We now show that DSP is mathematically equivalent to PSA, since it may not be entirely obvious from the foregoing description. We will consider only the forward-strand cases; the proof for the reverse-strand cases can be derived by a series of trivial substitutions in the proof below (see Exercise 8.25).

Theorem 8.2 *(Equivalence of PSA and DSP) Let S be a sequence and θ a set of model parameters for a GHMM. Then given parses $\phi^*{}_{PSA} = PSA(S, \theta)$ and $\phi^*_{DSP} = DSP(S, \theta)$ selected under the PSA and DSP decoding algorithms, respectively, we have $\phi^*_{PSA} = \phi^*_{DSP}$.*

Proof: To begin, we show by induction that the signal propagator $\pi_j[\omega]$ for signal s_j is initialized to the PSA inductive score $R_I(s_j, \omega)$. For the basis step, recall that the left terminus anchor signals were initialized to have zero scores in both PSA and DSP, regardless of whether a given signal began a coding or noncoding feature. In the case of coding features, substituting Eq. (8.26) into Eq. (8.30) yields:

$$\pi_j[\omega] \leftarrow \pi_i[(\omega - \Delta) \bmod 3] + R_D(s_i, s_j) + R_T(s_i, s_j) + R_S(s_j). \tag{8.31}$$

According to Eq. (8.17), this initialization will result in $\pi_j[\omega] = R_I(s_j, \omega)$ only if:

$$\pi_i[(\omega - \Delta) \bmod 3] = R_I(s_i, \gamma_i) + R_C(s_i, s_j, \omega), \tag{8.32}$$

where $\gamma_i = (\omega - \Delta) \bmod 3$ according to Eq. (8.15). At the time that signal s_j is instantiated by its signal sensor, π_i has been propagated up to $e = wpos(s_j) - 1$, the nucleotide just before the leftmost position of the context window for s_j. By the inductive hypothesis, $\pi_i[\gamma_i]$ was initialized to $R_I(s_i, \gamma_i)$. This initialization occurred at the time when the current DSP position was at the beginning of the predecessor's context window. Note, however, that π_i effectively began receiving updates at position $b = wpos(s_i) + wlen(s_i)$, the position immediately following the end of the signal's context window, at which point s_i graduated from its holding queue. Thus, $\pi_i[\gamma_i]$ will have accumulated content scores for positions b through e, inclusive. In order to establish Eq. (8.32), we need to show that these accumulations sum to precisely $R_C(s_i, s_j, \omega)$.

Substituting Eq. (8.23) into Eq. (8.21) we get the following formula describing propagator updates as if they came directly from content sensor M:

$$\forall_{0 \le \omega \le 2} \pi[\omega] \leftarrow \pi[\omega] + \log P_{M[(\omega + \Delta) \bmod 3]}(x_f), \tag{8.33}$$

where $\Delta = f - (pos(s_i) + len(s_i))$ is the distance between the rightmost end of signal s_i and the current position f in the DSP algorithm. Let us introduce the notation:

$$F(i, j, \omega) = \sum_{k=i}^{j} \log P_{M[(\omega + k) \bmod 3]}(x_k). \tag{8.34}$$

Using this notation, $\pi_i[\gamma_i]$ has since its initialization accumulated $F(b, e, \gamma_i - pos(s_i) - len(s_i))$; this can be verified by expanding this expression via Eq. (8.34) and observing that the result equals a summation of the log term in Eq. (8.33) over $f = b$ to e. The reader can easily verify that the effect of Eq. (8.10) will be that:

$$\alpha_{i,\gamma}[h] = \sum_{k=0}^{h} \log P_{M[(k+\gamma) \bmod 3]}(x_k) = F(0, h, \gamma). \tag{8.35}$$

According to Eq. (8.12), showing that $\pi_i[\gamma_i]$ has accumulated $R_C(s_i, s_j, \omega)$ is therefore equivalent to:

$$F(b, e, \psi) = F(0, wpos(s_j) - 1, \gamma) - F(0, wpos(s_i) + wlen(s_i) - 1, \gamma), \tag{8.36}$$

where $\psi = \gamma_i - pos(s_i) - len(s_i)$ and $\gamma = \omega - pos(s_j)$. Equivalently:

$$F(b, e, \psi) = F(0, e, \gamma) - F(0, b - 1, \gamma). \tag{8.37}$$

To see that $\psi \equiv \gamma (\bmod 3)$, observe that $pos(s_j) - (pos(s_i) + len(s_i)) = \Delta$, the length of the putative exon (possibly shortened by three bases, in the case where s_i is a start codon), and further that $\gamma_i - \omega \equiv -\Delta (\bmod 3)$ according to Eq. (8.15), so that $\psi - \gamma \equiv \Delta - \Delta \equiv 0 (\bmod 3)$. Thus, Eq. (8.37) is equivalent to:

$$F(b, e, \gamma) = F(0, e, \gamma) - F(0, b - 1, \gamma), \tag{8.38}$$

which can be established as a tautology by simple algebra after expansion with Eq. (8.34). This shows that the signal propagator for signal s_j is initialized to the PSA inductive score $R_I(s_j, \omega)$, and thus establishes the inductive step of the proof in the case of coding features.

To see that the above arguments also hold for noncoding features, note that Eq. (8.28) simplifies Eq. (8.32) to:

$$\pi_i[\omega] = R_I(s_i, \omega) + R_C(s_i, s_j), \tag{8.39}$$

that Eqs. (8.20) and (8.25) combine to simplify Eq. (8.33) to:

$$\forall_{0 \le \omega \le 2} \pi[\omega] \leftarrow \pi[\omega] + \log P_M(x_f), \tag{8.40}$$

and that Eq. (8.9) causes:

$$\alpha_i[h] = \sum_{k=0}^{h} \log P_M(x_k) = F_{NC}(0, h), \tag{8.41}$$

for $F_{NC}(i, j) = \sum_{k=i..j} \log P_M(x_k)$. We can thus reformulate Eq. (8.36) as:

$$F_{NC}(b, e) = F_{NC}(0, wpos(s_{cur}) - 1) - F_{NC}(0, wpos(s_{pred}) + wlen(s_{pred}) - 1), \tag{8.42}$$

or, equivalently:

$$F_{NC}(b, e) = F_{NC}(0, e) - F_{NC}(0, b - 1), \tag{8.43}$$

which is again a tautology. In the interests of brevity, we leave it up to the reader to verify that the above arguments still apply when the noncoding features are intergenic, thereby invoking Eq. (8.29) rather than Eq. (8.28) in formulating Eq. (8.38).

To see that the selection of optimal predecessors is also performed identically in the two algorithms, note that the PSA criterion given in Eq. (8.14) is equivalent to the $argmax(R_{CI} + R_D + R_T)$ criterion of DSP as long as $R_{CI}(s_i, \omega) = R_C(s_i, s_j, \omega) + R_I(s_i, \gamma_i)$ at the time the optimal predecessor is selected, which we have in fact already shown by establishing Eq. (8.32).

Thus, DSP and PSA build identical trellises; application of the same `traceback()` procedure will therefore produce identical parses and identical gene predictions. ❑

8.3.4 A DSP example

We now give a simplified example of the operation of the DSP algorithm on a single prospective parse, as illustrated in Figure 8.12. For simplicity, we will assume the signal sensor windows include only the signal consensus positions (e.g., the ATG for a start codon), and we will also ignore the issue of holding queues.

In the figure, the sequence is shown in tier 2; we assume that the sequence begins two bases before the leftmost ATG in that tier. Only the nucleotides for the signals in the prospective parse are explicitly given; other bases are shown as n's. Above the topmost tier we have annotated the coding and noncoding intervals. Tier 1 gives the frame, which by definition is 0 at the first base of the sequence and cycles by threes as we move in the 5′-to-3′ direction (left to right). The phase (tier 3) also cycles by threes, but it advances only in putative coding regions. Tier 4 provides a crude representation of the fact that the coding sensor is *three-periodic* – i.e., has a separate submodel for each of the three phases. Each dot in each of the three rows of that tier represents a single-residue score from a Markov chain. A small number of these dots are circled and attached to directed arrows to show which (logarithmic) values are summed into particular accumulator entries, as we will describe in greater detail shortly.

The accumulators and propagators are represented in the bottom two tiers; each accumulator or propagator is represented by a column of three boxes, where (in the case of a propagator π_i) the top box represents $\pi_i[0]$, the middle box represents $\pi_i[1]$, and the bottom box represents $\pi_i[2]$; and similarly for an accumulator α. Scores of $-\infty$ are denoted using the $\times$ symbol. Finally, below the bottom tier is shown a set of Δ values, which we will describe shortly.

The operation of the algorithm proceeds left-to-right across the sequence. At the extreme left we show a propagator (labeled LT) for a *left terminus* signal; recall that a left terminus is instantiated for each signal type. This propagator is then propagated along the sequence, taking occasional updates from the appropriate (i.e., signal-type-specific) accumulator; the second propagator from the left in the figure (also labeled LT) shows this same left terminus propagator after it has been propagated up to the ATG. At this point, a propagator is instantiated for the start codon signal s_j (shown as the third propagator from the left, and labeled ATG) and initialized so that $\pi_j[1] = \pi_j[2] = -\infty$ (since phases 1 and 2 are not valid phases for a start codon), while $\pi_j[0]$ is initialized via Eqs. (8.29, 8.30) to the phase 0 score of its chosen predecessor (the updated LT signal – the second propagator from the left). Since the putative feature preceding s_j is an intergenic region, the chosen predecessor is considered a *noncoding predecessor*. The components of Eq. (8.30) used to initialize $\pi_j[0]$ are the signal score $R_S(s_j)$ for the ATG, the duration score $R_D(s_i, s_j)$ which is based on the length of the preceding feature, the transition score $R_T(s_i, s_j)$, and the inductive score $R_{CI}(s_i, \omega)$ obtained from the propagator of the predecessor signal $s_i = LT$. The new signal s_j is then inserted into the initial-exon and single-exon queues (since start codons begin both initial exons and single exons).

As the algorithm progresses across the initial exon (the leftmost interval marked "coding" at the top of the figure), the accumulator α for the initial-exon queue (second accumulator from the left) accumulates coding content sensor scores (tier 4) in a phase-dependent manner, as defined by Eq. (8.23). In the figure, we explicitly show the five scores which are summed into the $\alpha[1]$; the other two elements of the accumulator are computed in a similar manner, but are omitted from the figure for clarity. The first position summed into $\alpha[1]$ is (zero-based) position 5, which corresponds to frame 2, since 5 mod 3 = 2. Equation (8.23) indicates that the model phase to be used in this position for $\alpha[1]$ is given by $(\omega + f) \bmod 3 = (1 + 2) \bmod 3 = 0$, and hence the circled model score in tier 4 for this position is in the top row of that tier, denoting model phase 0. The reader can verify that the other phase-frame relations are likewise properly tracked.

At position 10 in the sequence, the GT dinucleotide causes the donor-site signal sensor to instantiate a new donor signal. Before this new signal can be linked back to the ATG, however, all of the elements of the coding queues must be updated from the respective accumulators, so that comparisons performed during evaluation of prospective predecessors will be made based on scores that have all been propagated up to the same location in the sequence; otherwise a signal may be selected as a predecessor simply because it had accumulated fewer scores (which are negative, since we work in log space). Thus, the ATG's propagator is updated from the appropriate accumulator, in a phase-dependent fashion (according to Eq. 8.21), as shown in the figure by the arrows extending from the second accumulator from the left to the third propagator from the left. At this point, a propagator π_j for the newly encountered GT signal s_j can be instantiated and its components $\pi_j[\omega]$ initialized from the sum of the GT signal sensor score $R_S(s_j)$, the duration score $R_D(s_i, s_j)$, the transition score $R_T(s_i, s_j)$, and the scores $R_{CI}(s_i, \omega)$ of the ATG's propagator (Eq. 8.30), as shown near the $\Delta = 2$ mark in the figure. Here, Δ denotes the length of the preceding exon, mod 3, which works out to $\Delta = 8 \bmod 3 = 2$.

The Δ value is used in Eq. (8.26) to initialize the $R_{CI}(s_i, \omega)$ value for the GT signal in a phase-dependent fashion: $R_{CI}(s_i, \omega) \leftarrow \pi_i[(\omega - \Delta) \bmod 3]$.

Following the GT is a 3 bp intron (not counting the splice sites). Over the course of this intron, the phase 0 element of the intron queue's accumulator (the third accumulator from the left) accumulates the single-base scores from the intron content sensor (not shown). When the AG is reached, the phase 0 element of this accumulator is added to all three phase elements of the GT's propagator; however, because only the phase 2 element of that propagator was initialized to a finite number, the other two elements (phases 0 and 1) remain $-\infty$, since $-\infty + x = -\infty$ for any finite x. Next, a propagator for the AG is instantiated and initialized in a phase-independent manner from the GT's propagator – the phase-independence of this operation can be seen by the horizontal arrows between the GT's propagator and the AG's propagator; horizontal arrows obviously denote no change in phase, whereas sloped arrows denote a change of phase.

We leave it as an exercise for the reader to follow the operation of the algorithm across the remainder of the sequence.

8.3.5 *Shortcomings of DSP and PSA*

At the beginning of section 8.3 we stated two assumptions: that noncoding lengths follow a geometric distribution, and that the variable-length emission functions (i.e., content sensors) P_e are factorable. Both PSA and DSP make use of these assumptions in order to achieve reasonable memory and run-time efficiency. As we will see in section 8.4.2, the former assumption may be relaxed somewhat without reducing efficiency to an unacceptable level. However, the assumption of factorable emission functions does place a fairly restrictive constraint on the choice of submodel for the content sensors.

Recall that the notion of factorability of a content sensor was defined as:

$$\textit{factorable}(P_e) \Leftrightarrow \exists_f \, P_e(S) = \prod_i f(S, i) \tag{8.44}$$

for emission function P_e and putative feature S, where $0 \leq i < |S|$. Although Eq. (8.44) places no constraint on the form of $f(S,i)$, the very fact that the product $\prod_i f(S, i)$ places the f terms in one-to-one correspondence with the residues in the sequence effectively limits the range of functions which one may consider in practice for use in evaluating content regions, so that practical submodel implementations in GHMMs tend invariably to be based on Markov chains or their various extensions, such as IMMs – hence our emphasis on these model types in Chapter 7. Furthermore, the use of a nonprobabilistic scoring function f can result in significant biases and scaling issues which can result in a total failure of the gene finder to produce reasonable predictions unless the implementor takes additional steps to address these issues.

An additional shortcoming of the DSP algorithm is that it does not explicitly construct prefix sum arrays, which one could conceivably wish to use for some unforeseen application.

8.4 Higher-fidelity modeling

In the sections below we consider several methods for higher-fidelity modeling of various characteristics of genes. Our treatment is by no means exhaustive - we merely wish to illustrate that the GHMM architecture readily permits such specialization for the more detailed modeling of gene structures and their statistical properties. Additional results are given in section 11.3 from experiments involving the use of additional states for features such as *CpG islands* and *branch points*.

8.4.1 Modeling isochores

A number of studies have demonstrated that the human genome can be segmented into regions of different {G,C} density called *isochores* (Bernardi *et al.*, 1985), and that the genes occurring in these different isochores can have significantly different statistical properties. For this reason, a number of gene finders utilize different models specific to the individual isochore from which the input sequence originated. In GENSCAN (Burge and Karlin, 1997), four isochores are defined according to the {G,C} density:

$$\begin{aligned} &\text{I:} && \{\texttt{G,C}\}\text{ density} \in [0\%, 43\%], \\ &\text{II:} && \{\texttt{G,C}\}\text{ density} \in (43\%, 51\%], \\ &\text{III:} && \{\texttt{G,C}\}\text{ density} \in (51\%, 57\%], \\ &\text{IV:} && \{\texttt{G,C}\}\text{ density} \in (57\%, 100\%]. \end{aligned} \tag{8.45}$$

GENSCAN differentiates between these four isochores when computing the P_t terms for all transitions, and when computing the P_d term for introns and intergenic regions. In addition, when computing the P_e term, GENSCAN differentiates between two ranges of {G,C} density: 0–43%, and 43–100%.

Differentiating between isochores during gene prediction can be done very simply by scanning through the input sequence to measure the {G,C} density of the entire sequence, and then loading the appropriate submodel parameters from a file specific to the indicated isochore. An obvious drawback to this approach is that the input sequence may actually span more than one true isochore - this will increasingly be the case as improvements in genome sequencing and assembly techniques allow for longer contigs. One solution is to partition longer contigs into a number of shorter sequences to be provided to the gene finder on separate runs. The latter strategy has in fact been in practice at some genome annotation centers for some time, as a number of early gene finders were found incapable of handling sequences longer than a few hundred kilobases without exhausting the computer's

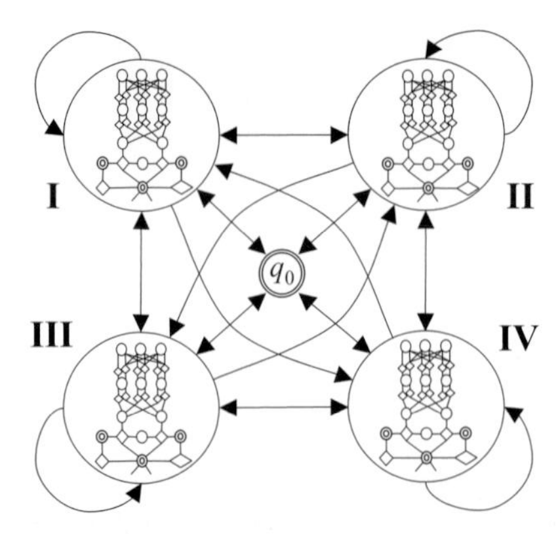

Figure 8.13 An HMM for isochore prediction. States are labeled with roman numerals denoting a range of {G,C} density. Within each HMM state is shown a GHMM, signifying that the submodels of our GHMM may be parameterized differently for different isochores.

memory. Of course, applying such a strategy requires some care in order to avoid splitting a contig in a place where there is a gene, thereby making it difficult to obtain complete, accurate predictions of these artificially segmented genes.

For gene finders capable of processing exceptionally long input sequences (such as those using a DSP-like decoding algorithm), more sophisticated methods for handling isochores are desired. One possibility is to preprocess the sequence to identify the likely isochore boundaries, and then to apply the appropriate isochore *profiles* (i.e., sets of GHMM submodels) to the different isochores. The use of isochore-specific submodels can be accommodated easily enough during decoding, by keeping all of the required models in memory and simply noting when a switch of the active submodel for each GHMM state is necessary, based on the previously identified isochore boundaries. This may be seen to constitute a crude form of *nonstationarity*, as described in section 7.2.6.

Identifying the likely isochore boundaries may be attempted by applying a separate HMM to the input sequence during a preprocessing stage. One may construct, for example, a 5-state HMM, with each of the four nonsilent states representing one of four isochores, as suggested by Figure 8.13. Given a set of contigs with known isochore boundaries, it is a simple task to train such an HMM using the labeled sequence training techniques presented in section 6.3. Applying the Viterbi algorithm would then allow us to obtain the most probable partitioning of an input sequence among the four states, thereby identifying the most probable isochore boundaries in the sequence, though a post-processing phase may be desired to enforce a minimum isochore length. Other methods for *de novo* isochore prediction have also been put forward (e.g., Oliver *et al.*, 2004).

One difficulty that afflicts all of these approaches to the modeling of isochores is that by partitioning our training data into isochore-specific subsets, we effectively

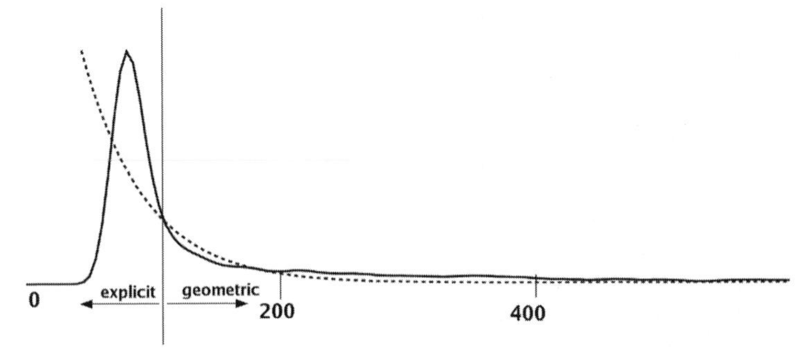

Figure 8.14 Explicit length modeling for noncoding features. The observed distribution of *Arabidopsis thaliana* intron lengths (solid line) is shown with a geometric distribution (dashed line). The geometric distribution is a reasonably good fit for lengths >100 bp. Below 100 bp we can use an explicit length distribution, as long as the intron queue stores all donor sites fewer than 100 bp away from the current position.

reduce the sample sizes available for the training of the individual submodels, thereby increasing the risk of sampling error and of negatively influencing the accuracy of the resulting gene finder. An ingenious solution to this problem was suggested by Stanke and Waack (2003), who implemented the approach in the gene finder AUGUSTUS. The key insight is that one need not partition the training set into smaller subsets; rather, we can train each isochore-specific submodel using the entire training set, with those training examples falling within the correct isochore weighted more highly than those falling in a different isochore. In AUGUSTUS, this is achieved by assigning each sequence S_i a weight $w_i = w_{\alpha,\beta}$ according to:

$$w_{\alpha,\beta} = \left\lceil 10e^{-200(\alpha-\beta)^2} \right\rceil \tag{8.46}$$

for α the desired {G,C} density of the submodel, and β the observed {G,C} density of sequence S_i. Each sequence S_i with weight w_i can then be treated as if it occurred w_i times in the training set, so that all frequency counts obtained from S_i are correspondingly multiplied.

8.4.2 *Explicit modeling of noncoding lengths*

In section 8.3 we made the assumption that the distribution of feature lengths for noncoding features such as introns and intergenic regions could be approximated reasonably well with a geometric distribution, and we saw that making this assumption permitted us to place an upper bound on the time complexity of GHMM decoding algorithms by keeping track of at most one potential noncoding predecessor, per phase, of each type during decoding. However, it is not difficult to find instances of noncoding length distributions for which at least some part of the distribution differs markedly from the geometric distribution used to model it. Such an example is shown in Figure 8.14.

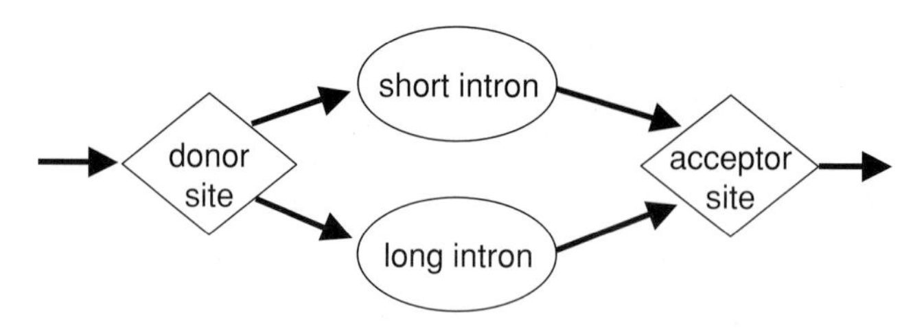

Figure 8.15 Explicit length modeling for introns. The short intron model is capable of generating introns of length up to some maximum, L_{short}, whereas the long intron model can generate only introns longer than this.

In the figure, it can be seen that the geometric distribution (dashed line) provides a reasonably good fit to the observed distribution of intron lengths in *Arabidopsis thaliana* for introns of length >100 bp. For introns shorter than 100 bp, however, the geometric distribution is a very poor fit.

One way of rectifying this problem without significantly impacting efficiency is to model short noncoding features differently from long ones. This is the approach taken by the gene finder AUGUSTUS (Stanke and Waack, 2003) in its modeling of intron lengths.

Figure 8.15 depicts a portion of a GHMM consisting of two intron models, one capable of generating introns of length up to some maximum L_{short}, and another capable of generating only introns longer than this. Referring back to Figure 8.14, we can use this two-intron scheme to model introns greater than 100 bp in length using an intron submodel based on a geometric distribution, and we can model introns of length 100 bp or less using an explicit length distribution as we do for coding features – though it should be noted that these two distributions must be renormalized so as to sum to 1 over the appropriate domain (i.e., $length > L_{short}$ for the geometric and $length \leq L_{short}$ for the explicit distribution). The actual emission probabilities of these two states would presumably be tied so as to use the same emission submodel, though one might consider using separate intron emission models for short and long introns, as suggested by Stanke and Waack (2003). The possibility of modeling splice sites differently for short and long introns has also been suggested (e.g., Lim and Burge, 2001).

Using an explicit length model for introns as described above requires a slight change to our decoding algorithm in order to avoid greedy behavior. Let us suppose that the maximum length for the short intron model is L_{short}. We must now relax our assumption that at most one putative noncoding predecessor of each type in each phase need be tracked during decoding. We will do this by allowing the noncoding signal queues to store an arbitrary number of potential noncoding predecessors, subject to the constraint that at most one signal further away than L_{short} will be retained in the queue. In this way, whenever we reach a putative acceptor site s_i, we will consider all putative donor sites s_j which are no more than L_{short} bp

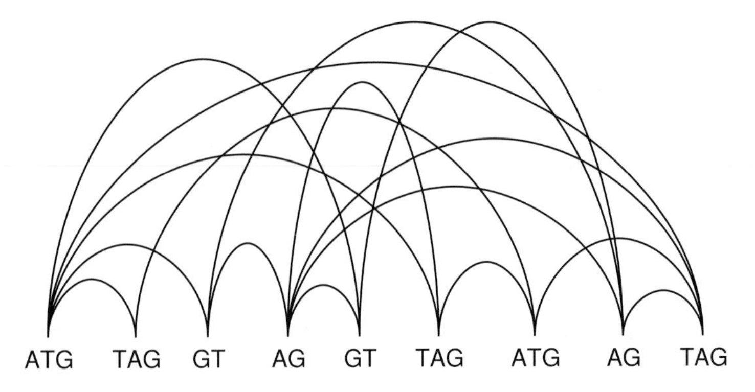

Figure 8.16 A hypothetical ORF graph for a contig. Vertices are shown at the bottom with labels indicating the type of signal represented. Each edge represents a putative exon, intron, or intergenic region, according to the edge type. Edges are implicitly directed from left to right. Vertices are associated with specific positions in a DNA sequence (omitted for clarity).

left of s_i, and we will consider at most one putative donor site further away than L_{short}, so that if the density of putative donor sites discovered by the signal sensor is not too great, and if L_{short} is reasonably small, then the amount of additional computation which needs to be performed to accommodate the explicit length modeling for short introns will not be too great.

Indeed, since the number of potential donor sites which could possibly occur within an L_{short} bp window must be less than some fixed maximum based on L_{short}, the asymptotic time complexity of our decoding algorithm remains unchanged. In practice, the run-time of the gene finder will slow by some constant factor, but we can mitigate this somewhat by imposing a threshold on our donor-site signal sensor, so that any putative donor site scoring less than the imposed threshold will be ineligible for further consideration as a donor site.

To see that the modified decoding algorithm is still optimal with regard to Eq. (8.6), simply note that for introns longer than L_{short}, the same arguments given in the proof for Theorem 8.1 still hold, since the inductive scores of any pair of potential predecessors at least L_{short} bp to the left of the current position will still decline at the same rate due to the geometric length distribution; for shorter introns, the argument does not hold, but since we consider all possible predecessors up to distance L_{short}, the algorithm remains optimal.

8.5 Prediction with an ORF graph

In section 3.4 we introduced the notion of an *ORF graph*, and we argued that the problem of predicting the structure of eukaryotic genes could be formulated as one of choosing the best path through the ORF graph for a sequence. A hypothetical ORF graph is depicted in Figure 8.16.

In formulating the decoding algorithms in the foregoing sections we showed how a subset of the ORF graph, known as a *trellis*, is commonly built by GHMM-based gene finders during decoding so that a simple traceback procedure can reconstruct the optimal parse. We will now consider the possibility of using the ORF graph in a more explicit manner for the purpose of gene finding and other related tasks.

8.5.1 *Building the graph*

Building an explicit representation of the underlying ORF graph for a sequence is a relatively simple task, given an existing GHMM decoding algorithm. We have already described the instantiation of trellis links which represent, for each putative signal encountered during decoding, the optimal predecessor for the current signal in each phase. To build the full ORF graph, we simply need to construct links back to *all* possible predecessors for each signal.

For those signals that can terminate a coding feature, this can be done by simply linking the current signal back to all of the signals that are currently present in the appropriate signal queue, since the signal queues are already maintained by the decoding algorithm so as to keep only those signals that have not yet been eclipsed by intervening stop codons in all three frames.

For noncoding predecessors, the problem is slightly more complicated, since the signal queues for such signals maintain at most one predecessor of each type in each phase. However, we saw in section 8.4.2 that it is possible to relax this discipline somewhat by allowing the noncoding queues to maintain multiple predecessors in each phase, subject to some constraint such as a maximum distance (e.g., the L_{short} used for explicit intron length modeling). In the gene finder *GeneZilla* the noncoding queues are constrained to hold no more than N signals for some N specified by the user in a configuration file. Each queue is represented using a data structure called a *priority queue* (see, e.g., Cormen *et al.*, 1992) in which the signals are kept in sorted order according to their inductive scores, with the lowest-scoring signal being dropped from the queue whenever a higher-scoring signal is added to a full queue. In this way, we can maintain the *N best* potential noncoding predecessors rather than simply the *N most recent* ones.

Limiting the number of noncoding predecessors in this way is essential for maintaining efficiency, since noncoding predecessors are not subject to the eclipsing phenomenon which affects coding predecessors. In the absence of any limits on the number of noncoding predecessors, the number of edges in an ORF graph would grow as a function of the square of the sequence length – i.e., $\mathbb{O}(|S|^2)$ – resulting in unmanageably large ORF graphs for long sequences; in the gene finder *Exonomy* (Majoros *et al.*, 2003), for example, building full ORF graphs for sequences longer than 50 kb routinely exhausted all computer memory. Using an *N best* strategy as described above allows us to retain the high-scoring parts of the ORF graph while keeping the size of the graph manageable – with the vertices and edges both obeying $\mathbb{O}(|S|)$, assuming fixed N. For a large enough N, we can be reasonably confident

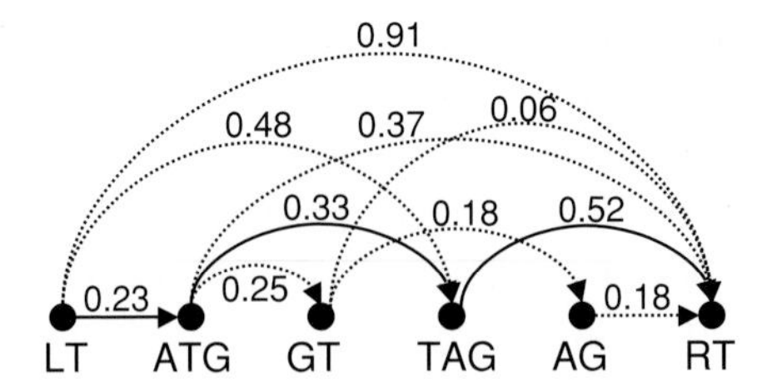

Figure 8.17 Operation of the `highestScoringPath()` algorithm for a small portion of a weighted ORF graph. The highest scoring path is shown in bold, and represents the optimal gene parse for this portion of the sequence. LT, left terminus; RT, right terminus.

that all of the portions of the graph that are consistent with the actual underlying gene structures are explicitly represented. A reasonable strategy is therefore to choose the largest N that produces reasonable run-times for the gene finder and other tools which use the emitted ORF graph for other tasks.

8.5.2 Decoding with a graph

Given an ORF graph G for a sequence S, it is a simple matter to produce a gene prediction for S by applying a standard algorithm from graph theory to extract the highest-scoring path. Indeed, we have already provided such an algorithm in the form of procedure `highestScoringPath(T,f,b,e)` from section 2.14. Since the vertices in the ORF graph are already implicitly ordered according to their positions within the sequence, the topological sort T can be replaced with the ordered vertex set V from the graph. The weighting function f is merely a mapping from the edges of the graph to the content scores assigned by the appropriate content sensor to the variable-length feature represented by a given edge, combined with the signal score for the preceding vertex. Finally, b and e represent the virtual anchor signals which were instantiated by the decoding algorithm at either terminus of the sequence (see Figure 8.17). Note that explicit phase tracking as defined in section 8.3 must also be performed (see Exercise 8.27).

In cases where one wishes to obtain the optimal gene parse for a specific interval of a sequence, we can apply the same procedure to just the subgraph extending from one end of the interval to another, so that the b and e vertices are now chosen to lie at either end of the interval. Ideally, a posterior decoding approach should be applied as in section 8.5.4 to evaluate the boundary conditions for vertices b and e – that is, to weight the edges incident on b by the probability, summed over all possible parses of the region to the left of b, of the vertex b being in the proper phase for a given edge; and similarly, to weight the edges incident on e by their corresponding right-context probabilities as defined by summing over all possible parses of the region to the right of e in the frame consistent with each such edge. We leave the details of this construction as an exercise for the reader (see Exercise 8.18).

8.5.3 *Extracting suboptimal parses*

To further motivate the reader's interest in the use of ORF graphs, we will make the following conjecture: that for most of the state-of-the-art gene finders currently available, when the gene finder does not produce a perfectly correct gene prediction, had that gene finder emitted an ORF graph rather than a single prediction, the correct gene structure would very likely have been one of the top few highest-scoring paths through the graph, and would therefore have been available for later discovery by human annotators or automated tools having access to additional forms of evidence. Because most gene finders do not emit ORF graphs, such later analyses are not possible as we have only an incorrect prediction and (in some cases) a set of gene or exon scores which are often expressed on a scale which is not easy to interpret relative to the predictions of other programs or to other forms of evidence which may be available (such as ESTs or cDNA alignments – see Chapters 1 and 9).

In other words, we posit that many gene finders very often come tantalizingly close to producing the correct gene prediction, only to be "deceived" – by slight inaccuracies in their statistical models, or sampling error due to short features in the input sequence – into emitting an erroneous prediction instead. We might think of this in terms of *discrimination power* at different *resolutions*: the statistical models of well-trained gene finders are clearly able to distinguish, in most cases, between the correct gene model and one which is so erroneous that it effectively reverses the coding and noncoding regions of the prediction; but as we consider pairs of gene predictions that are less dramatically different – i.e., as we attempt discrimination at a finer level of resolution – we find that the statistical models become more sensitive (in an adverse way) to the effects of "noise" or sampling error. Given the amount of computation that is performed by the gene finder in its analysis of the input sequence, it seems a monumental waste of effort to discard all but the highest-scoring path through the ORF graph, when we know that many times that path will not constitute the correct gene structure.

A number of tasks are made possible when we have access to the full ORF graph, or to an ORF graph retaining some reasonable percentage of its top-scoring edges – we term the latter a *pruned ORF graph*, since the low-scoring portions of the graph have been discarded, or "pruned away." One potentially useful task is to enumerate the top N highest-scoring parses in the graph, for some N. It should be evident that this is merely an extension of the decoding problem, and one that can be easily solved, as follows. Given a graph G, we apply the procedure for decoding with a graph, as given previously, except that each vertex may now have up to N trellis links in each phase. When we are selecting the trellis links to install into some vertex v_i, we consider all potential predecessors v_j of v_i. Each predecessor v_j will have up to N trellis links of its own (i.e., pointing back to its own predecessors), each with an inductive score that includes the signal score for v_j itself. We form the union T of all trellis links of all predecessors v_j of v_i, sort them according to their inductive scores at v_i (i.e., their inductive scores propagated up to v_i), and then select the top N edges leading into v_i as trellis links for v_i. In this way, the trellis links from a signal v_i point not to preceding *signals* but rather to specific trellis *links* belonging to those preceding signals (see Exercise 8.19).

8.5.4 *Posterior decoding for GHMMs*

A particularly useful feature of the GHMM-based gene finder GENSCAN (Burge, 1997) is that the scores assigned by the program to individual exon predictions have been found to be fairly reliable indicators of the quality or confidence of each putative exon. These scores represent the probability of a given exon being emitted by the GHMM in *any* parse, regardless of the other features emitted as part of the parse. This probability can be computed via *posterior decoding*, which we described in section 6.10 in the context of pure HMMs. The technique involves summing over all possible left contexts of a feature via a dynamic programming algorithm called the *Forward algorithm*, and similarly for the right contexts of the feature via the *Backward algorithm*. These can be generalized to the GHMM case as follows.

Given a weighted ORF graph for a sequence, we compute a generalized Forward variable f_i:

$$f_i = \begin{cases} 1 & \text{if } deg_{in}(y_i) = 0 \\ \sum\limits_{j \in pred(i)} f_j \lambda_{j \rightarrow i} P_e(s_i | y_i) & \text{otherwise} \end{cases} \tag{8.47}$$

where $\lambda_{j \rightarrow i}$ is the edge score for the edge connecting vertices (i.e., states) y_j and y_i in the ORF graph:

$$\lambda_{j \rightarrow i} = P_e(S_{(end(j), begin(i))} | y_{(j,i)}, \delta_{(j,i)}) P_d(\delta_{(j,i)} | y_{(j,i)}) P_t(y_i | y_j),$$

where *begin(i)* is the first nucleotide position of the signal s_i emitted by state y_i, *end(i)* is the position of the last nucleotide in s_i, and $d_{(j,i)}$ is the length of the feature represented by that edge (i.e., corresponding to the variable-length state $y_{(j,i)}$); *pred(i)* is the set of predecessors of y_i in the ORF graph; and $deg_{in}(y_i)$ is the number of predecessors of y_i in the graph. Note that we assume the ORF graph has been pruned so that all remaining nonterminal vertices have both nonzero in-degrees and nonzero out-degrees (see Exercise 8.16).

The generalized Backward variable b_i is computed as:

$$b_i = \begin{cases} 1 & \text{if } deg_{out}(y_i) = 0 \\ \sum\limits_{j \in succ(i)} b_j \lambda_{i \rightarrow j} P_e(s_j | y_j) & \text{otherwise} \end{cases} \tag{8.48}$$

for *succ(i)* the successors of y_i in the ORF graph and $deg_{out}(y_i) = |succ(i)|$. The posterior probability of a variable-length feature $S_{[begin(i),end(j)]}$ delimited by signals s_i and s_j generated by states y_i and y_j can then be written:

$$P(s_i, s_j, \delta|S) = \frac{P(s_i, s_j, \delta, S)}{P(S)} \tag{8.49}$$

for feature length δ. We can factor the numerator as:

$$\begin{aligned} &P(s_i, s_j, \delta, S) \\ &= P\left(S_{[0,end(i)]}, y_i\right) P\left(S_{(end(j),L)}|y_j\right) P_e\left(S_{(end(i),begin(j))}|y_{(i,j)}, \delta\right) P_t(y_j|y_i) P_d(\delta_{(i,j)}|y_{(i,j)}) P_e(s_j|y_j) \\ &= f_i \cdot b_j \cdot \lambda_{i\to j} \cdot P_e(s_j|y_j). \end{aligned} \tag{8.50}$$

The $P(S)$ term in Eq. (8.49) is given by:

$$P(S) = \sum_{\substack{right\\termini\\y_i}} P(S|y_i) = \sum_{\substack{right\\termini\\y_i}} f_i. \tag{8.51}$$

In practice, the Forward and Backward variables must also be indexed by phase (see Exercise 8.31). Note that any additional pruning of the ORF graph to reduce non-dead-end vertices (i.e., via "thresholding" during signal sensing) will render the computation only approximative, since the summation will not cover all possible paths; see Burge (1997) for a more exact procedure.

8.5.5 *The ORF graph as a data interchange format*

The very fact that the construction of an ORF graph and its subsequent use in decoding can be performed separately raises an interesting possibility. Imagining for the moment that all currently available gene finders could be modified to emit an ORF graph (in some standard file format) rather than a set of gene predictions, we could conceive of a single program D_U which, given a weighted ORF graph G, would extract the optimal gene prediction $\phi = D_U(G)$ from the graph using the methods described in section 8.5.2. D_U could then be called a *universal decoder*, since it could be used in conjunction with any gene finder emitting graphs in the standard format, and the need for an explicit decoding procedure in any of these gene finders would thus be eliminated.

Now let us imagine that the available comparative gene finders (Chapter 9) could be modified so as to take as input one or more weighted ORF graphs in the standard file format, and to produce as output a re-weighted ORF graph for the target genome. This re-weighted ORF graph could of course be subjected to decoding by the universal decoder D_U in order to extract a single optimal parse of the target genome, but since the comparative gene finder used to re-weight the ORF graph for the target genome presumably had access to additional information such as BLAST hits and/or ORF graphs for genomes of related organisms, the re-weighted ORF graph might be expected to produce greater gene-finding accuracy when provided to D_U for decoding than would the original graph.

In this way, not only would the construction and decoding of ORF graphs be separated into distinct programs, but also the methods used in their construction and re-weighting (in the case of comparative gene finding) could be decoupled, so that the end-user of a particular comparative gene finder F_C could choose the best *ab initio* gene finder F_A available for the organism of interest to use in constructing the initial graph. One could perhaps even conceive of cascades of general-purpose graph re-weighting programs which would sequentially process a graph and modify its weights according to some specialized computational method or evidence resource specific to that program; the job of the genome annotator would then be to identify the best graph generator and the best combination of re-weighting utilities to use in producing the final gene predictions.

A further benefit of such an (admittedly utopian) scheme would be that the extraction of suboptimal parses (as described in section 8.5.3) would also be greatly facilitated, as the availability of a generalized universal decoder $D_{subopt} : \mathcal{G} \times \mathbb{N} \mapsto \Phi$ (for $\mathcal{G}$ the set of all ORF graphs and Φ the set of all parses) capable of extracting the N best parses from a graph would allow the exploration of suboptimal parses within the context of any gene-finding system which emitted ORF graphs.

In this way, *ab initio* gene finders would be largely commoditized, allowing gene-finding researchers to effectively ignore mundane issues related to parsing and phase-tracking, etc., and to concentrate on the more important task of finding optimal methods for incorporating external evidence into the gene prediction process. Figure 8.18 illustrates some of these ideas.

While it may indeed be unrealistically utopian to imagine that all the authors of popular gene-finding programs will opt to modify their programs to conform to a standard graph-based interface (assuming such a standard could be designed and agreed upon by the gene-finding community), it should be clear that many (if not most) state-of-the-art *ab initio* eukaryotic gene finders could be modified with relative ease to produce ORF graphs, since nearly all such programs utilize some form of dynamic programming in which optimal predecessor signals are selected for linking into some form of trellis structure comprising the final gene prediction. At the stage where a number of potential predecessors are compared based on their inductive scores, a vertex for the current signal could be instantiated and linked back to all such potential predecessors, with the new edges weighted according to some function of a predecessor's inductive score. Thus, to the extent that most gene finders at least trace out, either explicitly or implicitly, an ORF graph (or some subgraph thereof) via their algorithmic processing of putative signals and their potential predecessors according to some numerical scoring function, modification of these programs to emit a weighted ORF graph should in most cases be almost trivial.

One seeming obstacle to the adoption of the ORF graph as a standard interchange format is the potentially large memory requirements for their storage. An unpruned ORF graph (i.e., one in which all potential noncoding predecessors are retained) can be expected to grow quadratically with the length of the input sequence (see

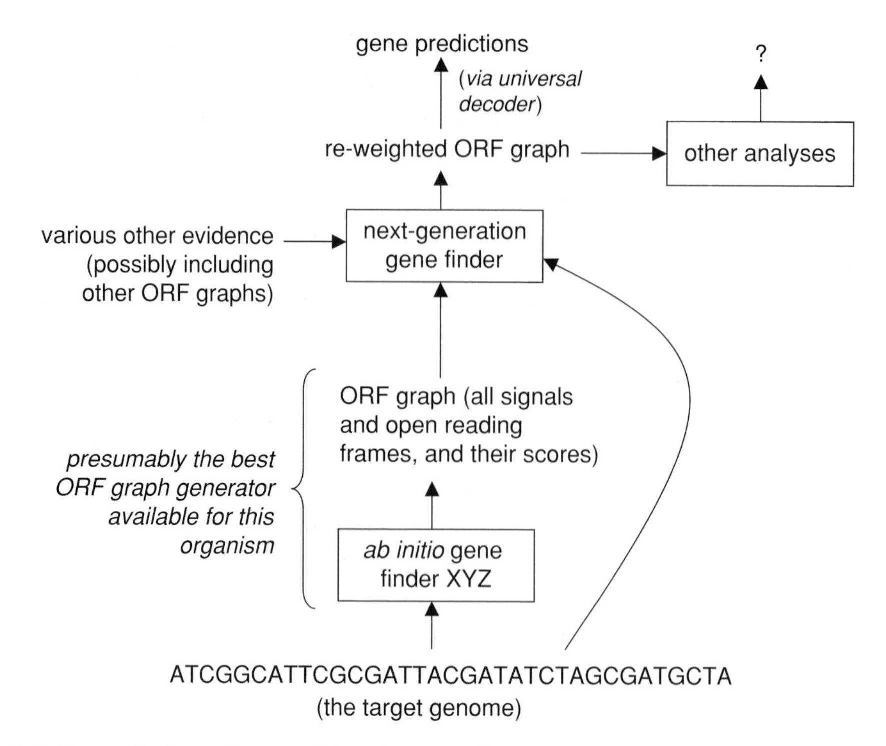

Figure 8.18 Proposed schema for a graph-based system of genome annotation.

Exercise 8.30). One solution to this problem is to permit programs that generate ORF graphs to prune their graphs before emitting them, using methods similar to those described previously (i.e., maintaining only the highest-scoring N noncoding predecessors at any signal during graph construction, for some reasonable N). For many purposes, a graph pruned with a reasonably large N could produce near-optimal results for the application at hand. Allowing for end-users to specify the value of N would allow for greater flexibility in downstream processing.

Another optimization involves representing the graph using a list of vertices and coding edges, with the noncoding edges being implicitly defined via a prefix sum array of noncoding content sensor scores. Such a representation would require effectively linear space, while still allowing for exploration of all suboptimal parses, though consumers of graphs in this representation would be burdened with the task of reconstructing noncoding edges as needed.

An additional advantage of the proposed paradigm for computational gene finding is the simplicity and elegance of graph-based algorithms versus their corresponding sequence-based analogs. Once an ORF graph has been produced by an *ab initio* gene finder, all syntactic and phase constraints relating to gene structure may be fairly simply abstracted into a reduced set of graph traversal operations that may be encoded in an appropriate *application programming interface (API)*. Because such an API could be implemented in a high-level programming language such as

Perl (e.g., via a package such as *BioPerl* – Stajich *et al.*, 2002) almost as efficiently as in a low-level systems language such as C/C++, the science of gene prediction could conceivably be opened to larger segments of the research community, including gene researchers with perhaps less of a computational background, but more intuition for the ultimate biological goals of accurate genome annotation.

As a final bit of speculation, we posit that the archiving of ORF graphs for every sequenced genome in some central public repository[1] might further promote annotation (and re-annotation) efforts by making available the precomputed syntactic analyses of the sequences. In much the same way as whole-genome alignments are now produced and archived at centers such as UCSC for easy download by bioinformaticians over the internet, ORF graphs aligned using methods such as those described in section 9.5 could likewise be archived for later re-processing by future tools. Just as archived genome alignments eliminate (or at least reduce) the need for the end-user to run costly alignment procedures on site, the existence of pre-computed ORF graph alignments between two or more genomes might considerably simplify the work of both end-users and the developers of next-generation comparative annotation tools.

8.6 Training a GHMM

We now consider the task of parameter estimation for GHMM-based gene finders. As argued in section 6.9 for pure HMMs, the ideal goal when training a gene finder is to find that parameter set θ^* which maximizes the expected accuracy of the program on unseen data. Assuming that we accept the training set as a reasonable surrogate for this "unseen data," then we may consider maximizing the joint probability of the gene parses in the training set T:

$$\begin{aligned}
\theta^* &= \underset{\theta}{argmax}\left(\sum_{(S,\phi)\in T} \log P(\phi|S,\theta)\right)\\
&= \underset{\theta}{argmax}\left(\sum_{(S,\phi)\in T} \frac{\log P(S,\phi|\theta)}{P(S|\theta)}\right)\\
&= \underset{\theta}{argmax}\left(\sum_{(S,\phi)\in T} \log \frac{\prod_{i=1}^{n+1} P_e(S_i|y_i,d_i)P_t(y_i|y_{i-1})P_d(d_i|y_i)}{P(S|\theta)}\right).
\end{aligned} \tag{8.52}$$

Unfortunately, the above equation is not known to have a simple closed-form solution. Historically, an alternative criterion based on *maximum likelihood estimation* has been optimized instead of Eq. (8.52), and we will begin by describing this approach. Following this we will explore a general gradient-ascent method

[1] Such as the NCBI – www.ncbi.nlm.nih.gov/.

for optimizing the accuracy of the gene finder on the training set, thereby pursuing a training protocol more consistent with the notion of *discriminative training* introduced in section 6.9 with regard to pure HMMs.

8.6.1 Maximum likelihood training for GHMMs

Maximum likelihood training of a GHMM is very straightforward. We have already described MLE training for each of the common signal and content sensors (Chapter 7), which implement the various emission functions P_e for the states of the GHMM. The duration functions P_d can be estimated by observing the length distributions of appropriate features in a training set, smoothing the resulting distributions as necessary, and then using either a histogram or a curve-fitting approach to implement the P_d function as described in section 2.7. The only remaining parameters are those of the P_t function, which can be estimated exactly as for HMMs, by observing how often each type of feature is followed in a training sequence by each of the other types of features.

Thus, the common MLE protocol for GHMMs optimizes the three likelihood terms in the numerator of Eq. (8.52), and thus optimizes $\sum_i \log P(S_i, \phi_i|\theta)$, subject to the standard Markov assumptions. However, because it does not optimize the entire ratio in Eq. (8.52), this procedure does not maximize $\sum_i \log P(\phi_i|S_i, \theta)$.

8.6.2 Discriminative training for GHMMs

It has long been observed (though not so commonly documented) by practitioners in the field that the accuracy of a GHMM-based gene finder trained using the above MLE techniques can often be significantly improved by tuning, or "tweaking," the parameters manually after MLE training. This hand-tuning can be thought of as a very crude form of *discriminative training* (section 6.9; see also section 12.4), since the tuning is typically performed iteratively while the practitioner observes the effect of each parameter change on the prediction accuracy on the training set – the practitioner effectively re-estimates the model parameters so as to optimize the expected accuracy on unseen data. In this way, we can conceive of a form of discriminative training for GHMMs which utilizes an automatic *gradient-ascent* procedure (see sections 2.5 and 10.15) to optimize the accuracy of the gene finder on the training set.

Such an approach is illustrated schematically in Figure 8.19. Under this scheme, we utilize the MLE-trained gene finder as a starting point and iteratively improve the accuracy of the gene finder on the training set by varying model parameters in some random or systematic way and observing which changes lead to accuracy improvements. Similar approaches might be attempted based on the use of *genetic algorithms* (section 10.12) or *simulated annealing* (section 10.13).

The total number of parameters that need to be estimated when training a typical GHMM-based gene finder is on the order of tens of thousands; the large

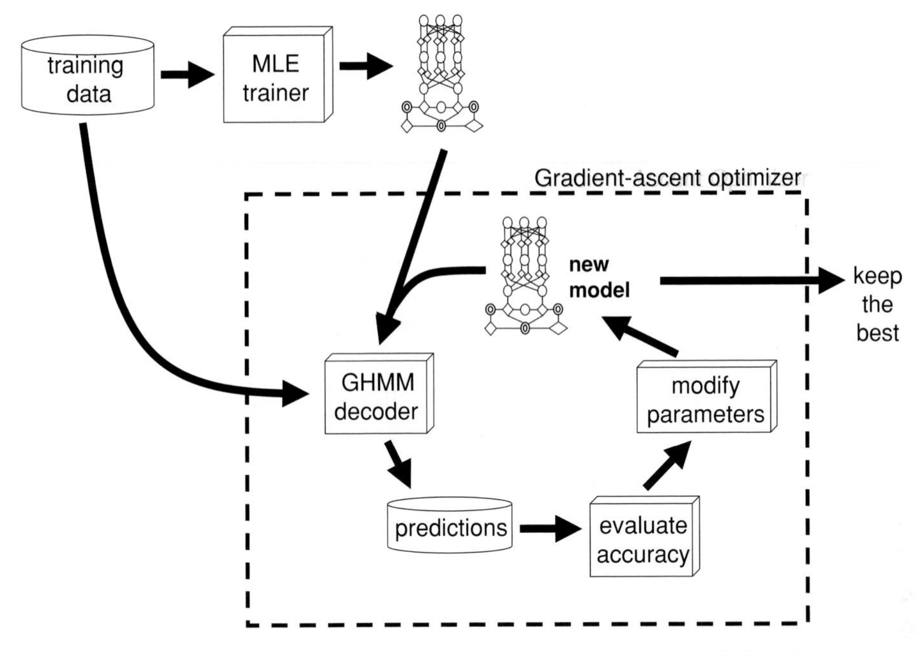

Figure 8.19 A gradient-ascent scheme for optimizing a GHMM-based gene finder. The process begins with an MLE-trained model and proceeds by perturbing the model's parameters and observing the effect of such perturbations on the accuracy of the resulting model when used to predict genes in the training set. Favorable perturbations are accepted and unfavorable ones are rejected.

bulk of these are the *n*-gram statistics comprising the content sensors (in the case of Markov chains and related models). However, some of these parameters exert a much stronger influence on the accuracy of the gene finder than do others, so that it is often possible to achieve significant gains in accuracy by considering only a very small subset of the parameters for modification. Some of the more useful parameters for this purpose are:

- mean intron, intergenic, and UTR lengths
- transition probabilities
- exon "optimism" (section 7.3.4)
- sizes of all signal sensor windows
- locations of consensus regions within signal sensor windows
- emission orders for Markov chains and other models
- sensitivity of signal thresholds (when thresholding is used – section 8.3.1)
- number of signal *boosting* iterations to utilize during signal training (section 11.1)
- *skew* and *kurtosis* of exon length distributions (section 2.6).

The exon "optimism" parameter is a constant which is multiplied by the probability for any transition leading into an exon state, and has been found anecdotally to be useful during manual tuning of GHMM parameters (i.e., "coding bias" in the

Table 8.3 Gradient ascent vs. maximum likelihood estimation for GHMM training. Both protocols were applied to 1000 training and 1000 (distinct) test genes from *Arabidopsis thaliana*

	Nucleotide SMC	Exon *F*-measure	Gene sensitivity
Gradient ascent	94%	81%	48%
MLE	90%	71%	33%

Source: Results from Majoros and Salzberg (2004).

terminology of Siepel and Haussler, 2004b). Signal thresholds are those that are mentioned in section 8.3.1 for deciding which consensus sequences may be safely ignored when searching for putative signals in DNA. The *skew* and *kurtosis* of a distribution were introduced in section 2.6. Signal "boosting" will be described in section 11.1.

To illustrate the potential utility of a gradient-ascent approach based on a subset of the parameters listed above, Table 8.3 shows the results of optimizing 29 parameters of a GHMM-based gene finder to improve the prediction accuracy on *Arabidopsis thaliana*; the optimizer also explored various *parameter tying* options (section 6.4). As can be seen from the table, a significant improvement in accuracy was observed at all levels.

In order for a gradient-ascent approach to be feasible in practice, one must identify an objective function which both reflects the criterion one wishes to optimize, and which can be efficiently evaluated. The more quickly we can evaluate the objective function, the more evaluations we can afford in a given amount of time, and therefore the more effectively we can explore the parameter space in search of an optimal parameterization.

A particularly convenient objective function is one based on running the gene finder on a test set T and evaluating the whole-gene-level accuracy of the predictions relative to the known gene structures. Such an accuracy measure is easy to obtain, and provides a reasonable heuristic for crudely approximating $\sum_{(S,\phi)\in T} \log P(\phi|S,\theta)$ by indicating roughly how often the current model θ would cause the correct parse ϕ to be predicted for training sequence S. In practice, we can utilize accuracy measurements at various levels of granularity, including the level of nucleotides, exons, or genes. One possible objective function combining all of these levels is:

$$f(\theta) = w_0 A_{nuc} + w_1 A_{exon} + w_2 A_{gene}, \tag{8.53}$$

where A_{nuc}, A_{exon}, and A_{gene} represent F-measures or similar metrics at the nucleotide, exon, and gene levels, respectively, and the w_i are weights. Using a combination of scores in this way helps to smooth the objective function – this can be seen by considering the use of A_{gene} alone, in which case the objective function would behave

Table 8.4 Prediction accuracy of a GHMM-based gene finder (TIGRscan) on 500 *Arabidopsis thaliana* genes, compared to the accuracy of our pure HMM model H_{95} (with fifth-order emissions) from Chapter 6, as well as UNVEIL, another pure HMM-based gene finder; all numbers are percentages

	Nucleotides			Splice sites		Start/stop codons		Exons			Genes
	Sn	Sp	*F*	Sn	Sp	Sn	Sp	Sn	Sp	*F*	Sn
H_{95}	98	97	98	90	81	66	62	74	67	70	27
UNVEIL	94	98	96	79	83	75	71	68	81	70	37
TIGRscan	94	96	95	87	89	77	74	79	80	80	45

very much like a *step function* (see Exercise 8.29) in which an entire gene either is or is not predicted correctly, but the proportion of the gene that is predicted correctly is not measured. For the purposes of gradient-ascent optimization, a smooth function is desired in order to provide maximal information to the optimization procedure regarding the gradient of the current location in the search space.

A number of decisions must be made about the way in which a gradient-ascent optimizer will operate on a given data set: the initial step size and the schedule for reducing that step size; the number of dimensions; the parameters to be optimized; whether to allow occasional score reduction as in *simulated annealing* (section 10.13); whether to optimize all parameters on every iteration, or to sample from them stochastically; how to seed the optimizer; these all must be decided before an optimizer can be applied. Many practitioners adopt the practice of making these choices in an ad-hoc fashion, repeating the optimization procedure for different experimental set-ups until a model is trained which appears to be optimal. Optimization methods in general have received much attention in the literature (e.g., Fletcher, 1980; Powell, 1981). An open-source software package for gradient-ascent optimization is described in section 10.15.

8.7 Case study: GHMM versus HMM

In Chapter 6 we quantified the effect of various improvements to our simple HMM-based gene finder by testing the accuracy of the models on a 500-gene *Arabidopsis thaliana* test set. Here we report the accuracy of a GHMM-based gene finder (*TIGRscan* – Majoros *et al.*, 2004) on the same test set. The results are shown in Table 8.4.

As can be seen from the table, the GHMM-based gene finder achieves significantly greater accuracy at the exon and whole-gene levels: 80% exon accuracy versus the 70% achieved by the HMMs, and 45% gene accuracy versus the 37% of the pure HMM-based program UNVEIL. Interestingly, the nucleotide accuracy is highest for the 5th-order model H_{95} and lowest for the GHMM-based system, suggesting that the different programs are finding somewhat different sets of genes in the test set. It is tempting to speculate that this ensemble of programs might therefore make a particularly useful combination of inputs to a *combiner* system such as those described in section 9.2.

EXERCISES

8.1 Explain why a fully general Viterbi-like solution for GHMM decoding requires that in each cell we consult potential predecessor cells arbitrarily far back in the matrix. Hint: you may want to make use of the fact that the duration distribution can be of arbitrary shape.

8.2 Continuing with the previous exercise, explain how the situation changes if we assume that all duration distributions are geometric.

8.3 Let M be a (finite) HMM capable of generating infinitely many different sequences. Prove that the length distribution induced by M has a geometric tail.

8.4 Explain how it is possible to enforce a nongeometric length distribution during decoding by modifying an HMM's transition probabilities and relaxing sum-to-1 constraints on those transition probabilities. Is it necessary to relax sum-to-1 constraints? Explain your answer.

8.5 Let M be an HMM having N states, and suppose that M is known to have generated a sequence S for which $|S| > N$. Prove that M can generate infinitely many different sequences. Hint: you may want to look up the *pumping lemma* from the theory of finite automata.

8.6 In Figure 8.3(b), observe that from any internal exon state $E_i (0 \leq i \leq 2)$ one can transition (indirectly, through a D state) to any of the three intron states I_j $(0 \leq j \leq 2)$, yet from \ , I_j one cannot transition to an arbitrary internal exon state (after first passing through an A state); i.e., of the internal exon states, only state E_j may immediately follow (after passing through an A state) an I_j intron in a run of this model. Explain why the model was designed in this way. What exactly does the i represent for exon state E_i? What exactly does the j denote for intron state I_j?

8.7 Explain how the assumption of *sparse* GHMM state topologies (i.e., with all in-degrees bounded by a small, fixed constant) reduces the time complexity of the decoding problem. Can we expect gene-finding GHMMs to be sparse in practice? Justify your answer. Hint: consider how the definition of eukaryotic gene syntax is likely to change over time.

8.8 Show that the decoding complexity of a GHMM can be reduced to $O(|Q|^2 \times |S|)$ by imposing a maximum feature length L.

8.9 Implement a program to generate random DNA sequences with the following nucleotide distributions. Use the program to estimate the density of stop codons in sequences having similar composition to these:

(a) A = 25% C = 25% G = 25% T = 25%
(b) A = 21% C = 27% G = 25% T = 28%
(c) A = 18% C = 31% G = 31% T = 20%
(d) A = 28% C = 27% G = 22% T = 23%
(e) A = 24% C = 19% G = 26% T = 31%

8.10 As illustrated in Figure 8.10, the recommended method of partitioning an input sequence during GHMM decoding is to evaluate a modestly sized window surrounding each putative signal using a signal sensor specific to that signal type, and to evaluate only the regions between windows using a content sensor such as a Markov chain. An alternative would be to reduce the signal sensor windows so as to cover only the consensus positions (e.g., ATG for a start codon), so that the regions evaluated by the content sensors can extend right up to the putative signal consensus. Explain why the former practice is preferable to the latter.

8.11 Give a proof of correctness for Algorithm 8.1.

8.12 Give a proof of correctness for the reverse-strand version of Algorithm 8.1.

8.13 Give a proof of correctness for the reverse-strand version of Algorithm 8.2.

8.14 Prove that any GHMM-based gene finder having the same state topology of GENSCAN will induce a geometric distribution on the number of exons.

8.15 Continuing with the previous exercise, show also that a geometric distribution is induced on the number of genes.

8.16 Give an algorithm for eliminating from an ORF graph any nonterminal vertex v_i having $deg_{in}(v_i) = 0$ or $deg_{out}(v_i) = 0$.

8.17 Give a proof of correctness for the algorithm designed in the previous exercise.

8.18 Specify in formal pseudocode the algorithm mentioned in section 8.5.2 for evaluating the probability of a portion of a gene parse within a specific interval of the input sequence via posterior decoding.

8.19 Formalize (in pseudocode) the algorithm sketched at the end of section 8.5.3 for extracting the N highest-scoring paths from an ORF graph. Describe how the ORF graph formalism must be extended to support this algorithm.

8.20 Derive a formula for the minimal ORF length L for which we can be P% confident that such an ORF is not merely due to chance (i.e., that it is not merely a noncoding region which happens by chance to lack stop codons). Assume equal nucleotide frequencies and independence of noncoding residues.

8.21 Building on your answer to the previous question, modify your formula to account for unequal nucleotide frequencies.

8.22 Give a formal algorithm for the procedure mentioned in section 8.3.1 for detecting a stop codon which straddles an intron. Include phase-tracking in your solution.

8.23 Give a proof that the use of likelihood ratios as given by Eqs. (8.18, 8.19) will not change the semantics of a PSA decoder, as long as a factorable model is used in the denominator.

8.24 Explain why the phase parameter to RC_I in Eq. (8.28) is necessary in order to avoid greedy behavior. Explain also how this issue is addressed in the PSA algorithm.

8.25 Extend the proof of Theorem 8.2 to address the reverse strand.

8.26 Describe how the PSA decoding procedure can be modified with an additional set of variables to keep track of the number of putative exons present in each prospective parse.

8.27 Modify the `highestScoringPath()` algorithm from section 2.14 to work with ORF graphs, and in particular, to enforce phase constraints.

8.28 Explain why training a GHMM by maximizing the likelihood of the full training sequences would not be useful for gene finding.

8.29 A step function is one in which changes in the *y*-value are discrete rather than continuous (think of a set of stairs viewed from the side). Explain why approximating an objective function with a step function, rather than a continuous function, can make gradient-ascent optimization more difficult.

8.30 Explain why an unpruned ORF graph for a typical genome may be expected to grow quadratically with the size of the genome. Which part of the graph grows quadratically?

8.31 Modify the equations in section 8.5.4 for posterior decoding to account for phase.

9

Comparative gene finding

The most promising of current methods for computational gene prediction are those that utilize *comparative* information by incorporating evidence from the genomes of one or more additional organisms, or from expression data such as ESTs or proteins, or even by comparing and combining the predictions of other gene-finding systems (Guigó *et al.*, 2006). In this chapter we review the various comparative methods which have been successfully employed to improve gene-finding accuracy, as well as several emerging techniques at the forefront of current gene-finding research.

In Figure 9.1 we summarize the major approaches to comparative gene finding that have been applied to date, and that we will be considering shortly. As it may be argued that all scientific and engineering endeavors tend to advance in an evolutionary manner – i.e., by modifying an existing theory or method so as to obtain an incremental improvement in descriptive, predictive, or manipulative power – we draw attention to the similarity of Figure 9.1 to a *phylogeny* or *family tree* (section 1.2), in which case we can view the paths through the figure as evolutionary lineages – whether lineages of software, or of ideas embedded in software. Furthermore, as it is certainly the case that comparative methods are at this point in time still evolving, the following exposition will necessarily be less dogmatic than our descriptions of *ab initio* methods given in the foregoing chapters, relying to a correspondingly greater degree on the presentation of case studies to illustrate the various methods that have been put forth. In order that the reader may better appreciate the evolutionary history of current state-of-the-art comparative methods, we present the major developments in roughly chronological order, so that our ordering of topics will be seen to constitute a *topological sort* (section 2.14) of the graph shown in Figure 9.1.

It will be seen shortly that the analogy with evolutionary patterns of descent is an appropriate one for another reason. Recalling from section 1.2 that evolution can be described most concisely as a form of *descent with modification* – with the modification being driven largely by the action of natural selection – it should

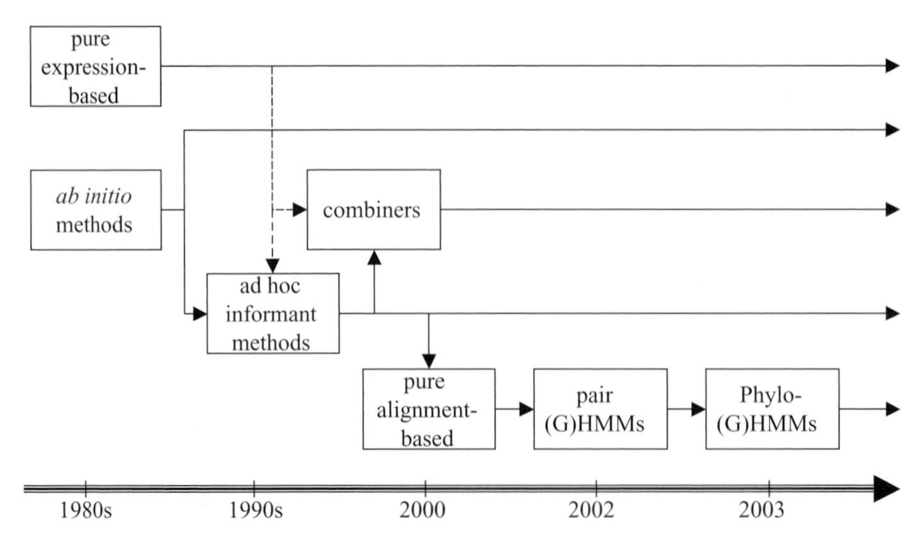

Figure 9.1 A timeline of computational advances in comparative gene finding.

seem reasonable to suppose that the patterns of evolutionary modification in a set of related genomes may inform the gene-finding process inasmuch as the selective pressures constraining evolutionary changes may be inferred based on our understanding of organismal and molecular biology. From the *central dogma* expounded in Chapter 1 we may deduce that the coding portions of eukaryotic genes should be subject to the strongest selective pressures – in particular, selective pressures against changes resulting in significantly altered proteins – so that the greatest tendency toward *conservation* (i.e., matching ancestral and descendent residues) should be evident in the CDS portions of genes (though exceptions certainly exist – see, e.g., Kamal *et al.*, 2006). Thus, one may envision that at some evolutionary distance, the random drift in noncoding regions which is allowed to proceed largely unchecked due to a relative lack of selective pressures in those regions will result in a very low level of nucleotide conservation, in contrast to the high level of amino acid conservation that would be observed in the encoded proteins of the respective coding regions.

In addition to evolutionary conservation, a number of comparative methods have made extensive use of *expression-based* evidence, including ESTs and full-length cDNAs synthesized from mRNA populations in the living cell, as well as the amino acid sequences of known proteins. Large databases of such expression data now exist, and are efficiently searchable via BLAST. Given a protein or an assembly of ESTs corresponding to a previously uncharacterized locus, it is in many cases possible to infer with high accuracy the exonic structure of the underlying gene; when used in combination with other sources of evidence, such approaches can be used to produce automatically a draft annotation of an entire genome (e.g., Allen *et al.*, 2006; Guigó *et al.*, 2006).

9.1 Informant techniques

Among the more popular of the early comparative gene finders were those that exemplified what we call the *informant-based* approach to comparative gene finding; these include programs such as *TWINSCAN* (Korf *et al.*, 2001), *GenomeScan* (Yeh *et al.*, 2001), *SGP-2* (Parra *et al.*, 2003), and a number of others. This approach involves the use of one or more *informant sequences* aligned to the *target genome* (the genome in which we wish to predict genes). These methods largely rely on the expectation that protein-coding sequences, because of their clear importance to the organism, will be more strongly conserved over evolutionary time through the action of natural selection than so-called "nonfunctional" DNA (e.g., introns, intergenic regions, and UTRs – excluding the various regulatory elements that may occur within those regions).

It is interesting to note that while humans and mice are thought to be separated by more than 65 million years of evolution, approximately 99% of mouse genes appear to have a direct counterpart in the human genome, and vice versa (Mouse Genome Sequencing Consortium, 2002). One may thus hope, by observing the level of conservation between a target region and a set of aligned informant sequences – in nucleotide and/or amino acid space – that selective pressures, and therefore the coding/noncoding status of a given stretch of DNA, will be the more easily discerned computationally, thereby improving gene prediction accuracy.

The use of expression-based evidence has likewise been harnessed by a number of these systems to improve prediction accuracy. In these cases, the details of the database search and the filtering, clustering, and aligning of the resulting database hits to the target genome are almost universally left to specialized programs such as BLAST for searching, *sim4* (Florea *et al.*, 1998), *BLAT* (Kent, 2002), or *GeneSequer* (Usuka *et al.*, 2000) for spliced alignment, or *PASA* (Haas *et al.*, 2003) or *CRASA* (Chuang *et al.*, 2003) for clustering and ambiguity resolution. In the following sections we will thus confine our attention to the algorithmic techniques used in integrating this evidence into the gene-prediction process, and refer the interested reader to the primary literature for the details of these auxilliary programs.

9.1.1 *Case study: TWINSCAN*

The program *TWINSCAN* (Korf *et al.*, 2001) consists of a GHMM-based gene finder (GENSCAN++) which has been modified to incorporate homology information in the form of BLASTN hits. The program requires as inputs both a target genome sequence and a set of nucleotide HSPs (see section 1.4) from a related genome. From this the program is able to infer a *conservation sequence* consisting of symbols from the 3-letter alphabet $\alpha_C = \{``|", ``:", ``."\}$ (*pipe*, *colon*, and *period*), which we will denote, respectively, $\{s_{match}, s_{mismatch}, s_{unaligned}\}$; the first denotes a match between the target and informant nucleotides, the second denotes a mismatch, and the

```
     GCTATCGGTCTTA
  ...|:||:||||:|:|...
ATCGGTAACGGTGTAATGC
```

Figure 9.2 Example of a conservation sequence. A target sequence S is shown at the bottom, with an HSP at the top. The conservation sequence, indicating matches (|), mismatches (:), and unaligned positions (.), is shown in between.

third denotes a position for which no informant HSP was found. An example of a conservation sequence is shown in Figure 9.2.

The decoding algorithm employed by TWINSCAN is an extension of the standard GHMM decoding algorithm, with the difference being the conservation term, $P(C\,|\phi)$:

$$\begin{aligned}
\phi^* &= \underset{\phi}{argmax}\ P(\phi|S,C) \\
&= \underset{\phi}{argmax}\ \frac{P(\phi,S,C)}{P(S,C)} \\
&= \underset{\phi}{argmax}\ P(\phi,S,C) \\
&= \underset{\phi}{argmax}\ P(\phi)P(S,C\,|\phi) \\
&= \underset{\phi}{argmax}\ P(\phi)P(S|\phi)P(C\,|S,\phi) \\
&= \underset{\phi}{argmax}\ P(\phi)P(S|\phi)P(C\,|\phi).
\end{aligned} \tag{9.1}$$

The first line in the derivation states that the optimal parse ϕ^* is the one which maximizes $P(\phi|S,C)$, the conditional probability of the parse given the target sequence S and the conservation sequence C. This conditional probability we then factor using the conditional probability rule (line 2), and then we discard the denominator since it has no effect on the *argmax* (line 3), just as in traditional GHMM decoding. The last three lines factor the joint probability $P(\phi,S,C)$ into the familiar term $P(\phi)P(S|\phi)$ from the GHMM decoding problem (section 8.1), with the additional factor $P(C\,|\phi)$. Note that in the final line we take the liberty of replacing $P(C\,|\phi,S)$ with $P(C\,|\phi)$; this was done largely for the purpose of simplifying the implementation.[1]

What remains is to specify how to evaluate $P(C\,|\phi)$; in TWINSCAN this is accomplished using a 5th-order Markov chain trained on the 3-letter alphabet of conservation sequences; e.g., invoking the term $P(s_{match}|s_{unaligned},s_{unaligned},s_{unaligned},s_{match},s_{mismatch})$ for the sixth position in the example sequence above, or $P(\text{“|”}|\text{“...|:”})$. Parameter estimation for this Markov chain is accomplished by including informant sequences in the training set, encoding these training HSPs

[1] I. Korf, personal communication.

into conservation sequences, and then training the Markov chain on the resulting conservation sequences. During decoding, only the "best" HSP (i.e., the highest scoring) is used in determining the conservation symbol covered by one or more HSPs at a given nucleotide position.

9.1.2 *Case study: GenomeScan*

GenomeScan (Yeh *et al.*, 2001) is somewhat similar to TWINSCAN, with the most salient difference being that BLASTX is used for detecting HSPs, rather than BLASTN as in TWINSCAN; that is, GenomeScan utilizes similarity to known *proteins* rather than similarity to DNA or RNA sequence. Recall from section 1.4 that BLASTX performs 6-frame translation of the target sequence as it searches a given protein database, so that HSPs are identified at the amino acid level, rather than at the nucleotide level as in BLASTN.

Rather than constructing an explicit "conservation sequence" as in TWINSCAN, GenomeScan reformulates the decoding objective function by factoring in a slightly different manner:

$$\begin{aligned}\phi^* &= \underset{\phi}{argmax}\ P(\phi|S, I)\\ &= \underset{\phi}{argmax}\ \frac{P(\phi, S|I)}{P(S|I)}\\ &= \underset{\phi}{argmax}\ P(\phi, S|I)\end{aligned} \tag{9.2}$$

where I represents a BLASTX hit. The probability of the parse ϕ given the target S and the HSP I is factored and then simplified to $P(\phi, S|I)$, which is then treated specially depending on whether the parse ϕ is *consistent* with I – i.e., whether the *centroid* of I (i.e., the highest-scoring region of the BLASTX hit) falls within a putative exon contained in ϕ:

$$P(\phi, S|I) = \begin{cases} \dfrac{P_I P(\phi, S)}{P(\Phi_I)} + P_I P(\phi, S) & \text{if consistent} \\ (1 - P_I)P(\phi, S) & \text{if inconsistent} \end{cases} \tag{9.3}$$

where Φ_I is the set of all parses consistent with I, and $P(\Phi_I)$ can be evaluated using a posterior decoding approach (see section 8.5.4). P_I is the probability that the BLASTX hit I is not *artifactual*, but instead represents a true homology relationship; this explicit modeling of the reliability of the HSP is another difference between GenomeScan and TWINSCAN. For parses inconsistent with I (i.e., line 2 in Eq. 9.3), one can therefore model $P(\phi, S|I)$ as being independent of the HSP – $P(\phi, S)$ – while also including a factor $(1 - P_I)$ to account for the artifactual nature of the HSP. The consistent cases are thus mutually exclusive and therefore include a P_I term – the logical complement to $(1 - P_I)$. The consistent cases are further divided into those in which the HSP is nevertheless artifactual, despite the parse being consistent with it – this is the $P_I P(\phi, S)$ term in the consistent case. The other term in the consistent case is derived from $P_I P(\phi, S|\Phi_I)$ by utilizing the conditional probability rule. In either case (consistent or inconsistent), the $(1 - P_I)$ term is modeled as being an integral root of the BLASTX score P_B: $(1 - P_I) = P_B^{1/r}$, for some integer r which is presumably chosen in a maximally discriminative manner with respect to some training set. Additional details can be found in Yeh *et al.* (2001).

9.1.3 *Case study: SGP-2*

The program *SGP-2* (Parra *et al.*, 2003) is yet another informant-based program which utilizes HSPs to modify exon probabilities. However, whereas TWINSCAN uses BLASTN to find nucleotide homology and GenomeScan uses BLASTX to find amino acid homology to known proteins in a protein database, SGP-2 utilizes TBLASTX to find amino acid homology between the target nucleotide sequence and an informant nucleotide sequence, by considering all 6-frame translations of the nucleotide sequences. SGP-2 also differs from the other two programs in that its underlying model is not a GHMM or HMM. The program instead relies on finding all high-confidence putative exons in a sequence and then, in a separate pass, assembling these into consistent gene structures using a dynamic programming algorithm (Guigó, 1998). Content scores for putative exons are still, however, computed via Markov chains. The underlying noncomparative gene finder used by SGP-2 is *GENEID* (Blanco *et al.*, 2002).

SGP-2 incorporates homology evidence into the GENEID exon-assembly procedure as follows. The final score $s(e)$ of an exon e used during exon assembly is defined as the sum of the raw GENEID score for the exon and a weighted conservation score:

$$s(e) = s_{GENEID}(e) + w \cdot s_{conservation}(e) \tag{9.4}$$

where the $s_{GENEID}(e)$ and $s_{conservation}(e)$ scores are log-likelihood ratios; hence the use of addition in combining them. The coefficient w is an *ad-hoc* weight which determines the influence of the homology evidence on the final exon score, and is presumably chosen so as to maximize gene finding accuracy on the training set. The conservation score $s_{conservation}(e)$ is simply a sum of the maximal TBLASTX scores for the HSPs overlapping a putative exon:

$$s_{conservation}(e) = \sum_{\substack{HSP \\ h}} s_{TBLASTX}(h) \cdot p_{overlap}(h, e). \tag{9.5}$$

That is, $s_{conservation}(e)$ is the weighted sum of HSP (log-likelihood) scores $s_{TBLASTX}(h)$ for maximal HSPs h partitioning the putative exon e, with the weight $p_{overlap}(h, e)$ being the proportion of the exon that is overlapped by the HSP. HSPs are thus projected onto the target genomic sequence in a mutually exclusive manner, so that, e.g., a higher-scoring HSP h_i will “eclipse” any lower-scoring HSP h_j overlapped by h_i (with the eclipsing process acting on those overlapped portions of HSPs in the case where one HSP overlaps only a portion of another).

9.1.4 *Case study: HMMgene*

A number of early *ab initio*, noncomparative methods were subsequently augmented to incorporate homology evidence from database matches, and were generally shown to be rendered more accurate in this way. One of these was the HMM-based gene finder *HMMgene* (Krogh, 2000). A number of evidence types were incorporated into the base gene finder, including protein matches, cDNA and EST alignments (including inferred exons and introns from the spliced alignment of transcripts to the target genome), and genomic repeats. These various evidence sources – here termed *database matches* – were "compressed" in a mutually exclusive manner according to an evidence hierarchy, resulting in a series of intervals describing matches and the gaps where no matches were found.

The HMM decoding procedure of the underlying gene finder was then augmented to incorporate a conservation term, $P(I|\phi)$:

$$\begin{aligned}
\phi^* &= \underset{\phi}{argmax}\ P(\phi|S, I) \\
&= \underset{\phi}{argmax} \frac{P(\phi, S, I)}{P(S, I)} \\
&= \underset{\phi}{argmax}\ P(\phi, S, I) \\
&= \underset{\phi}{argmax}\ P(\phi)P(S, I|\phi) \\
&= \underset{\phi}{argmax}\ P(\phi)P(S|\phi)P(I|S, \phi) \\
&= \underset{\phi}{argmax}\ P(\phi)P(S|\phi)P(I|\phi) \qquad (9.6)
\end{aligned}$$

for *evidence track* (i.e., "informant") I. The final step in the above derivation involves the assumption that $P(I|S, \phi) = P(I|\phi)$ – i.e., conditional independence between I and S, given ϕ. The alert reader will observe that this derivation closely parallels that given earlier for the program TWINSCAN. However, whereas TWINSCAN modeled an informant sequence C which explicitly represented the matches and mismatches (and unaligned positions) between the target sequence and an informant sequence, the homology-aware version of HMMgene modeled an evidence track I using intervals annotated only with a single match score (i.e., the "confidence" of the alignment) and a match type (e.g., EST, cDNA, protein, etc.). Each match score was then weighted by a fixed matrix indexed by match type and HMM state. In this way, the modeling of conservation patterns between the target sequence and the matched element (e.g., protein, cDNA, etc.) was somewhat coarser than in TWINSCAN. In contrast, we will see when we consider phylogenetic HMMs later in this chapter that an even more *finely* detailed model of conservation patterns may be employed in comparative gene finding.

9.1.5 Case study: GENIE

Another early gene finder that was subsequently modified to incorporate homology evidence was GENIE (Kulp and Haussler, 1997). Like GENSCAN, this program was one of the earliest implementations of a GHMM-based gene finder. The decoder utilized by the base gene finder utilized a pruned *ORF graph*, much like that described in section 3.4. Homology evidence included protein homology discovered via BLASTX. The use of protein homology evidence thus required the modeling of the translation process from the target DNA sequence S to its protein product, denoted here $\tau(S)$:

$$\begin{aligned}
(1)\ \phi^* &= \underset{\phi}{argmax}\, P(\phi, S|I) \\
(2)\quad &= \underset{\phi}{argmax}\, P(\phi|I)P(S|\phi, I) \\
(3)\quad &= \underset{\phi}{argmax}\, P(\phi)P(S|\phi, I) \\
(4)\quad &= \underset{\phi}{argmax}\, P(\phi)P(S, \tau(S)|\phi, I) \\
(5)\quad &= \underset{\phi}{argmax} \prod_{y_i \in \phi} P(y_i|y_{i-1})P(\tau(S_i)|y, I)P(S_i|\tau(S_i), y, I) \\
(6)\quad &= \underset{\phi}{argmax} \prod_{y_i \in \phi} P(y_i|y_{i-1})P(\tau(S_i)|I)P(S_i|\tau(S_i), y) \\
(7)\quad &= \underset{\phi}{argmax} \prod_{y_i \in \phi} \left(P(y_i|y_{i-1})2^{-blastscore(\tau(S),I)} \prod_{\substack{codons \\ c \in S_i}} P(c|amino(c)) \right).
\end{aligned} \tag{9.7}$$

The justification for this derivation is as follows. The second line follows simply from the conditional probability rule. The third line is obtained by assuming that the parse ϕ is independent of the evidence I; although this seems somewhat unreasonable, given that we wish to use the informant sequences I to help us more accurately select a parse, the term $P(S|\phi, I)$ will still allow informants to influence the scoring of a prospective parse via their joint influence (i.e., jointly with the parse) on the value of the term $P(S|\phi, I)$. In the fourth line we have utilized the fact that $P(S) = P(S)P(\tau(S)|S) = P(S, \tau(S))$ – omitting the other conditioning terms for clarity – due to the deterministic nature of the translation process from a DNA sequence to a polypeptide (i.e., since $P(\tau(S)|S) = 1$). However, in line 5 we then factor $P(S, \tau(S)|\phi, I)$ as $P(\tau(S)|\phi, I)P(S|\tau(S), \phi, I)$, where the $P(S|\tau(S), \phi, I)$ term does not evaluate to 1 since the reverse mapping (amino acid → codon) is not unambiguous, unlike the forward mapping (codon → amino acid); a *codon usage table* giving the frequencies of association between codons c and amino acids *amino*(c) for a given organism can be used to evaluate this latter term (ignoring the other conditioning terms). Line 5 also decomposes the parse ϕ into a product over the states y_i comprising the parse and their corresponding feature subsequences S_i. In line 6 we drop some conditioning terms, so that $P(\tau(S_i)|I)$ and $P(S_i|\tau(S_i), q)$ remain. In line 7 we replace $P(\tau(S_i)|I)$ with $2^{-blastscore(\tau(S_i),I)}$, where *blastscore(A,B)* is the score (in bits) returned by BLASTX for query *A* and matching sequence *B*; use of the formula $2^{-blastscore}$ results from the fact that BLASTX scores are given in bits (see Exercise 9.8).

Finally, note that the derivation above explicitly treats only coding states. For noncoding states (and for regions of putative coding segments having no database match) the $P(S|\phi, I)$ term is replaced with $P(S|\phi)$ and then decomposed in the usual way for a GHMM – i.e., using an appropriate content sensor such as a Markov chain.

9.2 Combiners

Whereas the informant-based approaches described above incorporate database matches into the gene prediction process via one or more similarity measures, a distinct (and highly successful) class of gene-finding program has emerged which integrates evidence from a number of disparate sources, including database matches as well as the predictions from other gene finders and related programs (e.g., splice-site predictors such as *GeneSplicer* – Pertea *et al.*, 2001). This approach to automated genome annotation is sometimes referred to as *integrative gene finding*, since it seeks to integrate heterogeneous evidence sources into a uniform framework for accurately inferring gene structures. As with the informant approaches, the methods for integrative gene finding are still a subject of active research, with no clear consensus, as of yet, as to the optimal way to formalize the problem, and so we again present the material as a series of case studies of current systems.

9.2.1 *Case study: JIGSAW*

One of the most successful integrative systems, as shown by the results of the EGASP gene-finding workshop (Guigó, *et al.*, 2006), is *JIGSAW* (Allen *et al.*, 2006), an extension of an earlier program known simply as "*Combiner*" (Allen *et al.*, 2004). We will briefly trace the evolution of the earlier program through to its most recent incarnation. In all versions of the program, a set of *evidence tracks* – i.e., database matches, gene predictions from one or more programs, splice site predictions, etc. – are made available to the program as sets of intervals in the target sequence, with (in most cases) an associated score for each evidence interval. In the case of gene-finder tracks, the beginning and ending points (i.e., *boundaries*) of these intervals are used by the combiner to establish the set of possible signals (i.e., start/stop codons and splice sites) for delimiting the exons of putative genes.[2] Exons are assembled by the combiner program into complete gene predictions by a dynamic programming algorithm similar to that described in (Guigó, 1998).

[2] Exactly which tracks are used to establish signals can be determined by the user via a configuration file (J. Allen, personal communication).

The earliest version of the program – called *LC1* (*Linear Combiner 1*) – formed a linear combination of the evidence interval scores:

$$P(\phi) = \prod_{I \in \phi} P(I),$$
$$P(I) = \sum_k w_k \cdot s(I, k) \tag{9.8}$$

for evidence types k and intervals I in a putative parse ϕ, where the evidence-type-specific weights w_k were fixed at 1 for LC1 (i.e., LC1 is a *voting combiner*). Only gene-finder tracks were considered by this version of the program. The scores $s(I, k)$ were a function of the number of positions in I predicted to be of the appropriate type for interval I (i.e., predicted to be coding for putative exon intervals; noncoding for putative intron and intergenic intervals).

The *LC2* version (*Linear Combiner 2*) of this program extended the earlier version by allowing explicit weights w_k and by also allowing database matches as evidence tracks. With each evidence interval is associated a vector $\mathbf{v} = [v_i]$ of evidence scores for the interval, with the ith dimension of the vector determined by the ith evidence track. The evidence scores are simply the alignment scores in the case of database hits, or scores assigned by a gene finder in the case of prediction tracks (with a default of 1 for programs that do not provide scores with their predictions). The score $s(\mathbf{v})$ of an individual evidence vector $\mathbf{v}$ is defined as a length-normalized dot product of the evidence vector $\mathbf{v}$ with a fixed weight vector $\mathbf{w}$:

$$s(\mathbf{v}) = \frac{\langle \mathbf{w}, \mathbf{v} \rangle}{|I|} \tag{9.9}$$

where $\langle \mathbf{w}, \mathbf{v} \rangle$ denotes the *inner product* $\sum_i \mathbf{v}_i \mathbf{w}_i$ (section 2.1), and $|I|$ is the length of interval I. The weight vector is determined during training so as to maximize predictive accuracy. A dynamic programming algorithm based on the one used by LC1 finds the optimal parse as a series of intervals under the scoring function $s(\mathbf{v})$.

LC2 was supplanted by *SC* (the *Statistical Combiner* program), which utilizes the same dynamic programming algorithm as LC2 but with a different formulation for the $P(I)$ term in Eq. (9.8). Because the intervals $I \in \phi$ in a prospective parse ϕ have associated with them putative feature types $type(I)$ – i.e., exon, intron, intergenic – the $P(I)$ term can be rewritten $P(type(I)|E)$ for evidence vector E, and if we assume conditional independence on the evidence falling in intervals that are not adjacent to interval I, we can refine this to $P(type(I_k)|E_{k-1}, E_k, E_{k+1})$, where E_k is the evidence vector for interval I_k. These probabilities are computed by a *decision tree* (section 10.6), which classifies a novel evidence tuple $V_k = (E_{k-1}, E_k, E_{k+1})$ into one of several disjoint groups G_i of training tuples $G_i = \{W_j | 0 \leq j < n\}$, where the groups G_i reside at the n leaves of the tree (one group per leaf), and are distinguished via a series of predicates applied to evidence attributes (i.e., elements of the evidence vectors in the tuple to be classified) on a path from the root of the tree to a leaf. The training tuples W_j in any group may consist of positive or negative examples, with respect to the posited *type*(I) – i.e., some support the posited type and some support a different type for I. The desired probability $P(type(I)|E)$ is then estimated simply as the fraction $|pos(W_j)|/|G_i|$ of positive elements in the group G_i to which the novel tuple V_k has been assigned by the decision tree.

The newest incarnation of this lineage of combiners is the program *JIGSAW*, which incorporates transition probabilities into the evaluation of prospective parses, as well as an intron model which infers an intron evidence track from the existence of consecutive exon predictions in the gene finder tracks. The resulting program was found to perform exceptionally well on the *EGASP* human gene-finding exercise (Guigó *et al.*, 2006). The interested reader is referred to Allen and Salzberg (2005) and Allen *et al.* (2006) for additional details.

9.2.2 *Case study: GAZE*

Although a number of other combiner programs exist (e.g., *ExonHunter* – Brejová *et al.*, 2005; *Eugène* – Foissac and Schiex, 2005; *GLEAN* – Mackey, 2005), perhaps the most flexible of these is the program *GAZE* (Howe *et al.*, 2002), which provides a reprogrammable framework for integration of gene evidence. Nearly all aspects of the gene-finding task can be externally programmed by the user without the need to modify the GAZE source code; this includes even the definition of gene syntax and phase tracking. The program does not work directly with sequence data, but rather makes its predictions based entirely on the evidence tracks provided as inputs in *GFF* format (see section 4.1). The system provides a rich language for the specification of the scoring strategy to be used in integrating evidence to form gene predictions. The system has been successfully applied to gene finding in *Caenorhabditis elegans*, and promises to see use in other annotation projects as well, due to its great flexibility.

9.3 Alignment-based prediction

Another class of comparative gene-finding methods can be best described as *alignment-based*, since the programs in this category perform their computation exclusively on pairs of pre-aligned genomes. While the combiner and informant techniques described above also make use of alignments, this class of system is more similar to the *pair HMM* systems described in the next section, and indeed may be considered the direct forerunners of the latter.

The approach is conceptually simple: given a pair of pre-aligned genomes G_1 and G_2, we note all conserved splice site consensuses (i.e., those which are paired with one another in the alignment) and conserved start/stop codon consensuses in the two sequences, and then use simple *ab initio* methods to perform prediction based on these conserved signals. Evaluation of content scores can proceed by observing only one of the two sequences, possibly modified by a conservation score reflecting the quality of the alignment covering a putative gene feature. Prediction of genes in one of the two genomes thus constitutes prediction in the other genome as well, since the coordinates of predicted gene features in one genome can be directly mapped to homologous features in the other genome, though, of course, extra work must be done to ensure that the prediction in the second genome satisfies frame constraints (i.e., CDS length divisible by 3 and no in-frame stop codons except at the end of the CDS).

The input to an alignment-based gene finder is a whole-genome alignment computed by a program such as *MUMmer* (Delcher *et al.*, 1999b), *GLASS* (Batzoglou *et al.*, 2000), *AVID* (Bray *et al.*, 2003), or the like. The output consists of a set of gene predictions for one or both genomes. We illustrate the method with a set of case studies.

9.3.1 Case study: ROSETTA

The alignment-based gene finder *ROSETTA* (Batzoglou *et al.*, 2000) exemplifies one of the simplest – but in some cases highly effective – approaches to comparative gene prediction. The program was found to perform remarkably well compared to a number of other systems with more sophisticated statistical modeling.

The approach taken by ROSETTA is to identify conserved splice sites and start/stop codons which may constitute valid signals in a gene parse. These signals are linked into the chosen parse via a dynamic programming algorithm according to the following scoring scheme. Splice sites are evaluated using an extension of the standard MDD/WAM methods employed in GENSCAN, with the extension serving to reward observed *directionality biases* (see Pachter *et al.*, 1999). Codon usage scores are computed for each pair of aligned putative exons, using a simple sum of log odds ratios from a published codon frequency table (Delphin *et al.*, 1999). The signal and codon usage scores for the two genomes are summed into a single logarithmic parse score. The amino acids encoded by the codons of putative genes are also scored using a similarity matrix (Dayhoff *et al.*, 1978), with the resulting similarity score being incorporated directly into the parse score. Finally, the lengths of putative paired (i.e., aligned) exons are scored according to length histograms for the two genomes, with a penalty term reflecting differences in exon lengths (i.e., to reflect expected constraints on evolutionary divergence of exon lengths from the most recent common ancestor).

9.3.2 Case study: SGP-1

SGP-1 (Wiehe *et al.*, 2001), the predecessor to the program SGP-2 described in section 9.1.3, is somewhat more sophisticated than ROSETTA. The program takes as input an alignment between two genomes, which may be at the nucleotide or amino acid

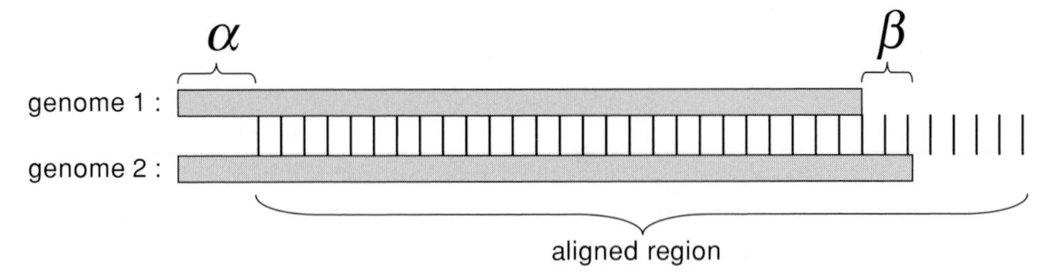

Figure 9.3 So-called "fuzziness" of exon alignments in SGP-1. Parameters control how far an alignment may be extended to include exon boundaries (α), as well as how far apart a pair of homologous signals may be (β). (Adapted from Wiehe et al., 2001.)

levels. This alignment is processed to identify high-confidence HSPs, which then serve as the substrates for prediction. Potential signals (start/stop codons, splice sites) are identified in the two genomic sequences, and the potential exons delimited by these signals are identified independently in the two sequences. These exon candidates from the two genomes are then paired off according to their positions in the alignment. Putative exons having conserved splice sites across the alignment are obviously retained, since they are clearly promising candidates for the final gene prediction. Exons that are paired by an HSP, but which do not have precisely matching endpoints, however, may be retained, depending on a "fuzziness" parameter β which specifies the maximum distance allowed between paired signals (see Figure 9.3).

Furthermore, if a pair of candidate exons are covered in part by an HSP, but one or more of their endpoints are not covered by the alignment, then the alignment may be "extended" by at most α positions (another user-specified parameter) in hopes of including the putative exon boundaries. These latter accommodations clearly make SGP-1 more flexible than ROSETTA, which requires that paired splice sites be strictly aligned in the precomputed whole-genome alignment. Exon scores are computed as a linear combination of splice site scores (via WMM-like profiles) and a similarity score computed by summing entries from an amino acid similarity matrix such as BLOSUM80 (Henikoff and Henikoff, 1992). Finally, a dynamic programming algorithm similar to that used in SGP-2 is applied independently to the remaining candidate exons in each genome to arrive at a final set of gene predictions for the two genomes.

9.3.3 Case study: CEM

Our final example of an alignment-based gene finding system is *CEM* (the *conserved exon method* – Bafna and Huson, 2001), which may be seen in a number of ways to constitute a forerunner to the *pair HMM* (*PHMM*) approach which is described in the next section. The primary difference, as will become apparent to the reader after we introduce PHMMs (and in particular, after we describe practical PHMM-based

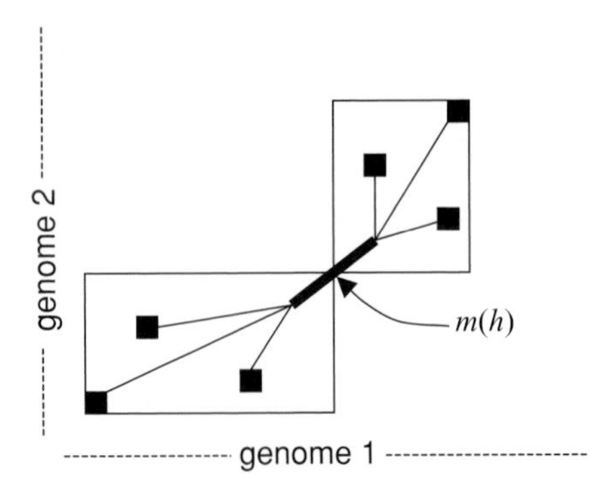

Figure 9.4 The two dynamic programming subproblems for an HSP in CEM. The thick segment in the middle represents an HSP. The two large rectangles bound the dynamic programming submatrices. Black boxes are cells in the dynamic programming matrices corresponding to paired signals (e.g., a splice site present in both genomes). Thin lines represent alignment paths. The two matrices meet at $m(h)$, the middle of the HSP. (Adapted from Bafna and Huson, 2001.)

gene-finding systems) is the lack of a content scoring model or state transition model.

The program takes as input a set of HSPs as computed by TBLASTX or some whole-genome alignment program. The program scans the DNA sequences for the two genomes using a set of species-specific WMMs to identify potential splice sites and start/stop codons. WMM scores are used only to accept or reject potential signal sites by applying a predefined threshold to the scores. As in SGP-1, candidate exons are then identified, and a dynamic programming algorithm is later applied to assemble candidate exons into a gene prediction. The system is considerably more sophisticated than SGP-1, however, in the way that it identifies paired signals and scores candidate exon pairs.

Consider an HSP h as shown in Figure 9.4. The midpoint $m(h)$ is identified and its projection onto the two genomic DNA sequences is found. Any candidate coding exon which contains this projected midpoint is used to extend the dynamic programming matrices indicated by the rectangles in the lower-left and upper-right portion of the figure. These matrices are used by a pair of dynamic programming algorithms similar to the Needleman–Wunsch alignment algorithm (Needleman and Wunsch, 1970) at the amino acid level (i.e., the DNA sequences are translated in the appropriate frames as determined by the given HSP) to find the optimal alignment paths connecting each signal pair to the midpoint $m(h)$ of the HSP in the dynamic programming matrix.

Candidate exons are represented as subgraphs in which each signal pair and each HSP midpoint is a vertex. The dynamic programming algorithm which is used to assemble candidate exons into complete gene predictions therefore computes a shortest-path-like function on the weighted graph in order to identify the highest-scoring gene parse. The weights of the edges in the graph are determined

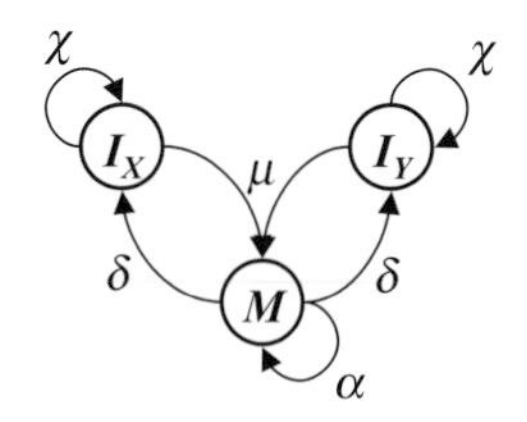

Figure 9.5 A simple pair HMM. State M represents a match or mismatch, state I_X represents an insertion, and state I_Y represents a deletion. The silent start/stop state is omitted for clarity. (Adapted from Durbin *et al.*, 1998.)

by combining the HSP (protein) alignment scores with an intron penalty and a frameshift penalty.

9.4 Pair HMMs

Of the various proposals that have been put forward to date for the incorporation of homology evidence into the gene prediction process, one stands out as being particular elegant in its formulation – even if it does not in its idealized form lend itself to efficient implementation generally. This is the notion of the *pair HMM*, or *PHMM* as we will call it.

Consider the state-transition diagram shown in Figure 9.5. This simple 3-state model operates just as an HMM, except that each state emits a *pair* of symbols, rather than a single symbol, at each time unit. In particular, the state M emits pairs of symbols from the standard DNA alphabet $\alpha = \{\texttt{A},\texttt{C},\texttt{G},\texttt{T}\}$, with the first symbol being emitted into the first sequence, and the second being emitted into the second sequence. In contrast, the state I_X emits pairs from the set $\alpha \times \{-\}$, where the dash symbol ("$-$") can be thought of as a blank space, or as a gap in an alignment. Similarly, the state I_Y emits pairs from $\{-\} \times \alpha$, so that I_X effectively models gaps in the one sequence and I_Y models gaps in the other sequence. Thus, state M generates matching or mismatching pairs of nucleotides, whereas I_X and I_Y generate *gap positions* in one or the other sequence. In this way it can be seen that PHMMs emit alignments, and as we will see, the problem of decoding with a PHMM is analogous to the problem of *pairwise sequence alignment* (e.g., Needleman and Wunsch, 1970; Smith and Waterman, 1981).

More generally, any HMM can be transformed into a PHMM by employing three copies of all nonsilent states from the original HMM, with the first set of states (the "match" states) emitting pairs of symbols (whether matching or mismatching) from α into the two output sequences, and the other two sets of states ("insertion" states and "deletion" states) emitting a single symbol into one of the two output sequences, as in the simple example above. It has been noted often in the literature that noncoding match states (so-called *conserved noncoding sequence states*, or *CNS states*) are as important for the predictive accuracy of these systems as are the coding match states, due to the existence of conserved noncoding elements in genomic DNA, such as transcription factor binding sites and other functional noncoding elements (Alexandersson *et al.*, 2003).

Formally, we define a PHMM for DNA sequences as a 7-tuple:

$$M = (q^0, Q_M, Q_I, Q_D, \alpha, P_t, P_e), \tag{9.10}$$

for state set $Q = Q_M \cup Q_I \cup Q_D \cup \{q^0\}$, DNA alphabet α, transition distribution $P_t : Q \times Q \mapsto \mathbb{R}$, silent initial/final state q^0, and emission distribution $P_e : Q \times \alpha_+ \times \alpha_+ \mapsto \mathbb{R}$, for the augmented alphabet $\alpha_+ = \{\texttt{A}, \texttt{C}, \texttt{G}, \texttt{T}, \texttt{-}\}$.[3] As before, all emission probabilities in q^0 are 0. All the state sets Q_M, Q_I, Q_D, and $\{q^0\}$ are disjoint. States in Q_M are referred to as *match states*, and are subject to:

$$P_e(q \in Q_M, s \in \alpha_+, -) = P_e(q \in Q_M, -, s \in \alpha_+) = 0, \tag{9.11}$$

though it should be clear that these so-called "match" states can emit nonmatching pairs of symbols as well as matching pairs. States in Q_I are known as *insertion states*, and satisfy:

$$P_e(q \in Q_I, s \in \alpha_+, s \in \alpha) = P_e(q \in Q_I, -, -) = 0, \tag{9.12}$$

while states in Q_D are known as *deletion states*, and satisfy

$$P_e(q \in Q_D, s \in \alpha, s \in \alpha_+) = P_e(q \in Q_D, -, -) = 0, \tag{9.13}$$

so that insertion states can emit only pairs in $\alpha \times \{-\}$ while deletion states emit only pairs in $\{-\} \times \alpha$. Note that we forbid the emission of $(-, -)$ by any state, as in traditional sequence alignment: $P_e(-, -|q \in Q) = 0$. It is also sometimes desirable to forbid transitions between insertion and deletion states:

$$P_t(q_i \in Q_I | q_j \in Q_D) = P_t(q_j \in Q_D | q_i \in Q_I) = 0. \tag{9.14}$$

We will denote the emitting states as $Q_E = Q - \{q^0\}$.

Decoding with a PHMM is somewhat more complicated than decoding with a simple HMM. Recall that the Viterbi decoding algorithm for simple HMMs operated via the systematic computation of a two-dimensional dynamic programming matrix, with the two matrix dimensions corresponding to the sequence length and the state set of the HMM. In the case of a PHMM, a three-dimensional dynamic programming matrix is required, where two of the dimensions correspond to positions in the two output sequences, and the third dimension corresponds to the set of states (see Figure 9.6).

[3] Technically, since the alignment is *unobservable*, the emission alphabet should not include the gap character. For our present purposes, however, this is the most convenient formulation.

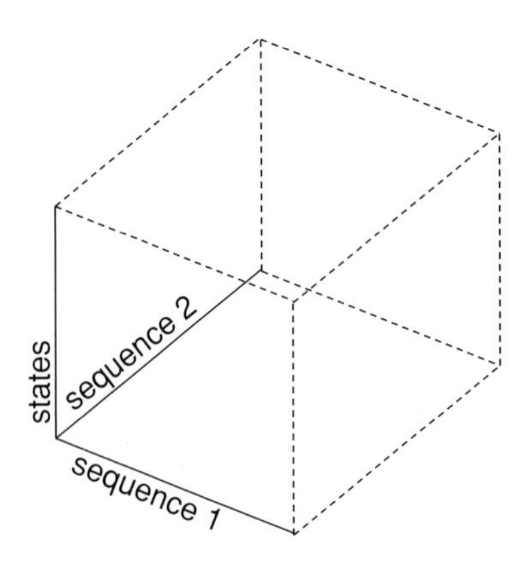

Figure 9.6 A three-dimensional dynamic programming matrix for a PHMM. Two axes correspond to the two output sequences, and the third axis corresponds to the states of the machine.

The objective function for PHMM decoding can be stated as:

$$\phi^* = \underset{\phi=\{y_0,\ldots,y_{m-1}\}}{argmax} \; P_t(q^0|y_{m-2}) \prod_{i=1}^{m-2} P_e(a_{i,1}, a_{i,2}|y_i) P_t(y_i|y_{i-1}) \tag{9.15}$$

for $y_i \in Q$, where $a_{i,j}$ is the ith symbol in the jth sequence ($j \in \{1, 2\}$) of the resulting alignment $A(\phi)$. It should be clear that these $a_{i,j}$ are unambiguously defined for any given parse $\phi = \{y_0, \ldots, y_{m-1}\}$, since the succeeding states in the parse can each be classified as belonging to Q_M, Q_I, or Q_D, so that the precise sequence of (mis)matches, insertions, and deletions comprising $A(\phi)$ can be thereby determined (see Exercises 9.19 and 9.20).

A dynamic programming recursion for PHMM decoding can be devised by observing the following discipline. A valid parse must start in cell $V_{0,0,0}$, since the machine always begins in state q^0. For any cell $V_{i,j,k}$, the optimal predecessor will be one of $V_{i-1,j,h}$ (for insertions), $V_{i,j-1,h}$ (for deletions), or $V_{i-1,j-1,h}$ (for matches or mismatches), with the optimal predecessor state q_h chosen so as to maximize each of these terms, subject to the constraint that insertions, deletions, and matches/mismatches occur only when the machine transitions into a state in Q_I, Q_D, or Q_M, respectively. We thus arrive at the following recursion formulae:

$$V_{i,j,k} = \begin{cases} \max\limits_h V_{i-1,j-1,h} P_t(q_k|q_h) P_e(s_{i-1,1}, s_{j-1,2}|q_k) & \text{for } q_k \in Q_M \\ \max\limits_h V_{i-1,j,h} P_t(q_k|q_h) P_e(s_{i-1,1}, -|q_k) & \text{for } q_k \in Q_I \\ \max\limits_h V_{i,j-1,h} P_t(q_k|q_h) P_e(-, s_{j-1,2}|q_k) & \text{for } q_k \in Q_D \\ 0 & \text{for } q_k = q^0 \end{cases}$$

$$T_{i,j,k} = \begin{cases} \underset{(i-1,j-1,h)}{argmax} V_{i-1,j-1,h} P_t(q_k|q_h) P_e(s_{i-1,1}, s_{j-1,2}|q_k) & \text{for } q_k \in Q_M \\ \underset{(i-1,j,h)}{argmax} V_{i-1,j,h} P_t(q_k|q_h) P_e(s_{i-1,1}, -|q_k) & \text{for } q_k \in Q_I \\ \underset{(i,j-1,h)}{argmax} V_{i,j-1,h} P_t(q_k|q_h) P_e(-, s_{j-1,2}|q_k) & \text{for } q_k \in Q_D \end{cases} \tag{9.16}$$

for $0 < i \leq |S_1|$, $0 < j \leq |S_2|$, and $0 \leq k < |Q|$, where the following initializations are assumed:

$$
\begin{aligned}
V_{0,0,k} &= \begin{cases} 1 & \text{for } q_k = q^0 \\ 0 & \text{otherwise} \end{cases} \\
V_{i>0,0,k} &= \begin{cases} \max\limits_{h} V_{i-1,0,h} P_t(q_k|q_h) P_e(s_{i-1,1}, -|q_k) & \text{for } q_k \in Q_I \\ 0 & \text{otherwise} \end{cases} \\
V_{0,j>0,k} &= \begin{cases} \max\limits_{h} V_{0,j-1,h} P_t(q_k|q_h) P_e(-, s_{j-1,2}|q_k) & \text{for } q_k \in Q_D \\ 0 & \text{otherwise} \end{cases} \\
T_{i>0,0,k} &= \begin{cases} \underset{(i-1,0,h)}{argmax}\, V_{i-1,0,h} P_t(q_k|q_h) P_e(s_{i-1,1}, -|q_k) & \text{for } q_k \in Q_I \\ NIL & \text{otherwise} \end{cases} \\
T_{0,j>0,k} &= \begin{cases} \underset{(0,j-1,h)}{argmax}\, V_{0,j-1,h} P_t(q_k|q_h) P_e(-, s_{j-1,2}|q_k) & \text{for } q_k \in Q_D \\ NIL & \text{otherwise} \end{cases} \\
T_{0,0,k} &= NIL
\end{aligned}
\tag{9.17}
$$

where V is a three-dimensional real matrix of size $|Q| \cdot (|S_1| + 1) \cdot (|S_2| + 1)$, and $s_{i,j}$ is the ith symbol in the jth input sequence – these are distinct from the $a_{i,j}$ described above, as the index i is relative to the beginning of the raw sequence, not the current partial alignment. Note that the matrix is wider and deeper than the respective lengths of the two sequences. The matrix T, of the same dimensions as V, stores *traceback pointers* (i, j, h) to facilitate reconstruction of the optimal parse, as in traditional Viterbi decoding:

$$
\begin{aligned}
C_0 &= \underset{T_{|S_1|,|S_2|,h}}{argmax}\, V_{|S_1|,|S_2|,h} P_t(q^0|q_h) \\
C_i &= \begin{cases} T_{C_{i-1}[0],C_{i-1}[1],C_{i-1}[2]} & \text{if } C_{i-1} \neq NIL \\ NIL & \text{otherwise} \end{cases} \\
L &= \max i \ni C_i \neq NIL \\
\forall_{0 \leq i \leq L}\, R_i &= C_i[2]
\end{aligned}
\tag{9.18}
$$

where each C_i is a triple (x, y, z) specifying a cell in V or T, and the optimal parse $\phi^* = \{R_L, R_{L-1}, \ldots, R_0, q_0\}$ is formed by reversing the order of the elements in R; $C_i[2]$ evaluates to the state associated with the ith cell found during traceback.

Because a three-dimensional matrix (rather than a two-dimensional matrix in the case of simple Viterbi decoding) must be computed while decoding with a PHMM, the latter model can be seen to incur a greater computational cost than does a simple HMM. While the greater expense incurred by the PHMM might be thought to be worthwhile when the accuracy of the resulting gene predictions are significantly improved, for sufficiently long sequences the cost in terms of time complexity and memory requirements can in many cases quite easily exceed the available resources. Under such circumstances, appropriate heuristics are instead desired which can provide near-optimal gene prediction accuracy while satisfying the constraints of a realistic computing environment. A number of such heuristics which have been employed in practice are described below.

Once the optimal path has been determined, a pair of gene predictions can be extracted – one prediction per genome – in a way directly analogous to the method employed for HMM-based gene finders described in Chapter 6 (see Exercise 9.22).

The training of a PHMM can be performed in much the same way as for a pure HMM when labeled sequence data are available. If a set of sequence alignments are available, a simple 3-state PHMM such as the one shown in Figure 9.5 can be trained by segmenting the alignments into contiguous regions consisting of gapless regions (corresponding to state M), regions with only gaps in genome 2 (corresponding to state I_X), and regions with only gaps in genome 1 (corresponding to state I_Y). From these training examples, maximum likelihood estimates of emission and transition probabilities for the appropriate states can be estimated in much the same way as for a standard HMM (see Exercise 9.23).

Up to this point we have considered the idealized case in which the PHMM consists of a reasonably small number of states and in which the sequences to be subjected to decoding are reasonably short. In cases where these assumptions do not hold, there are a number of obvious heuristics which can be employed to reduce the computational cost of PHMM decoding. The most common heuristic involves the use of an efficient tool such as BLAST to detect regions of significant homology between the two input sequences. These HSPs can then be mapped into the coordinate space of the full dynamic programming matrix for a particular PHMM and used to *band* (i.e., eliminate from consideration the less promising regions of) the decoding matrix. In particular, it is common practice to evaluate only those cells in the matrix which occur within a fixed distance (measured in numbers of intervening cells) of an HSP, or within submatrices defined by pairs of nearby HSPs much as in the CEM method described earlier, and to then prohibit during decoding any path that passes through cells which have not been selected for evaluation. In this way, the portion of the matrix that needs to be evaluated can be significantly reduced, often to such a degree that decoding becomes computationally tractable once again. An obvious drawback of this approach is that the correct parse may actually pass through a nonevaluated region of the matrix, in which case the decoder cannot possibly produce the correct prediction.

9.4.1 *Case study: DoubleScan*

A well-known PHMM implementation for comparative gene prediction is *DoubleScan* (Meyer and Durbin, 2002). The model consists of 54 states representing gene structures (exon, intron, and intergenic regions – including introns occurring within

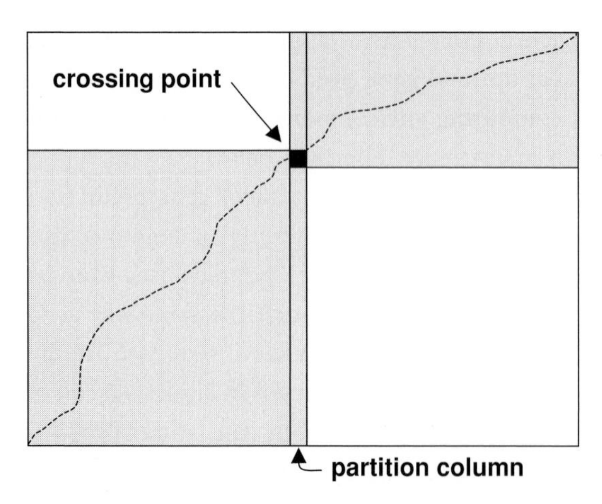

Figure 9.7 A crude illustration of the Hirschberg algorithm. A first pass over the matrix identifies the crossing point (black square) of the optimal path (dashed line) through the partition column. The two gray submatrices are then evaluated recursively using the same strategy.

UTRs,[4] but no promoters or polyadenylation sites) on a single DNA strand. Insertions and deletions within exons are modeled at the whole codon level, and pairs of codons emitted by the match state are scored according to their amino acid similarity (e.g., Henikoff and Henikoff, 1992). At the higher level of gene syntax, DoubleScan models insertions and deletions of whole exons or introns, effectively modeling evolutionary changes to gene structure such as the fusing or splitting of whole exons. Note that the input to the program must be a pair of likely orthologous DNA sequences; a preprocessing step is thus necessary to identify apparent "co-linear" regions of two genomes to which the PHMM decoder can then be applied.[5]

As indicated above, evaluation of the full dynamic programming matrix for PHMM decoding can become intractable for long sequences. DoubleScan employs two clever tricks to reduce the time and space complexity of decoding. The first is the *Hirschberg algorithm* (Hirschberg, 1975) for sequence alignment, which reduces the space requirements of a standard alignment algorithm from $O(n^2)$ to $O(n)$ while leaving the time complexity $O(n^2)$, for sequence lengths n. It accomplishes this via a recursive procedure in which a "memoryless" decoding pass is made over the two halves of the matrix to determine the crossing point of the optimal path through the middle column. Once this has been accomplished, the matrix can be partitioned in half at the crossing point, and the two remaining subproblems can then be recursively solved in the same manner (with some additional pruning – see Figure 9.7). The interested reader is encouraged to consult Hirschberg (1975) for

[4] Note, however, that the current version of the program does not emit UTR predictions as part of its output (I. Meyer, personal communication).

[5] I. Meyer, personal communication.

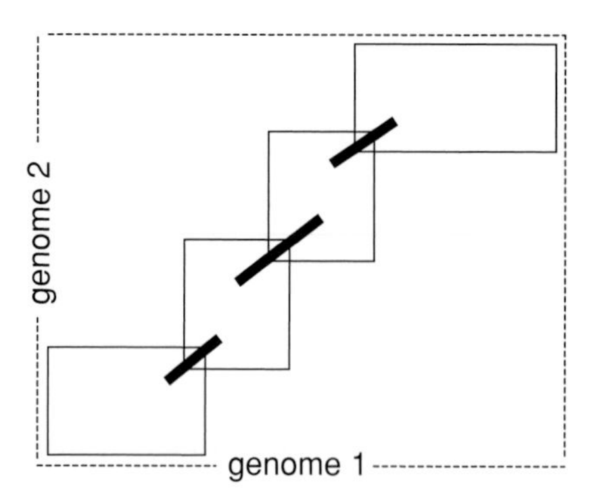

Figure 9.8 Rough illustration of the stepping-stone algorithm employed by DoubleScan. A set of consistent HSPs (thick black line segments) are joined by slightly overlapping submatrices to form a wide path through the full matrix. Each submatrix can be evaluated using the (extended) Hirschberg algorithm.

the details of this ingenious algorithm. The extension of this algorithm for use in PHMM decoding obviously requires that the partition column be generalized to a column-like volume in the three-dimensional decoding matrix (see Exercise 9.32).

The second clever "trick" that is employed in DoubleScan to reduce the time and space complexity of the decoding process is the so-called *stepping-stone algorithm*, which effectively reduces the portion of the matrix which has to be evaluated in a manner very reminiscent of a technique used in the comparative gene finder CEM, which we described earlier.

The algorithm begins by identifying all promising regions of the matrix via BLASTN, and then orders the most promising HSPs using a procedure similar to the *longest increasing subsequence* (*LIS*) algorithm (see, e.g., Cormen *et al.*, 1992). A set of submatrices is then formed by placing their corners at the midpoints of successive HSPs in the LIS sequence (see Figure 9.8). These submatrices are then evaluated using an approach very similar to the Hirschberg algorithm to reduce the amount of memory required. The result is a decoding heuristic which, while not guaranteed to find the optimal parse, has been found to perform very well in practice, while achieving quite tolerable space and time efficiency.

The DoubleScan system has been modifed by its authors to produce a related program, called *Projector*, which solves the simpler problem of transferring annotations from one genome to another, using a PHMM in which the path of the optimal parse is constrained in the one genome to match the known gene structure (Meyer and Durbin, 2004). A similar approach was taken by Seneff *et al.* (2004).

9.5 Generalized pair HMMs

Just as we generalized HMMs into GHMMs by abstracting the generative capabilities of individual states in the model, we can generalize the notion of a PHMM into that of a *generalized PHMM* – or *GPHMM* – in which each state may generate not just one symbol into each of two genomes, but rather a pair of arbitrary-length subsequences into both genomes (Pachter *et al.*, 2002). As with GHMMs, this generalization allows us to model more flexibly the content sensors and duration distributions of the emitted sequences.

Formally, we denote a GPHMM as a 6-tuple $(Q, q^0, P_t, P_d, P_e, \alpha)$ in which all elements are as defined for GHMMs except that now P_d is a joint distribution of paired durations $P_d : \mathbb{N} \times \mathbb{N} \times Q \mapsto \mathbb{R}$ conditional on state, and P_e is now a joint distribution of paired emissions $P_e : \alpha^* \times \alpha^* \times Q \times \mathbb{N} \times \mathbb{N} \mapsto \mathbb{R}$ conditional on state and two durations. Upon entering each state q_i the GPHMM first selects a pair of durations $d_{i,1}$ and $d_{i,2}$ according to $P_d(d_{i,1}, d_{i,2}|q_i)$ and then emits subsequences $S_{i,1}$ and $S_{i,2}$ into genomes 1 and 2, respectively, according to the joint emission distribution $P_e(S_{i,1}, S_{i,2}|q_i, d_{i,1}, d_{i,2})$, where $d_{i,j} = |S_{i,j}|$.

Formulated in this way, gene prediction with a GPHMM can be accomplished by finding the most probable parse ϕ^* according to:

$$\phi^* = P_t(q^0|y_{n-2})\, \underset{\phi}{argmax} \prod_{i=1}^{n-2} P_e(S_{i,1}, S_{i,2}|y_i, d_{i,1}, d_{i,2}) P_t(y_i|y_{i-1}) P_d(d_{i,1}, d_{i,2}|y_i) \tag{9.19}$$

for features $S_{i,1}$ and $S_{i,2}$ of lengths $d_{i,1}$ and $d_{i,2}$, respectively, and where the concatenation $S_{0,j} \ldots S_{n-1,j}$ of individual features forms the input sequence S_j, $j \in \{0, 1\}$. A parse $\phi = \{(y_i, d_{i,1}, d_{i,2})|0 \leq i < n\}$ is now a series of states (y_i) and corresponding pairs of durations. The joint emission probability term P_e can be decomposed via conditional probability as:

$$P_e(S_{i,1}, S_{i,2}|y_i, d_{i,1}, d_{i,2}) = P_e(S_{i,1}|y_i, d_{i,1}) P_{cons}(S_{i,2}|S_{i,1}, y_i, d_{i,2}) \tag{9.20}$$

(if we ignore any dependence of $S_{i,1}$ on $d_{i,2}$ and $S_{i,2}$ on $d_{i,1}$ – this is reasonable, given that P_d is in general able to penalize differences between $d_{i,1}$ and $d_{i,2}$), where $P_e(S_{i,1}|y_i, d_{i,1})$ can be estimated in the same way as for a GHMM, and the conservation term P_{cons} can be estimated by aligning $S_{i,1}$ and $S_{i,2}$ and mapping the resulting alignment to a conditional probability based on the alignment properties of a set of known orthologs (see below). Similarly, the joint duration distribution P_d can be estimated from known orthologs, if such are available, though in practice one may have difficulty obtaining large enough sample sizes to accurately model the full joint distribution. One solution is to factor the duration probability as follows:

$$\begin{aligned} P_d(d_{i,1}, d_{i,2}|y_i) &= P(d_{i,1}|y_i) P(d_{i,2}|d_{i,1}, y_i) \\ &\approx P(d_{i,1}|y_i) P(\Delta_d|y_i) \end{aligned} \tag{9.21}$$

where $\Delta_d = |d_{i,1} - d_{i,2}|$. The term $P(d_{i,1}|y_i)$ can be estimated using a simple histogram of feature lengths as in a standard GHMM, leaving the term $P(\Delta_d|y_i)$, which can also be estimated using a simple histogram.

Optimal decoding with a GPHMM is theoretically more computationally costly than with a PHMM, due to the variable duration of states. It is possible, however, to accelerate the decoding process by making a set of more-or-less reasonable assumptions, much as was done for GHMMs. In particular, we can assume that noncoding lengths are geometrically distributed and that content functions are factorable, allowing us to use one of the GHMM decoders from section 8.3 as a foundation for building a GPHMM decoder. Since the latter will unfortunately have to evaluate a three-dimensional matrix of size $\mathbb{O}(|S_1| \times |S_2| \times |Q|)$, we will also wish to prune the matrix in some way, as was done in the PHMM case described earlier. One approach to doing this is illustrated in the following case study.

9.5.1 *Case study: TWAIN*

The open-source comparative gene finder *TWAIN* (Majoros *et al.*, 2005b) is based on a GPHMM. The system consists of two programs: *ROSE* (*Region-Of-Synteny Extractor*) and *OASIS* (*Optimal Alignment of Structures in Synteny*). Whereas OASIS implements the GPHMM decoding algorithm, ROSE acts as a preprocessor for OASIS, identifying regions of contigs on which to run the GPHMM decoder and also providing a set of precomputed "guide" alignments for OASIS to use as a basis for estimating P_{cons} terms.

The top-level system structure of TWAIN is illustrated in Figure 9.9. A pair of genomes is supplied to the whole-genome alignment program *MUMmer* (Delcher *et al.*, 2002; Kurtz *et al.*, 2004), which produces a set of HSPs. ROSE selects the most promising MUMmer HSPs (based on scores supplied by MUMmer) and groups them into probable whole genes using ad-hoc heuristics; a set of *ab initio* gene predictions for the target genomes may optionally be provided to help ROSE identify the probable gene regions. Once ROSE has identified a probable gene, a region including the gene (with a suitable margin on either side) is provided to OASIS. OASIS performs GPHMM decoding to produce a pair of gene predictions for each region. ROSE automatically reverse-complements the sequences as necessary before providing them to OASIS, to render the putative genes on the forward strand; OASIS thus only needs to predict genes on one strand.

An important feature of TWAIN is that it models sequence conservation at both the nucleotide and the amino acid levels. This roughly corresponds to the use in other GPHMM implementations of conserved noncoding as well as conserved coding states. Recall that the primary justification for cross-species comparative gene finding is the assumption that the different nature of selective pressures operating on the coding versus the noncoding regions of a genome will result in differential conservation patterns due to the operation of these different selective pressures over evolutionary time. It is thus intuitively very reasonable to expect that methods which observe only nucleotide-level conservation patterns will be somewhat handicapped, so that the explicit use of both nucleotide and amino acid alignments in a comparative gene finder is highly recommended.

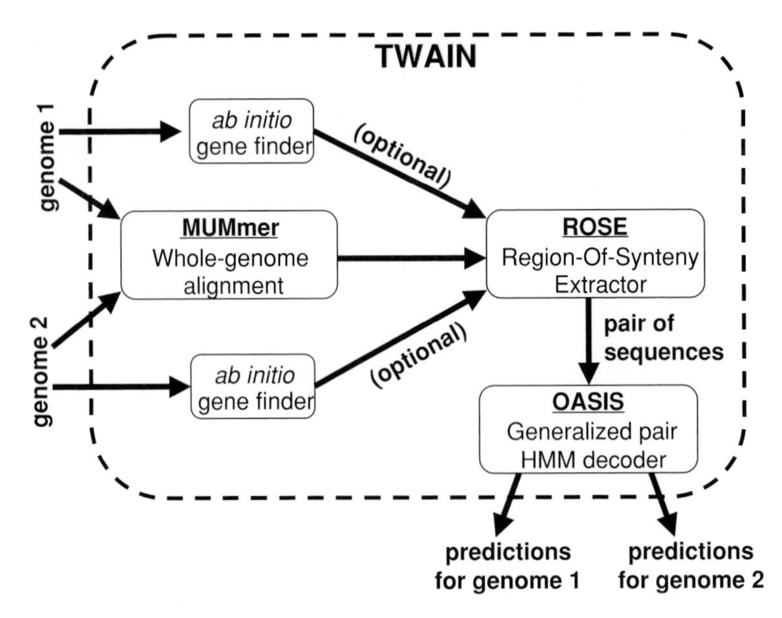

Figure 9.9 Top-level system structure of TWAIN. Two genomes are aligned by MUMmer, and the most promising HSPs are selected by ROSE and grouped into probable gene regions. The GPHMM decoder OASIS is applied to each probable gene region separately.

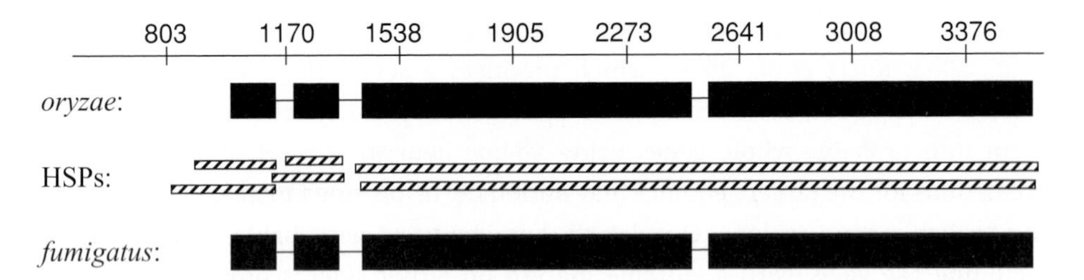

Figure 9.10 A pair of *Aspergillus* (*A. oryzae* vs. *A. fumigatus*) orthologs and their corresponding PROmer HSPs. The annotated gene structures are shown at top and bottom as black rectangles (exons) connected by line segments (introns). PROmer HSPs (in several frames) are shown as thin, hatched rectangles in the middle row.

The MUMmer program which is used by TWAIN provides both nucleotide alignments (via the *NUCmer* program) and amino acid alignments (via the *PROmer* program). Both NUCmer and PROmer alignments are utilized by ROSE and passed on to OASIS to inform the decoding process. An example pair of *Aspergillus* orthologs and their corresponding PROmer HSPs are illustrated in Figure 9.10.

From the figure it can be seen that PROmer HSPs can be very informative indicators of the location of exons, though the ends of the HSPs often extend some distance beyond the edge of the corresponding exons, and they often extend across short introns, so that identification of exon boundaries based solely on the predictions of PROmer or similar programs such as TBLASTX will not in general be very

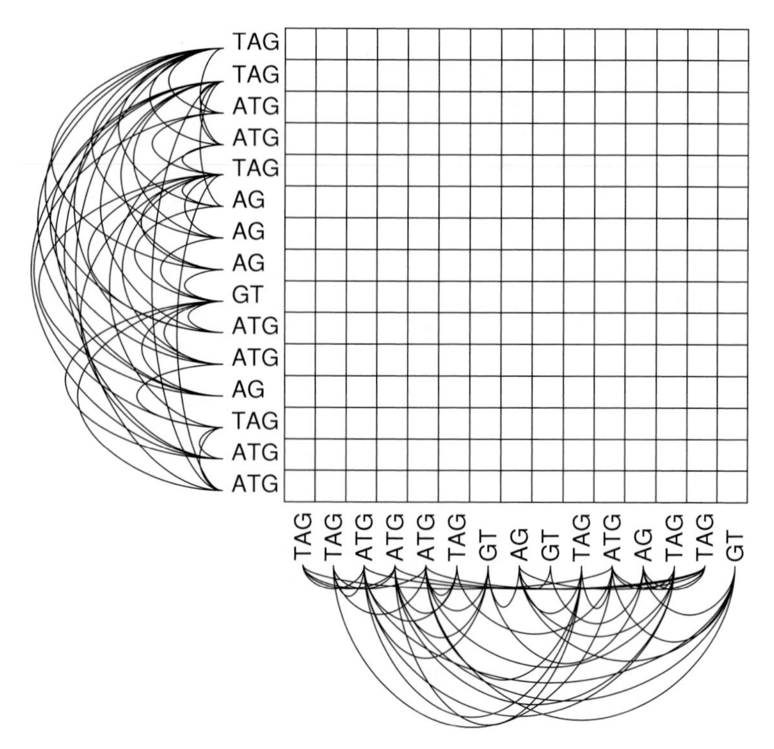

Figure 9.11 Alignment of ORF graphs. The ORF graph for each sequence in an alignment is placed along one axis of the matrix. Matrix rows and columns correspond to signals in the ORF graphs. Alignment of the ORF graphs produces an isomorphism between subgraphs which denote respective gene parses in the two genomes.

accurate. Nevertheless, for appropriately related genomes, predicted amino acid similarity information for homologous DNA sequences is clearly of great value to a comparative gene finder.

Because each state in a GPHMM parse ϕ corresponds to a pair of features, the OASIS algorithm can be conceptualized as an alignment algorithm between gene parses, as illustrated in Figure 9.11.

To formalize the notion of aligning gene parses, define a set of *signal types* $\sigma = \{\mathrm{ATG}, \mathrm{TAG}, \mathrm{GT}, \mathrm{AG}\}$ denoting start codon, stop codon, donor site, and acceptor site, respectively, of any appropriate consensus (e.g., TAG $\equiv$ {"TAG", "TGA", "TAA"}). Recall from section 3.4 our definition of an *ORF graph*. We denote an ORF graph $G = (V, E)$ for vertex set $V \subset \sigma \times \mathbb{N}$ and directed edge set $E \subset V \times V$, so that each vertex $v = (s, x)$ corresponds to a putative signal of type $s \in \sigma$ at position x in one of the two genomes, and each directed edge corresponds to a putative exon, intron, or intergenic region. In TWAIN, the following edge types are explicitly represented: ATG $\rightarrow$ TAG, ATG $\rightarrow$ GT, GT $\rightarrow$ AG, AG $\rightarrow$ GT, AG $\rightarrow$ TAG, and TAG $\rightarrow$ ATG.

The GHMM-based gene finder *TIGRscan* (Majoros *et al.*, 2004) is used as a subroutine by OASIS to construct an ORF graph for each of the two genomes. The

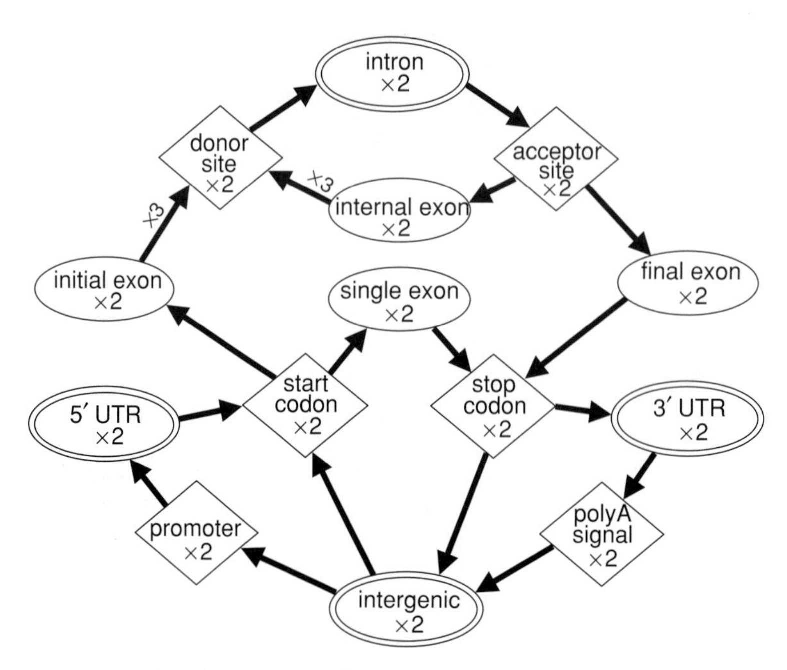

Figure 9.12 State-transition diagram for the GPHMM modeled by OASIS.

state-transition diagram of TIGRscan (omitting the reverse-strand states) is shown in Figure 9.12, where each state is additionally labeled with a "×2" to denote the fact that in TWAIN each state matches a pair of genomic features.

TIGRscan constructs its graph left-to-right by sliding its signal sensors (WMM, WAM, or MDD models) along the sequence and allocating a vertex at each high-scoring position, as described for the DSP algorithm in section 8.3.2. Nonconsensus splice sites are permitted. Each new vertex is attached via an edge back to any previous vertex of the appropriate type (unless eclipsed in all frames by stop codons). Only the top-scoring r noncoding predecessor edges are kept at each vertex, for some user-defined r, as well as all coding edges. Vertices having an in-degree or out-degree of zero are discarded. This heuristic tends to retain only the high-scoring subgraphs, and was found to be invaluable for maintaining speed and space efficiency. Each edge is annotated by TIGRscan with three scores associated with the corresponding feature in the three possible phases (or a single phase for intergenic edges). These scores comprise the $P_e(S_{i,1}|y_i, d_{i,1})P_t(y_i|y_{i-1})P_d(d_{i,1}|y_i)$ term corresponding to Eqs. (9.19, 9.20).

The vertices in these two graphs are then placed along the axes of a two-dimensional dynamic programming matrix as illustrated in Figure 9.11. Each cell in this matrix corresponds to a pair of vertices – one in each ORF graph. Because this matrix can get very large in practice, OASIS employs a sparse representation of the matrix by eliminating all but the most promising regions from consideration. In particular, each PROmer alignment is mapped into the nucleotide-based coordinate

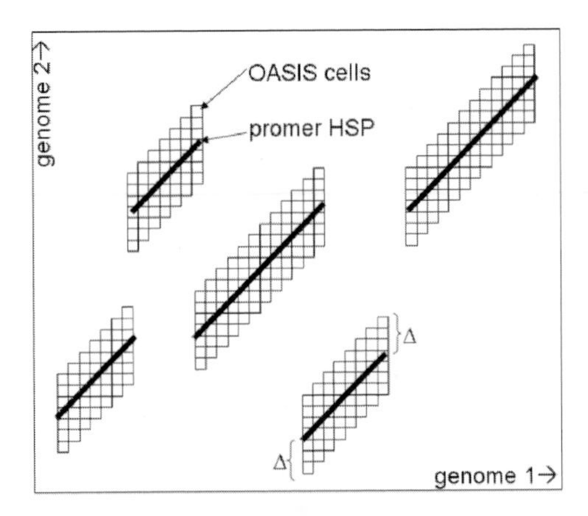

Figure 9.13 Mapping PROmer alignments into the nucleotide-based coordinate space of the OASIS dynamic programming matrix. The matrix is extensively pruned so that only regions close to a PROmer HSP (shown) or a NUCmer HSP (not shown) are represented.

system of the OASIS matrix (using the genome coordinates of the putative signals in the two ORF graphs as a frame of reference), as depicted in Figure 9.13. Each HSP is maximally extended in a gapless fashion at either end until an in-frame stop codon is encountered in either genome. Only those OASIS cells within a user-specified nucleotide distance Δ of a PROmer or NUCmer alignment path are retained, and only those for which the two corresponding signals are of the same signal type (e.g., GT–GT, ATG–ATG, TAG–TGA, etc.). The result is typically a rather sparse matrix with very reasonable memory requirements.

NUCmer alignments are treated differently from PROmer alignments. Whereas the individual PROmer HSPs are all retained separately, the NUCmer HSPs are combined by ROSE into a single global nucleotide alignment, so that OASIS is given potentially many PROmer HSPs but only one NUCmer alignment, where the latter extends from the lower-left corner of the OASIS matrix to the upper-right corner. The task of the decoding algorithm is then to find the highest-scoring path through the unpruned portion of the matrix, where coding features may only be predicted in the vicinity of a PROmer hit, but noncoding features may be predicted anywhere. Scoring of alignments may be performed using *prefix sum arrays* (see below).

Once the OASIS matrix has been pruned around the PROmer and NUCmer hits, each OASIS cell is attributed with a set of predecessor links (see Figure 9.14) formed by taking the cartesian product $E_1 \times E_2$ of the outgoing edges on the two vertices corresponding to that cell. In this way, each OASIS link has associated with it two TIGRscan edges, one from each ORF graph for the two genomes. The resulting set of OASIS cells and their predecessor links form a *trellis*. It should be evident that a path through the trellis outlines an isomorphism (see section 2.14) between subgraphs of the two ORF graphs, and that these subgraphs outline corresponding

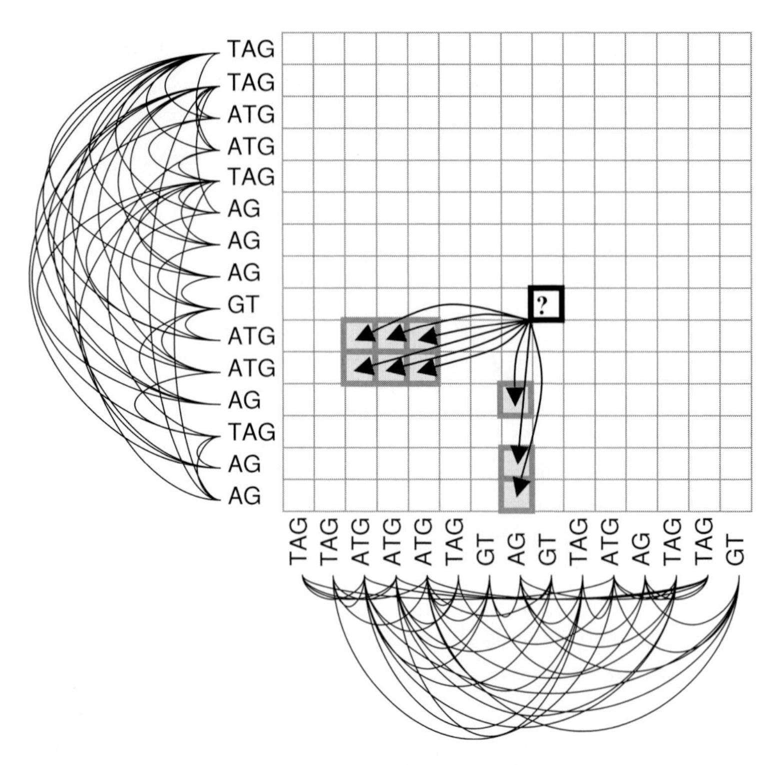

Figure 9.14 Building the OASIS trellis. Each cell in the dynamic programming matrix associated with a pair of signals of the same type (e.g., ATG–ATG) is linked back to all possible predecessor cells by forming the cartesian product of the backward-pointing edges in the two ORF graphs, and considering only cells which have not been pruned from the matrix.

gene predictions (although ensuring that those gene predictions are well formed requires tracking of phase constraints using the usual *mod 3* arithmetic employed in GHMM decoding, as described in section 8.3).

Evaluation of the matrix is then performed according to Eq. (9.19). Each coding link in the trellis is annotated with nine scores (or one for nonphased links), corresponding to the cartesian product of the three possible phases of the associated features in the two genomes (i.e., phase ω_1 in genome 1 and phase ω_2 in genome 2). Storing all nine scores separately is necessary for phased edges in order to avoid greedy behavior in the GPHMM decoding algorithm.

The dynamic programming recurrence for this algorithm is as follows:

$$\beta(i,j) = \max_{(h,k)\in pred(i,j)} \begin{pmatrix} \beta(h,k)P_e(S_{h:i,1}|y_{h:i},d_{h:i})\cdot \\ P_e(S_{i,1}|y_i)P_{cons}(S_{k:j,2}|S_{h:i,1},y_{h:i},d_{h:i})\cdot \\ P_d(d_{h:i}|y_{h:i})P_t(y_i|y_h) \end{pmatrix} \tag{9.22}$$

where (i, j) and (h, k) address cells in the OASIS matrix, $pred(i, j)$ is the set of links to predecessor cells in the trellis, $S_{i,1}$ is the fixed-length sequence in the immediate vicinity of signal i in the ORF graph for genome 1, $S_{h:i,g}$ is the sequence between signals h and i in genome $g \in \{0, 1\}$, y_i is the state corresponding to signal i, and $y_{h:i}$ is the state corresponding to $S_{h:i,g}$. The product $P_e(S_{h:i,1}|y_{h:i}, d_{h:i})P_e(S_{i,1}|y_i)$ $P_d(d_{h:i}|y_{h:i})P_t(y_i|y_h)$ is cached in each edge of ORF graph 1 so that these terms need not be re-evaluated for each cell in the OASIS matrix; these terms are evaluated by TIGRscan, as previously noted. The $P_d(d_{i,1}, d_{i,2}|y_i)$ term has been very crudely approximated using $P_d(d_{h:i}|y_{h:i})$, TIGRscan's duration term for genome 1. $\beta(h, k)$ is the inductive score which is stored separately for each of the nine phase pairs (omitted from Eq. (9.22) for clarity) in each cell of the OASIS matrix. Note that the phases of putative orthologous exons are *not* required to match; phase is tracked only to enforce phase constraints separately in the two genomes, as in a typical GHMM-based gene finder.

The $P_{cons}(S_{k:j,2}|S_{h:i,1}, y_{h:i}, d_{h:i})$ term is approximated in TWAIN very simply as:

$$(P_{match})^{P_{ident}d_{h:i}}(1 - P_{match})^{(1-P_{ident})d_{h:i}} \tag{9.23}$$

where P_{ident} (*percent identity* for nucleotide alignments, or *percent similarity* for amino acids – see Delcher *et al.*, 2002) is estimated using a rudimentary but highly efficient approximate alignment procedure, as described below. P_{match} is a parameter to the GPHMM which is estimated during training to reflect the probability of a single residue match in a coding or noncoding alignment (as per the coding/noncoding status of state $y_{h:i}$ – thus, two P_{match} parameters are required by the program: $P_{match,nuc}$ and $P_{match,amino}$). A pointer to the predecessor (h, k) selected by the *max* term in Eq. (9.22) is stored in one of the nine slots at each cell (according to phase). Once the matrix has been evaluated, the highest-scoring path through the trellis is found using a standard traceback procedure with the usual phase constraints for coding regions and introns.

The P_{ident} term in Eq. (9.23) is estimated using an approximate alignment procedure (see Figure 9.15), using *prefix sum arrays*, as follows. In each cell of a PROmer or NUCmer alignment is stored two values: μ, the cumulative number of matches from the beginning of the alignment to the current cell; and λ, the length of the alignment up to the current cell. For PROmer alignments, a BLOSUM matrix (Henikoff and Henikoff, 1992) score greater than zero is counted as a match. Which BLOSUM matrix to use is specified in a configuration file loaded at run-time. Because a pair of OASIS cells A and B might not fall directly on the PROmer or NUCmer alignment, additional indel terms δ_A and δ_B are assessed by finding the nearest alignment cells z and y to OASIS cells A and B, respectively, using a binary search, and then simply assessing the distance from each OASIS cell to the corresponding alignment cell in nucleotide space (see Figure 9.15). In this way, P_{ident} can be estimated using:

$$P_{ident} = \begin{cases} \dfrac{\mu_z - \mu_{y-1}}{\lambda_z - \lambda_{y-1} + \delta_A + \delta_B} & \text{if } \mu_z - \mu_{y-1} > 0 \\ 0 & \text{otherwise} \end{cases} \tag{9.24}$$

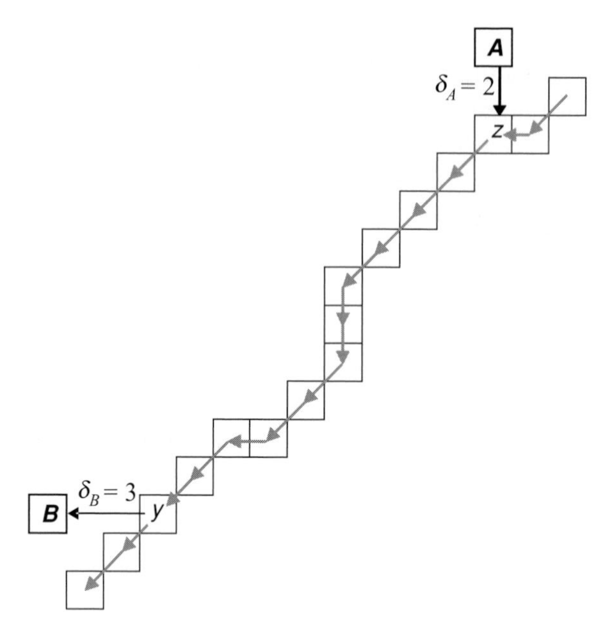

Figure 9.15 A crude method for computing approximate alignment scores. A guide alignment (MUMmer HSP) is shown extending through cells z and y. The OASIS cells A and B, between which an alignment and its score are desired, do not fall directly on the guide alignment, so the shortest path from each cell to the guide alignment is constructed. The alignment score between z and y is rapidly obtained using a prefix sum array (PSA), with additional indel penalties corresponding to distances δ_A and δ_B being incorporated into the approximate score.

where the putative feature pair extends from OASIS cell B to A. When no PROmer or NUCmer alignment is near enough to A and B to give a nonnegative value for the numerator of Eq. (9.24), we set $P_{ident} = 0$.

Figure 9.16 shows that these approximate alignment scores (x-axis) correlate fairly well with full Needleman–Wunsch (Needleman and Wunsch, 1970) alignment scores (y-axis), though the approximate scores appear to underestimate the degree of conservation for the less-conserved features, and there is a large amount of variance.

The alert reader will have inferred from the foregoing description that since OASIS considers only those cells in its dynamic programming matrix corresponding to signal pairs of matching types, the software in its current form is unable to predict pairs of homologous genes having different numbers of exons – i.e., allowing for exon fusion, splitting, insertion, or deletion events in one of the two lineages. Such evolutionary changes are known to occur with some frequency; a somewhat extreme example of two orthologs in *Aspergillus oryzae* and *A. fumigatus* is shown in Figure 9.17.

Although it would be quite simple to modify the OASIS algorithm to allow for the prediction of orthologs with different numbers of exons, to do so would

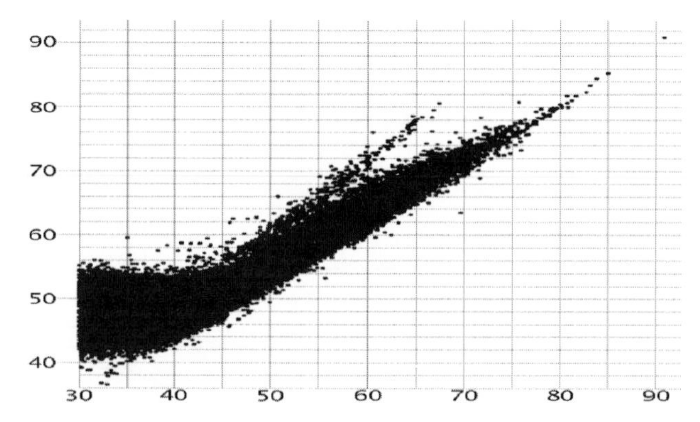

Figure 9.16 Graph of approximate alignment score (x-axis) versus full Needleman–Wunsch alignment score (y-axis). Percent identity was used for both alignment scores in the graph. Sequences were selected randomly from *A. fumigatus* and *A. nidulans* ORFs and filtered so as to consider only those which differed by no more than 5% in length and which overlapped a PROmer HSP. Only nucleotide sequences were aligned.

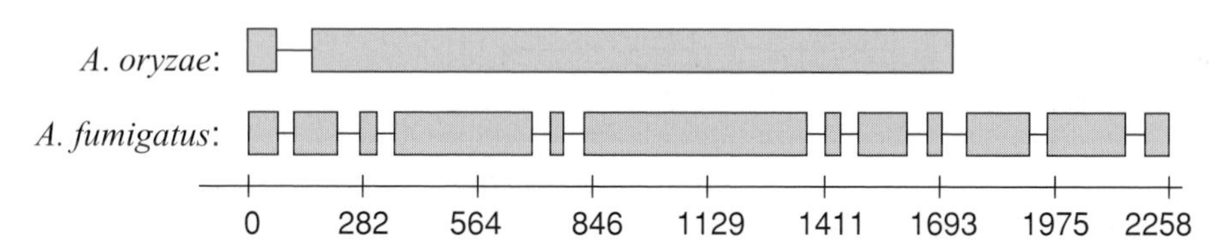

Figure 9.17 A pair of *Aspergillus* orthologs showing evidence of exon fusion and/ or splitting. Gray rectangles represent exons, and the thin line segments connecting them denote introns. Despite their radically different appearance, these genes produce identical proteins.

significantly increase the time and space complexity of the algorithm, due to the much larger number of dynamic programming cells that would need to be considered during decoding (see Figure 9.18). While none of the available GPHMM-based gene finders currently models orthologs with differing exon counts (including SLAM – Alexandersson *et al.*, 2003), the (nongeneralized) PHMM-based gene finder DoubleScan described earlier does do this.

As was mentioned earlier, TWAIN constructs a single, global nucleotide alignment over the input sequences by combining NUCmer HSPs, and utilizes this single nucleotide alignment in assessing approximate alignment scores for putative noncoding features. An alternative strategy is suggested by Figure 9.19, which depicts an *HSP graph* consisting of NUCmer HSPs linked left-to-right and bottom-to-top across the matrix – i.e., with coordinates that strictly increase in both dimensions of the matrix for every pair of adjacent HSPs in the graph. Given two endpoints for a putative noncoding feature in the OASIS matrix, one may then conceive of constructing an optimal path representing the putative feature – i.e., from one endpoint to the nearest HSP, then along some number of links in the HSP graph, and then from the

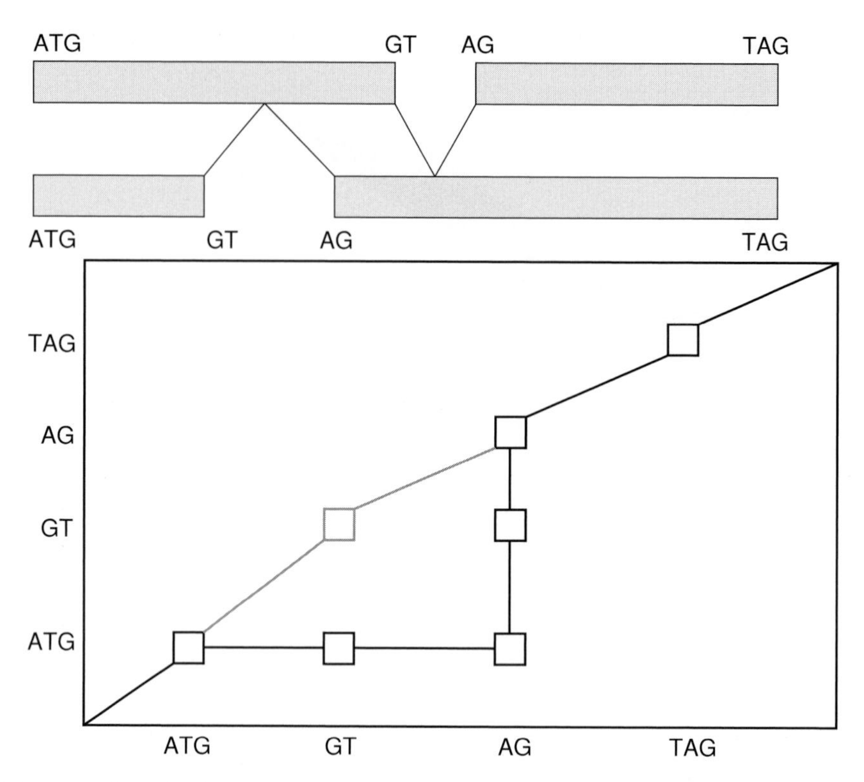

Figure 9.18 Intron insertion in two gene lineages. (a) A pair of aligned orthologs in which an intron has been independently inserted in each lineage. (b) A dynamic programming matrix in which the cells for the relevant exon boundaries have been retained. Allowing for all possible insertions of introns in either genome would clearly interfere with aggressive matrix pruning, since the correct path through the matrix may wander arbitrarily far from the diagonal.

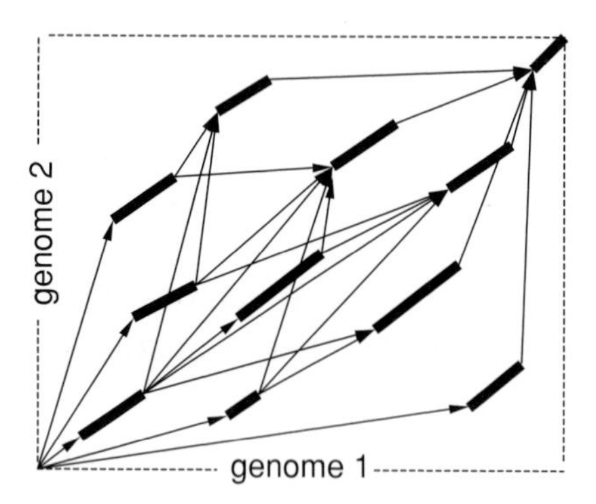

Figure 9.19 Crude illustration of an HSP graph. Thick diagonal lines represent HSPs. Directed arrows indicate possible paths between HSPs. A full path from the lower-left corner to the upper-right corner outlines a global alignment; other paths may be constructed between the endpoints of putative noncoding features, for the assessment of approximate alignment scores for those features.

Table 9.1 Gene-finding accuracy of a GPHMM versus its underlying (non-homology-aware) GHMM, on 147 *Aspergillus fumigatus* × *A. nidulans* orthologs

	Nucleotide SMC	Exon *F*	Gene Sn
GHMM	99%	75%	54%
GPHMM	99%	87%	74%

Majoros *et al.* (2005b).

final HSP in the path to the far endpoint of the putative feature. An approximate alignment score may then be obtained by combining the scores of the individual HSPs on the path through the HSP graph, with adjustments being made for the gaps between successive HSPs on the path, as well as the gaps from each of the endpoints (i.e., the two OASIS cells delimiting the putative feature) to their nearest HSPs which terminate the HSP path.

Such HSP graphs are similar to *Steiner graphs* and *Manhattan networks*, which have been studied in the context of global sequence alignment, and have also been proposed for use in gene finding with a GPHMM (Lam *et al.*, 2003), though efficient algorithms for the latter application remain to be fully described in the gene-finding literature.

Table 9.1 illustrates the value of homology information in the case of a GPHMM. The accuracy of the homology-aware model (GPHMM) is much elevated over that of the non-homology-aware model (a GHMM). Both models used the same signal and content sensors; the GHMM was in fact employed as a subroutine by the GPHMM.

9.6 Phylogenomic gene finding

When a relatively large amount of conservation evidence from different organisms is available, a key difficulty in accurate comparative gene finding is avoiding biases caused by the nonindependence of informant sequences, due to their commonality of evolutionary descent. This is the problem which *phylogenomic (or phylogenetic) HMMs* (*PhyloHMMs*; also known as *evolutionary HMMs*) attempt to address. These methods are also sometimes referred to as *phylogenetic footprinting*.

9.6.1 *Phylogenetic HMMs*

A PhyloHMM can be interpreted as a straightforward extension of the notion of a pair HMM in which our model alternates between emitting a single sequence (our target sequence S) and emitting stretches of aligned sequences (S and one or more informants $I^{(1)}, \ldots, I^{(n)}$). In this way, we can conceive of modeling the process as a "noisy" N-ary HMM ($N = n + 1$) in which all but one of the output channels (i.e., the informants) are prone to malfunction (i.e., to cease emitting) for periods of time in a semi-coordinated fashion. Periods of inactivity in an output channel can be modeled as *missing data*, as we will describe later.

We define a PhyloHMM as a 6-tuple $M = (Q, P_e, P_d, P_t, \alpha, \Psi)$ which operates like a GHMM having alphabet α, states $Q = \{q_i | 0 \leq i < N_{states}\}$, transition distribution P_t, emission distribution P_e, and duration distribution P_d, except that in addition to emitting a *target sequence* S, M also emits a set of *informant sequences* $I^{(1)}, \ldots, I^{(n)}$ according to the state-specific evolution models $\psi_q \in \Psi$. Given a suitable parameterization of the model M, the problem is then one of finding the most probable series of states and state durations $\phi^* = \{(q_i, d_i) | 0 \leq i < m\}$ whereby M could have generated the target and informant sequences $S = S_0 S_1 \ldots S_m$ and $I^{(j)} = I_0^{(j)} I_1^{(j)} \ldots I_m^{(j)}$:

$$\begin{aligned}
\phi^* &= \underset{\phi}{argmax}\ P(\phi | S, I^{(1)}, \ldots, I^{(n)}) \\
&= \underset{\phi}{argmax}\ \frac{P(\phi, S, I^{(1)}, \ldots, I^{(n)})}{P(S, I^{(1)}, \ldots, I^{(n)})} \\
&= \underset{\phi}{argmax}\ P(\phi, S, I^{(1)}, \ldots, I^{(n)}) \\
&= \underset{\phi}{argmax}\ P(\phi) P(S, I^{(1)}, \ldots, I^{(n)} | \phi) \\
&= \underset{\phi}{argmax}\ P(\phi) P(S | \phi) P(I^{(1)}, \ldots, I^{(n)} | S, \phi).
\end{aligned} \tag{9.25}$$

The term $P(\phi)P(S|\phi)$ is the familiar joint probability term from the GHMM decoding problem, which we factor in the usual way:

$$P(\phi)P(S|\phi) = P_t(q^0 | y_n) \prod_{i=1}^{n} P_t(y_i | y_{i-1}) P_d(d_i | y_i) P_e(S_i | y_i, d_i). \tag{9.26}$$

The standard GHMM decoding methods of section 8.3 suffice to evaluate this expression, so that what remains is only to evaluate the joint probability of the informant sequences, given the target sequence and the parse:

$$P(I^{(1)}, \ldots, I^{(n)} | S, \phi) \tag{9.27}$$

which it can be seen is reminiscent of the P_{cons} term for GPHMMs as given in Eq. (9.20). Let us ignore ϕ for the moment, and suppose the sequences $S, I^{(1)}, \ldots, I^{(n)}$ are related by a *phylogeny*, or *phylogenetic tree*, such as the one shown in Figure 9.20. Present in the tree are the target species, the informant species, and a number of ancestral species whose purpose is to root each *clade* (subtree) in the tree. It is also possible to attach a *substitution matrix* (i.e., a matrix of probabilities for specific mutation events between pairs of residues) to each edge of the tree to model the rate of evolutionary change in each lineage. These rates are known to be variable across lineages (and even across sites within a lineage – see Yang, 1994; Felsenstein and Churchill, 1996).

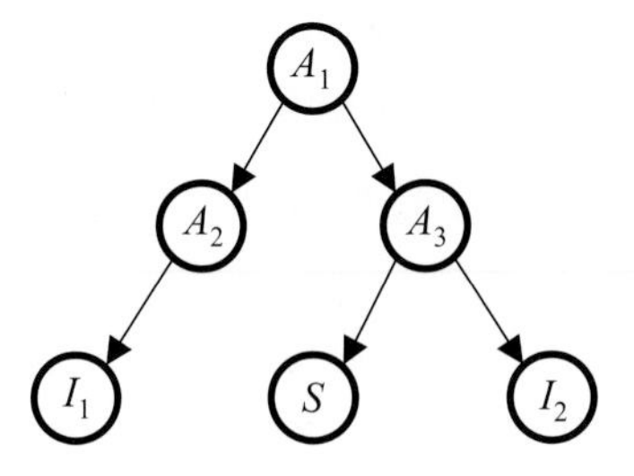

Figure 9.20 A simple phylogenetic tree. The extant species are shown as leaves at the bottom, and the ancestral species are shown as internal nodes. If S is our target species, then I_1 and I_2 are considered informants.

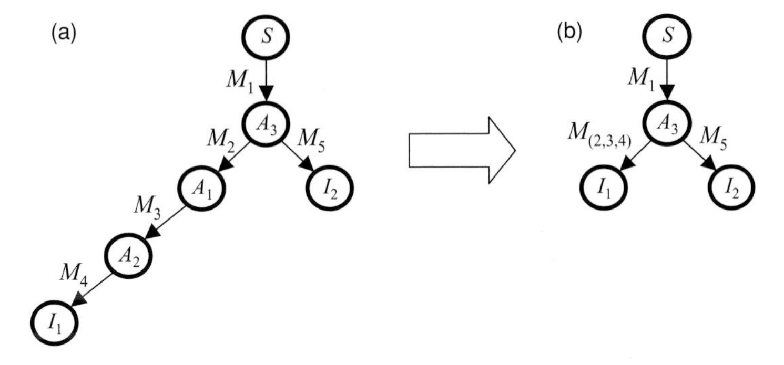

Figure 9.21 (a) A re-rooted version of the phylogeny shown in Figure 9.20. The target species S is now at the root of the tree, and only the informants occur at the leaves. Substitution matrices have also been attached to the branches. (b) Nonbranching internal nodes below the root have been eliminated. (Adapted from Gross and Brent, 2005.)

It is a well-known fact from the study of phylogenetics that we can re-root this tree using any vertex, and any *reversible* substitution models (see section 9.6.4) along the lines of descent of the phylogeny remain valid (Durbin *et al.*, 1998; Siepel and Haussler, 2005). We will define reversibility shortly. Since we wish to condition the informant sequences on the target sequence S according to Eq. (9.27), it will be useful to root the tree with the target species. The re-rooted tree for the above example is shown in Figure 9.21(a).

Note that the informant species in Figure 9.21(a) are all located at the leaves of the tree and the target sequence is at the root. In part (b) of the figure, we show that it is also possible to eliminate nonbranching internal nodes below the root, combining their incident branches with a new branch having a length which is the sum of the lengths for the replaced branches; the effect of this modification on the substitution models of these branches will become clear later when we give a more thorough treatment of substitution models.

Returning for the moment to the unmodified tree in Figure 9.21(a), we can use the structure of the tree to factor Eq. (9.27) as a product of conditional probabilities in which each sequence is conditioned only on its unique parent (and in which we marginalize over the ancestors, since they do not appear in Eq. (9.27)):

$$\sum_{A_1,A_2,A_3} P(I_1|A_2)P(A_2|A_1)P(A_1|A_3)P(I_2|A_3)P(A_3|S) \tag{9.28}$$

or, for the tree in Figure 9.21(b):

$$\sum_{A_3} P(I_1|A_3)P(I_2|A_3)P(A_3|S) \tag{9.29}$$

where the summation is over all possible nucleotide sequences of the appropriate length for the ancestral taxa. This factorization is possible because of the common (and fully justifiable) assumption of conditional independence in traditional phylogenetic models – i.e., that any feature of species X is conditionally independent of all speciation events involving any of its aunts, uncles, siblings, nephews, nieces, and grandparents (arbitrarily far removed), given its own speciation event (i.e., given its parent) (e.g., McAuliffe *et al.*, 2004).

More generally,

$$P(I^{(1)}, \ldots, I^{(n)}|S) = \sum_{\mathit{unobservables}} \left(\prod_{\substack{\mathit{nonroot} \\ v}} P(v|\mathit{parent}(v)) \right) \tag{9.30}$$

where we have summed over all possible assignments to the *unobservables* (ancestral vertices) as in Eqs. (9.28, 9.29).

Given a precomputed multi-alignment of the target and informant sequences, and assuming independence between *sites* (columns of the alignment), Eq. (9.30) can be efficiently computed on a per-column basis using a recursion known as *Felsenstein's pruning algorithm*:

$$L_u(a) = \begin{cases} \delta(u, a) & \text{if } u \text{ is a leaf} \\ \prod_{c \in C(u)} \sum_{b \in \alpha} L_c(b)P(c = b|u = a) & \text{otherwise} \end{cases} \tag{9.31}$$

(Felsenstein, 1981, 2004), for $C(u)$ the children of node u, $\delta(u, a)$ the *Kronecker* match function (section 2.1), and the augmented DNA alphabet $\alpha = \{\mathtt{A}, \mathtt{C}, \mathtt{G}, \mathtt{T}, \mathtt{-}\}$, where the dash "–" denotes missing information or a gap in the alignment (e.g., Siepel and Haussler, 2004c); $a \in \alpha$. $L_u(a)$ denotes the likelihood of the subtree rooted at node u, given that the value of u is the symbol a. Actual evaluation of the recursion should be performed during a *postorder* traversal of the tree (section 2.14), to take advantage of the efficiency benefits of dynamic programming.

Using Felsenstein's recursion, the conditional likelihood for a single column j of the alignment is given by

$$P(I^{(1)}[j], \ldots, I^{(n)}[j]|S, \theta) = L_r(S[j]|\theta) \tag{9.32}$$

where r is root node of the tree, $S[j]$ denotes the jth symbol in the S track of the alignment (e.g., the "human" row in Figure 9.23), and we have made explicit that

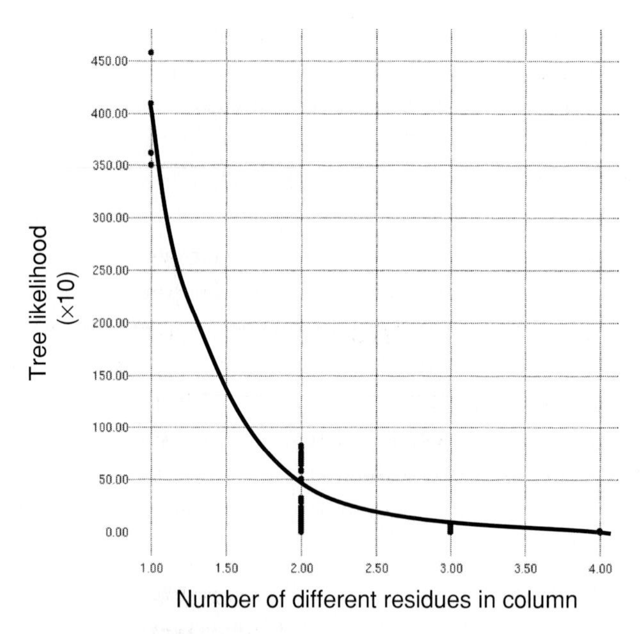

Figure 9.22 Likelihood of a tree for five species, for 5000 columns of an alignment.

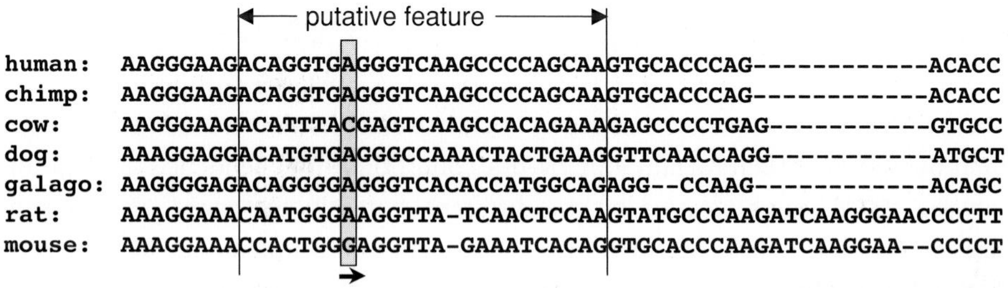

```
                 |<-- putative feature -->|
human:  AAGGGAAGACAGGTGAGGGTCAAGCCCCAGCAAGTGCACCCAG------------ACACC
chimp:  AAGGGAAGACAGGTGAGGGTCAAGCCCCAGCAAGTGCACCCAG------------ACACC
cow:    AAGGGAAGACATTTACGAGTCAAGCCACAGAAAGAGCCCCTGAG-----------GTGCC
dog:    AAAGGAGGACATGTGAGGGCCAAACTACTGAAGGTTCAACCAGG-----------ATGCT
galago: AAGGGGAGACAGGGGAGGGTCACACCATGGCAGAGG--CCAAG------------ACAGC
rat:    AAAGGAAACAATGGGAAGGTTA-TCAACTCCAAGTATGCCCAAGATCAAGGGAACCCCTT
mouse:  AAAGGAAACCACTGGGAGGTTA-GAAATCACAGGTGCACCCAAGATCAAGGAA--CCCCT
```

Figure 9.23 Evaluation of a putative feature spanning a given interval in the target sequence (human in this example). Felsenstein's pruning algorithm can be applied to individual columns within the interval. Different evolution models ψ can be employed for the different feature types, thereby providing some basis for feature discrimination in unannotated DNA.

the likelihood depends on the model parameters θ. Figure 9.22 shows the likelihood of a particular tree for five species, for 5000 columns, showing that the likelihood function is distinctly nonlinear.

Evaluating the conditional likelihood of the entire alignment (still assuming independence between sites) can be accomplished via multiplication:

$$P(I^{(1)}, \ldots, I^{(n)}|S, \theta) = \prod\nolimits_{0 \leq j < L} L_r(S[j]|\theta) \tag{9.33}$$

where L is the length of the alignment. This is illustrated in Figure 9.23 for an interval of an alignment corresponding to a putative gene feature.

Alternatively, we may consider evaluating $L_r(S_i)$ for an entire subsequence S_i of S, in which case we need not assume site independence within S_i; we will return to the issue of site independence later. In the case of individual sites, we can evaluate $P(c = b|u = a) = P(c|parent(c))$ from Eq. (9.31) by accessing the appropriate entry of the substitution matrix **P** attached to the branch of the phylogeny connecting c to its parent u. **P** is parameterized by the length t of this branch, since the propensity for nucleotide substitutions over evolutionary time is clearly dependent on the length of time that has elapsed since the speciation event separating the parent and child species. We will consider the time-parameterization of **P** in section 9.6.3. Recall from section 9.5.1 that substitution matrices for amino acids are also possible (e.g., Henikoff and Henikoff, 1992).

It is worth noting at this point that the re-rooted phylogeny can be interpreted more generally as a *Bayesian network* (Gross and Brent, 2005; see also section 10.4), in which case we are then free to employ any of the well-known inference algorithms that have been developed for these structures (e.g., Pearl, 1991; Jensen, 2001), whether purely probabilistic or otherwise. We will revisit this issue.

Now let us attend to ϕ. Recall that for the $P(\phi)P(S|\phi)$ term we partitioned ϕ into a series of states and their durations, as shown in Eq. (9.26). We can do the same within the context of Eq. (9.27), to produce:

$$P(I^{(1)}, \ldots, I^{(n)}|S, \phi) = \prod_{(y_i, d_i) \in \phi} P(I_i^{(1)}, \ldots, I_i^{(n)}|S_i, y_i, d_i) \qquad (9.34)$$

where $I_i^{(j)}$ is the subsequence emitted by y_i (the ith state in ϕ) into the $I^{(j)}$ track of the alignment, and we have again employed a conditional independence assumption between the features emitted by the different states in the parse. This decomposition by state allows us to utilize a different evolution model for different feature types, which is the key advantage of using phylogenetic models in gene finding. Because we expect the patterns of nucleotide substitution to vary significantly between the functional elements within a genome (due to the different selective pressures at work within those elements), the use of separate evolution models within each state (i.e., for each feature type) can provide significant discriminatory power for predicting the boundaries of gene features, as shown in Figure 9.24.

Combining Eq. (9.26) together with Eq. (9.34) gives us an expression for the *maximum a posteriori* parse, ϕ^*:

$$\phi^* = \underset{\phi}{argmax} \prod_{(y_i, d_i) \in \phi} P(S_i|y_i, d_i)P(y_i|y_{i-1})P(d_i|y_i)P(I_i^{(1)}, \ldots, I_i^{(n)}|S_i, y_i, d_i) \qquad (9.35)$$

$$= \underset{\phi}{argmax} \prod_{\substack{(y_i, d_i) \in \phi, \\ y_i \neq q_0}} P(S_i|y_i, d_i)P(y_i|y_{i-1})P(d_i|y_i) \prod_{b_i \leq j \leq e_i} L_r(S[j]) \qquad (9.36)$$

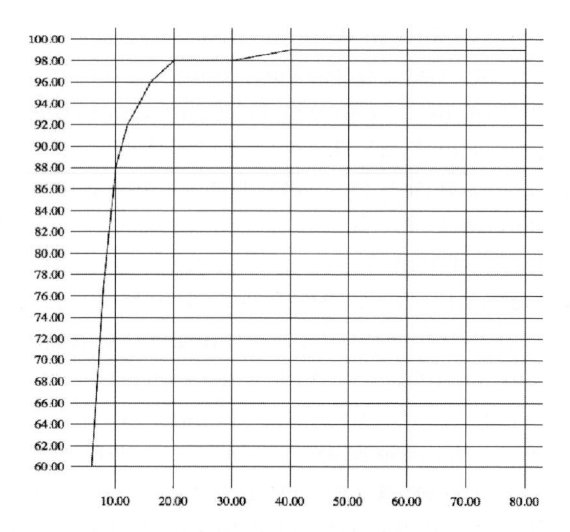

Figure 9.24 Classification accuracy (*y*-axis, percentages) of a PhyloHMM for an exon identification task. Equal numbers of coding and noncoding segments were independently evolved over a simulated phylogeny, using a Jukes–Cantor model (section 9.6.4) with variable substitution rates, and then classified via likelihood ratio. Coding substitution rate was fixed at 5%; noncoding substitution rate was varied from 6% to 80% (*x*-axis). Increasing the noncoding substitution rate relative to the coding rate quickly enabled the PhyloHMM to achieve reliable discrimination between coding and noncoding elements.

where the output of the ith state in ϕ spans the interval $[b_i, e_i]$ in the alignment, and we have omitted the transition to state q^0 for simplicity. This optimization problem can be solved using an extension of the standard GHMM decoding algorithms. We describe this next.

9.6.2 *Decoding with a PhyloHMM*

As detailed in section 8.3, there are two distinct approaches currently in use for decoding with a GHMM: the *prefix sum arrays* (PSA) algorithm, and the more efficient *dynamic score propagation* (DSP) algorithm. Our discussion will be equally applicable to both, though we will borrow an important concept from the former for incorporating homology evidence into the decoding process. In particular, we require the notion of a *prefix sum array*. Define a prefix sum array A for a function $f(x)$ and a sequence S of length L to be a $1 \times L$ dimensional array in which the ith element is defined by:

$$A[i] = \begin{cases} \log f(0) & \text{for } i = 0 \\ A[i-1] + \log f(i) & \text{for } i > 0 \end{cases} \tag{9.37}$$

for $0 \leq i < L$. Such an array allows one to compute the product $\prod_{b \leq x \leq e} f(x)$ for an arbitrary interval $[b, e]$ within the sequence by simple subtraction followed by exponentiation:

$$e^{A[e]-A[b-1]}. \tag{9.38}$$

Because this operation can be performed in constant time, the use of prefix sum arrays is very fast. The price of this speed is the cost of storing the array, which can become significant if a large number of lengthy arrays are required (especially for double-precision floating-point numbers). This cost can be considerably reduced through the use of sparse arrays (Exercise 9.26); using arrays of integers rather than real values is also possible (Exercise 9.29).

Before describing how to decode a sequence with a PhyloHMM, we briefly review the steps of the DSP and PSA algorithms for GHMM decoding. In either case, the process consists of a single 5′-to-3′ (left-to-right) pass over the input sequence, where at each position we evaluate our (factorable) content sensors and use the resulting single-nucleotide values to update various data structures (prefix sum arrays in the case of PSA; *partial parse* scores in the case of DSP). Whenever a putative signal is encountered, a signal record is instantiated, linked to potential predecessor signals according to the grammar of legal genes, and added to the appropriate queues. Upon reaching the end of the sequence, the trellis of links between putative signals is traversed right-to-left to outline (in reverse) the most probable parse ϕ^*.

The attentive reader will now see that it is relatively straightforward to modify either decoding algorithm to incorporate homology evidence. In either case we may first compute a set of prefix sum arrays A_i – one for each feature type $\Gamma_i \in \{exon,\ intron,\ etc.\}$ – for the function $L_r(S[j])$, and using Eq. (9.38) evaluate the log-likelihoods of the informant subsequences for a putative feature spanning interval $[b, e]$ via $A_i[e] - A_i[b-1]$. The prefix sum arrays A_i can all be computed during a single pass over the sequence prior to the decoding pass. Alternatively, the conditional tree likelihoods L_r may be computed during the (one and only) DSP pass over the sequence. In either case, the appropriate values ($\log L_r(S[j])$ for DSP, $A_i[e] - A_i[b-1]$ for PSA) can be added to the inductive scores of putative signals before they are inserted into their respective queues (and, in the case of DSP, during propagator updates – see section 8.3.2), so that the final traceback procedure for the GHMM decoding algorithm will then select automatically the parse that scores maximally under Eq. (9.36), as can be verified by the reader. This constitutes a complete decoding procedure for a PhyloHMM.

9.6.3 *Evolution models*

We now consider the evolution models ψ employed by the PhyloHMM. Each $\psi = (\tau, \beta, \mathbf{Q})$ consists of a tree topology τ, a set $\beta = \{t_i | 0 \leq i < n\}$ of branch lengths, and a substitution *rate matrix* **Q**. The tree topology may be inferred using a standard algorithm such as *UPGMA* (see below), which is essentially equivalent to a hierarchical clustering procedure. The branch lengths and the single rate matrix together allow us to establish substitution rates along each of the branches of the (re-rooted) phylogeny, as we will show. Our treatment closely follows that of Clote and Backofen (2000).

A simple algorithm for phylogeny reconstruction is as follows. We will provide only an informal definition, since the procedure is such a trivial one; interested readers are referred to Durbin *et al.* (1998) or a similar reference for a more detailed treatment. The procedure, called *UPGMA* (Sneath and Sokal, 1973), begins with a set of degenerate binary trees in which each tree consists of a single root corresponding to each of the genomes to be included in the phylogeny. The procedure iteratively combines subtrees in this population until only one tree remains, which is the phylogeny to be returned by the procedure. Each step of the iteration chooses the two nearest subtrees according to a distance matrix D, which is initialized to contain pairwise edit distances d_{ij} provided by a suitable alignment algorithm such as Needleman–Wunsch (Needleman and Wunsch, 1970). Upon combining trees T_i and T_j, a combined tree T_k is formed by creating a new root R having as children the roots of T_i and T_j. T_i and T_j are removed from the population and replaced with T_k, and the pairwise distances between T_k and all other active subtrees T_h in the population are computed via:

$$d_{kh} = \frac{d_{ih}v_i + d_{jh}v_j}{v_i + v_j} \tag{9.39}$$

for v_i the number of vertices in tree T_i. The procedure terminates when only a single tree remains; this is the phylogeny returned by the procedure.

An alternative procedure which tends to be more accurate in practice is the method which is known *as neighbor joining*, or *NJ* (Saitou and Nei, 1987); the following concise description is adapted from Durbin *et al.* (1998). This algorithm follows the same logic as UPGMA, but with a different distance formula:

$$D_{ij} = d_{ij} - r_i - r_j \tag{9.40}$$

where d_{ij} is as defined for UPGMA. Nearest trees are chosen according to this D_{ij}, where r_i is defined as:

$$r_i = \frac{1}{|L| - 2} \sum_{k \in L} d_{ik}. \tag{9.41}$$

Updating of distances for a new subtree T_k is performed via:

$$\forall_{m \in L} \left[d_{km} = \frac{1}{2}(d_{im} + d_{jm} - d_{ij}) \right]. \tag{9.42}$$

Branch lengths for the children (i, j) of new node k are given by:

$$d_{ki} = \frac{1}{2}(d_{ij} + r_i - r_j) \tag{9.43}$$

and

$$d_{kj} = \frac{1}{2}(d_{ij} + r_j - r_i). \tag{9.44}$$

As an example, consider the randomly generated phylogeny shown in Figure 9.25. The root sequence labeled "ancestor_4" was evolved stochastically over this phylogeny to produce the five leaf sequences labeled "species_1" through "species_5." The UPGMA algorithm was then applied to these five leaf species in an attempt

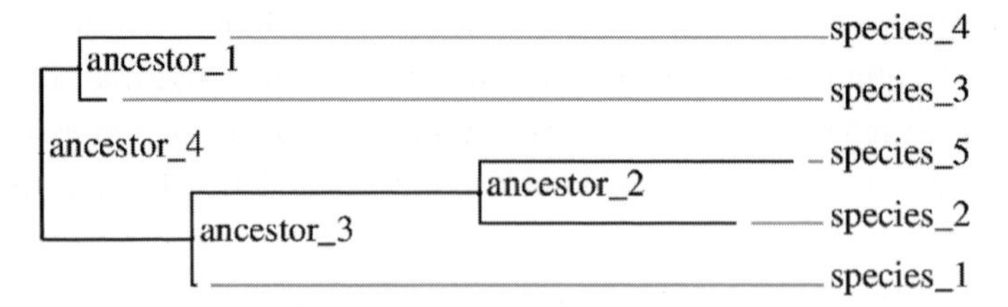

Figure 9.25 A phylogenetic tree. Evolution is modeled as proceeding from left to right across the branches of the tree. Tree branches are shown in black; the gray lines associate labels with the leaves of the tree. Branch lengths indicate the amount of evolutionary change in a lineage, which may even vary over equal spans of time.

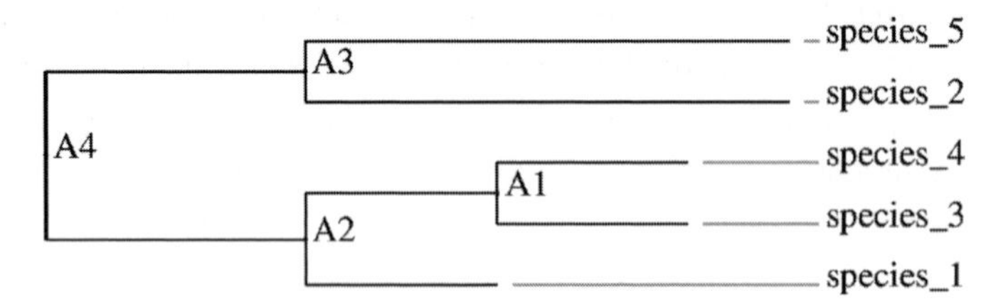

Figure 9.26 A phylogenetic tree reconstructed via UPGMA from the five genomic (leaf) sequences shown in Figure 9.25.

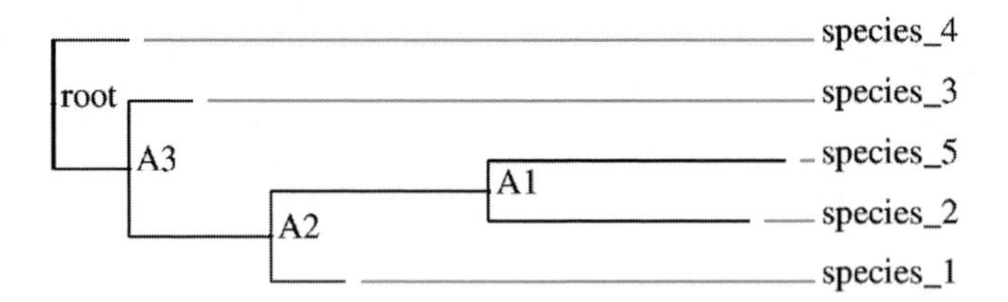

Figure 9.27 A phylogenetic tree reconstructed via neighbor joining. The reconstructed tree more closely resembles the original in Figure 9.25 than does the tree inferred through UPGMA (Figure 9.26).

to reconstruct the original, unknown phylogeny. Figure 9.26 gives the phylogeny inferred via UPGMA. While the UPGMA algorithm has correctly joined species 5 and 2 into a clade and species 4 and 3 into a separate clade, it can be seen from Figure 9.26 that the branch lengths within these clades do not accurately represent those in the correct phylogeny.

Figure 9.27 shows the result of applying the neighbor-joining algorithm instead of UPGMA to the leaf sequences. Although NJ does not normally identify a root for the synthesized tree, we have arbitrarily rooted this tree at the last step of the neighbor-joining process by calling the last-introduced ancestral node the root. As can be seen from the figure, the branch lengths appear to have been more accurately reconstructed, as is evident from the species_5/species_2 clade.

There is a large literature devoted to the subject of accurate phylogeny reconstruction; we assume the foregoing will suffice to illustrate the problem. We progress now to the problem of modeling the evolutionary changes occurring along individual branches of the phylogeny. Our interest in this problem is motivated entirely by our need to evaluate Eq. (9.34), via Eq. (9.31). Models describing such evolutionary changes are called *substitution models*, since they describe the propensity for one residue X in some descendant sequence to *substitute for*, or *replace*, a residue Y at the corresponding position in an ancestral species, given the elapsed time between speciation events for the ancestral and descendant genomes (or between a speciation event and the present).

Substitution models are typically based on *continuous-time Markov chains*, which are similar to the discrete-time Markov models which we have broadly expounded in this book, but with the obvious difference that time is modeled continuously rather than in discrete units. Let us denote by $\mathbf{P}(t)$ the matrix of substitution probabilities over a time lapse $t \in \mathbb{R}$ – *i.e.*, $\mathbf{P}(t)_{a,b}$ denotes the probability of observing residue b at a particular site, given that residue a occupied that site at precisely t time units into the past, for real number t. In practice, t is typically scaled to the expected number of substitutions per site per unit time (Liò and Goldman, 1998).

The so-called *Markov property* for continuous-time Markov chains states that:

$$\mathbf{P}(t + s) = \mathbf{P}(t)\mathbf{P}(s) \tag{9.45}$$

for time lengths t and s. That is, the probability of a given substitution is insensitive to the absolute position along the time axis (i.e., the substitution rate is *stationary*), so that time-dependent substitution rates are simply compounded via matrix multiplication. From this we can derive an *instantaneous rate matrix* $\mathbf{Q}$ from $\mathbf{P}(t)$, where we make use of the obvious fact that $\mathbf{P}(0) = \mathbf{I}$:

$$\begin{aligned}\frac{d\mathbf{P}(t)}{dt} &= \lim_{\Delta t \to 0} \frac{\mathbf{P}(t + \Delta t) - \mathbf{P}(t)}{\Delta t} \\ &= \lim_{\Delta t \to 0} \frac{\mathbf{P}(t)\mathbf{P}(\Delta t) - \mathbf{P}(t)\mathbf{I}}{\Delta t} \\ &= \mathbf{P}(t) \lim_{\Delta t \to 0} \frac{\mathbf{P}(\Delta t) - \mathbf{P}(0)}{\Delta t} \\ &= \mathbf{P}(t)\mathbf{Q}\end{aligned} \tag{9.46}$$

(Clote and Backofen, 2000), giving us the differential equation:

$$\frac{d\mathbf{P}(t)}{dt} = \mathbf{P}(t)\mathbf{Q} \tag{9.47}$$

which can be solved by *Taylor expansion* (see, e.g., Mizrahi and Sullivan, 1986):

$$\mathbf{P}(t) = e^{\mathbf{Q}t} = \sum_{n=0}^{\infty} \frac{\mathbf{Q}^n t^n}{n!} = I + t\mathbf{Q} + \frac{\mathbf{Q}^2 t^2}{2!} + \ldots \tag{9.48}$$

where the notation $e^{\mathbf{Q}t}$ is known as the *matrix exponential*. In practice, the matrix $\mathbf{P}(t)$ is generally solved via *spectral decomposition* (i.e., *diagonalization*):

$$\mathbf{P}(t) = \mathbf{G} \begin{bmatrix} e^{t\lambda_1} & & & \\ & e^{t\lambda_2} & & \\ & & \ddots & \\ & & & e^{t\lambda_n} \end{bmatrix} \mathbf{G}^{-1} \tag{9.49}$$

(Liò and Goldman, 1998), where λ_i is the ith *eigenvalue* of **Q**, and **G** contains the *eigenvectors* (section 2.1) of **Q** as its columns; the off-diagonal entries of the expanded matrix in this formula are all zeros. Thus, given an evolution model $\psi_q = (\tau, \beta, \mathbf{Q})$ for state q in our PhyloHMM, we can derive from β and **Q** the substitution matrices $\mathbf{P}(t)$ along each of the branches in the phylogeny, for corresponding branch length $t \in \beta$. We will see that constraining all the substitution matrices in a phylogeny to be instantiations $\mathbf{P}(t)$ of **Q** allows for a reduction in the number of parameters in the evolution model, thereby reducing the potential for overfitting.

9.6.4 *Parameterization of rate matrices*

The number of parameters in an evolution model can be further reduced by constraining the form of the rate matrix **Q**. For mathematical reasons, the diagonal entries of **Q** are constrained so as to make each row in the matrix sum to zero (Liò and Goldman, 1998). Thus, for the 4-letter DNA alphabet, the matrix will have at most 12 free parameters (i.e., the number of off-diagonal entries), though simpler models with fewer free parameters are typically used.

Two key features of rate matrices important for phylogenetic gene finding are *reversibility* and *transition–transversion modeling*. A *reversible* model is one in which

$$\forall_{ij} \pi_i \mathbf{P}_{ij}(t) = \pi_j \mathbf{P}_{ji}(t), \tag{9.50}$$

where π_i is the background frequency of base i. One advantage of reversible models is that their eigenvector decomposition (e.g., Eq. 9.49) is guaranteed to produce only real-valued (as opposed to *imaginary*) eigenvalues (Keilson, 1979). Another advantage is that they allow us to re-root the phylogeny, as noted previously. *Transition–transversion* modeling involves parameterizing the model in such a way that *transition* (R ↔ R, Y ↔ Y, for R = purines A and G, and Y = pyrimidines C and T) and *transversion* (R ↔ Y) rates are differentially expressible; this is important biologically, as it is known that transition and transversion rates are often drastically different, with transitions typically much more frequent than transversions.

We adopt the convention that substitution matrices $\mathbf{P}(t)$ are oriented so that the rows correspond to the parental residues and the columns correspond to the descendant residues:

$$\mathbf{P}(t) = \overbrace{\begin{array}{cccc} A & C & G & T \end{array}}^{\textit{parent}} \left[\begin{array}{cccc} - & P_{A\to C} & P_{A\to G} & P_{A\to T} \\ P_{C\to A} & - & P_{C\to G} & P_{C\to T} \\ P_{G\to A} & P_{G\to C} & - & P_{G\to T} \\ P_{T\to A} & P_{T\to C} & P_{T\to G} & - \end{array} \right] \left.\begin{array}{c} A \\ C \\ G \\ T \end{array}\right\} \textit{child}$$

The simplest substitution model is that due to Jukes and Cantor (1969):

$$Q_{JK} = \begin{bmatrix} - & \alpha & \alpha & \alpha \\ \alpha & - & \alpha & \alpha \\ \alpha & \alpha & - & \alpha \\ \alpha & \alpha & \alpha & - \end{bmatrix}$$

in which all nonidentical substitutions are modeled as having equal rate α. Diagonal elements are shown as "–" and are assumed as stated above to bring the row sums to zero. At *equilibrium* (i.e., in the limit as $t \to \infty$), all bases under this model would have equal frequencies, and so this model is reversible, though it clearly neglects any differences in transition–transversion rates.

The *Kimura 2-parameter model* (Kimura, 1980) satisfies both our requirements of reversibility and transition–transversion modeling:

$$Q_{K2P} = \begin{bmatrix} - & \beta & \alpha & \beta \\ \beta & - & \beta & \alpha \\ \alpha & \beta & - & \beta \\ \beta & \alpha & \beta & - \end{bmatrix}$$

but it also results in equal nucleotide equilibrium frequencies. This latter fact can be demonstrated with the following UNIX command:

```
/usr/bin/perl -e '$p=0.6;$r=0.4;$a=0.7;
$b=0.3;for($i=0;$i<10;++$i){print "$i
$p\n";($p,$r)=($p*$a+$r*$b,$r*$a+$p*$b)}'
```

by graphing the value of `$p` (purine frequency) over time, to show that in an analogous discrete system, purines and pyrimidines become equally frequent. A refinement of this program can show that the individual base frequencies also become equal (see Exercise 9.30). Note that α and β need not sum to 1; indeed, their sum determines the overall rate of evolution under this model.

Felsenstein's 4-parameter model (Felsenstein, 1981) is reversible and allows for nonuniform nucleotide equilibrium frequencies, but no transition–transversion modeling:

$$Q_{FEL} = \begin{bmatrix} - & \alpha\pi_C & \alpha\pi_G & \alpha\pi_T \\ \alpha\pi_A & - & \alpha\pi_G & \alpha\pi_T \\ \alpha\pi_A & \alpha\pi_C & - & \alpha\pi_T \\ \alpha\pi_A & \alpha\pi_C & \alpha\pi_G & - \end{bmatrix}.$$

Here, α is the mutation rate and $(\pi_A, \pi_C, \pi_G, \pi_T)$ are the equilibrium frequencies.

The so-called *HKY* model of (Hasegawa, Kishino, and Yano, 1985) satisfies our requirements of reversibility and transition–transversion modeling, while also allowing nonuniform equilibrium frequencies:

$$Q_{HKY} = \begin{bmatrix} - & \beta\pi_C & \alpha\pi_G & \beta\pi_T \\ \beta\pi_A & - & \beta\pi_G & \alpha\pi_T \\ \alpha\pi_A & \beta\pi_C & - & \beta\pi_T \\ \beta\pi_A & \alpha\pi_C & \beta\pi_G & - \end{bmatrix}$$

HKY has five free parameters, including the equilibrium nucleotide frequencies (three of which are free, and the fourth of which is constrained to bring their sum to 1).

The most general reversible model, which has nine free parameters, is:

$$Q_{REV} = \begin{bmatrix} - & \beta\pi_C & \alpha\pi_G & \chi\pi_T \\ \beta\pi_A & - & \kappa\pi_G & \omega\pi_T \\ \alpha\pi_A & \kappa\pi_C & - & \tau\pi_T \\ \chi\pi_A & \omega\pi_C & \tau\pi_G & - \end{bmatrix}.$$

Considering that the bulk of the free parameters in a large phylogeny are comprised of the $2(n-1)$ branch lengths, the use of the 9-parameter model does not seem unreasonable, and indeed, will allow for the closest fit to the training data in the general case. Note that Q_{FEL}, Q_{HKY} , and Q_{REV} are nonsymmetric, which limits somewhat the available methods for their eigenvector decomposition; fortunately, the open-source library *TNT* (*Template Numerical Toolkit* – Pozo, 1997) provides routines for eigenvector decomposition of arbitrary real matrices.

To appreciate the reduction in parameters that can be achieved by employing parametric substitution models as described above, consider that employing one nucleotide substitution matrix per edge in a phylogeny would require $24 \cdot (n-1)$ free parameters for n extant species, whereas using one rate matrix for the entire tree and a single branch length per edge requires only $2n + 10$ parameters in the case of unconstrained matrices, and fewer still for any of the rate matrix families described above. For a tree with seven species, these work out to 144 parameters for the former, but only 24 parameters for the latter, and an even smaller number can be achieved if the model is suitably constrained. In practice this may translate into different tendencies toward overtraining of the model.

Given enough training data, matrices may be devised to encode higher-order relations – e.g., $P(v = \text{ATCG}|parent(v) = \text{ATGG})$ – so that the site-independence assumption may be (somewhat) relaxed (Siepel and Haussler, 2004c), although this will have the effect of increasing the number of parameters in the model. We will return to this issue shortly.

9.6.5 Estimation of evolutionary parameters

Given a set of training alignments, the goal (in a maximum likelihood setting) of parameter estimation for a PhyloHMM is to maximize the likelihood of the training alignments, given the model. That is, we would like to jointly optimize the rate matrix $\mathbf{Q}$ and the branch lengths $\beta = \{t_i | 0 \leq i < n\}$ of a given topology τ so as to maximize the likelihood of the given multiple sequence alignment A:

$$(\mathbf{Q}, \{t_i\})^* = \underset{(\mathbf{Q}, \{t_i\})}{argmax}\, P(A_{S,I^{(1)},\ldots,I^{(n)}} | \mathbf{Q}, \{t_i\}, \tau). \tag{9.51}$$

In practice, we may instead decide to first infer the rate matrix from the tree, using appropriately rescaled branch lengths from UPGMA or NJ, and then to re-estimate the optimal branch lengths given the tree and the rate matrix. Felsenstein (1981) gives *expectation maximization* (*EM*) update equations for optimizing a single branch length while holding the others constant, assuming a 4-parameter rate matrix of the form $\mathbf{Q}_{FEL}$ described above (i.e., without explicit transition–transversion modeling). Siepel and Haussler (2004a, b) report success in using the *BFGS algorithm* (Press *et al.*, 1992) to simultaneously optimize the rate matrix and all the branch lengths for a given topology, assuming an arbitrary reversible rate matrix form; open-source software for the general-purpose BFGS optimization algorithm is available in the *GNU Scientific Library* (*GSL* – see section 10.15). This procedure requires all first partial derivatives of the objective function, which Siepel and Haussler (2004a, b) evaluated via *differencing* – i.e., by simply computing the full tree likelihood via Felsenstein's algorithm (Eq. 9.31) at two nearby points, and taking the difference via:

$$\partial L(x)/\partial x \approx (L(x + dx) - L(x))/dx \tag{9.52}$$

for some small dx, where the approximation is asymptotically accurate for diminishing dx. The same authors in a subsequent work (Siepel and Haussler, 2004c) provide EM update equations for simultaneous optimization of the rate matrix and branch lengths, and report that this converges faster than BFGS. Pedersen and Hein (2003) report using the *Powell algorithm* (also known as *DFP*, or *Davidson–Fletcher–Powell*) for optimizing the rate matrix and branch lengths for their system; Powell does not require explicit gradients (Press *et al.*, 1992), and can therefore be expected to converge more slowly than even BFGS, due to the larger number of function evaluations required. The reader should note that the efficiency of the estimation procedure becomes increasingly more critical as the order of the model is increased (see below).

Evaluating the gradient of the tree likelihood can be done analytically for the branch lengths as follows:

$$\frac{\partial L_u(a)}{\partial t_{x,y}} = \begin{cases} 0 & \text{for } u \text{ a leaf} \\ f_= & \text{for } u = x \text{ internal} \\ f_{\neq} & \text{for } u \neq x \text{ internal} \end{cases} \tag{9.53}$$

for:

$$f_= = \left(\sum_{b\in\alpha} L_y(b) \frac{\partial \mathbf{P}_{a,b}(t_{x,y})}{\partial t_{x,y}}\right)\left(\sum_{b\in\alpha} L_{other(y)}(b)\mathbf{P}_{a,b}(t_{x,other(y)})\right) \tag{9.54}$$

and:

$$
\begin{aligned}
f_{\neq} = & \left(\sum_{b\in\alpha} L_{\,left}(b)\mathbf{P}_{a,b}(t_{u,left})\right)\left(\sum_{b\in\alpha} \frac{\partial L_{\,right}(b)}{\partial t_{x,y}}\mathbf{P}_{a,b}(t_{u,right})\right) \\
& + \left(\sum_{b\in\alpha} L_{\,right}(b)\mathbf{P}_{a,b}(t_{u,right})\right)\left(\sum_{b\in\alpha} \frac{\partial L_{\,left}(b)}{\partial t_{x,y}}\mathbf{P}_{a,b}(t_{u,left})\right)
\end{aligned} \tag{9.55}
$$

where $t_{x,y}$ is the branch length for the branch connecting node y and its parent x, and:

$$
\frac{\partial \mathbf{P}(t_{x,y})}{\partial t_{x,y}} = \mathbf{P}(t_{x,y})\mathbf{Q} \tag{9.56}
$$

by Eq. (9.46). This can be evaluated using a dynamic programming algorithm similar to Felsenstein's algorithm for Eq. (9.31). The gradient of the likelihood function with respect to some rate-matrix parameter π is given by:

$$
\frac{\partial L_{\,u}(a)}{\partial \pi} = \begin{cases} 0 & \text{for } u \text{ a leaf} \\ g_{int} & \text{for } u \text{ internal} \end{cases} \tag{9.57}
$$

where g_{int} is given by:

$$
\begin{aligned}
g_{int} = & \left(\sum_{b\in\alpha} L_{\,left}(b)\mathbf{P}_{a,b}(t_{u,left})\right) \\
& \times \sum_{b\in\alpha}\left(\frac{\partial L_{\,right}(b)}{\partial \pi}\mathbf{P}_{a,b}(t_{u,right}) + L_{\,right}(b)\frac{\partial \mathbf{P}_{a,b}(t_{u,right})}{\partial \pi}\right) \\
& + \left(\sum_{b\in\alpha} L_{\,right}(b)\mathbf{P}_{a,b}(t_{u,right})\right) \\
& \times \sum_{b\in\alpha}\left(\frac{\partial L_{\,left}(b)}{\partial \pi}\mathbf{P}_{a,b}(t_{u,left}) + L_{\,left}(b)\frac{\partial \mathbf{P}_{a,b}(t_{u,left})}{\partial \pi}\right).
\end{aligned} \tag{9.58}
$$

As recently shown by Schadt and Lange (2002), we have:

$$
\frac{\partial \mathbf{P}(t)}{\partial \pi} = \mathbf{G}\left[\mathbf{F} \circ \left(\mathbf{G}^{-1}\frac{\partial}{\partial \pi}(\mathbf{Q}t)\mathbf{G}\right)\right]\mathbf{G}^{-1} \tag{9.59}
$$

for componentwise ("Hadamard") matrix product operator $\circ$, where $\mathbf{G}$ is the matrix of (column) eigenvectors of $\mathbf{Q}$, and where the elements of matrix $\mathbf{F} = [f_{a,b}]$ are given by:

$$
f_{a,b} = \begin{cases} te^{\lambda_a t} & \text{if } \lambda_a = \lambda_b \\ \dfrac{e^{\lambda_a t} - e^{\lambda_b t}}{\lambda_a - \lambda_b} & \text{otherwise} \end{cases} \tag{9.60}
$$

for eigenvalues $\lambda_c, c \in \{\mathtt{A}, \mathtt{C}, \mathtt{T}, \mathtt{G}, -\}$ (Siepel and Haussler, 2004c). The term $\partial/\partial\pi(\mathbf{Q}t) = t\partial\mathbf{Q}/\partial\pi$ in Eq. (9.59) requires knowledge of the functional form of the rate matrix **Q** so that its partial derivative may be computed:

$$\frac{\partial \mathbf{Q}_{JK}}{\partial \alpha} = \begin{bmatrix} -3 & 1 & 1 & 1 \\ 1 & -3 & 1 & 1 \\ 1 & 1 & -3 & 1 \\ 1 & 1 & 1 & -3 \end{bmatrix}$$

$$\frac{\partial \mathbf{Q}_{K2P}}{\partial \alpha} = \begin{bmatrix} -1 & 0 & 1 & 0 \\ 0 & -1 & 0 & 1 \\ 1 & 0 & -1 & 0 \\ 0 & 1 & 0 & -1 \end{bmatrix}, \quad \frac{\partial \mathbf{Q}_{K2P}}{\partial \beta} = \begin{bmatrix} -2 & 1 & 0 & 1 \\ 1 & -2 & 1 & 0 \\ 0 & 1 & -2 & 1 \\ 1 & 0 & 1 & -2 \end{bmatrix}$$

$$\frac{\partial \mathbf{Q}_{FEL}}{\partial \alpha} = \begin{bmatrix} -\pi_C - \pi_G - \pi_T & \pi_C & \pi_G & \pi_T \\ \pi_A & -\pi_A - \pi_G - \pi_T & \pi_G & \pi_T \\ \pi_A & \pi_C & -\pi_A - \pi_C - \pi_T & \pi_T \\ \pi_A & \pi_C & \pi_G & -\pi_A - \pi_C - \pi_G \end{bmatrix}$$

$$\frac{\partial \mathbf{Q}_{HKY}}{\partial \alpha} = \begin{bmatrix} -\pi_G & 0 & \pi_G & 0 \\ 0 & -\pi_T & 0 & \pi_T \\ \pi_A & 0 & -\pi_A & 0 \\ 0 & \pi_C & 0 & -\pi_C \end{bmatrix},$$

$$\frac{\partial \mathbf{Q}_{HKY}}{\partial \beta} = \begin{bmatrix} -\pi_C - \pi_T & \pi_C & 0 & \pi_T \\ \pi_A & -\pi_A - \pi_G & \pi_G & 0 \\ 0 & \pi_C & -\pi_C - \pi_T & \pi_T \\ \pi_A & 0 & \pi_G & -\pi_A - \pi_G \end{bmatrix}$$

$$\frac{\partial \mathbf{Q}_{REV}}{\partial \alpha} = \begin{bmatrix} -\pi_G & 0 & \pi_G & 0 \\ 0 & 0 & 0 & 0 \\ \pi_A & 0 & -\pi_A & 0 \\ 0 & 0 & 0 & 0 \end{bmatrix}, \quad \frac{\partial \mathbf{Q}_{REV}}{\partial \beta} = \begin{bmatrix} -\pi_C & \pi_C & 0 & 0 \\ \pi_A & -\pi_A & 0 & 0 \\ 0 & 0 & 0 & 0 \\ 0 & 0 & 0 & 0 \end{bmatrix},$$

$$\frac{\partial \mathbf{Q}_{REV}}{\partial \chi} = \begin{bmatrix} -\pi_T & 0 & 0 & \pi_T \\ 0 & 0 & 0 & 0 \\ 0 & 0 & 0 & 0 \\ \pi_A & 0 & 0 & -\pi_A \end{bmatrix},$$

$$\frac{\partial \mathbf{Q}_{REV}}{\partial \kappa} = \begin{bmatrix} 0 & 0 & 0 & 0 \\ 0 & -\pi_G & \pi_G & 0 \\ 0 & \pi_C & -\pi_C & 0 \\ 0 & 0 & 0 & 0 \end{bmatrix}, \quad \frac{\partial \mathbf{Q}_{REV}}{\partial \tau} = \begin{bmatrix} 0 & 0 & 0 & 0 \\ 0 & 0 & 0 & 0 \\ 0 & 0 & -\pi_T & \pi_T \\ 0 & 0 & \pi_G & -\pi_G \end{bmatrix},$$

$$\frac{\partial \mathbf{Q}_{REV}}{\partial \omega} = \begin{bmatrix} 0 & 0 & 0 & 0 \\ 0 & -\pi_T & 0 & \pi_T \\ 0 & 0 & 0 & 0 \\ 0 & \pi_C & 0 & -\pi_C \end{bmatrix}$$

and similarly for other matrix families.

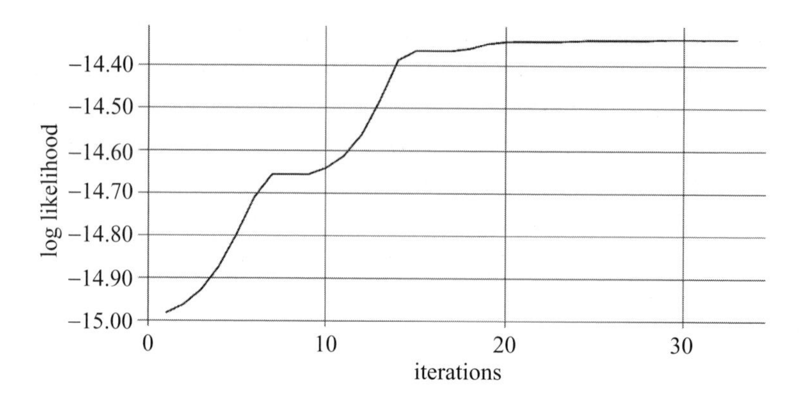

Figure 9.28 Convergence graph for a training run of a PhyloHMM. Log likelihood ($\times 10^3$) is shown on the *y*-axis and number of iterations of gradient ascent is shown in the *x*-axis. The model consisted of nine parameters (eight branch lengths and one matrix parameter). Training data consisted of an alignment of a 10 Mb segment from each of five genomes.

Because these gradients need to be computed over all columns of the alignment, we have:

$$\frac{\partial L}{\partial x}(all\ columns) = \sum_{\substack{columns \\ i}} \left(\prod_{\substack{columns \\ j \neq i}} L(col_j) \right) \frac{\partial L}{\partial x}(col_i) \tag{9.61}$$

where L is the likelihood function implemented by Felsenstein's pruning algorithm, and col_i is the ith column in the alignment used for training.

It should be evident that evaluation of the above gradients is asymptotically no faster than applying the differencing method (Eq. 9.52), although the former may be more accurate in practice. Note also that the gradient-based and EM-based methods both suffer from being local optimization techniques; they are not guaranteed to find the global optimum, and it may therefore be worthwhile to run them a number of times from random starting points.

Figure 9.28 illustrates the convergence behavior of a gradient-ascent optimizer during an example training run; nine parameters were estimated from approximately 50 Mb of training data.

9.6.6 *Modeling higher-order dependencies*

Relaxing of the site-independence assumption can be (partly) achieved by incorporating higher-order dependencies into the evolution model, so that, e.g., $P(\mathbf{X}[j]|S[j])$ is replaced with $P(\mathbf{X}[j]|S[j], \mathbf{Y}[j-n], \ldots, \mathbf{Y}[j-1])$, for $\mathbf{X}[j]$ the jth column of the alignment excluding the target sequence, and $\mathbf{Y}[j]$ the jth column of the alignment including the target sequence, where n is the order of the model. Thus, an nth-order model assumes that column probabilities are explicitly dependent only on the preceding n columns. Obviously, the order n must be reduced for the leftmost n columns in the alignment; we will gloss over this fine point in what follows.

Formulating higher-order models is a simple matter of recoding the alphabet into a higher-order alphabet by consider all n-grams of some fixed order $n+1$, with appropriate renormalization to produce conditional probabilities. For example, in a 2nd-order model, the matrix would have dimensions $4^3 \times 4^3 = 64 \times 64$, with the rows and columns corresponding to triples of nucleotides $X_1X_2X_3$; an individual cell $(X_1X_2X_3, Y_1Y_2Y_3)$ of the matrix would correspond to the conditional probability $P(Y_1Y_2Y_3|X_1X_2X_3)$, for ancestral 3-gram $X_1X_2X_3$ and descendant 3-gram $Y_1Y_2Y_3$. Estimating parameters for such matrices clearly involves a significantly greater amount of computational effort than for 0th-order models (e.g., Siepel and Haussler, 2004c).

Pedersen and Hein (2003) utilized only 1st-order models in most states of their (nongeneralized) PhyloHMM *EvoGene*, though their codon state emitted triples. However, as they note, it is difficult to ensure that the generic alignment preprocessing step does not introduce frameshifts in (or between) coding regions of informant sequences. They applied a simplistic heuristic for addressing this issue, but were still unable to match the performance of the noncomparative gene finder GENSCAN, surprising as that may seem.

Siepel and Haussler (2004a, b, c) model emissions at higher orders in all states of their (nongeneralized) PhyloHMM *ExoniPhy*, though doing so significantly complicates their parameter estimation process, increasing the CPU time required for training. Though they do not explicitly state so, their solution would also seem to be vulnerable to misalignment problems due to the fact that the alignment preprocessing stage aligns only in nucleotide space.

Another program, called *N-SCAN* (Gross and Brent, 2005), utilizes 1st-order emission models for the informant columns of the alignment over a 6-letter alphabet including the four DNA bases plus a gap and "unaligned" symbol, so that 170 free parameters are required for its partially reversible substitution model. Unlike EvoGene and ExoniPhy, N-SCAN utilizes a generalized HMM (GHMM) rather than a simple HMM as its underlying noncomparative model. Thus, we could call their solution a *PhyloGHMM*.

A simpler method for implementing higher-order models is to condition each descendent base b on both the ancestral base a and the n-gram preceding b in the root's track of the alignment: $P(b_j|r_{j-n}\ldots r_{j-1}a_j)$ for column j of the alignment. This can be accomplished very simply by maintaining 4^n rate matrices (one for each possible n-gram in the ancestral sequence, not counting gaps) over the standard 4-letter DNA alphabet, so that the need for large rate matrices associated with higher-order alphabets would be eliminated. Though this would effect a significant reduction in the number of parameters to be estimated during training, a thorough comparison of these and other methods of modeling higher-order dependencies in a PhyloHMM (especially in the presence of frameshifts due to misalignment) remains to be performed. One might also consider some sort of interpolation scheme, similar to those described in section 6.8.3, for the reduction of sampling error at higher orders, giving rise to what might be termed an *interpolated PhyloHMM*, or *PhyloIMM*. Such a model would presumably make use of full contextual information when sample sizes permit, and rely on lower-order models (suitably interpolated) when the potential for sampling error is detected.

Note also that for coding regions the evolutionary model should ideally (in the absence of higher-level modeling such as at the level of whole amino acids) be partitioned into three separate models for the three phases; given that the underlying GHMM upon which a Phylo(G)HMM is to be built would already possess the appropriate phase-tracking machinery, the extra effort needed to model phases separately in the phylogenetic portion of the model would be essentially trivial, though a reduction in sample size during training would of course be unavoidable. Applying phylogenetic modeling techniques to the implied amino acid sequences of putative orthologous coding sequences is also conceivable, though work remains to be done to refine the available methods in the presence of frameshifts due either to misalignments or to evolution.

9.6.7 *Enhancing discriminative power*

In implementing the PhyloHMM-based gene finder N-SCAN, the program's authors found it necessary to introduce an "arbitrary constant called the conservation score coefficient," which serves no theoretical role in the generative model but was observed empirically to improve the discriminative accuracy of the gene finder when assigned a value between 0.3 and 0.6 (Gross and Brent, 2005). Similarly, the authors of the phylogenetic gene finder ExoniPhy (Siepel and Haussler, 2004b) describe the use of an extra "tuning parameter" called "coding bias" which was intended to account for the unknown density of exons in a genome, but likely serves the additional purpose of weighting the conservation evidence, just as the "conservation score coefficient" employed by N-SCAN. These "fudge factors" help to improve the discriminatory power of what would otherwise be a purely generative model and would therefore likely offer only suboptimal parsing accuracy (see section 12.4), though the degree to which this effect is due to inadequate training data versus the need for some form of discriminative training remains to be quantified. For the present, the use of extra "tuning parameters" such as those mentioned above seems to be necessary for maximizing predictive accuracy. An explicitly discriminative framework for phylogenetic modeling in a gene-finding context has been very recently investigated by Vinson *et al.* (2006).

9.6.8 *Selection of informants*

One question which has received some attention recently is that of the optimal set of informants which are to be used in a PhyloHMM, or more generally in any comparative gene finder. The simpler form of the question is the optimal evolutionary distance between a target species and an informant species in the case of single-informant methods. Interest in this question arises from the observation that at the two extremes – i.e., informant genomes which are virtually identical to the target genome, and informants which are so diverged from the target that they have no detectable homology – no useful information can be gleaned from the informant to aid the gene-prediction process. From this observation it is intuitively clear that the evolutionary distance between the informant and target will influence the amount of useful information available to the gene finder for accurate gene prediction. The ideal case would seem to be one in which the target and informant are virtually identical in their coding segments, but fully divergent (i.e., with no significant HSPs) in their noncoding regions. Simulation results by Zhang *et al.* (2003) suggest that reasonable prediction accuracy may be obtained using informants at a range of evolutionary distances; in the case of human gene finding, the latter authors suggest that an informant genome somewhat more divergent from human than mouse (i.e., >75 million years) would seem to be ideal. Additional research in this vein is also needed.

9.7 Auto-annotation pipelines

In this volume we have chosen to concentrate on methods of gene prediction based on coherent, formal mathematical models such as HMMs, as well as various other constructs from the field of machine learning, such as Bayesian networks and decision trees (see Chapter 10 for a survey of other machine-learning methods). At the higher level of whole-genome annotation, it is often necessary to take into account various additional information, such as the existence of genomic repeats and pseudogenes, or explicit knowledge regarding the reliability of a given database of ESTs or proteins, or even regarding the quality of the genome assembly. While these issues are usually addressed by human annotators in those cases where a full manual curation of the target genome is to be performed, for many genome projects there is, unfortunately, insufficient funding and/or manpower for such a manual curation effort.

In such cases, a draft annotation of the target genome must be produced through wholely automated means. This is the task of the so-called *auto-annotation pipelines* such as Celera's *Otto* system (Venter *et al.*, 2001), TIGR's *EGC* system (Wortman *et al.*, 2003), and the Sanger Center's *Ensembl* system (Birney *et al.*, 2006). These systems typically employ many complicated heuristics meant to simulate the logic of a human annotator in disambiguating the various forms of evidence available, such as the conflicting predictions of different gene finders which are fed as inputs into the pipeline. Other evidence such as existing annotations of known genes, EST assemblies, RepeatMasker predictions (Smit and Green, 1996), and *synteny* information (e.g., Mural *et al.*, 2002) are typically integrated using a set of rules encoded by

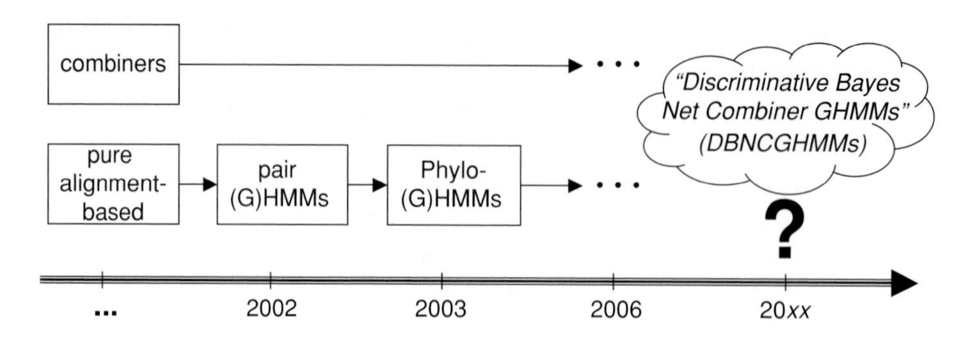

Figure 9.29 Extrapolation of the timeline of computational advances for comparative gene finding.

experienced annotators for the hierarchical sorting and disambiguating of evidence at the whole-genome or whole-contig level.

Unfortunately, the ad-hoc nature of these rule-based systems renders them difficult to compare in a rigorous manner, so that it will likely be some time before a coherent, principled theory of whole-genome annotation will emerge which allows the formal characterization of optimal methods for auto-annotation. In the meantime, existing systems such as Ensembl and its several competitors continue to play an invaluable role in whole-genome annotation.

9.8 Looking toward the future

It is always tempting upon reviewing the evolutionary trajectory of one's field to speculate as to its future directions, and in particular to try to guess the form of the next "paradigm shift." In Figure 9.29 we recap the major developments in comparative gene finding since the opening of the millennium, in which it can be seen that the rate of advance has been fairly rapid; to add additional context to this figure, we remind the reader that the initial draft sequence of the human genome was published in 2001 (International Human Genome Sequencing Consortium, 2001; Venter *et al.*, 2001).

Based on the various strengths and weaknesses of known comparative techniques, as described in this chapter, we will venture to predict that the next computational advance will minimally involve a combination of the following elements:

(1) A GHMM-like framework for efficient decoding.
(2) Incorporation of arbitrary evidence using a combiner-like approach.
(3) Utilization of a Bayesian network for weighting evidence in a principled manner.
(4) Some form of discriminative training to ensure optimal accuracy.

The first point follows both from our observation that most (successful) comparative systems to date have been based on extensions of an underlying *ab initio* HMM-based or GHMM-based model, and also from our belief that the GHMM decoding architectures described in section 8.3 provide sufficiently flexible frameworks for efficient sequence parsing to allow for their continued use in more sophisticated comparative settings. The second point merely reflects the fact that multiple sources of evidence are typically available and that no evidence of sufficient quality should be ignored. In the fourth point we suggest that the strength of Bayesian networks in accounting for the nonindependence of evidence sources will be essential in mitigating biases that would otherwise negatively influence the effective integration of available evidence. Finally, we reaffirm our belief that none of the existing methods – comparative or otherwise – has yet proven to provide optimal prediction accuracy without additional "tweaking" of the model parameters so as to maximize the discriminative power of the model.

Another significant shortcoming of most (though not quite all) comparative gene finders is that they rely on precomputed alignments, with those alignments having been computed by general-purpose alignment programs – i.e., alignment programs which do not take signal or content cues for coding or noncoding states into account. In theory, pairHMMs attend to this issue by allowing the alignment and gene prediction processes to proceed simultaneously, though in practice the use of aggressive banding and/or "pruning" of the dynamic programming matrix around precomputed "guide" alignments often negates this feature. In the case of PhyloHMMs, which seem to enjoy the greatest popularity at present, the use of precomputed alignments is effectively universal, so that any alignment errors that occur in the precomputed alignments have the potential to significantly mislead the gene finder. Additional work is clearly needed to address this issue.

EXERCISES

9.1 Give a formal justification for each step of derivation (9.1). Refer to material from Chapter 2 as needed.

9.2 Consider the conservation sequence used by TWINSCAN to represent matches, mismatches, and unaligned positions. How many symbols would be required to represent all possible combinations of conservation between three sequences (i.e., one target and two informants) rather than two as in TWINSCAN?

9.3 Generalize your answer to the previous question to give the alphabet size for a conservation sequence between N sequences (i.e., one target sequence and $N - 1$ informants).

9.4 Explain how a PSA-type decoder (section 8.3.1) can be generalized to efficiently incorporate conservation sequence as in TWINSCAN. Explain how this will impact the time and space complexity of the decoder.

9.5 As shown in derivations (9.1) and (9.2), TWINSCAN and GenomeScan factor the objective function $P(\phi|S, I)$ for informant I in a slightly different manner. Explain the significance of this difference.

9.6 Download a genome and set of annotations for an organism of your choice, from the NCBI (www.ncbi.nlm.nih.gov). Extract the CDSs, translate them into amino acid sequences, and then construct a codon usage table giving the probability $P(C|A)$ that the codon coding for a particular amino acid A in some peptide is C.

9.7 Explain how a 3-state PHMM for sequence alignment such as the one shown in Figure 9.5 may be reduced to 2 states (plus q_0). How will this change affect the resulting alignments?

9.8 Explain why, if a score s is given in bits, the quantity 2^{-s} can be interpreted as a probability. What is it the probability of?

9.9 Suppose you are tasked with implementing a combiner program. Given a set of putative exon boundaries from the available evidence tracks, devise an efficient dynamic programming algorithm to assemble exons into the highest-scoring parse of the target (input) sequence. You may ignore the issue of phase (i.e., CDSs need not have length divisible by 3) for this exercise.

9.10 Extend your solution to the previous exercise by modeling phase constraints.

9.11 Extend your solution to the previous exercise to ensure that no in-frame stop codons exist in any predicted CDS.

9.12 Explain why it is necessary in Eq. (9.9) to divide by the length $|I|$ of the interval I.

9.13 Give a geometric interpretation of Eq. (9.9). You may ignore the denominator for this exercise.

9.14 Referencing the primary literature on one of the whole-genome alignment programs mentioned in the text, explain the technique used by the selected program to align very long sequences.

9.15 In our description of ROSETTA we stated that a penalty term was incorporated to reflect differences in length between two putatively homologous exons. Devise a formula for such a penalty term, based on probability theory.

9.16 Explain why the "fuzziness" of splice site alignments in SGP-1 is useful.

9.17 Compare and contrast the approaches taken by the GPHMM-based program TWAIN and the CEM program. What minimal additions to CEM would be necessary to transform the latter program into a GPHMM-based program like TWAIN?

9.18 In what way is CEM more sophisticated than the other alignment-based approaches? Under what circumstances might this increased level of sophistication result in more accurate gene predictions?

9.19 Give a formal algorithm for determining the $a_{i,j}$ terms in Eq. (9.15) denoting the ith symbol in the jth sequence of the alignment $A(\phi)$ resulting from a PHMM decoding run.

9.20 Give a pair of algorithms $A_{a,i}$ and $A_{i,a}$ to convert an alignment to a sequence of insertions, deletions, and (mis)matches ($A_{a,i}$) and to convert a sequence of insertions, deletions, and matches (via the symbols `I`, `D`, `M`) back to an alignment ($A_{i,a}$).

9.21 Implement a PHMM-based alignment program in your (or your instructor's) favorite programming language. All model parameters should be loaded from a configuration file specified by the user.

9.22 Give a formal algorithm for extracting a pair of gene predictions from a PHMM parse ϕ.

9.23 Give a formal algorithm to estimate the maximum likelihood parameters of a 3-state PHMM for alignment, given a set of training alignments. Hint: consider recoding the paired emission symbols into a higher-order alphabet and treating the model as an ordinary HMM.

9.24 In section 9.4.1 we refer to the Hirschberg algorithm as being "memoryless." After reading the original paper describing this algorithm, explain what is meant by "memoryless." How is decoding possible if the underlying algorithm is "memoryless"?

9.25 Give a justification for Eq. (9.61). Hint: this requires some knowledge of calculus.

9.26 Explain how the memory requirements of a PhyloHMM's prefix sum arrays may be reduced by using sparse arrays. Hint: consider that the only elements of the prefix sum arrays accessed during decoding are those corresponding to a putative signal.

9.27 Explain how the P_{cons} term in TWAIN could be evaluated in a more rigorous fashion using an appropriate substitution matrix.

9.28 Why is a phylogeny a good choice for a Bayesian network decomposition of the joint sequence probability of a multi-alignment? Could a Bayesian network that differed in some substantial way from the true phylogeny of a set of organisms conceivably produce a more accurate gene finder? Explain your answer.

9.29 Explain how arrays of integers may be used to reduce the memory requirements of a prefix sum array implementation of a PhyloHMM decoder.

9.30 Refine the UNIX command given in section 9.6.4 to show that the individual base frequencies of the Kimura 2-parameter model approach uniformity in the limit.

9.31 Download the *GNU Scientific Library* and implement an optimizer for the function $f(x) = 3x^2 + \cos(x) - \sin(1-x)$. Show your code and the output.

9.32 Extend the Hirschberg algorithm for use in PHMM decoding. You may wish to refer to the original description of the algorithm in Hirschberg (1975).

9.33 Give an efficient algorithm for finding the optimal path in an HSP graph, given the beginning and ending HSPs in the path, as described in section 9.5.1. Hint: consider a depth-first search procedure on DAGs, with the recursion being terminated whenever the x-coordinate or y-coordinate of the current HSP in the search exceeds the corresponding coordinate for the ending HSP.

10

Machine-learning methods

Quite a few of the techniques described in the foregoing chapters could be said to qualify as *machine-learning* methods. In this chapter we consider a number of other popular machine-learning algorithms and models which either have seen limited use in gene finding, or would seem to offer possible avenues for future investigation in this arena. Most of the methods which we describe are relatively easy to implement in software, and nearly all are available in open-source implementations (see Appendix). While the current emphasis in the field of gene prediction seems to be on Markovian systems (in one form or another), an expanded role for other predictive techniques in the future is not inconceivable.

10.1 Overview of automatic classification

Perhaps the most typical setting for machine-learning applications is that of N-way classification (Figure 10.1). In this setting, a *test case* (i.e., a novel object) is presented to a *classifier* for assignment to one of a fixed number of discrete categories. The test case is typically encoded as a vector of real-valued or integer-valued *attributes* (i.e., *random variables* – section 2.6), though since $\mathbb{N} \subset \mathbb{R}$ we will generally treat all attributes as being real-valued; thus the attributes of a single test case are drawn from $\mathbb{R}^m$, for some integer m. The categories to which test cases are to be mapped are typically encoded as integer values in the range $\mathbb{N}_N = \{0 \leq i < N\}$. Thus, the *training* problem, or *learning* problem, for automatic classification can be stated as that of finding a function $f : \mathbb{R}^m \rightarrow \mathbb{N}_N$ to serve as the desired classifier. Ideally, we would like to find the classifier f which achieves maximal classification accuracy on unseen test cases (i.e., on the set of test cases which will be encountered by the classifier when it is deployed for use in the real world). In practice, we might settle for maximizing the classification accuracy on a training set T, though in doing so we must accept that the level of accuracy achieved on the training set may not be representative of the classifier's accuracy on novel cases, due to the phenomenon of *overtraining* (see below).

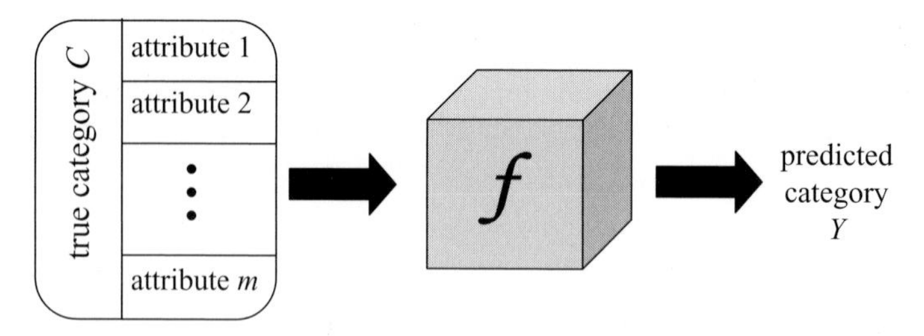

Figure 10.1 The typical classification paradigm. A novel object is presented (at left) to a classifier *f*, which must predict a category *Y* for the object. The true category *C* of the object is not visible to the classifier, and may or may not be known to the researcher.

We thus define a *learning algorithm* $\mathcal{A}$ as one which takes as input a training set $T \subset \mathbb{R}^m \times \mathbb{N}_N$ and emits a classifier $f \in \mathcal{C}$, for $\mathcal{C} = \{f \mid f : \mathbb{R}^m \rightarrow \mathbb{N}_N\}$, where the *dimensionality* of the training set T is defined as m. In practice, $\mathcal{C}$ is typically highly constrained, since practitioners generally limit their search for f to their favorite family of models $\mathcal{F}$ (such as *neural networks* or *decision trees*) corresponding to some subset of classifiers $\mathcal{C}_\mathcal{F} \subset \mathcal{C}$. The learning process may therefore be described as the task of inducing the optimal parameter set θ^* for a model, resulting in a classifier f_{θ^*}:

$$\theta^* = \underset{\theta \in \Theta(\mathcal{F})}{argmax} \sum_{c \in T} \delta(f_\theta(c), class(c)) \tag{10.1}$$

for Kronecker match function δ (section 2.1), where θ is a vector of model parameters, *class*(c) is the correct class of c, and Θ is the set of all possible parameterizations (i.e., vectors of parameters) for models in model family $\mathcal{F}$. In the case of neural networks, for example, $\mathcal{F}$ would represent the set of all neural networks, and f_θ would represent a particular neural network having a topology and synaptic weights specified by θ (perhaps also including other relevant details germane to the operation of a neural network – see section 10.5). Once the classifier has been trained, classification of a novel test case c can be accomplished via $f_{\theta^*}(c)$.

For the purposes of computational gene finding, we may envision that the test cases are intervals of a DNA sequence and the corresponding categories of those test cases are $\{0 = \textit{is not an exon}, 1 = \textit{is an exon}\}$. Thus, if we merely wish to identify all intervals of a sequence that are transcribed and spliced into protein-coding exons in cells of some type, and we are unconcerned with exactly which exons in a possibly alternatively spliced locus actually combine to produce functional proteins (section 12.1), then the problem is one which seems to fit quite well into the classification paradigm.

Although early attempts at gene prediction were often formulated in traditional machine-learning terms (e.g., Snyder and Stormo, 1993), most of the current state-of-the-art systems rely solely on HMMs and their various extensions, with the Viterbi decoding process enforcing a one-locus-one-isoform discipline. Recent revelations regarding the prevalence of alternative splicing in eukaryotic genomes (section 12.1) suggest, however, that the single-isoform paradigm enforced by traditional HMM-based methods deserves to be relaxed by future gene-finding systems. It is reasonable to suppose that the classification paradigm of traditional machine learning may provide an effective means to achieve the desired generalization of the gene-finding problem. Furthermore, it may be argued that the need for *discriminative training* of HMMs and GHMMs (see section 12.4) suggests that purely Markovian models are perhaps not perfectly suited to the problem of gene parsing, and that alternative methods may be worth exploring (e.g., Culotta *et al.*, 2005; Vinson, 2006). Additional support for the latter view is given by the growing body of evidence that there is significant interaction between the processes of splicing, 5′ capping, and 3′ polyadenylation (Yan and Marr, 2005), and that compensatory effects manifested in correlated signal strengths between, e.g., the donor and acceptor sites bounding an intron or an exon, can influence the splicing process and should therefore be modeled by gene finders (Zhao *et al.*, 1999; Clark and Thanaraj, 2002; Fairbrother *et al.*, 2002), though they typically are not modeled in (G)HMM-based implementations due to the assumed independence between signal scores (i.e., due to the *Markov assumption*).

Thus, in this chapter we describe in detail a number of classification and learning algorithms which the reader should find reasonably easy to implement and utilize within the context of exon prediction. Open-source software implementing many of these algorithms can be downloaded through the book's website (see Appendix).

As illustrated in Figure 10.1, these methods all require as input a set of attribute values. These might take the form of hexamer frequencies, signal or content sensor scores, putative feature lengths, or other continuous or discrete measures taken from a putative feature in a DNA sequence. At the end of this chapter (section 10.17) we present results of experiments in which signal and content sensors were provided as inputs to a number of different classifiers and the resulting exon predictions were scored as to their classification accuracy.

A central issue in machine learning is that of *overtraining*. As mentioned earlier (section 4.1), the accuracy of any predictive model as measured on the training set may not reflect the eventual accuracy of the model when it is deployed for use on unseen cases; that is, the model's knowledge about how to classify accurately the training examples may not *generalize* to knowledge of how to accurately classify unseen test cases. A model that performs well on the training set but fails to generalize to unseen cases is said to be *overtrained*. For example, consider the following classifier:

$$f_\theta(d) = \begin{cases} class(c) & \text{if } d = c \text{ for some } c \in T \\ random(N) & \text{otherwise.} \end{cases} \tag{10.2}$$

This function classifies any case d which happens to be identical to some training case c into the same *class* (i.e., category) as c, but classifies all others randomly.

This function will achieve 100% accuracy on the training set (assuming no inconsistencies in the training set), but will obviously perform very poorly on unseen test cases.

A number of techniques have been investigated for the avoidance of overtraining, though we lack the space to cover all of them here. One approach is to jointly minimize the complexity of the model while maximizing its accuracy on the training set (e.g., *minimum description length*, or *MDL* – Rissanen, 1978). The *regularization* method (section 10.11) described later for logistic regression is another approach. An ad-hoc approach which is often used is to fit a model to the training data and then to iteratively explore various simplifications of the resulting model, accepting only those simplifications that do not significantly reduce the accuracy of the model on the training set. For example, if the model is a neural network, we might find the optimal network M for the training set T and then iteratively consider removing a single neuron from the network and retraining the reduced network, reinserting the deleted neuron back into the network only if the retrained model achieves a significantly lower accuracy when applied to the training set.

Another common technique is that of *early stopping*, which is applicable to learning algorithms that are iterative in nature. These algorithms typically modify a model's parameters in discrete steps, following a gradient-ascent approach to maximize the performance of the model on the training set. An early stopping approach partitions the training set T into disjoint subsets $T = T_1 \cup T_2$ and then proceeds to train a model M on T_1, with the accuracy of the current model M_i at each iteration i measured both on T_1 and T_2. The accuracy of M_i on T_1 is used as the objective function for gradient ascent, as usual. However, if a model M_i is found to perform more poorly on T_2 than did M_{i-1}, then the gradient-ascent is terminated, even if M_i performs as well as or better than M_{i-1} on T_1. The model is then typically re-trained from scratch on all of T, with the gradient ascent process limited to at most $i - 1$ iterations, since the results of the early stopping run suggested that i iterations would result in overtraining.

The foregoing should suffice to prepare the reader to implement the methods which follow and train them to achieve respectable performance on many real-world problems. A more thorough description of these and other methods can be found in the primary machine-learning literature; many excellent monographs and reviews of the field exist as well (e.g., Mitchell, 1997; Duda *et al.*, 2000; Hastie *et al.*, 2003) to which the interested reader is referred.

10.2 *K*-nearest neighbors

Perhaps the simplest and most intuitive method for classification of novel objects is via the *K-nearest-neighbors* approach (*KNN* – Cover and Hart, 1967). In this method, a novel object presented for classification is compared to a number of *exemplar* objects (i.e., *training cases*) for which the correct classification is known. In the case of 1-nearest-neighbors (i.e., $K = 1$), the classification rule is simple: if an exemplar E turns out to be the nearest exemplar (according to some *distance metric*) to the novel object R, then we conclude that R should be classified to category $C = class(E)$, the known category of training case E. For $N > 1$, we can instead take the set $G = \{E \mid E$ is one of the nearest N exemplars to $R\}$, and then simply assign R to the most common category represented in G (with ties resolved in some arbitrary manner).

Since the test case and the exemplars all take the form of m-dimensional vectors of real or integral values, an obvious distance metric is the *Euclidean distance*:

$$d\left(X^{(i)}, X^{(j)}\right) = \sqrt{\sum_{\substack{attributes\\k}} \left(X_k^{(i)} - X_k^{(j)}\right)^2} \tag{10.3}$$

where $X^{(i)} = R$ is the object to be classified, and $X^{(j)} = E$ is an arbitrary exemplar; $X_k^{(j)}$ denotes the kth attribute of exemplar $X^{(j)}$, and similarly for test case $X^{(i)}$.

A more sophisticated distance metric is the *Mahalanobis distance*:

$$D\left(X^{(i)}, X^{(j)}\right) = \sqrt{(X^{(i)} - X^{(j)})^T \mathbf{V}^{-1} (X^{(i)} - X^{(j)})} \tag{10.4}$$

where $\mathbf{V}^{-1}$ is the inverse of the *covariance matrix* $\mathbf{V} = [c_{hk}]$:

$$C_{hk} = \frac{\sum_{i=1}^{n} \left(X_h^{(i)} - \overline{X}_h\right)\left(X_k^{(i)} - \overline{X}_k\right)}{n - 1}. \tag{10.5}$$

$\overline{X}_h$ denotes the mean of the hth attribute over the training cases, and n is the number of training cases. The Mahalanobis distance tends to be superior to the Euclidean distance for many applications because it accounts for the correlation structure between the attributes (Manly, 1994). That is, if two attributes are correlated, their combined influence on the resulting classification will be mitigated to control for the fact that the evidence provided by these two attributes is not independent, and is therefore not simply additive (see section 10.16). The Mahalanobis distance can be applied within the framework of other classification algorithms, though we will not explicitly consider it here.

10.3 Naive Bayes models

As most of the techniques presented in this book have been probabilistic in nature, it is natural to consider probabilistic solutions to the classification problem. The *naive Bayes* method is one such probabilistic approach. The naive Bayes strategy simply classifies each novel object $X = (x_1, x_2, \ldots, x_m)$ into the category Y which is most probable, given the attributes of the test case. That is, we maximize $P(Y_i|X)$ over the possible categories Y_i (i.e., *MAP* classification – section 2.8). The question then arises as to how we should evaluate this probability. One possibility is to apply Bayes' theorem to invert $P(Y_i|X)$ into its *Bayesian inverse*, $P(X|Y_i)$:

$$P(Y_i|X) = \frac{P(X|Y_i)P(Y_i)}{\sum_j P(X|Y_j)P(Y_j)}. \tag{10.6}$$

Since the denominator is invariant with respect to category Y_i, it suffices to compute:

$$Y^* = \underset{Y_i}{argmax}\ P(X|Y_i)P(Y_i). \tag{10.7}$$

Here, $P(Y_i)$ is the *prior probability* of category Y_i, which is typically taken to be the proportion of the training examples which fall into category Y_i. What remains is to evaluate $P(X|Y_i)$, the *posterior probability* of test case X given that it belongs to category Y_i. This we can approximate as follows:

$$P(X|Y_i) \approx P(X_1 = x_1|Y_i) \cdot P(X_2 = x_2|Y_i) \cdot \ldots \cdot P(X_m = x_m|Y_i) \tag{10.8}$$

using a separate *histogram* (section 2.7) for attribute variable X_j to assess the probability of (discretized) attribute value x_j conditional on category Y_i (i.e., one histogram per variable-category pair). Construction of these histograms, together with the assessment of prior probabilities $P(Y_i)$, comprises the training of the model.

It will be noticed that Eq. (10.8) assumes independence of the attributes; that is, we assume that each attribute X_j is conditionally independent (given the category Y_i) of the other attribute values of the test case – the so-called "naive Bayes" assumption. This is generally not a valid assumption, and in the next section ("Bayesian networks") we will show one way in which this unreasonable assumption can be relaxed. It should be noted, however, that naive Bayes is often applied in practice for real classification problems in industry and research, and its performance can in fact be quite good, especially if the attributes are largely uncorrelated. Ensuring that the attributes are uncorrelated can be accomplished by some form of *feature selection* or by an appropriate *dimensionality reduction* procedure, such as *principal components analysis (PCA)* (see section 10.16).

10.4 Bayesian networks

The independence assumption of naive Bayes can be relaxed somewhat by explicitly modeling dependencies between attributes; this is the purpose of *Bayesian networks*. The first task in this approach is to establish a *dependency structure* among the attributes – i.e., a directed acyclic graph (section 2.14) – in which each edge denotes a dependence between the attributes represented by the vertices at its two endpoints. For example, if $v_A \rightarrow v_B$ is an edge in the graph, then we assume that attribute A is dependent on attribute B. A common heuristic for inducing a dependency structure for a set of attributes is to apply a χ^2 *test of independence* (section 2.9) to all pairs

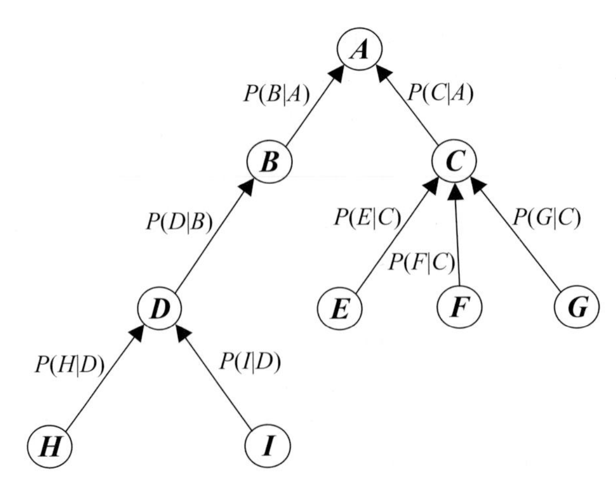

Figure 10.2 A Bayesian network. Variables are represented by vertices in the graph. Edges correspond to conditional probabilities between dependent variables. This particular Bayesian network is a tree.

of attributes, and to establish edges between attribute vertices for each statistically significant test statistic. An alternative method is to compute *Pearson correlation coefficients* (section 2.6) and then to establish edges only between those attribute vertices having a correlation higher than some given threshold. Yet another alternative is to rank the edges based on *mutual information* (section 2.10) and to again apply some threshold. Additional steps must be taken to establish a directionality for the edges, and then to ensure that the resulting directed graph is acyclic.

An especially convenient form of dependency structure is the tree, since trees are always acyclic. A reasonable method for inducing the tree is to compute all pairwise dependence scores (via some function of the χ^2 statistic, or some correlation measure) between attributes and then to compute a *maximal spanning tree* on the complete graph induced by those pairwise scores. Section 2.14 provides an algorithm for computing a minimal spanning tree; this can be easily modified to produce a maximal spanning tree instead, by choosing the largest edge weights instead of the smallest. Once the tree has been produced, we can factor the probability $P(X|Y_i)$ according to the individual edges, where each edge $v_A \rightarrow v_B$ gives rise to a term $P(X_A|X_B, Y_i)$:

$$P(X|Y_i) \approx P(X_0|X_{parent(0)}, Y_i) \cdot P(X_1|X_{parent(1)}, Y_i) \cdot \ldots \cdot P(X_{m-1}|Y_i) \tag{10.9}$$

where $X_{parent(i)}$ is the attribute whose vertex in the tree is the parent node of the vertex for attribute X_i; note that we have adopted the convention that the edges in the tree point upward, as illustrated in Figure 10.2. The last term, $P(X_{m-1}|Y_i)$, corresponds to the root of the tree, which has no parent; we assume for convenience that the attributes have been renumbered so that the root corresponds to attribute X_{m-1}.

Although this formulation can be expected to perform better than naive Bayes in many cases, it should be clear that we are still making some assumptions – in particular, we make conditional independence assumptions, in which we assume that X_i is conditionally independent of any other attribute outside the subtree rooted at *parent(i)*, given the value of that parent attribute. Just as naive Bayes can work well for problems in which attributes are uncorrelated, a tree-based Bayesian network can work well for problems in which the true dependency structure among attributes is tree-like. For problems having more complicated dependency structures, more sophisticated methods are required; we refer the interested reader to the large literature on Bayesian networks (e.g., Heckerman *et al.*, 1994; Jordan *et al.*, 1999).

An example use of Bayesian networks for splice site prediction was given in section 7.3.8 when we described the PTM model (Cai *et al.*, 2000). Bayes networks have also been used for evidence weighting in gene finders (e.g., Zhang *et al.*, 2003); indeed, as we noted in section 9.6, the use of probabilistic phylogenetic trees in a PhyloHMM constitutes such a use as well.

10.5 Neural networks

Neural networks, first introduced in a simpler form by Rosenblatt (1958), are perhaps the most celebrated machine-learning method. They have been used for many applications and have also been developed into a number of specialized varieties. We will consider only *feedforward* networks.

A *feedforward neural network* is a machine denoted by a 5-tuple $M = (X, Q, W, \xi, Y)$ where $X = \{x_0, x_1, \ldots, x_n\}$ is a set of input variables, $Q = \{q_0, q_1, \ldots, q_m\}$ is a set of *neurons*, $W{:}Q \times Q \to \mathbb{R}$ is a set of weighted *synapses* connecting pairs of neurons, $\xi : Q \times \mathbb{R} \to \mathbb{R}$ is a neuron-specific *activation function*, and $Y = \{y_0, y_1, \ldots, y_k\}$ is a set of output variables.

An example of a neural network is depicted in Figure 10.3, in which the vertices denote neurons and the edges connecting neurons denote synapses. Each neuron in the network comprises a simple computational unit having enough memory to store a single numeric value, called the *activation value* of the neuron. A neuron may also have a finite number of synaptic inputs and/or synaptic outputs. In Figure 10.3 we have drawn the network so that the inputs to a neuron are shown on the left and the outputs are on the right. Synapses should ideally be drawn as directed edges; we have omitted the arrowheads for clarity (they should be directed left-to-right in this example). Note that a feedforward network must be acyclic.

The machine operates as follows. A set of numeric values is supplied externally in the form of the input variables $x_0, x_1, \ldots, x_n$, which are copied directly (after rescaling them to the range [−1,1]) into the neurons in the input layer (shown on the left in Figure 10.3). The activation values stored in these neurons are then

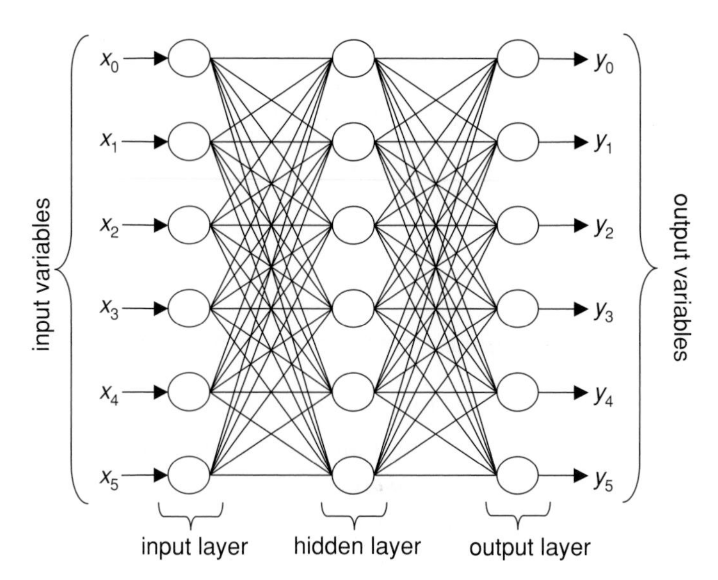

Figure 10.3 A three-layer feedforward neural network. Input variables are shown on the left and output variables on the right. The values of the input variables are copied into the input neurons, and then the activation values are propagated across the synaptic connections to activate the neurons in the next layer. Once the output layer has been activated its activation values are copied into the output variables and the network becomes stable again until a new set of inputs is supplied.

propagated across their output synapses (as described below) and into the next layer of neurons, thereby initializing these neurons' activation values. This propagation process is repeated between each successive pair of layers in the network, until finally the activation values of the output neurons have been determined. At this point, the activation values of the output neurons can be copied into the output variables $y_0, y_1, \ldots, y_k$, thereby completing the computation of the machine. In this way, a network can be viewed as a set of functions $f_0, f_1, \ldots, f_k$ mapping the input variables to a set of output variables via $f_i(x_0, x_1, \ldots, x_n) = y_i$.

The topology of a feedforward network can vary considerably. The numbers of neurons in each layer can vary independently of the other layers, and the total number of layers can also vary. For networks with $L > 2$ layers, all layers but the first and last are considered *hidden layers*. The pattern of synaptic connections can also vary; for example, we might omit one or more synapses between a pair of layers, or we might consider introducing synapses that connect neurons in nonadjacent layers. The interested reader is referred to the many books available on neural networks (e.g., Fausett, 1994).

The rules of propagation for a feedforward network are fairly simple. Let us consider the problem of initializing the activation value for some neuron q_k in a noninput layer of the network. Figure 10.4 illustrates the problem.

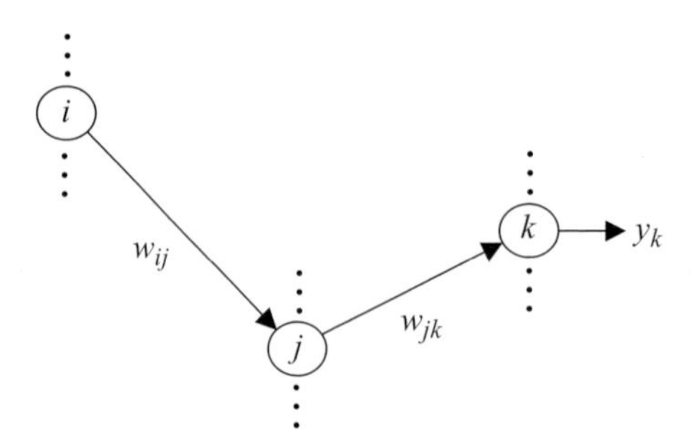

Figure 10.4 A path through a neural network. Synapses are shown labeled with their weights.

If we let $pred(q_k) = \{j \,|\, (q_j, q_k) \in dom(W)\}$ denote the indices of all the neurons in a previous layer having a synaptic connection to q_k, then we can define the *input value* σ_k as the sum of weighted inputs to q_k:

$$\sigma_k = \sum_{j \in pred(q_k)} w_{jk} \alpha_j \tag{10.10}$$

where α_j denotes the activation value of neuron q_j. Given this weighted sum[1] of inputs to a neuron q_k, we can compute q_k's activation value via the neuron-specific activation function $\xi_k(\sigma_k)$. A very commonly used activation function is the *standard sigmoid*:

$$\alpha_k = \xi_k(\sigma_k) = \frac{1}{1 + e^{-\sigma_k}}. \tag{10.11}$$

We will assume hereafter that this standard activation function is used for all neurons, and so we will drop the subscript k from ξ. Note that the activation function is sometimes referred to as a *transfer function*.

We now describe the training of a feedforward network. Let us suppose we have a training set $T \subset \mathbb{R}^{n+k+2}$ in which each training case $c = (x_0, x_1, \ldots, x_n, t_0, t_1, \ldots, t_k)$ prescribes a vector of expected outputs $t_0, t_1, \ldots, t_k$ for a given input vector $x_0, x_1, \ldots, x_n$. If we initialize the synaptic weights of the machine to have random values in the interval [−1,1], then we can expect that the output values $y_0, y_1, \ldots, y_k$ emitted by the machine for the training case $x_0, x_1, \ldots, x_n$ will differ markedly from the desired outputs $t_0, t_1, \ldots, t_k$. We can quantify the total *error* in the emitted outputs – i.e., the degree to which the emitted outputs differ from the expected outputs – via the error function E:

$$E(M, T) = \frac{1}{2} \sum_{c \in T} \sum_{i=0}^{k} (t_i - y_i)^2 \tag{10.12}$$

(Tveter, 1998) where the summation is over all outputs y_i of the network M and over all training cases $c = (x_0, x_1, \ldots, x_n, t_0, t_1, \ldots, t_k)$ in T.

Training can now be accomplished by *propagating* the measured error backward from the output layer to each preceding layer, adjusting the synaptic weights

[1] Inputs may be combined in other ways besides a weighted sum, but this formulation works well in many cases.

between each pair of layers in accordance with the amount of error which is propagated across each synapse. In this way, we effectively assign different levels of "blame" to the different components of the network in proportion to their contributions to the observed error of the network. This process is called *backpropagation* and is only one of a variety of training procedures available for feedforward neural networks.

In the backpropagation approach to training, synaptic weights are adjusted according to the derivative $\partial E/\partial w_{jk}$ of the error function:

$$\Delta w_{jk} = -\eta \frac{\partial E}{\partial w_{jk}} \tag{10.13}$$

where η is a constant known as the *learning rate*, and is typically set to a small positive real value (Tveter, 1998). The rationale for Eq. (10.13) is as follows. We wish to adjust each synaptic weight w_{jk} in such a way as to reduce the overall error of the network on the training set. The derivative $\partial E/\partial w_{jk}$ specifies the gradient of the error function and therefore indicates the direction (i.e., increase or decrease) in which we should modify w_{jk}; we take the negative of the gradient since the gradient points in the direction of increasing error, whereas we wish to decrease the error. Finally, the learning rate η adjusts the magnitude of the change to w_{jk}, and as such acts as a *step size* much like the step size used in the *gradient ascent* method described in section 2.5; indeed, backpropagation is effectively a form of gradient descent over the error function $E(M, T)$, so that the effect is to (locally) minimize the error of the network on the training set.

What remains is to derive the actual update formulas using the derivative of the error function. The following is adapted from Tveter (1998). For the output layer, assuming the standard sigmoid activation function given by Eq. (10.11), the derivation is as follows:

$$\frac{\partial E}{\partial w_{jk}} = \frac{\partial E}{\partial \alpha_k}\frac{\partial \alpha_k}{\partial \sigma_k}\frac{\partial \sigma_k}{\partial w_{jk}} = -(t_k - \alpha_k)\alpha_k(1-\alpha_k)\alpha_j \tag{10.14}$$

so that the update formula for output neuron q_k is:

$$\Delta w_{jk} = \eta(t_k - \alpha_k)\alpha_k(1-\alpha_k)\alpha_j \tag{10.15}$$

where t_k is the expected output for neuron q_k for a particular test case c in T. We can sum these Δw_{jk} terms over all the training cases and then make a single update to w_{jk}, or we can apply the updates from each training case separately, before evaluating the next training case – note, however, that these two strategies can give different training behaviors; in our experience, the former strategy seems to work reasonably well much of the time.

For the middle layer of the example three-layer network shown in Figure 10.4, the $\partial E/\partial w_{ij}$ term can be derived as follows:

$$\begin{aligned}\frac{\partial E}{\partial w_{ij}} &= \sum_{j\to k}\frac{\partial E}{\partial \alpha_k}\frac{\partial \alpha_k}{\partial \sigma_k}\frac{\partial \sigma_k}{\partial \alpha_j}\frac{\partial \alpha_j}{\partial \sigma_j}\frac{\partial \sigma_j}{\partial w_{ij}}\\ &= \sum_{j\to k} -(t_k - \alpha_k)\alpha_k(1-\alpha_k)w_{jk}(1-\alpha_j)\alpha_i\end{aligned} \tag{10.16}$$

where the summation is over all synapses $j \rightarrow k$ leading out of neuron q_j. More generally, for any layer containing neuron q_j in an N-layer network:

$$\begin{aligned} \Delta w_{ij} &= \eta \delta_j \alpha_i, \text{ where:} \\ \delta_j &= (t_j - \alpha_j)\alpha_j(1-\alpha_j) && \text{for the output layer} \\ \delta_j &= \alpha_j(1-\alpha_j)\sum_{j \rightarrow k} w_{jk}\delta_k && \text{for any hidden layer} \end{aligned} \tag{10.17}$$

(Tveter, 1998). A neural network with $k+1$ outputs $\{y_0, y_1, \ldots, y_k\}$ may be used for $(k+1)$-way classification by simply classifying a test case X into that category c_i corresponding to the output variable y_i having the greatest magnitude; this is known as a 1-out-of-k representation, in which the identity of the output neuron with the highest activation value is taken to be the network's answer to the question which has been posed to it in the form of test case X. Training of a network for 1-out-of-k classification may be accomplished simply by coding the training data such that for training case $X = (x_0, x_1, \ldots, x_n, t_0, t_1, \ldots, t_k)$ having category c, the only output variable t_i having a nonzero value is $t_c : t_i = \delta(i, c)$, for Kronecker function δ (section 2.1).

Neural networks have been employed in a number of gene-finding tasks, including both the prediction of full gene parses (e.g., Uberbacher and Mural, 1991; Snyder and Stormo, 1993) and also more specialized tasks such as promoter prediction (Reese and Eeckman, 1995; Bajic *et al.*, 2002; Burden *et al.*, 2005) and splice site detection (e.g., Reese and Eeckman, 1995). The following case study gives a more detailed example.

10.5.1 Case study: GRAIL

The gene finder *GRAIL* (Uberbacher and Mural, 1991) employs a neural network to integrate the scores of seven content sensors, in order to predict the likelihood of a putative exon. The seven content sensors are evaluated in 99 bp windows; all of them produce some measure of coding potential, such as the frequencies of each of the four nucleotides in each putative phase, the overall base composition of the window, the dinucleotide *fractal dimension*, relative hexamer frequencies (i.e., log-likelihood ratios of hexamer frequencies based on coding versus noncoding status), and measures for detecting repetitive DNA. These measures are integrated over a 99 bp window to produce a measure for each nucleotide position (i.e., the position in the center of the window). At each position in the sequence, the scores from the seven content sensors are fed as inputs into a four-layer neural network. The input layer obviously consists of seven neurons – one for each content sensor. The two middle layers contain 14 and five neurons, respectively; each of these is connected to at most four neurons in the previous layer (i.e., has an *in-degree* of at most 4) and at most four neurons in the following layer (i.e., has an *out-degree* of at most 4). The network has a single output neuron, which is interpreted as the

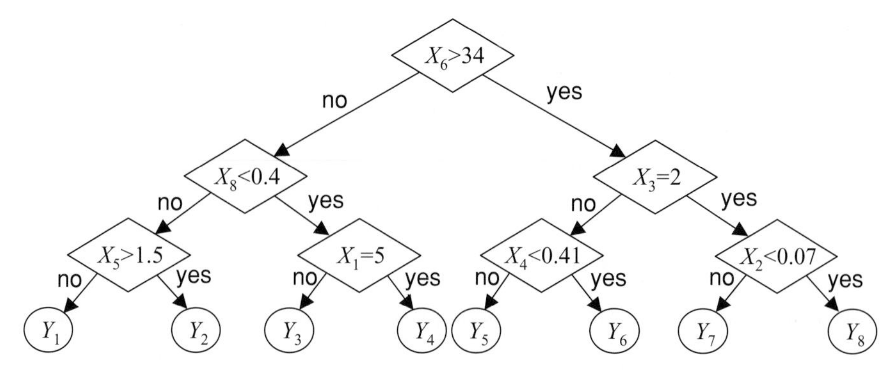

Figure 10.5 An example decision tree. Predicates (internal nodes) are shown as diamonds, and leaf nodes are shown as circles at the bottom. A test case is classified by starting at the root (top) and descending through the appropriate path. Classification of the test case occurs when a leaf node (Y_i) is reached.

likelihood that the current position is within the coding portion of an exon. These individual nucleotide-level predictions are then coalesced to produce exon-level predictions through the application of heuristic, window-based rules. Later versions of GRAIL have considerably extended this early version in various ways (e.g., Xu and Uberbacher, 1996), and some of these were used quite heavily in Celera's annotation of the human genome (Venter *et al.*, 2001).

It is interesting to note that in GRAIL the DNA sequence is not directly fed as an input to the network; instead, the sequence is first processed by the content sensors in order to extract local statistical properties of the sequence. The framework is an attractive one, since one can easily conceive of other content sensors which might be integrated into the system by merely adding another input neuron to the network (and possibly additional hidden nodes) followed by appropriate retraining.

Other systems using neural networks for gene and/or exon prediction were developed by Lapedes *et al.* (1989) and Reese *et al.* (1997), the latter system being the well-known GENIE program which has been used for annotation of the *Drosophila* and human genomes (see also the GENIE case study in section 9.1.5).

10.6 Decision trees

We have already seen the use of decision trees in the context of *maximal dependence decomposition* (*MDD* – section 7.3.7). Recall that a decision tree is a data structure in which binary or *N*-ary decisions decorate the internal nodes of the tree, providing a means of navigating from the root of the tree to an appropriate leaf node in search of a solution to a given computational problem (see Figure 10.5).

Here we will consider the general-purpose decision tree algorithm known as *C4.5* (Quinlan, 1993). This approach is based on the notion of *entropy reduction*. Recall from section 2.10 that the entropy $H = -\sum_i p_i \log p_i$ measures the "evenness" of a

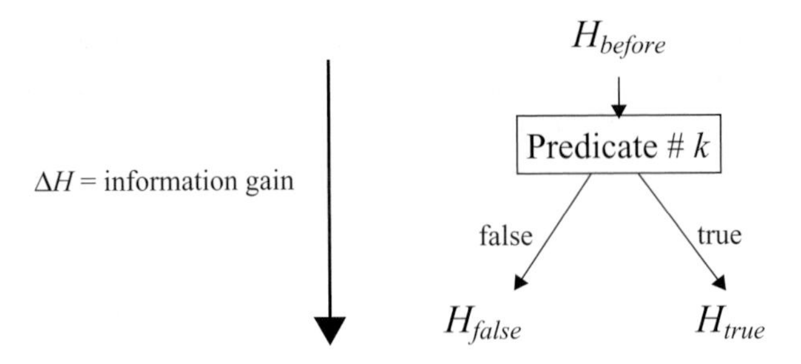

Figure 10.6 Information gain, or reduction of uncertainty, at a bifurcation in a decision tree. The classification entropy before the bifurcation is compared to a weighted sum of the entropies after the bifurcation.

distribution, and can also be interpreted as the amount of uncertainty in anticipating the next draw from a random variable governed by the given distribution.

In the C4.5 method, a decision tree is recursively constructed from the root down to the leaves, where at each recursive step we select the bifurcation which maximally reduces the *classification entropy* (i.e., the entropy of the distribution of categories among the training cases in a set – see below) of the training cases which flow down to this bifurcation during classification (see Figure 10.6). That is, if we define H_{before} to be the classification entropy of the cases flowing down to this bifurcation, H_{false} to be the classification entropy of that subset of cases for which the predicate at this bifurcation evaluates to false, and H_{true} to be the classification entropy of the subset of cases evaluating to true under the predicate, then the average reduction in entropy across this predicate is given by:

$$\Delta H = H_{before} - \left(\frac{n_{false}}{n} H_{false} + \frac{n_{true}}{n} H_{true} \right). \tag{10.18}$$

The term in parentheses is simply a weighted average of the H_{false} and H_{true} scores; n_{false} is the number of cases for which the predicate evaluates to false, and similarly n_{true} is the number for which it evaluates to true; $n = n_{false} + n_{true}$. At each point in the recursive tree construction procedure, we evaluate ΔH for each prospective predicate, and choose the predicate R_k having the highest ΔH value. This predicate is then assigned to the current bifurcation, and the procedure recurses to the two nascent children of this node in the tree.

The notion of *classification entropy* can be defined more precisely as:

$$H = -\sum_c \frac{n_c}{n} \log \frac{n_c}{n}$$

where the summation is over all categories c; the number of training cases flowing down to the current node in the tree is n, with n_c of these having category c. At the root of the tree we obviously have $n = |T|$.

The recursion can obviously be terminated when all the training cases reaching a point in the tree share the same category. However, it is often preferable (in order to avoid overtraining) to terminate prior to this point using some other heuristic, such as when the sample size n drops below some threshold, or when the category distribution becomes sufficiently dominated by one category. In the latter cases the leaves can then be used to assign probabilities according to the proportion of training cases of each class at a leaf, rather than assigning a single category to the test case; the combiner-type program *JIGSAW* uses such an approach (see section 9.2.1), as does the eukaryotic gene finder *GlimmerM* described in the next section.

The predicates which are considered during the tree-building process are drawn from the set of all (*attribute*, *value*) pairs present in the training set. For continuous attributes, values must be discretized in some manner, or the operators $<$ and $>$ may be used. A number of discretization methods have been explored, including methods which are themselves entropy-based (e.g., Kohavi and Sahami, 1996). A simpler heuristic is to *bin* the values for a particular variable using fixed-size bins – i.e., to construct a histogram, as described in section 2.7 – and then to recode the variable using the bin numbers. Another complexity that arises in implementing a decision tree system is in deciding whether to restrict the tree to a bifurcating structure, or whether to allow N-ary decisions to be made at each internal node of the tree. Other very promising approaches to building decision trees have been proposed, including the *CART* (Breiman *et al.*, 1984) and *OC1* (Murthy *et al.*, 1994) methods.

10.6.1 Case study: *GlimmerM*

The eukaryotic gene finder *GlimmerM* (Pertea and Salzberg, 2002) utilizes an OC1 decision tree in a novel way. The tree is evaluated for each putative feature (exon, intron, intergenic region) in a sequence, but the result of evaluating the tree is a probability rather than a discrete category. Because the tree is pruned in such a way that the training cases flowing to individual leaves need not all share the same category, it is possible to associate with each leaf a probability for each category, using the proportion of training cases at that node having a given category as an estimate for the associated probability as suggested in the previous section.

The attributes provided to the tree-building procedure consist of: (1) a WAM score for the signal at the 5′ end of the putative feature, (2) a WAM score for the signal at the 3′ end of the feature, (3) a length probability for the putative feature, based on a histogram of lengths of training features, and (4) a log-likelihood score computed from the *hexamers* (strings of six nucleotides) observed in the sequence:

$$s_{hex} = \sum_{\substack{hexamers \\ H}} \log \frac{P(H|\Gamma)}{P(H)} \tag{10.19}$$

where Γ is the feature type for the putative feature, and $P(H)$ is a background distribution estimated from long stretches of unannotated DNA.

10.7 Linear discriminant analysis

A method originally introduced by the great statistician and geneticist R. A. Fisher in the 1930s is *linear discriminant analysis*, or *LDA* (Fisher, 1936), also sometimes referred to as *DFA* – *discriminant function analysis*. In the case of binary classification, the goal of LDA is to find a *linear combination* (i.e., $a_1x_1 + a_2x_2 + \cdots + a_nx_n$) of attributes which maximizes the ratio of between-group variance to within-group variance, where a group is defined as those training cases sharing the same category. Thus, the goal is to maximize the separability of the test cases via an appropriate linear transformation. The variance ratio can be measured using an *F-ratio* defined as $MS_{between}/MS_{within}$, where *MS* denotes the *mean square* term from an *analysis of variance* (*ANOVA* – see, e.g., Zar 1996); because we will be describing an *eigenvector decomposition* method for maximizing the *F*-ratio, we will omit a full description of ANOVA and mean squares.

Let us define a matrix $\mathbf{W} = [w_{rc}]$ of *within-group sums of squares and cross products*:

$$w_{rc} = \sum_{j=1}^{m}\sum_{i=1}^{n_j}\left(X_{j,r}^{(i)} - \overline{X}_{j,r}\right)\left(X_{j,c}^{(i)} - \overline{X}_{j,c}\right) \tag{10.20}$$

(Manly, 1994) where j iterates over the m categories and i iterates over the n_j training cases in category C_j, so that $X_{j,r}^{(i)}$ denotes the rth attribute of the ith training case in category j and $\overline{X}_{j,r}$ denotes the mean of the rth attribute for category j. Thus, w_{rc} denotes the entry at the rth row and cth column (corresponding to attributes r and c) of matrix $\mathbf{W}$.

Similarly, we can compute a matrix $\mathbf{T} = [t_{rc}]$ representing the *total* (i.e., not within-group nor between-group, but across the entire training set) *sums of squares and cross products*:

$$t_{rc} = \sum_{j=1}^{m}\sum_{i=1}^{n_j}\left(X_{j,r}^{(i)} - \overline{X}_r\right)\left(X_{j,c}^{(i)} - \overline{X}_c\right) \tag{10.21}$$

(Manly, 1994) for $\overline{X}_r$ the mean of the rth attribute over the entire training set. We can then obtain the *between-groups sums of squares and cross products* matrix $\mathbf{B}$ via simple matrix subtraction: $\mathbf{B} = \mathbf{T} - \mathbf{W}$. A useful fact from the field of multivariate statistics is that the dominant *eigenvector* V (i.e., the one associated with the largest *eigenvalue*; see below) of $\mathbf{W}^{-1}\mathbf{B}$ is the linear combination which maximizes the *F*-ratio, and can therefore be used as a linear function for optimally separating the two groups in the training set (Manly, 1994). That is, if we collect the attributes of a training case into a vector X, then the dot product $V \cdot X$ will result in a real value d which can then be subjected to a *K*-nearest-neighbors (KNN) algorithm to perform the classification. For multi-way classification ($m > 2$) we can apply the top $r = m - 1$ dominant eigenvectors to each incoming test case via linear combination, effectively mapping the test case to an r-dimensional space, and then we can again use KNN to classify the test case by comparing it to all the similarly transformed

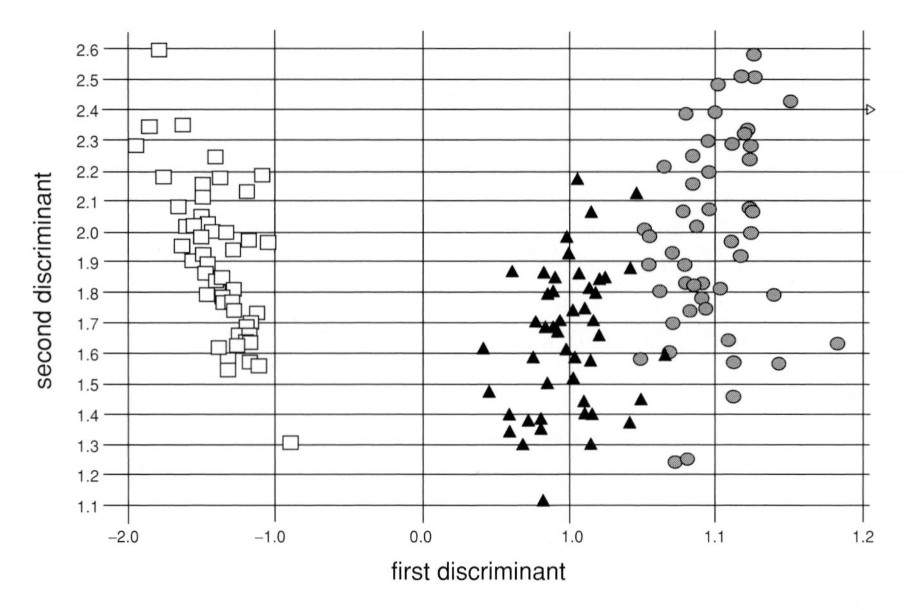

Figure 10.7 Linear discriminant analysis of morphometric attributes of three species of *Iris*. The first discriminant function (x-axis) achieves quite good separation of classes; the second (y-axis) adds virtually nothing to the discriminatory power of the model.

training cases in the induced r-dimensional space. This technique is similar to using the Mahalanobis distance metric with KNN, and can often produce accurate classifications on practical data sets (Manly, 1994). Note that in classical LDA the attributes are generally assumed to be normally distributed for the purposes of significance testing, though in practice the resulting classifier may still work well under nonnormality in the absence of extreme outliers – see, e.g., Friedman (1989) for other conditions on the optimality of LDA.

Unfortunately, the need for robust computation of eigenvectors in software can be somewhat of a nuisance, since many eigenvector decomposition methods constrain the form of the input matrix to be, e.g., symmetric or hermitian. A robust eigenvector decomposition procedure is included in the *TNT* toolkit (*Template Numerical Toolkit* – Pozo, 1997), an open-source C++ library developed by the U. S. National Institute for Standards and Technology. The *GNU Scientific Library* (*GSL* – Galassi *et al.*, 2005) also provides procedures for eigenvector decomposition, but at present they are not general purpose.

A classic example of the use of LDA is the problem of separating three species of iris (*Iris setosa*, *I. versicolor*, and *I. virginica*) based on morphometric attributes such as petal length and sepal length. Figure 10.7 shows the results of applying LDA to this data set. The discriminants found by LDA were:

$$d_1(X) = -0.21x_0 - 0.39x_1 + 0.55x_2 + 0.71x_3$$
$$d_2(X) = 0.01x_0 + 0.60x_1 - 0.23x_2 + 0.78x_3.$$

As can be seen from the figure, the first linear discriminant (d_1; x-axis) was able to achieve a fairly clear separation between the classes, whereas the second discriminant (d_2; y-axis) appears to contribute virtually nothing to the discriminatory power of the model.

An example of the application of LDA to gene finding is the program *FGENEH*, which uses linear discriminants to score possible exons which are then evaluated into complete gene structures using a dynamic programming algorithm (Solovyev *et al.*, 1995). Other uses include the identification of core promoters (Hannenhalli and Levy, 2001; Solovyev and Shahmuradov, 2003; see also section 12.3).

10.8 Quadratic discriminant analysis

A weakness of the LDA method is that the resulting classifier is constrained to using linear decision boundaries in discriminating between test cases in different groups. *Quadratic discriminant analysis*, or *QDA*, relaxes this constraint, allowing nonlinear decision boundaries. Unfortunately, the use of nonlinear decision boundaries can sometimes lead to overfitting of the decision model to the training data, and can therefore compromise classification accuracy on unseen data. Nevertheless, for some problems nonlinear methods can at times be more effective than a simple linear classifier.

Here we describe a crude heuristic which, while less sophisticated than "canonical" QDA as defined in the statistics literature, is extremely easy to implement and can sometimes produce improved classification over LDA. The procedure involves merely computing new attributes consisting of squares and cross products of the original attributes, and providing these to LDA. That is, if the original set of attributes is $A = \{x_0, \ldots, x_{n-1}\}$, then we simply add to this set all terms $x_i x_j$ for $0 \leq i \leq j < n$. These terms are computed for each test case, and the test case is correspondingly enlarged to include these new attributes. This expansion is performed both on the training set prior to training the linear discriminant, and also on unseen test cases that are presented for classification during deployment.

Although the resulting classifier is still linear in the augmented set of variables, it is quadratic in the original set of attributes. Any of the other machine-learning algorithms described in this chapter can also be applied to these higher-order attributes. More sophisticated nonlinear discriminant methods are amply described elsewere (e.g., Sakai *et al.*, 1998; Roth and Steinhage, 1999).

An existing application of QDA to gene finding is the *MZEF* program (Zhang, 1997), which uses a quadratic discriminant function to evaluate possible exons; although the program does not predict entire genes, the output of this program can be used as inputs to a dynamic programming algorithm, such as that described by Guigó (1998), for assembling exons into complete gene predictions.

10.9 Multivariate regression

Many readers will be familiar with linear regression; here we describe the use of multivariate linear regression to perform *N*-way classification. First we describe the process of multivariate regression.

Suppose that we form a matrix **X** in which the columns correspond to attributes and the rows correspond to the training cases. We can thus populate the matrix with attribute values from the training set. The categories of the training cases can then be collected into a column vector Y, with the indices of Y corresponding to rows of **X**. The problem of multivariate regression is to find the vector A of coefficients which can be used to form a linear combination with any test case (i.e., vector of attribute values) $Z : c = A \cdot Z$, where the errors between c and the known category C for each training case are minimized in some way (typically in the form of their squared differences, $\sum_i (c_i - C_i)^2$). A simple method which often works in practice is given by:

$$A = (\mathbf{X}^T\mathbf{X})^{-1}\mathbf{X}^T Y \tag{10.22}$$

where $\mathbf{X}^T$ is the *transpose* of **X** and $\mathbf{X}^{-1}$ is the *inverse* of **X** (see section 2.1); both of these operations are easily performed using routines such as those provided by the open-source *GNU Scientific Library* (*GSL* – Galassi *et al.*, 2005). For problems with small numbers of uncorrelated variables, this method can be very effective. Unfortunately, the existence of *multicollinearity* – strong correlations between variables – in the **X** matrix can result in the matrix $\mathbf{X}^T\mathbf{X}$ being noninvertible, so that A cannot be obtained via Eq. (10.22). Performing *dimensionality reduction* (see section 10.16) prior to regression can often solve this problem.

Assuming that the vector A is computable, this regression technique provides a way to classify unseen cases into two categories (i.e., *binary classification*), by forming the linear combination $c = A \cdot Z$ for test case Z and rounding the real value c to the nearest integer in $\{0,1\}$, where we assume the categories of the training examples were given as 0 or 1. For *N*-way classification ($N > 2$) we can decompose the problem into $N(N-1)$ binary classification problems between pairs of categories. For a given test case Z, we can perform all $N(N-1)$ binary classifications, tabulate the number of times Z was classified into each category C_i, and then select that C_i to which Z was most often classified. Although somewhat laborious (and susceptible to ties), this method can work surprisingly well for some problems.

10.10 Logistic regression

Whereas linear regression fits a linear function (i.e., linear combination) to the data, with the *response variable* (c in the previous section) representing the predicted

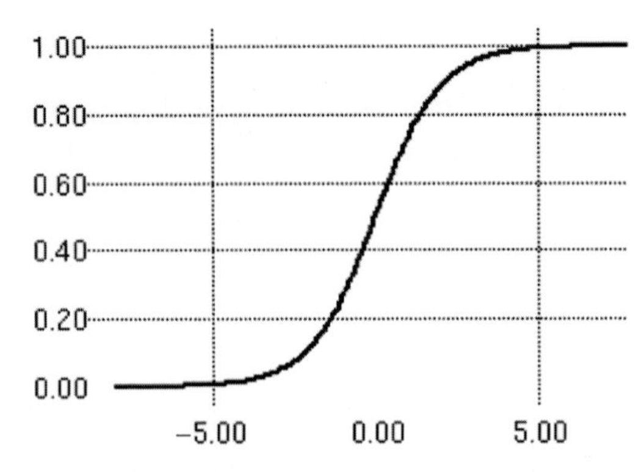

Figure 10.8 A logistic curve for a single-attribute problem. The attribute value is shown along the x-axis and the probability $P(X \in C_1)$ is shown along the y-axis.

category of a given test case, logistic regression fits a nonlinear curve to the relationship between the attributes of a training case and the probability that the case belongs to one of the two categories in a binary classification problem. That is, we find the best fit of the parameters $\theta = \{b_0, b_1, \ldots, b_n\}$ to the function:

$$P_X = P(X \in C_1) = \frac{1}{1 + e^{-(b_0 + b_1 x_1 + b_2 x_2 + \cdots + b_n x_n)}} \tag{10.23}$$

where P_X represents the probability that test case $X = (x_1, x_2, \ldots, x_n)$ belongs in category C_1; since we consider only binary classification problems in logistic regression, the probability that X belongs to category C_0 is simply $1 - P_X$. For problems having only a single attribute, this function forms an S-curve, as shown in Figure 10.8.

Fitting the parameters can be accomplished with a general-purpose gradient-ascent algorithm (see section 10.15), where the objective function for the gradient ascent optimization is the log-likelihood L of the training data:

$$L = \sum_{X \in C_1} \log P_X + \sum_{X \notin C_1} \log(1 - P_X). \tag{10.24}$$

The gradient of this function in any parameter dimension b_i can be computed simply by taking the first partial derivative for that parameter:

$$\frac{\partial}{\partial b_i} L = \sum_{X \in C_1} \frac{x_i \lambda}{(1+\lambda)^2 P_X} - \sum_{X \notin C_1} \frac{x_i \lambda}{(1+\lambda)^2 (1 - P_X)} \tag{10.25}$$

which can be easily derived given:

$$\frac{\partial}{\partial b_i} P_X = \frac{x_i \lambda}{(1+\lambda)^2} \tag{10.26}$$

where:

$$\lambda = e^{-b_0 - b_1 x_1 - \cdots - b_n x_n} \tag{10.27}$$

For the b_0 term we obviously take x_0 to be 1, since we defined $X = (x_1, x_2, \ldots, x_n)$. Given the estimated parameters of the logistic model, we can then classify a test case X by evaluating $P_X = P(X \in C_1)$ and assigning X to category C_1 if $P_X \geq 0.5$ and assigning it to category C_0 if $P_X < 0.5$.

10.11 Regularized logistic regression

For problems having many more dimensions (attributes) than training cases, the potential for overfitting can be very large for many classification algorithms. In the case of logistic regression, there is a simple way to mitigate this effect, by penalizing the use of model parameters with large absolute values:

$$L = \sum_{X \in C_1} \log P_X + \sum_{X \notin C_1} \log(1 - P_X) - \alpha \sum_X R(\theta; X). \tag{10.28}$$

Here we have augmented the formula for logistic regression with the term $-\alpha \sum_X R(\theta; X)$, where the summation is over all training cases X. The influence of this last term can be tuned by modifying the α parameter. The $R(\theta; X)$ term is a *regularization term*, which is a function of the other parameters $\theta = \{b_0, b_1, \ldots, b_n\}$. Two common regularization functions are R_{L1} and R_{L2}:

$$R_{L1} = \sum_{i=0}^{n} |b_i| \tag{10.29}$$

$$R_{L2} = \sum_{i=0}^{n} b_i^2. \tag{10.30}$$

Incorporation of the regularization term affects the gradient:

$$\frac{\partial L}{\partial b_i} = \sum_{X \in S} \frac{x_i \lambda}{(1+\lambda)^2 P_X} - \sum_{X \notin S} \frac{x_i \lambda}{(1+\lambda)^2 (1 - P_X)} - \alpha \sum_X \frac{\partial R}{\partial b_i}(\theta; X) \tag{10.31}$$

with the last term depending on the form of regularization used:

$$\frac{\partial R_{L1}}{\partial b_i} = -sign(b_i) \tag{10.32}$$

where:

$$sign(b_i) = \begin{cases} -1 \text{ if } b_i < 0 \\ 0 \text{ if } b_i = 0 \\ 1 \text{ if } b_i > 0 \end{cases} \tag{10.33}$$

for L_1 regularization, and:

$$\frac{\partial R_{L2}}{\partial b_i} = -2b_i \tag{10.34}$$

for L_2 regularization. Regularization effectively penalizes more extreme parameter values; the technique is more fully described elsewhere (e.g., Tikhonov, 1963; Hastie *et al.*, 2003).

10.12 Genetic programming

One of the more interesting machine-learning approaches, from a biological perspective, is *genetic programming* (Koza, 1992), a particular form of *evolutionary computation* (Fogel, 2005) or *genetic algorithm* (Goldberg, 1989). Genetic programming, or *GP*, involves the use of an evolutionary simulation in which a population of algorithms is subjected to a form of natural selection over some number of generations. Because we wish to solve a particular problem (a classification problem in our case), the form of natural selection that we employ should reflect the goals of the problem. In particular, if we view the algorithms in the evolving population as classifiers, then our goal is to evolve the most accurate classifier for the problem at hand, and so we need to implement our natural-selection process in a way which reflects the accuracy of the evolving classifiers on the target problem. We can do this by simply applying each classifier in the population to the training cases, measuring the classification accuracy, and then allowing each individual classifier in the population to survive and/or reproduce in proportion to the accuracy of that classifier. In this way, successive generations of classifiers should tend toward increasing average *fitness*, or classification accuracy. Thus, the strategy employed in GP is to run the simulation for a sufficient number of generations to increase the average classification accuracy of the population to an acceptable level, and then to select the most accurate classifier from the final population for deployment.

More formally, let $f_i : \mathbb{R}^n \rightarrow \mathbb{N}_m$ be an individual classifier mapping test cases $X \in \mathbb{R}^n$ to categories $Y \in \mathbb{N}_m$. We begin with a *population* $P_0 = \{f_i | 0 \leq i < N, f_i \in \mathcal{F}\}$ of N random functions f_i, for *population size* N, where $\mathcal{F}$ is the set of all possible algorithms expressible in some particular representation. The functions can be represented using *abstract syntax trees* (e.g., Aho *et al.*, 1986), which are simply trees in which each internal node is an *operation* (such as addition, multiplication, exponentiation, sin, or cos) and each leaf node is a variable or constant. The operations are drawn from an *operator set*, which is also known as the set of *nonterminals*. The leaves are drawn from a set of *terminals*, which as we have said will consist of variables (i.e., attributes) and randomly generated constants. The trees are generally limited to some maximal depth, for practical reasons (i.e., speed and space efficiency). $\mathcal{F}$ is thus taken to include not all algorithms, but rather the subset of all algorithms which can be represented by a syntax tree of acceptable height, over the given sets of terminals and nonterminals. Figure 10.9 depicts an example syntax tree. Clearly, more sophisticated algorithms than that illustrated in the figure can be expressed using the formalism of abstract syntax trees. The purpose of the figure is to illustrate intuitively how an algorithm can be represented using a syntax tree, and how an algorithm can be modified through appropriate *mutations* to the algorithm's syntax tree.

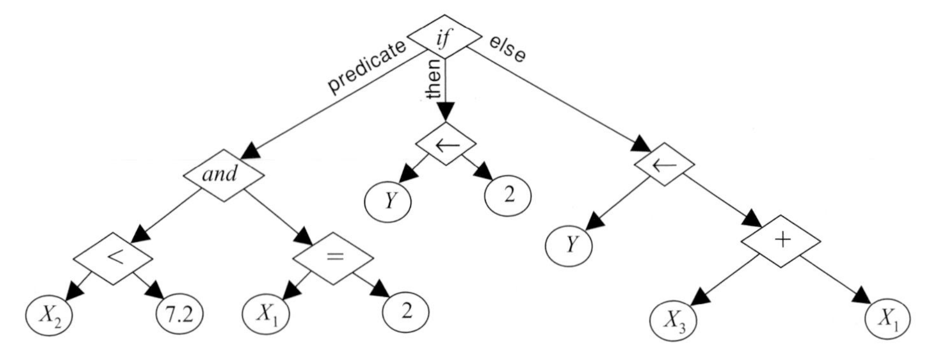

Figure 10.9 An example of an abstract syntax tree. The root is shown at the top, and the leaves, which contain the terminals, are shown at the bottom. The algorithm represented by the tree can be stated as: $\text{if}(X_2 < 7.2 \text{ and } X_1 = 2)$ then $Y \leftarrow 2$ else $Y \leftarrow X_3 + X_1$. Assignment to Y constitutes classification of the test case $C = (X_0, X_1, \ldots, X_n)$.

We define a *mutation operator* $M : \mathcal{F} \mapsto \mathcal{F}$ on classifiers in $\mathcal{F}$ as a procedure which perturbs a syntax tree in some way. Common mutation operators are *point mutation* (replacing a single terminal or nonterminal with another drawn from the appropriate set) and *subtree mutation* (replacing an entire subtree with a randomly generated subtree of acceptable depth). In addition, it is possible to define a *crossover operator* $R : \mathcal{F} \times \mathcal{F} \mapsto \mathcal{F} \times \mathcal{F}$ which takes two trees f_1 and f_2 and swaps subtrees between them, so that a new pair (f_3, f_4) of classifiers is produced which resemble the original "parent" classifiers in their syntax trees, but which also differ from them in some regards. This is similar to the genetic recombination which occurs in sexually reproducing species (see, e.g., Lewin, 2003). Yet another operator is the *cloning operator* $I : \mathcal{F} \mapsto \mathcal{F}$, which maps a function f back to itself.

A single step of the GP process consists of evaluating all the members of the current population P_i under the objective function (also called a *fitness function*) $Q : \mathcal{F} \to \mathbb{R}$, selecting some subset V of the population having the highest scores (or *fitnesses*) under Q, and then using V to populate the next generation P_{i+1} via the mutation, crossover, and cloning operators, so that P_{i+1} will contain individuals which resemble – to varying degrees – the best-scoring members of P_i. This process is repeated some number of times G, and then the highest-scoring individual f^* is selected from the final generation P_{G-1} and emitted by the GP; the selected individual f^* is taken to be the trained classifier which is deployed for the problem at hand.

For classification problems, the objective function Q is simply the classification accuracy on the training set (or some subset of the training set – see Koza, 1992). Utilizing a tree f for classification is a simple matter of binding the variables in the terminal set $(X_0, X_2, \ldots, X_{n-1})$ to the values of a given test case $c = (x_0, x_2, \ldots, x_{n-1})$:

$$X_0 \leftarrow x_0, X_1 \leftarrow x_1, \ldots, X_{n-1} \leftarrow x_{n-1},$$

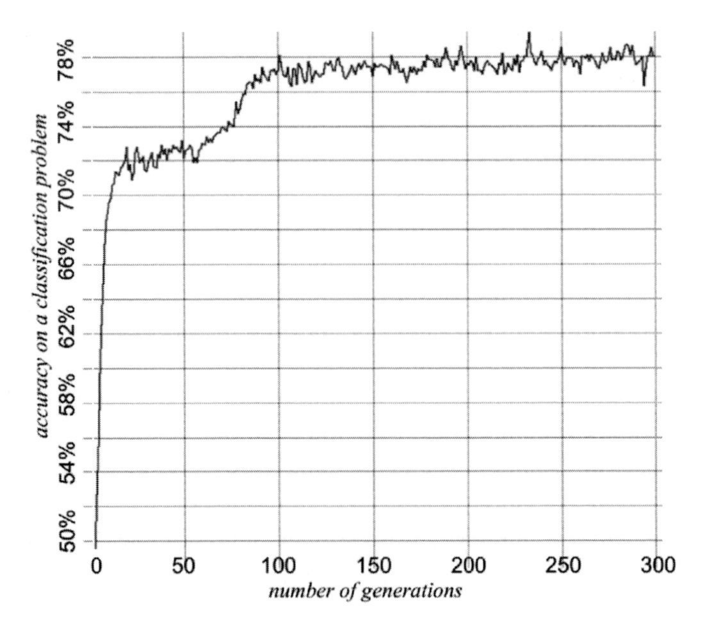

Figure 10.10 Accuracy of genetic programming on an example classification task, as a function of generation number. The average fitness increases rapidly at first, then quickly levels off.

then executing the algorithm denoted by the tree to produce a real-valued result r, and then interpreting r as a category C, via $C = round(r) \bmod m$.

The key to successfully applying GP is in crafting a set of nonterminals which are most effective for the target problem. This can be a difficult task; furthermore, a fair number of other parameters must be set by the user, such as the number G of generations of evolution to simulate, the size of the set V which are allowed to contribute their genetic material (i.e., portions of their syntax trees) to the next generation, the determination of which mutation and/or crossover operators should be used and in what proportion, what maximum depth for the parse tree to enforce, etc. In addition to having to select all of these parameters in an ad-hoc fashion, a user of GP may have to tolerate lengthy training times, depending on the computer hardware used. In addition, it is common practice to run the GP a number of times, taking the best individual from all runs as the trained classifier. Despite these difficulties, for some problems GP can perform very well, and is therefore a valuable tool an any machine-learning toolbox.

As a side-note, GP is an interesting framework for simulating and learning about evolutionary processes in an idealized setting. An instructive exercise is to monitor the average fitness over successive generations; Figure 10.10 shows an average-fitness curve for a particular GP run on an example problem.

An example of the application of GP to gene prediction is given by (Saetrom *et al.*, 2005), who applied the technique to the prediction of noncoding RNA genes in *Escherichia coli*. A somewhat similar population-based machine-learning approach, *estimation of distribution algorithms* (*EDA*), has been applied to the problem of splice site prediction (Saeys *et al.*, 2004).

10.13 Simulated annealing

A method which is in some ways very similar to genetic programming, but which can be much more efficient, is *simulated annealing*, or *SA*. We consider the special case of simulated annealing in which the solution which is sought is an abstract syntax tree representing a classifier, as in our formulation of GP. Whereas GP maintains an entire population of classifiers, SA maintains only one classifier at each step. We begin with a single randomly generated classifier f and apply a mutation operator M to f to obtain $f' = M(f)$. We again employ an objective function Q which measures the classification accuracy of a classifier on the training set. If the fitness of f' is greater than that of f – i.e., if $Q(f') > Q(f)$ – then we discard f and replace it with f'. This is similar to the application of natural selection in GP. If, however, $Q(f') < Q(f)$, we may still decide to replace f with f', even though f' performs more poorly than f.

In particular, we maintain during the SA run a *temperature* variable, t, which decreases at each step according to some *cooling schedule*. A typical cooling schedule sets $t_i = d \cdot t_{i-1}$ for some cooling factor $d < 1$, over successive steps i, until either the temperature drops to a designated stopping value or a prescribed number of steps have elapsed, at which time the process ends and the current classifier (or, alternatively, the best classifier seen so far) is emitted. In the case where $Q(f') < Q(f)$, we replace f with f' according to the following probability:

$$P = e^{-\Delta E/kt} \tag{10.35}$$

where $\Delta E = Q(f') - Q(f)$ and $k \approx 1.380\,65 \times 10^{-23}$ is the *Boltzmann constant* from statistical physics. In practice, a different value for k may be used in order to modify the convergence behavior of the algorithm. Figure 10.11 shows a fitness curve similar to that given previously for GP. As with GP, the fitness tends to increase very dramatically early in the process, and then to level off toward a final value. Like GP, simulated annealing is motivated by natural processes – in this case, the changing energy levels of molecules as the temperature drops.

10.14 Support vector machines

A relatively new model which has become especially popular in recent years is the *support vector machine*, or *SVM*. The usual formulation of the SVM is as a binary classifier in which positive cases are to be classified as +1 and negative cases as −1. The idea behind SVMs is to find an optimal hyperplane for separating positive and negative cases in a high-dimensional space, where the notion of optimality for a hyperplane is defined in terms of the margin between the hyperplane and the

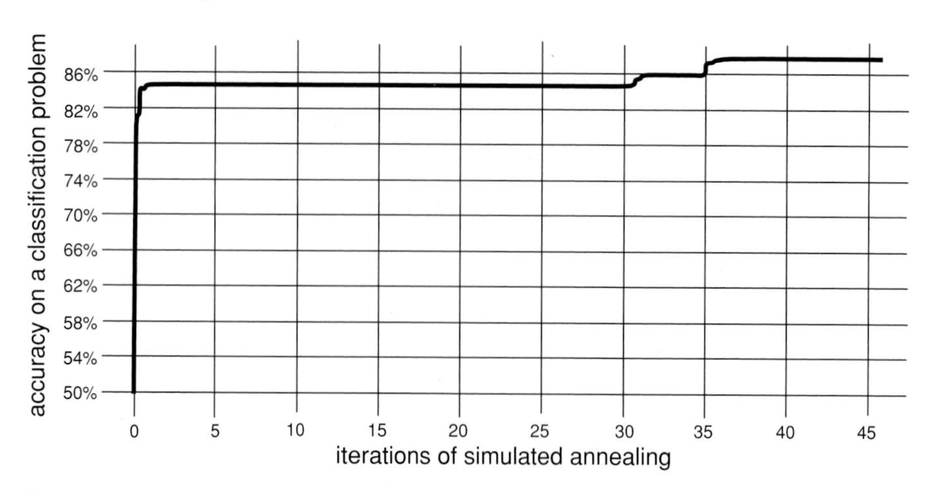

Figure 10.11 Accuracy of a simulated annealing process on an example classification task as a function of time. $k = 2.8 \times 10^{-10}$, t decreased from 100 to 1, and the cooling factor was 0.9999.

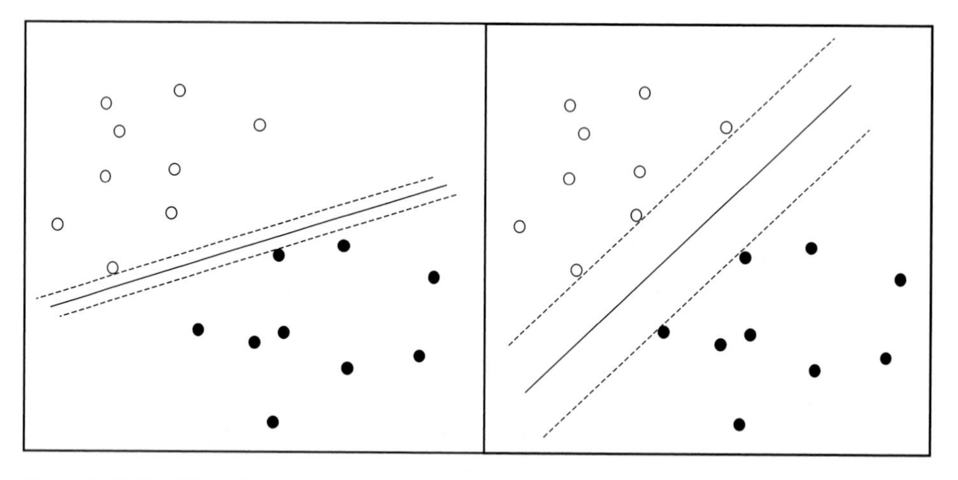

Figure 10.12 The idea of maximum margin hyperplanes. Left: a hyperplane (solid line) separating objects of two classes (white circles, black circles), with a relatively narrow margin (distance between solid and dashed line). Right: a hyperplane with a wider margin, for the same classification problem.

nearest positive and negative examples ("support vectors") in the training set, as crudely illustrated in two dimensions by Figure 10.12.

The justification for maximizing the margin is based on the notion of *risk minimization*, in which one tries to reduce the risk of overtraining and of misclassifying unseen objects. Actually finding the optimal hyperplane involves a *quadratic programming* problem to find an optimal set of parameters α_i, $0 \leq i < |T|$, for training set $T = \{(\mathbf{x}_i, y_i)\}$:

$$\max_{\alpha_i} \sum_i \alpha_i - \frac{1}{2} \sum_{i,j} \alpha_i \alpha_j y_i y_j K(\mathbf{x}_i, \mathbf{x}_j)$$

subject to:

$$\forall_i 0 \leq \alpha_i \leq C$$
$$\sum_i \alpha_i y_i = 0$$

for $\alpha_i \in \mathbb{R}$, *kernel function* $K(\cdot\,,\cdot)$, and training cases $x_i \in \mathbb{R}^m$ having category $y_i \in \{-1, 1\}$ (Burges, 1998). C is a user-provided parameter which is necessary in order to bound the optimization. The kernel function may be thought of as a distance function, with the vector *dot product* (section 2.1) providing a default (linear) implementation. Methods for solving the above quadratic programming problem are ably described elsewhere (Osuna *et al.*, 1997; Kaufman, 1998; Platt, 1998).

Once the α_i have been learned via the above optimization, classification of a novel case $\mathbf{x}$ may be achieved by computing the function:

$$f(\mathbf{x}) = \sum_i \alpha_i y_i K(\mathbf{x}, \mathbf{x}_i)$$

and observing the rule:

$$class(x) = \begin{cases} -1 & \text{if } f(\mathbf{x}) < 0 \\ +1 & \text{otherwise.} \end{cases}$$

The derivation of SVMs involves the use of *Lagrangian multipliers* and several results from the study of *reproducing kernel Hilbert spaces* (*RKHSs*) and is therefore beyond the scope of this text. The interested reader may consult the appropriate literature (Burges, 1998; Vapnik, 1998) for the derivation and other important modifications, such as for the scenario where the training data are not fully separable by a single hyperplane.

SVMs have been used in solving a number of problems related to gene finding. Just one example is the use by Dror *et al.* (2004) of an SVM with 228 features to predict *exon skipping* (a form of *alternative splicing* – see section 12.1) in eukaryotic genes. Among the 228 features were quantities reflecting sequence conservation, exon lengths, residues near donor sites, trinucleotide frequencies, and other information. That study utilized a publicly available implementation of SVMs called *SVMlight*; another popular SVM implementation is *Gist*.[2]

Other example uses of SVMs include the identification of individual exons (Jaakkola and Haussler, 1999), the prediction of start codons (Zien *et al.*, 2000), and the prediction of core promoters (Gangal and Sharma, 2005).

10.15 Hill-climbing with the GSL

A number of the methods described in this chapter require that parameters be estimated in some way so as to maximize the accuracy of the classifier on a training

[2] SVMlight is at http://svmlight.joachims.org and Gist is at http://microarray.cpmc.columbia.edu/gist/.

set. A common way to do this is via *hill-climbing*, or gradient ascent. In section 2.5 we gave a simple procedure for one-dimensional gradient ascent, in which the gradient was only implicitly used. In practice, rather larger numbers of parameters are present. A number of multivariate gradient-ascent algorithms have been developed over the years, including the popular *conjugate gradient* methods.

Implementations of a number of these methods are available in the open-source C-language *GNU Scientific Library* (*GSL* – Galassi *et al.*, 2005). In order to use these methods, the user must define an objective function $f(\bullet)$ and a gradient function $df(\bullet)$, as well as a function *fdf*($\bullet$) which computes both of these quantities:

```
double f(const gsl_vector *,void *params);
void df(const gsl_vector *,void *params,
       gsl_vector *df);
void fdf(const gsl_vector *,void *params,double *f,
       gsl_vector *df);
```

Once these have been defined and a small number of preparatory steps have been taken via GSL function calls, gradient ascent can be accomplished via repeated calls to the iteration function:

```
int status=gsl_multimin_fdfminimizer_iterate(s);
```

where `s` is a structure which encapsulates the current state of the optimization, and the returned status value indicates whether convergence has been reached. Currently, the following optimization algorithms are available in the GSL:

```
FLETCHER_REEVES
POLAK_RIBIERE
BFGS
STEEPEST_DESCENT
SIMPLEX
```

These predefined algorithm identifiers can be simply passed into a GSL initialization routine, so that switching between optimizers is trivial; this allows the user to try all the available optimizers and select the one which is observed to work best for the task at hand.

Computation of the gradient may be performed either analytically by taking the derivative of the objective function, or by using the *differencing* method described in section 9.6.5.

10.16 Feature selection and dimensionality reduction

As mentioned previously, correlations between attributes can in some instances and for some learning algorithms limit the accuracy of the resulting classifier, and methods for removing these correlations can sometimes result in substantially improved classifiers. This can be accomplished by performing *feature selection*, in which the original set of attributes for a problem is replaced by a subset of the original attributes, or *dimensionality reduction*, in which the original set of attributes is replaced by a new set of attributes which are functions of the original attributes. Some of the more common feature selection methods include:

- *F*-ratio
- *PCA*
- *LDA*
- *Mutual information*
- *Information gain.*

We have already encountered several of these earlier in this chapter. Linear discriminant analysis (LDA) employs the *F*-ratio $MS_{between}/MS_{within}$ of between-group variance to within-group variance to find one or more linear functions that maximize this ratio – i.e., that maximize the *separability* of the transformed test cases. These linear combinations of the original variables can be used to compute a new set of attributes to be used for classification by any general-purpose classifier. Alternatively, one can compute the *F*-ratio explicitly for each attribute and then select some subset of the original attributes having the highest *F*-ratios. The interested reader is referred to any suitable text on *analysis of variance* (*ANOVA*) (e.g., Zar, 1996) for instructions on computing *F*-ratios.

The *mutual information* $M(X, Y)$ between random variables X and Y (section 2.10) can be used to assess the *redundancy* between X and Y, with pairs of highly redundant variables being collapsed into one or the other variable of the pair; alternatively, taking X to be an attribute and Y to denote the category of a training case, we can rank the variables X_i according to their mutual information (or *correlation* in the case of binary classification and continuous attributes) scores and discard all but the top-scoring N variables for some N. Another information-based approach to variable selection utilizes the notion of *information gain* (Quinlan, 1993) from entropy-based decision trees to select those variables which maximally reduce the classification entropy of the test set when used in an appropriate predicate within a decision tree node. A number of methods for the *stepwise* selection of attributes in different orders have also been developed (e.g., Aha and Bankert, 1996).

Another very popular method of dimensional reduction is *principal components analysis*, or *PCA*. The method consists simply of computing the **T** matrix described previously for LDA – the matrix of sums of squares and cross products across all groups – scaling the entries of the matrix by $1/(n-1)$ for n the number of training cases, and then extracting the eigenvectors of the resulting matrix. These eigenvectors are then used to compute linear combinations (i.e., the dot product $V \cdot X$ for eigenvector V and test case X) of the attributes of test cases in order to produce the new set of transformed attributes – one attribute per eigenvector. A number of enhancements to this basic technique exist, which the interested reader may wish to pursue (e.g., Dunteman, 1989).

10.17 Applications

In an attempt to illustrate very crudely the relative merits of some of the foregoing methods, we present the results from several studies in which a number of machine-learning algorithms were applied to various classification problems.

The first problem is that of classifying an arbitrary interval of DNA as being coding or noncoding. The genomes of *Toxoplasma gondii*, *Aspergillus fumigatus*, *Arabidopsis thaliana*, *Plasmodium falciparum* (malaria), *Mus musculus* (mouse), and *Homo sapiens* (human) were all subjected to a feature-extraction procedure similar to that used by GlimmerM (section 10.6.1), in which four features were extracted for each CDS in a randomly selected subset of the annotated genes for the organism:

(1) The WAM score for the signal at the 5′ end of the ORF
(2) The WAM score for the signal at the 3′ end of the ORF
(3) The histogram-based length probability of the ORF
(4) The *hexamer score* S_{hex} of the ORF, which was computed via Eq. (10.36):

$$S_{hex} = \sum_{\substack{hexamers \\ H}} \log \frac{P(H|coding)}{P(H)}. \tag{10.36}$$

A roughly equal number of noncoding ORFs were also selected at random and used to produce negative training and test cases.

The resulting accuracy scores are shown in Table 10.1; algorithm abbreviations are defined in the table legend. From this table it can be seen that the ET (entropy-based decision trees) and naive Bayes models scored most highly on this particular problem, when averaged over all the organisms included in the study:

$$\text{ET} = \text{BAYES} > \text{KNN} > \text{REG} > \text{NET} > \text{GP}.$$

However, the reader should avoid drawing any firm conclusions from this single study, as it has often been noted in the machine-learning field that different methods may excel on different problems, so we cannot in general predict a priori which model type or learning algorithm will prove most effective on a novel problem, or even a problem previously studied but with a different set of available attributes.

To reinforce this point, consider the following ordering of classifier performance averaged over a set of 38 standard classification problems from the *UCI*

Table 10.1 Results from an exon classification task

	ET	NET	REG	KNN	BAYES	GP
Toxoplasma	91	91	82	91	**92**	85
Aspergillus #1	89	89	68	88	**91**	76
Aspergillus #2	89	86	63	90	**91**	72
Arabidopsis	91	86	84	**93**	**93**	84
Human	91	80	83	**93**	92	81
Mouse	**94**	83	85	**94**	**94**	83
Plasmodium	89	76	81	79	85	78
Average	**91**	84	89	90	**91**	80

Key: ET, entropy-based decision trees; NET, feedfoward neural networks; REG, multivariate linear regression; KNN, K-nearest neighbors for any $K < 20$; BAYES, naive Bayes; GP, genetic programming.

machine-learning repository (Murphy and Aha, 1994):

$$\text{NET} > \text{KNN} > \text{GP} > \text{ET} > \text{LDA} > \text{BAYES}.$$

Here, neural nets performed the best and naive Bayes the worst. On a set of 1123 time-series prediction tasks, the following ordering (after averaging over all problems) was observed:

$$\text{GP} > \text{NET} > \text{KNN} > \text{ET}$$

which it can be seen is essentially the opposite ordering as obtained for the exon classification task. Finally, the following three orderings were obtained from a literature-mining task, using three different forms of feature selection:

$$\text{KNN} > \text{GP} > \text{NET} = \text{ET} = \text{BAYES} > \text{LDA}$$
$$\text{GP} > \text{KNN} > \text{NET} > \text{ET} > \text{BAYES} > \text{LDA}$$
$$\text{ET} > \text{KNN} > \text{BAYES} > \text{GP} > \text{NET} > \text{LDA}$$

where it can be seen that while LDA was consistently the worst and KNN tended to be among the best, the specific ordering among algorithms changed between all three problems.

EXERCISES

10.1 Explain why in section 10.4 we insist that the dependency structure must be acyclic. Your answer should be based on probability theory.

10.2 Propose a solution for computing conditional probabilities in a Bayesian network when a vertex has more than one parent. Justify your solution using probability theory.

10.3 What changes need to be made to the MST procedure of section 2.14 in order to compute maximal spanning trees rather than minimal spanning trees? Give at least two distinct solutions.

10.4 The MST procedure produces unrooted trees. Explain (a) how to obtain a rooted tree from an unrooted tree, and (b) how the choice of root will affect the computation for the tree likelihood for a Bayesian network.

10.5 Provide a formal proof that the derivation of backpropagation given in section 10.5 is correct.

10.6 Design a neural network topology for the prediction of the next value in a sequence of real numbers, given the previous three values.

10.7 Expanding on your answer to the previous question, explain how the network can be used for predicting the next nucleotide in a DNA sequence, given the previous three nucleotides.

10.8 Explain how the network in the solution to the previous exercise can be used to approximate a 4th-order Markov chain. What is the difference between the function computed by the Markov chain and that computed by the neural network? Which would be better for incorporation into a gene finder? Explain your answer.

10.9 Verify that Eq. (10.17) works for the hidden layers of a four-layer neural network.

10.10 In the formula for the hexamer score utilized by GlimmerM, explain why the logarithms of the likelihood ratios are summed rather than the ratios themselves.

10.11 Download the *Template Numerical Toolkit (TNT)* over the internet, and verify that the eigenvector decomposition procedures work correctly, using the formula $Av = \lambda v$, for matrix A, eigenvector v, and eigenvalue λ. Perform the verification on at least 100 different matrices A.

10.12 Provide a derivation for the linear regression formula given by Eq. (10.22).

10.13 Describe what a logistic curve would look like for a two-attribute problem.

10.14 Derive an expression for the second derivative of the likelihood function for logistic regression.

10.15 Derive the second derivative of the likelihood function for regularized logistic regression.

10.16 Draw an abstract syntax tree for the MST algorithm given in section 2.14.

10.17 Explain why the temperature is decreased in simulated annealing.

10.18 Explain informally why simulated annealing might be expected to perform as well as genetic programming in many cases. Explain why genetic programming might perform better than simulated annealing in some cases.

10.19 Download the GNU Scientific Library (GSL) and use the FLETCHER_REEVES optimizer to maximize the function $f(x) = 6x^3 - 2x^2 + \cos(x)/2 - \sin(x/4)$.

10.20 Explain why it makes sense to maximize the between-groups variance while also minimizing the within-group variance. Why should these be performed simultaneously?

10.21 Implement a KNN classifier in your favorite programming language.

10.22 Implement a naive Bayes classifier in your favorite programming language.

10.23 Download the C4.5 decision tree system and apply it to a sample problem from the UCI machine-learning repository. Report the accuracy on a held-out test set.

11

Tips and tricks

In this chapter we describe a number of heuristics which we have found useful during the implementation, training, and/or deployment of practical gene finding systems for real genome annotation tasks.

11.1 Boosting

A well-known trick from the field of machine learning is *boosting*. This technique has been applied to the training of gene finders in the following way, with modest accuracy improvements being observed in a number of cases.

Suppose that while training a signal sensor for a GHMM-based gene finder we notice that a number of positive examples are assigned relatively poor scores by the newly trained sensor. One approach to boosting involves duplicating these examples in the training set and then re-training the sensor from scratch. The duplicated, low-scoring examples will now have a greater impact on the parameter estimation process due to their being present multiple times in the training set, so that the re-trained sensor is more likely to assign a higher score to those examples. Assuming that the low-scoring examples are not mislabeled training features, improvements to the accuracy of the resulting gene finder might be expected when the gene finder is later deployed on sequences having genes with similar characteristics to the duplicated examples. Care must be taken to avoid overtraining, however. To the extent that a gene finder with optimal genome-wide accuracy is desired, it is important that boosting not be allowed to bias the gene finder in way that is significantly inconsistent with the actual frequency of these difficult signals in the genome.

A more laborious (but possibly more rewarding) form of boosting for signal sensors would involve applying the gene finder to the full genomic contigs containing the training genes and identifying those training signals that are missing from the resulting predictions. As before, these example signals can be duplicated in the training set. Upon re-training the signal sensors on the modified training

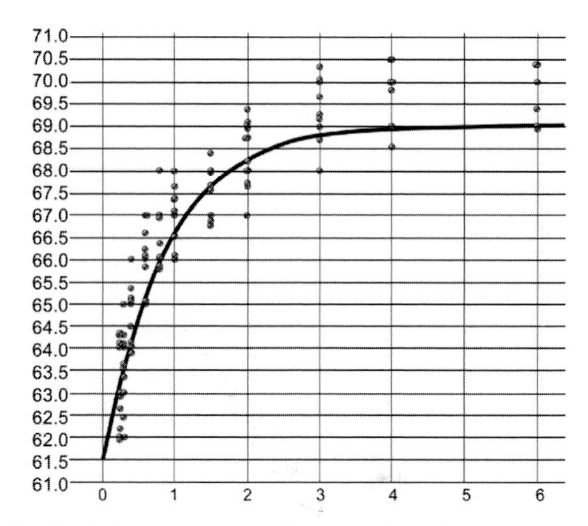

Figure 11.1 Predictive accuracy as a function of training set size. Numbers of human genes in a training set ($\times$1000) are shown on the *x*-axis; exon-level accuracy (*F*-score) of a GHMM-based gene finder on a held-out test set is shown on the *y*-axis. (Adapted from Allen et al., 2006.)

set, the accuracy of the re-trained gene finder may be observed to increase. This latter form of signal boosting in gene finders is reminiscent of the approach to discriminative training described in section 8.6.2, in which repeated training and evaluation runs are performed in order to find an optimal set of model parameters. In this case, such an iterative procedure would instead be used to find the optimal *weightings* for the training examples, where the duplication of an example effectively increments its weight. Nonintegral weights might also be used, somewhat like the isochore weightings used in *AUGUSTUS* (Stanke and Waack, 2003; see section 8.4.1). The use of boosting for content sensors is another possibility that remains to be investigated.

More sophisticated and rigorous boosting methods are detailed in the machine-learning literature (e.g., *AdaBoost* – Freund and Schapire, 1995); a rigorous characterization of their effect on gene prediction accuracy has yet to be performed (see Exercise 11.1).

11.2 Bootstrapping

It is an unfortunate fact that for many novel genome sequencing efforts, too few training data are available for robust training of a gene finder (e.g., Eisen *et al.*, 2006). Figure 11.1 illustrates the effect of sample size on gene-finding accuracy for a particular human gene finder.

In this example, a GHMM-based gene finder was trained many times on N human RefSeq genes, where $N/1000$ is shown on the *x*-axis. The resulting exon-level accuracy on a hold-out set is shown on the *y*-axis. Although the curve rapidly becomes flat above ~3000 genes, the slope is quite large below ~2000 genes. Assuming that the human genome contains at least 20 000 genes (International Human Genome Sequencing Consortium, 2004), and assuming that these results generalize to other organisms, this suggests that training a gene finder on fewer than 10% of an organism's full gene set could be expected to result in suboptimal predictive accuracy.

In cases where insufficient data are available for robust training of a gene finder, a simple procedure may be applied to effectively increase the amount of training data available. Suppose that we are given a relatively modest set T of training genes and a set C of contigs, both from the target genome G. From T we can train a preliminary gene finder M_1, which due to sampling error may be expected to perform relatively poorly on the target genome. If certain features are missing entirely from the training data (e.g., if our gene finder includes a promoter model, but no example promoters are present in the training set T) we can train the corresponding submodels from examples drawn from other genomes (preferably those of closely related species) – a practice termed *parameter mismatching* (Korf, 2004). We can then apply our preliminary gene finder M_1 to the contigs C to produce a set P of predictions, which can then be used to augment the training set T, producing a new training set $T_2 = T \cup P$. When incorporating the predictions P into T_2, we may consider weighting the predictions in various ways, to give the higher-confidence examples from T somewhat greater influence during the next round of training. The next round of training utilizes T_2 as the training set, to produce a new gene finder M_2, which it may be hoped will have suffered less from sampling error during its training than did M_1, since the augmented training set T_2 will generally be larger than T. Although one might consider repeating this process a number of times, the potential effects of doing so on gene finding accuracy have not been fully characterized as of yet, while the use of a single round of bootstrapping has been applied in practice and shows some promise of having a beneficial effect (e.g., Korf, 2004).

An alternative method for obtaining additional training data is the use of long ORFs, as mentioned briefly in sections 4.1 and 7.5. We show in Figure 11.2 that for ORFs of length $\geq$300 bp, the probability of such an ORF occurring in random DNA having equal nucleotide frequencies rapidly approaches zero with increasing ORF length, so that ORFs at least this long in real genomic DNA of similar nucleotide composition are likely to be real protein-coding genes – i.e., to owe their lack of in-frame stop codons to the action of natural selection. For {G,C} densities less than 50% this curve drops even more precipitously (see Exercise 11.3). By taking all ORFs in a genome longer than some reasonably chosen minimal length we can thus obtain a fairly high-confidence training set for the coding model of a gene finder.

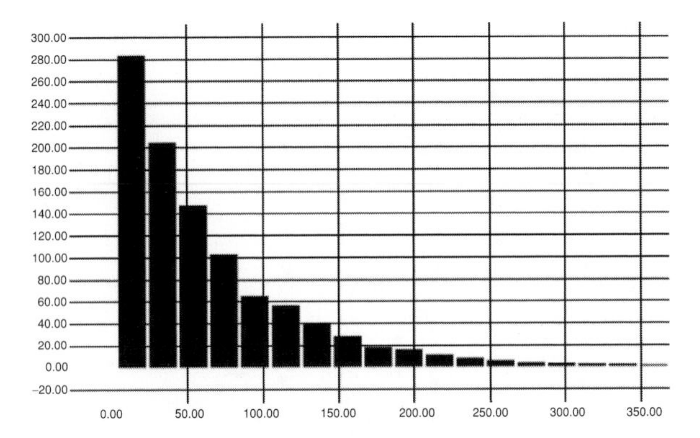

Figure 11.2 Length–frequency histogram for ORFs in random DNA. Ten megabases of random sequence were generated, using equal nucleotide frequencies. All ORFs were identified and their lengths (x-axis) collected into the given histogram, which shows relative frequency ($\times 10^{-3}$) on the y-axis.

11.3 Modeling additional gene features

In the HMM-based and GHMM-based gene-finding models that we have considered, the set of gene feature types which have been explicitly represented by corresponding states in the model has been relatively limited, including only states for exons, introns, intergenic regions, and the signals such as start and stop codons and splice sites which delimit those variable-length features. A number of other features are present in eukaryotic genes, however, which one might consider modeling in a gene finder. These include *promoter* elements such as TATA-boxes and CCAAT-boxes (Wingender *et al.*, 1997), the *branch points* (Hall *et al.*, 1988) which occur upstream of acceptor sites, *CpG islands* (Larsen *et al.*, 1992), *signal peptides* (Gierasch, 1989), polyadenylation signals, 5′ and 3′ UTRs, and various other elements having known or unknown biological significance, but which (with the exception of promoters, which we consider in more detail in section 12.3) have as of yet received relatively little attention from the gene-finding community.

Here we describe a set of experiments in which a number of these feature types were incorporated into a GHMM-based gene finder and the resulting effects on predictive accuracy were measured (Allen *et al.*, 2006). Our apparatus for these experiments consisted of the GHMM-based *ab initio* gene finder *GeneZilla*, previously known as *TIGRscan* (Majoros *et al.*, 2004). The augmented state-transition diagram for the forward strand of the model is shown in Figure 11.3.

The following states (shown in gray in the figure) were investigated:

- CpG islands (CpG)
- CAP sites (CAP)
- branch points (b)

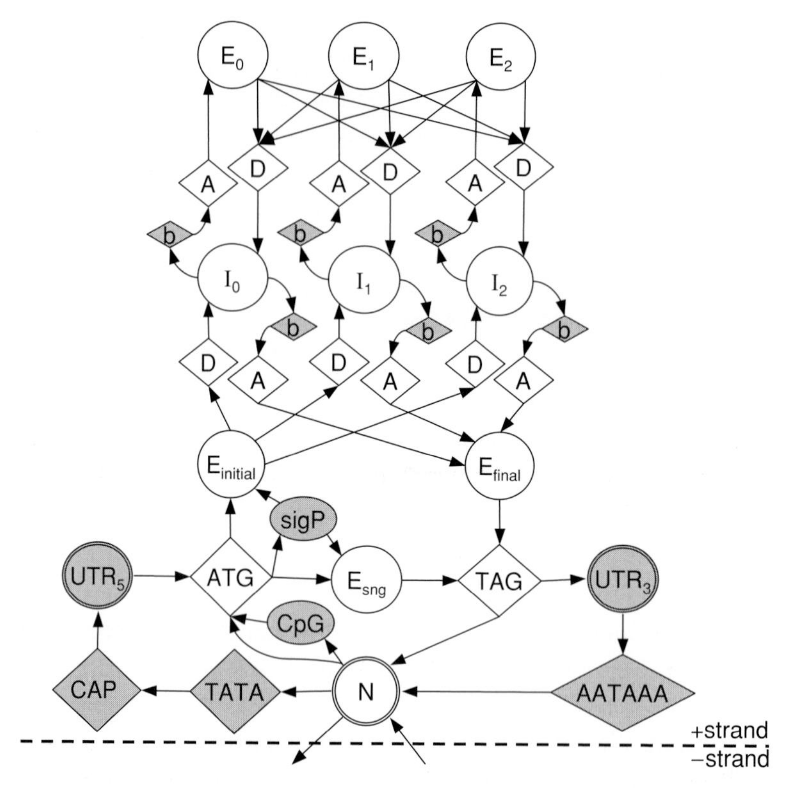

Figure 11.3 A version of the GHMM-based gene finder GeneZilla augmented with additional feature states (gray). E, exon; ATG, start codon; TAG, stop codon (including TGA and TAA forms); CAP, 5′ cap site; TATA, TATA-box; AATAAA, polyadenylation signal; D, donor site; A, acceptor site; b, branch point; sigP, signal peptide; CpG, CpG island; UTR, untranslated region; I, intron; N, intergenic.

- signal peptides (sigP)
- phase-specific introns (I_n)
- promoters (TATA → CAP → UTR_5)
- polyadenylation signals (UTR_3→AATAAA)

which we will describe individually below. In addition, we considered the explicit modeling of *isochore* boundaries (Bernardi *et al.*, 1985) and the use of UTR-trained parameters for the UTR models (versus the use of intergenic parameters for those states).

The base gene finder (not including the above added states) was trained on 8259 human RefSeq genes (Pruitt *et al.*, 2005) rendered nonredundant via BLASTN, so that no two genes were more than 80% identical over 80% of the gene length at the nucleotide level. Genes known to have multiple *isoforms* (see section 12.1) were also removed prior to training, since this gene finder currently predicts only one form for each putative gene.

CpG islands – regions in which the *methylation* of cytosine (C) residues into thymine (T) residues (when the cytosine is immediately followed by a guanine) is suppressed – have been observed to occur preferentially in the upstream regions of human genes (e.g., Venter *et al.*, 2001), especially *housekeeping genes* (i.e., those which are expressed in many different cell types). We applied the method of Larsen *et al.* (1992) to identify putative CpG islands during a preprocessing phase and then implemented the CpG island state as a simple matching state on the 5′ ends of these precomputed features – i.e., a fixed-length signal state having length 0 bp.

Polyadenylation signals were described in section 1.1; these are the motifs which are believed to signal the cleavage of the 3′ end of the mRNA prior to *polyadenylation* (i.e., the appending of a polyadenine tail). The polyadenylation signal state ("polyA") was implemented by a 16 bp 2nd-order WAM trained on 10 046 examples labeled as "polyA_signal" features in human *GenBank* (Benson *et al.*, 2005) entries. Two consensus sequences were allowed for this signal: AATAAA and ATTAAA. Only one isochore was modeled for this feature because the range of {G, C} densities for the example sequences were mostly <43%. Because the WAM was trained via simple maximum likelihood and is therefore not guaranteed to provide optimal discrimination power for the gene finder as a whole, we incorporated two additional parameters related to this state and explored a broad range of values for these parameters in an attempt to discover a maximally discriminative parameterization. The additional parameters were $R_{3'}$, a multiplicative factor which adjusts the existing $L_{3'}$ (mean 3′ UTR length) parameter; and O_{poly} ("polyA optimism"), another multiplicative factor which is applied to the (pre-logarithm) WAM score. Larger window sizes were investigated for the WAM but were found to provide no advantage over the 16 bp window.

The *promoter* state ("TATA") was implemented using a simple model very similar to the one used in *GENSCAN* (Burge and Karlin, 1997), consisting of a TATA-box WMM followed by a CAP site WMM with a variable 14–20 bp "spacer" region between. The TATA-box WMM was trained on 548 examples extracted from human "TATA_signal" elements in GenBank. WAMs of up to 5th order were also investigated for the TATA-box state, but no advantage was seen, so a WMM was used instead. The spacer region was modeled using simple 0th-order intergenic nucleotide frequencies. The CAP site WMM was obtained from the TRANSFAC 3.2 library of *transcription factor binding sites* (Wingender *et al.*, 1997; see also section 12.3). As with the polyA state, two additional parameters related to the promoter state were incorporated and tuned so as to maximize accuracy: $R_{5'}$, a multiplicative factor for the mean 5′ UTR length, and O_{prom} ("promoter optimism") which is multiplied by the promoter model score. Note that the tuning of these extra parameters was performed on the first of two test sets; to avoid undesirable post-hoc effects as a result of "peeking" at the test set, our final results were measured on a second, unseen, test set (described

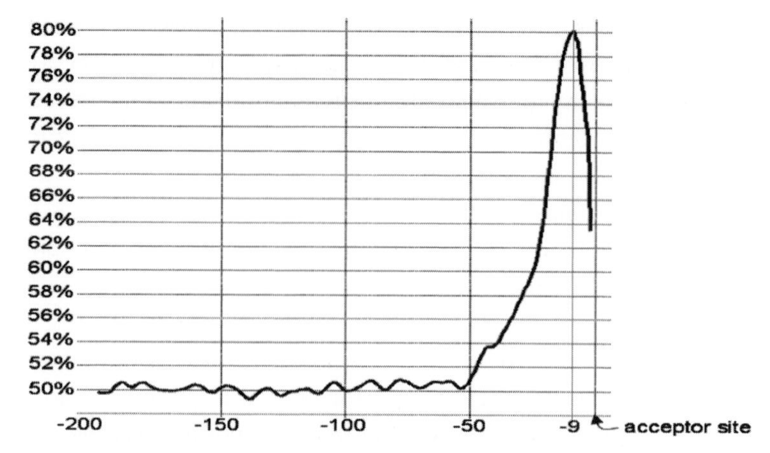

Figure 11.4 Pyrimidine density in polypyrimidine tracts (PPTs) upstream from acceptor sites. Frequencies are averaged over 2983 human acceptor sites.

below). It should also be noted that this promoter model is an exceptionally crude one, compared to the models that have been utilized in more rigorous promoter prediction experiments (see section 12.3). Nevertheless, its close similarity to the promoter model employed by the successful gene finder GENSCAN suggested that it might be of some value.

Signal peptides are specific amino acid sequences occurring near the C-terminus of a newly translated protein; these signals are believed to direct the protein to specific locations in the cell, where they are then cleaved from the peptide so as not to interfere with the subsequent functioning of the protein. Putative signal peptide sequences S were evaluated by the signal peptide model M_{sp} via:

$$P(S|M_{sp}) = \prod_{\substack{codons \\ c \, in \, S}} P(amino(c)|M_{sp})P(c|amino(c)) \tag{11.1}$$

where $amino(c)$ is the amino acid encoded by codon c. $P(amino(c)|M_{sp})$ was estimated by observing frequencies of amino acids in the set of training signal peptides; $P(c|amino(c))$ was estimated by observing the codon usage statistics of the training genes. Training data for this state consisted of 1048 "sig_peptide" features extracted from human GenBank entries.

The significance of the *branch point* for splicing was mentioned in section 1.1. Although precise identification of the branch point is difficult, it is well known that human acceptor sites are generally preceded by a *polypyrimidine tract* (*PPT*) of significantly elevated pyrimidine density (Figure 11.4). The branch point model used for our experiments consisted of a 19 bp WWAM (section 7.3.3) trained on intervals located at a fixed distance upstream from acceptor sites in the training set, and was thus intended to capture the pyrimidine bias of the PPT.

The 5′ and 3′ UTR states were trained on 18 432 and 19 977 untranslated regions, respectively, extracted from GenBank. These states were also retrained from scratch using pooled intergenic sequences, and the differences in accuracy resulting from this change were recorded.

A final experiment involved the use of precomputed isochore boundaries using an HMM structured like that shown in Figure 8.13. All signal and content sensors were trained separately on four isochores (G,C density: I = 0–43%, II = 43–51%, III = 51–57%, IV = 57–100%), treating the models trained from a particular isochore as a "profile" that could be dynamically loaded into the GHMM's states at any time during a gene-finding run. As the GHMM decoder reached a predicted isochore boundary, the appropriate isochore profile was loaded into the GHMM. It was hoped that this more fine-grained modeling of isochores (as opposed to applying a single isochore profile to the entire input sequence, as is typically done) would enhance predictive accuracy.

The test sets for the experiments consisted of 458 and 481 individual human genes selected randomly from the set of all nonredundant RefSeq genes available at the beginning of the study, with a margin of 1000 bp retained before and after the CDS portion of each gene when segmenting the sequence for input to the gene finder. This was done because we wished to test the ability of the gene finder to accurately predict the structure of genes, rather than to assess the false-positive rate for entire genes. However, for experiments targeting the utility of the polyA, promoter, and UTR states, a margin of 50 kb was instead used, since most UTRs in the training set were seen to be shorter than 50 kb in length. Under these latter conditions the test sets each comprised 62 Mb of sequence, or roughly 2% of the genome. Likewise, for the isochore-switching experiments we selected margin sizes so that each test chunk was $\geq$300 kb in length, as per the commonly accepted definition of isochores (e.g., Bernardi *et al.*, 1985).

The remaining parameters of the GHMM were initially trained via maximum likelihood estimation from the 8259 RefSeq training genes, and then a handful of the parameters (including transition probabilities, WMM and WAM sizes, WAM and Markov chain emission orders, and mean intron and intergenic lengths) were tuned by hand so as to maximize accuracy on the first of the two test sets. We report our results only on the second, unseen test set.

As can be seen in Table 11.1, none of the enhancements to the gene finder significantly improved its accuracy on the test set, and some of the enhancements (particularly the use of UTR training features) actually decreased the accuracy, even though some attempt was made to tune the relevant parameters to improve discriminatory power on the first test set (i.e., a primitive form of *discriminative training*). Additional experiments investigating various combinations of these feature states produced similar results (not shown). Informal discussions with the

Table 11.1 Results of incorporating additional feature states into a GHMM-based gene finder

	Nucleotide *F*-score	Whole exon *F*-score	Whole gene sensitivity
PolyA	−1%	+1%	+2%
TATA+CAP	+0%	+0%	+1%
TATA (no CAP)	+0%	+0%	+0%
CAP (no TATA)	+0%	+0%	+0%
Branch point	+1%	+1%	+1%
Signal peptide	−1%	−1%	+0%
Intron phase	+1%	+1%	+0%
CpG islands	+0%	+0%	−1%
Isochore switching	+0%	+1%	+1%
UTR (trained on UTR)	−15%	−30%	−7%

authors of several other gene finders lend additional support for these results; in addition, published results by Lim and Burge (2001) showed only a very modest (~1%) improvement due to modeling branch points. Thus, it appears that simply adding states into a GHMM-based gene finder to model features such as those described above may not provide any appreciable advantage in terms of predictive accuracy. Additional work is needed, however, to corroborate these results and to investigate whether some other computational framework or modeling technique may allow the explicit modeling of these additional features to improve predictive accuracy.

11.4 Masking repeats

A common practice before applying a gene finder to a contig is first to *mask* any repeat sequences that are found to be present in the sequence. Repeats can take a number of forms, with the most common being transposable repeat elements and low-complexity repeats. The former elements are known to be moved and/or copied into different places in the genome over evolutionary time, and in some cases in much smaller timescales such as between generations, creating unique patterns within the genomes of individual members of a species (such as those making up the *VNTRs – variable number of tandem repeats* – which are used in *DNA fingerprinting*; Jeffreys *et al.*, 1985). Several programs are available for identifying potential repeats in a sequence, the most popular being *RepeatMasker* (Smit and Green, 1996).

A repeat-masking program scans the input sequence for evidence of repeats, typically relying on a database of known transposons for a given species or family. Repeats can be *hard-masked*, in which case they are replaced by some fixed symbol such as "N", or they can be *soft-masked*, in which case the typographical case is changed – i.e., uppercase to lowercase, or lowercase to uppercase. The gene finder can then be applied to the masked sequence. In the case of hard-masking, it is typically the case that the gene finder's signal and content sensors will all reject "N" characters, except for the noncoding submodels, so that repeat regions can only be predicted to occur in introns, intergenic regions, or UTRs. Unfortunately, some types of putative repeats – especially low-complexity repeats – can occur in coding regions of genes, and even complex repeats such as transposons are known to occur in CDSs with low frequency (Sorek *et al.*, 2002), so that aggressive repeat masking can in some cases actually degrade accuracy rather than improve it. One common practice is to soft-mask low-complexity repeats and to hard-mask all others; many gene finders automatically convert input sequences to uppercase or lowercase letters, so that the identification of low-complexity repeats does not aid – nor hinder – the gene finder in any way. Additional work on the effects of repeat masking prior to gene finding is needed.

EXERCISES

11.1 Implement the AdaBoost algorithm for training a start codon sensor. Compare the performance of the boosted model to the nonboosted model.

11.2 Download an *ab initio* gene finder via www.geneprediction.org, and also a training and test set of genes for some organism. Apply n iterations of bootstrapping, for $0 \leq n \leq 5$. Graph the accuracy of the resulting gene finder as a function of n. What conclusions can you draw from this curve? Optionally, replicate this exercise for another *ab initio* gene finder. Did the second experiment cause you to change your conclusions in any way?

11.3 Construct a curve similar to Figure 11.2 for {G,C} densities of:
   (a) 35%
   (b) 40%
   (c) 60%
   (d) 65%.

11.4 Explain why {G,C} bias should be correlated with the mean length of random ORFs (e.g., ORFs occurring in random DNA).

11.5 Implement an acceptor-site signal sensor for *Homo sapiens*. Measure the accuracy (F-score) on the entire genome. Then augment your sensor by modeling the PPT upstream from the acceptor site. Test your model on the entire genome. Report the magnitude of the improvement due to the modeling of the PPT.

11.6 Implement the method described in Larsen *et al.* (1992) for finding CpG islands in the human genome. Apply your algorithm to the entire genome, compute the correlation coefficient between the annotated gene starts according to RefSeq (Pruitt *et al.*, 2005) and your predicted CpG islands (counting only overlaps between the gene start and any CpG island), and report the correlation.

11.7 Download an *ab initio* gene finder via www.geneprediction.org, and obtain human chromosome 22 from the NCBI website (www.ncbi.nlm.nih.gov/). Apply the gene finder to this chromosome both before and after application of RepeatMasker (Smit and Green, 1996). Score the accuracy of the gene finder relative to the RefSeq annotations (Pruitt *et al.*, 2005). Report your results.

12

Advanced topics

We conclude with a brief exposition of current research topics in computational gene prediction. As these all are areas of active research in which the optimal computational methods have yet to be identified, our treatment will necessarily be provisional. It may be hoped that future revisions of this text will be able to treat these topics in a more decisive manner due to the fruits of current research.

12.1 Alternative splicing and transcription

It may reasonably be argued that most current attempts at computational gene finding – at least those based on Markov models and Viterbi decoding – are fundamentally flawed. The paradigm of gene finding, as we have formulated it, consists of computing the single optimal parse ϕ of a genomic contig, and then predicting the set of nonoverlapping gene structures present in this "optimal" parse. As briefly mentioned in section 1.4, however, it is known that most human genes are subject to the phenomenon of *alternative splicing*, whereby a single locus in the genome may give rise to several different mature mRNAs, and therefore (in many cases) to distinct proteins with potentially divergent biological functions. Some estimates place the incidence of alternative splicing among human loci as high as 80% (e.g., Matlin *et al.*, 2005). Thus, the problem of alternative splicing is one of vital importance to a biological understanding of the human genome.

Figure 12.1 illustrates some of the possibilities for the alternative splicing of a locus. In the top tier is shown a diagram of potential exonic structures at the level of genomic DNA. Potential donor and acceptor splice signals are shown, as well as potential polyadenylation signals (pA_n) within putative 3′ UTRs. Below this are shown several possible splicing patterns, or *isoforms*, for this hypothetical gene.

As suggested by the figure, the phenomenon of alternative splicing may take several forms. The most obvious is that of alternative donor/acceptor sites, in which an exon is correspondingly shortened or lengthened (with respect to the "reference form" of the gene) through the use of an alternative donor or acceptor site during

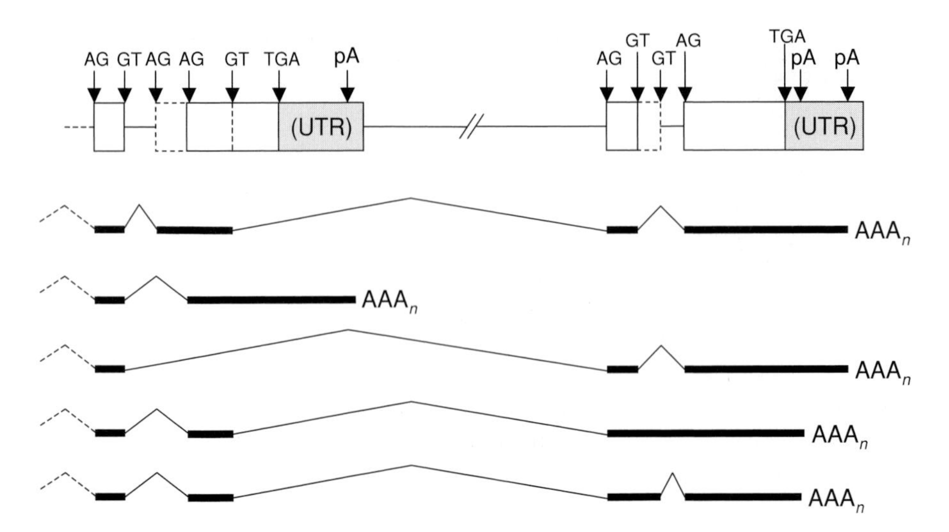

Figure 12.1 Several possibilities for the alternative splicing of a multi-exon gene. In the top tier, boxes represent exons and thin line segments represent introns or intergenic regions. In the lower tiers, thick horizontal lines represent exons and inverted-V structures represent introns. Polyadenylate tails of mature mRNAs are denoted AAA$_n$.

splicing. A more drastic form involves the wholesale omission of an exon from the mature mRNA – described in the literature as *exon skipping*, or the use of *cassette exons*. Intron skipping is also possible, via the differential recognition of a donor site, though the phenomenon is typically referred to as *intron retention*. Finally, the potential for differential recognition of polyadenylation signals and the differential cleavage of the mRNA which results is referred to as *alternative polyadenylation*. *Alternative promoter recognition* – possibly giving rise to alternative transcription start sites and potentially also to alternative translation initiation sites – is also plausible biologically, and has recently begun to receive more attention (e.g., Landry *et al.*, 2003; Carninci *et al.*, 2006).

Figure 12.2 demonstrates the prevalence of just one form of alternative transcription in the human genome – alternative polyadenylation, which appears to affect over half of all human loci. While alternative polyadenylation does not directly affect the CDS of a coding gene, through its effect on the length of the 3′ UTR it can influence the regulatory events which control the rate of transcription or translation of the mRNA, and is therefore equally important to our understanding of genes and their transcriptional dynamics.

Of the methods for alternative splicing prediction that have been investigated computationally to date, most are based on the spliced alignment of inconsistent ESTs to genomic sequence, with or without homology evidence, using largely ad-hoc approaches. The utility of observing inconsistent patterns in spliced ESTs should

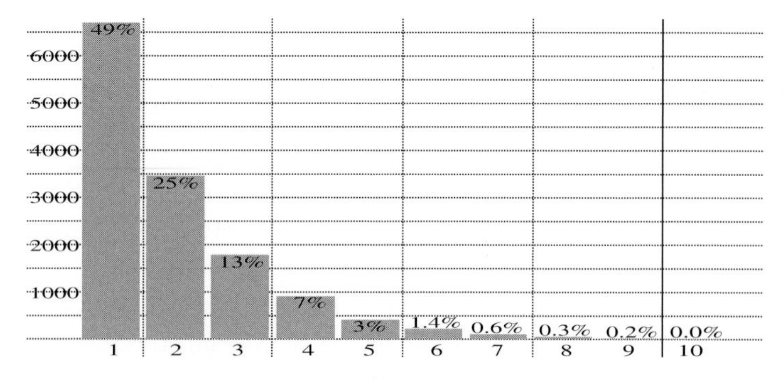

Figure 12.2 Histogram showing numbers of genes (y-axis) having the given number of polyadenylation sites (x-axis) in the human genome, as annotated in the database PolyA_DB (Zhang *et al.*, 2005). Represented are 27 180 sites in 13 664 genes. Mean sites/gene was 1.99 ± 1.33 (range: 1–13).

be abundantly clear: evidence for mutually exclusive splicing patterns within a single locus constitutes prima facie evidence of alternative splicing of transcripts. Two prominent systems employing an EST-based approach are *Eugene* (Foissac and Schiex, 2005) and *PASA* (Haas *et al.*, 2003). In the former program a graph-theoretic representation reminiscent of the *ORF graph* formalism promulgated in this volume is utilized in conjunction with EST evidence to score alternatively spliced variants of a predicted gene. Yet another system utilizing a similar approach is *AIR* (Florea *et al.*, 2005). Heber *et al.* (2002) refer to such a graph-theoretic representation as a *splice graph*, and examine some of its useful properties for predicting and representing alternative splicing patterns; an alternative formulation is described by Lee *et al.* (2002). It should be noted that the use of splice graphs alone can, unfortunately, lead to overprediction of isoforms, since the alternative splicing events for different exons and different splice sites need not be independent (see, e.g., Wojtowicz *et al.*, 2004).

A non-EST-based approach is exemplified by the program *UNCOVER* (Ohler *et al.*, 2005), which utilizes a pairHMM to model skipped exons. The program is applied to known pairs of orthologous introns, in hopes of finding conserved, unannotated, alternative exons residing within the introns. The program can also predict retained introns. Because the implementation computes the full dynamic programming matrix for the PHMM (i.e., without employing any banding or matrix "pruning" methods such as those described in Chapter 9), the speed of the program is somewhat affected, though the memory requirements are kept in check through use of a modified Hirschberg algorithm (section 9.4.1). A number of other classifier-based approaches have been proposed for discriminating between exons that are constitutive and those that are sometimes skipped (e.g., Dror *et al.*, 2004; Yeo *et al.*, 2005). Some of these latter approaches make heavy use of comparative information (see Chapter 9), though a few noncomparative studies have reported some success as well (e.g., Rätsch *et al.*, 2005).

In the purely noncomparative setting, two relatively obvious approaches to the prediction of alternative splicing spring to mind. The first involves the identification of some number n of suboptimal paths $\Phi_n = \{\phi_i | 0 \leq i < n\}$, for $P(\phi_i|S) \leq P(\phi_{i-1}|S)$ over $1 \leq i < n$, where $\phi_0 = \phi^*$ constitutes the optimal prediction from the gene finder. The identification of suboptimal parses of a sequence under an HMM or GHMM formulation has been considered previously (sections 6.2.1 and 8.5.3). While all parses in Φ_n might not actually occur in the living cell, the identification of some reasonable number n of such suboptimal parses in an automatic fashion may be expected to greatly assist the human annotator charged with identifying likely alternatively spliced forms of genes in the genome of interest. The incorporation of EST evidence via comparative methods as described in Chapter 9 may enable purely computational approaches to alternative splicing prediction to achieve elevated levels of accuracy, particular specificity, through the principled re-weighting of putative gene structures based on available (and possibly conflicting) spliced-EST evidence. A universally accepted means of accomplishing this in a principled manner has yet to emerge.

A second fairly obvious approach to the prediction of alternative splicing patterns in protein-coding genes involves the reformulation of the gene-prediction problem as one of *classification* rather than parsing. Traditionally, the notion of *parsing* in computer science has been equated with the identification of the "optimal" (under some scoring function) partitioning of an entire input sequence – whether utterances from a human speaker of some natural language or residues inferred by a sequencing machine – into a series of nonoverlapping features satisfying a predetermined grammar in some hierarchy of syntactic formalisms (e.g., the so-called *Chomsky hierarchy* mentioned in section 2.15). In fully automated annotation pipelines, the formulation of the problem as one of identifying suboptimal parses as described above seems a reasonable one, so long as some principled means of establishing the desired number of predicted spliced forms per gene may be adduced.

When human annotators are available, however, there may be greater utility in simply providing a set of high-confidence exon predictions, where the assembly of nonoverlapping exons into phase-consistent gene parses is left entirely up to the human annotator. Unfortunately, the tedious process of manual phase-tracking is further complicated in the case of alternative splicing due to the fact that a significant proportion of alternative splicing events in the coding portions of human genes are known to affect the reading frame (Matlin *et al.*, 2005). Thus, the widespread adoption of a more classification-oriented approach to exon/gene prediction will require greater support from gene-structure-aware graphical user interfaces (such as the *UCSC browser* mentioned in section 1.4) for managing the complexity of large numbers of high-scoring exons in different phases and their many possible splicing patterns.

To the extent that the ultimate consumer of gene prediction software is the human annotator, the very definition of the prediction problem should ideally derive from the needs and constraints of the manual annotation process, with the current capabilities of the genome browser tools having a significant impact on the definition of the annotator's requirements. Adopting such a view of the gene-finding problem may complicate the automatic evaluation of gene-finding accuracy and possibly limit our ability to organize large-scale "bake-off" competitions (e.g., *GASP* – Reese *et al.*, 2000; *EGASP* – Guigó *et al.*, 2006), but would hopefully benefit the ultimate goal of accurate (manual) genome annotation and thereby better promote the goals of genome science.

The problem of independently predicting high-confidence exons differs fundamentally from that of parsing. One tool that would likely prove useful for this task is the general *posterior decoding* procedure, which we have presented in Chapters 6 and 8 in the context of HMM- and GHMM-based models of gene structure. To the extent that posterior exon scores provide a reliable confidence rating for putative exons, a simple list of exon predictions for a region, sorted by posterior probability, may prove most useful to the human annotator. More fundamentally, the revised problem may be seen as one of *classification*, in which an arbitrary interval of DNA may be submitted to some appropriate machine-learning algorithm (such as those presented in Chapter 10) to be classified as being either *exonic* (class 1) or *nonexonic* (class 0). Those intervals deemed by the classifier to be of class 1 would then be made available to the human annotator for possible inclusion in an annotation for the locus under consideration.

Conversely, in the case of genomes for which a manual annotation phase is not feasible, the ultimate consumers of computational gene predictions are the combiner-type programs (section 9.2) and auto-annotation pipelines (section 9.7) which are typically employed instead. In these cases, the utility of emitting suboptimal parses and/or predicting exons independently will depend on the capabilities and sophistication of the downstream software being targeted.

12.2 Prediction of noncoding genes

According to the *central dogma of molecular biology* outlined in section 1.1, genes give rise to messenger RNAs (*mRNAs*), which in turn give rise to the proteins that by and large mediate all the major functions of the cell. Various noncoding RNAs (*ncRNAs*) are known, however, which play essential roles in cell biology, including the *transfer RNAs* (*tRNAs*) and *ribosomal RNAs* (*rRNAs*) utilized during translation of mature transcripts. In recent years the importance of noncoding RNAs has been elevated somewhat through the characterization of a class of RNAs known as *microRNAs* (*miRNAs*), which upon transcription by the cell are cleaved into small RNA molecules capable of exerting considerable influence on the expression patterns of other genes (typically through binding to the 3′ UTRs of the regulated genes – see Eddy, 2002; Bartel, 2004).

It is an unfortunate fact that noncoding RNAs, because they are never translated and thus never subjected to the action of natural selection at the protein level, do not exhibit the characteristic codon statistics of protein-coding genes. This renders these genes especially difficult to identify computationally, at least using the content- and signal-based methods detailed in the preceding chapters. At present the most useful forms of evidence for the computational identification of this special class of genes include their *secondary structure* (i.e., two-dimensional folding patterns) and their conservation patterns across related species (i.e., suggesting methods similar to those described in Chapter 9). It may also be noted that the use of secondary structure information has been put forward as a potentially useful form of evidence for protein-coding gene prediction, since the conformation patterns of both the DNA and the mRNA can affect the ability of molecules such as the polymerase and the eukaryotic spliceosome to interface with the substrate on which they are to act (Buratti and Baralle, 2004).

There are two popular methods for predicting RNA secondary structures. The first is based on thermodynamics, by finding the secondary structure that minimizes the *Gibbs free energy* of the RNA molecule (i.e., minimizing the potential for the molecule to spontaneously reconfigure itself into a different structure). A number of well-known programs (e.g., *mfold* – Zuker *et al.*, 1999; *RNAfold* – Hofacker *et al.*, 1994) are available for efficiently performing this optimization (in $O(n^3)$ time and $O(n^2)$ space, for an RNA sequence of length n) via dynamic programming, and many of the ncRNA prediction programs in current use treat these programs as "black box" subroutines, effectively offloading the problem of secondary structure prediction onto these well-established programs. These latter folding programs return one or more predicted structures and their associated free-energy scores. In the case of miRNA gene finding, one typically examines the top N structures reported by the folding program and selects the structure that most closely resembles the prototypical structure of a miRNA, which consists of a hairpin-like structure comprising a helical "stem" with a loop at one end (shown in idealized form in Figure 12.3). Similar approaches for other families of RNAs have been applied (e.g., Lowe and Eddy, 1999).

The thermodynamic approach to RNA secondary structure prediction is based on the *second law of thermodynamics*, which states that the entropy of a nonequilibrium closed system tends overwhelmingly to increase; the equilibrium state of the system is thus taken to be the state with maximum entropy (which in statistical mechanics is denoted S rather than H, though the two entropies are effectively analogous). In the case of RNA secondary structures, the Gibbs free energy of any

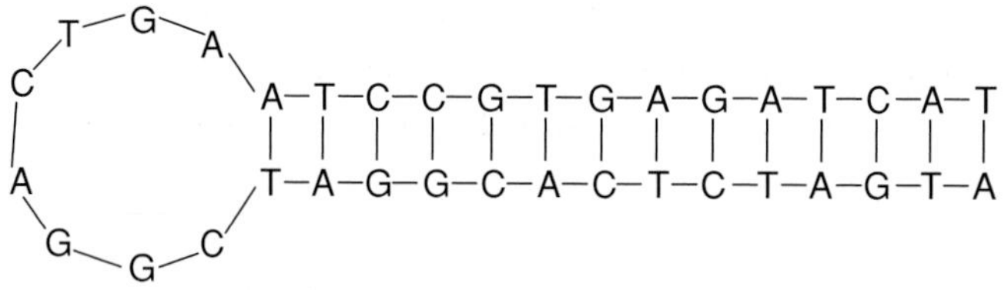

Figure 12.3 An idealized miRNA secondary structure, consisting of a loop (at left) attached to a helical stem (shown flattened into the plane, at right). Bulges and mismatches seen in typical miRNA structures are omitted for the sake of simplicity.

particular secondary structure is inversely proportional to the entropy, under constant environmental conditions:

$$G = U + PV - TS$$

for *internal energy* U, pressure P, volume V, temperature T, and entropy S, so that the second law of thermodynamics may be interpreted in this scenario to say that the equilibrium state of the molecule is the structure having minimal free energy. Finding the equilibrium structure thus equates to minimizing the free-energy over all possible secondary structures.

The free-energy computation for RNA secondary structures which is commonly employed in molecular modeling is not exact, however, being based on imperfect empirical measurements taken from known substructures (such as *loops* and *stems* – Figure 12.3), with these imperfect energy estimates typically being summed over the corresponding structural features of a novel structure to arrive at what is commonly acknowledged to be a crude estimate of the true free energy for the structure (Ding, 2006). Furthermore, there is evidence that for many RNA molecules, the *energy landscape* near the optimum is shaped such that many suboptimal structures have free energies which are numerically very close to that of the optimal structure, so that within the error of approximation used to compute free energies, it may not be possible to accurately identify the true optimum with any accuracy (McCaskill, 1990), and of course for nonconvex energy landscapes we are potentially also vexed by the problem of multiple local minima when gradient-ascent methods are employed (e.g., section 2.5).

An interesting technique aimed at rectifying these problems is one that utilizes the notion of an *ensemble* from statistical mechanics. In studying a closed system of particles we may distinguish between the macroscopic properties of the system (e.g., temperature, pressure, volume) and the microscopic state (or *microstate*) of the system as defined by the positions and momenta of all the particles in the system. In our case the microstates are the individual *conformations* (i.e., secondary structures) of the RNA molecule, and the macroscopic quantity of primary interest is the total free energy, G. An *ensemble* in our system is merely the set of all molecular conformations possible under some set of constraints, such as the constraint that the free energy of the system must lie in some interval $[G_0 - \Delta, G_0 + \Delta]$, or the constraint that a particular substructure such as a loop occupying a particular interval in the RNA must be present. Given an expression for the free energy G_i of structure S_i, it is well known that the probability of state S_i is given by:

$$P(S_i) = \frac{e^{\frac{G_i}{kT}}}{\sum_j e^{\frac{G_j}{kT}}}, \tag{12.1}$$

where T is the temperature and k is the *Boltzmann constant* introduced in section 10.13 in the context of simulated annealing. The denominator is referred to as a *partition function*, and is in practice typically evaluated using a number of simplifying assumptions and dynamic programming "tricks" in order to bound the computational complexity. Given the above formulation, the problem of finding the minimal free-energy secondary structure becomes one of finding the most *probable* structure. More importantly, by reformulating the partition function in Eq. (12.1) it is possible to compute the probabilities of specific structural features – for example, the probability $P(S_{(i,j)})$ that any given pair (i, j) of bases in the RNA are associated via hydrogen bonds in the equilibrium structure, allowing all other features of the structure to vary. The alert reader will notice that the latter construction is equivalent to the *posterior decoding* problem from sections 6.10 and 8.5.4, in which we summed the probabilities of all parses in which a given nucleotide (or feature) b was labeled with a particular label q. In the case of RNA secondary structure prediction, this construction gives us a confidence measure for individual base pairings in a putative structure, and may conceivably be used in a variety of ways for more accurate prediction of noncoding RNAs, though additional work remains to explore these avenues.

The other popular method for predicting RNA secondary structure is through the use of *stochastic context-free grammars (SCFGs)*, which model the *probability* of a structure rather than its *free energy* (though as demonstrated by Eq. (12.1), the probability and the free energy of a structure are closely related). Like finite automata, *context-free grammars (CFGs)* describe a class of formal languages in the *Chomsky hierarchy* (section 2.15). In terms of their ability to apply syntactic discrimination to input sequences, however, CFGs are strictly more powerful than finite automata, and as a corollary we state without proof (see Exercise 12.1) that SCFGs are therefore more powerful than HMMs.

Formally, we define a *context-free grammar* as a 4-tuple $G = (V, \alpha, S, R)$ comprising a *terminal alphabet* α, a set V of *nonterminal symbols* distinct from the symbols in α, a special *start symbol* $S \in V$, and a set R of rewriting rules called *productions*. The productions in R are rules of the form:

$$Y \to \lambda$$

for $\lambda \in (V \cup \alpha)^*$; such a production denotes that the nonterminal symbol Y may be *rewritten* (in a sense which we define below) by the expression λ, which may consist of zero or more terminals and nonterminals.

The computations of a CFG are referred to as *derivations*. A derivation consists of a series of *sentential forms*:

$$S \Rightarrow F_1 \Rightarrow F_2 \Rightarrow \cdots \Rightarrow F_n,$$

or, using a more compact notation,

$$S \Rightarrow^* F_n,$$

where each sentential form F_i is an ordered list of symbols from $V \cup \alpha$. All derivations begin with the start symbol S and end with a sentential form $F_n \in \alpha^*$ consisting only of symbols from α; all the other sentential forms $F_{i<n}$ in a derivation must contain at least one nonterminal symbol. Finally, each step $F_i \Rightarrow F_{i+1}$ in a derivation must be of the form:

$$\beta Y \gamma \Rightarrow \beta \lambda \gamma$$

where $Y \to \lambda \in R$ is a production which *rewrites* a single nonterminal Y into an expression λ; $\beta \in \alpha^*$ and $\gamma \in (V \cup \alpha)^*$ may obviously be empty strings. The productions $\{Y \to \lambda_0, Y \to \lambda_1, \ldots, Y \to \lambda_n\}$ may be more compactly written using the notation:

$$Y \to \lambda_0 | \lambda_1 | \ldots | \lambda_n.$$

An SCFG extends this framework by associating a probability with each production: $G = (V, \alpha, S, R, P_p)$ for $P_p : R \to \mathbb{R}$, where $\forall_{Y \in V} [\sum_{Y \to \lambda} P_p(Y \to \lambda) = 1]$. Given an SCFG G, we can compute the probability of any particular derivation $S \Rightarrow^* F_n$ by forming the product of probabilities for the productions utilized in the derivation. In the event that multiple derivations give rise to the same terminal sequence F_n, we may sum the probabilities of the alternate derivations to arrive at a probability:

$$P_G(F_n) = \sum\nolimits_{S \Rightarrow^* F_n} P(S \Rightarrow^* F_n).$$

As an example, consider $\mathcal{G} = (V_{\mathcal{G}}, \alpha, S, R_{\mathcal{G}}, P_{\mathcal{G}})$, for $V_{\mathcal{G}} = \{S, L, N\}$, $\alpha = \{\mathtt{A}, \mathtt{C}, \mathtt{G}, \mathtt{T}\}$ and R_{G} the set consisting of:

$$S \to \mathtt{A}\,S\,\mathtt{T} | \mathtt{T}\,S\,\mathtt{A} | \mathtt{C}\,S\,\mathtt{G} | \mathtt{G}\,S\,\mathtt{C} | L$$

$$L \to N\ N\ N\ N$$

$$N \to \mathtt{A} | \mathtt{C} | \mathtt{G} | \mathtt{T}$$

where $\forall_\lambda P_\mathcal{G}(S \rightarrow \lambda) = 0.2$, $P_\mathcal{G}(L \rightarrow NNNN) = 1$, and $\forall_\lambda P_\mathcal{G}(N \rightarrow \lambda) = 0.25$. Then the probability of the sequence ACGTACGTACGT is given by:

$$P(\text{ACGTACGTACGT}) =$$
$$P(S \Rightarrow \text{A}S\text{T} \Rightarrow \text{AC}S\text{GT} \Rightarrow \text{ACG}S\text{CGT} \Rightarrow \text{ACGT}S\text{ACGT} \Rightarrow$$
$$\text{ACGT}L\,\text{ACGT} \Rightarrow \text{ACGT}NNNN\text{ACGT} \Rightarrow \text{ACGTA}NNN\text{ACGT} \Rightarrow$$
$$\text{ACGTAC}NN\text{ACGT} \Rightarrow \text{ACGTACG}N\,\text{ACGT} \Rightarrow \text{ACGTACGTACGT}) =$$
$$0.2 \times 0.2 \times 0.2 \times 0.2 \times 0.2 \times 1 \times 0.25 \times 0.25 \times 0.25 \times 0.25 =$$
$$1.25 \times 10^{-6}$$

since this sequence has only one possible derivation, and it should be obvious that:

$$P(\text{ATGCATGCATGC}) = 0,$$

since grammar $\mathcal{G}$ cannot derive this terminal sequence.

For the purposes of ncRNA prediction, the final expression F_n is constrained to be the putative RNA sequence, and the probability $P_G(F_n)$ is taken as a measure of the propensity for F_n to fold into a shape typical of the class of RNAs being modeled. Given a null model $P_{null}(F_n)$ which evaluates to the probability of encountering the sequence F_n in the genome at random, we can use the log-likelihood ratio $LL(F_n) = P_G(F_n)/P_{null}(F_n)$ as a decision rule – i.e., whether $LL(F_n) > 0$ – to decide whether to emit F_n as a predicted ncRNA (Rivas and Eddy, 1999).

Identifying all possible derivations for a sequence F_n given a grammar G may be accomplished using a modified version of the well-known *CYK* parsing algorithm for (nonstochastic) CFGs (Kasami, 1965; Younger, 1967; Cocke and Schwartz, 1970) in *Chomsky normal form* (*CNF* – see Exercises 12.13, 12.14), where a CNF grammar is one in which all productions are of the form:

$$A \rightarrow BC$$

or:

$$A \rightarrow a$$

for nonterminals B and C and terminal a. Given a grammar $G = (V, \alpha, S, R)$ in CNF, the CYK dynamic programming algorithm begins by initializing its dynamic programming matrix V such that:

$$\forall_{0 \leq i < n} V_{i,i} = \{A | A \rightarrow x_i \in R\}$$

for the input sequence $I = x_0 x_1 \ldots x_{n-1}$. The remainder of the dynamic programming matrix is then computed column-by-column so that:

$$V_{i,j} = \{A | A \rightarrow BC \in R, \text{for some } B \in V_{i,k} \text{ and } C \in V_{k+1,j}, i \leq k < j\}$$

for $0 \leq i < j < n$. Given such a matrix, the reader should be able to verify quite easily (see Exercise 12.3) that the grammar G is capable of deriving the given input string I iff $S \in V_{0,n-1}$. Pseudocode specifications for this algorithm may be readily found elsewhere (e.g., Baker, 1979; Durbin *et al.*, 1998). Modification of this algorithm to compute the probability of a sequence involves associating a probability with each element in $V_{i,j}$, where for each nonterminal A we can multiply the probabilities associated with B and C in applying production $A \to BC$, while summing the probabilities associated with different productions for A and different values of k (see Exercise 12.4). Performing an *argmax* operation rather than summing the latter probabilities allows one to identify the single highest-scoring parse (see Exercise 12.5).

The advantage of SCFGs over HMMs lies in their ability to model arbitrary runs of matching pairs of elements, known to computer programmers as the problem of matching pairs of parentheses:

$$\cdots((((((((\cdots))))))))\cdots$$

where the number of opening parentheses on the left must match the number of closing parentheses on the right. When the number of matching pairs is unbounded, a finite-state model such as a DFA or an HMM is inadequate to enforce the constraint that all left elements must have a matching right element. In contrast, the modeling of matching pairs of elements can be readily achieved in a CFG, as illustrated by the example grammar $\mathcal{G}$ given earlier, which, the reader may have observed, models Watson–Crick base-pairing via the $S \to xSy|ySx$ productions, for $y = compl(x)$. This ability to model arbitrary runs of base pairings is obviously of great utility in secondary structure prediction. A potential use for SCFGs in the prediction of protein-coding genes is suggested by a recent report of apparent palindromic structures in a large proportion of alternatively spliced genes (Lian and Garner, 2005).

Parameter estimation for SCFGs can be accomplished using an algorithm called the *Inside–Outside algorithm*, which is similar to the Baum–Welch "Forward–Backward" algorithm used to train an HMM (section 6.6.3); the interested reader is again referred to the abundant literature on this topic (e.g., Baker, 1979; Durbin *et al.*, 1998).

Unfortunately, secondary structure alone seems to be insufficient for accurately locating noncoding RNAs in genomic DNA (Rivas and Eddy, 2000). The incorporation of comparative information has been shown to hold some promise for improving predictive accuracy for ncRNAs (e.g., Rivas and Eddy, 2001), though additional work is clearly needed.

12.3 Promoter prediction

The transcription of a gene – whether protein-coding or otherwise – present in genomic DNA is generally dependent on the presence of specific signals at positions located somewhat upstream of the actual gene. The importance of these signals appears to derive from their use as binding sites by regulatory proteins such as *transcription factors (TFs)*, and they are therefore termed *transcription factor binding sites (TFBSs)*. The interval containing these proximal signals immediately surrounding the *transcription start site (TSS)* of a gene is referred to as the *core promoter* of the gene.[1] In the case of a eukaryotic protein-coding gene, the relative positioning of these elements with respect to the CDS is given by:

$$\text{core promoter} \rightarrow 5'\ \text{UTR} \rightarrow \text{ATG} \rightarrow \text{CDS},$$

where the arrows indicate the 5′-to-3′ direction along the DNA.

Accurate identification of TSSs – and therefore of core promoters, which appear to occur largely within approximately 40 bp of the TSS (Zhang, 2003) – remains an important (and unfulfilled) goal of functional genomics, for a number of reasons. Identification of multiple core promoters for a gene may suggest multiple isoforms of the gene, whereas the lack of a convincing core promoter region upstream from a putative locus may suggest that the putative gene is not in fact transcriptionally active in the cell. Furthermore, the accurate delineation of the 5′ UTR and core promoter region for a gene are important prerequisites for regulatory studies of genes and gene networks, since the mechanisms for switching a gene between the transcriptionally active and transcriptionally inactive states are presumed to rely in large part on events involving the binding of regulatory proteins to the upstream regions of genes.

Like ncRNAs, promoters are difficult to predict via standard gene-finding methods because they do not directly code for proteins and therefore do not exhibit the strong codon statistics characteristic of many protein-coding genes. Thus, the standard content-sensor methods emphasized in earlier chapters will not suffice to identify promoter regions with the same degree of accuracy as for coding exons. The problem is further aggravated by species-specific differences in the presence and arrangement of core promoter motifs (Ohler *et al.*, 2002). Nevertheless, a number of methods relying on the same families of signal and content sensors introduced in earlier chapters have been investigated for eukaryotic promoter prediction, though not with overwhelming success. The most popular of these sensors is the weight matrix (WMM), with the majority of matrices in currently documented promoter prediction systems deriving from the popular *TRANSFAC* database of transcription factor binding sites (Wingender *et al.*, 1997). These weight matrices together with HMMs, interpolated Markov models, and a handful of machine-learning algorithms effectively span the computational approaches investigated to date for *ab initio*

[1] In addition to these core promoters located proximal to transcription start sites are the more distant *trans*-acting regulatory elements and upstream enhancer regions, which we will not consider here (see Lewin, 2003).

prediction of core promoters. Homology and expression (i.e., EST) evidence have also been used, though not as extensively.

A prime example of a promoter prediction system is *McPromoter* (Ohler *et al.*, 2000, 2001), which assumes the following structure for the typical class of promoter regions:

$$\text{UPR} \rightarrow \text{TATA} \rightarrow \text{Spacer} \rightarrow \text{Inr} \rightarrow \text{DPE}$$

where UPR = *upstream promoter region*, Inr = *initiator* (a region adjacent to the actual TSS), and DPE = *downstream promoter element*; TATA-boxes were mentioned in section 11.3. These elements are represented variously by weight matrices and 5th-order IMMs, with the overall structure being embodied in a basic GHMM. This GHMM is structured so as to permit the omission of some number of these elements, since it is known that, e.g., not all core promoters possess a TATA and Inr element. Because CpG islands (section 11.3) have been found to associate strongly with many mammalian promoters, modeling of CpG islands has also been profitably employed by McPromoter as well as other promoter-prediction programs, though it is well known that many promoters are not associated with CpG islands. Thus, the problem of promoter prediction has often been dichotomized into the twin problems of CpG-promoter prediction and non-CpG-promoter prediction, with the latter appearing to constitute somewhat of a greater challenge than the former (Bajic *et al.*, 2004).

In terms of the overall computational framework, the problem of promoter prediction as defined in the context of existing software systems has been formulated almost exclusively as a classification problem rather than a parsing problem. Thus, rather than computing the optimal parse ϕ of the entire input sequence into alternating regions consisting of putative promoters and intervening regions, the problem is more typically approached using a sliding window approach in which every possible interval (i, j) of reasonable size in the input sequence is classified as being more likely to be a core promoter region or a nonpromoter region, with a fixed scoring threshold typically serving to decide which intervals are to be classified as likely promoter regions. While it may seem obvious to some readers that the combination of a sophisticated promoter model with a global model of genes and intergenic regions might very well offer significant advantages over the standard threshold-based approach for promoter models used in isolation, few attempts at such an approach have been reported to date – rare examples include reports by Brown *et al.* (2005) and Davuluri *et al.* (2001). Unfortunately, the problem of *alternative transcription start sites* (section 12.1) may reduce the effectiveness of such parsing-based approaches, just as in the case of alternative splicing.

Early machine-learning approaches to promoter prediction considered only a handful of classifier methods including neural networks and linear and quadratic discriminant analysis (sections 10.7 and 10.8). More recently, SVMs and *RVMs* (*relevance vector machines* – Tipping, 2001) have been employed, with somewhat promising results (Down and Hubbard, 2002; Bajic *et al.*, 2004; Sonnenburg *et al.*, 2006). The application of *phylogenetic footprinting* approaches (section 9.6) has begun to receive some attention as well (e.g., Loots *et al.*, 2002; Sinha *et al.*, 2003; Corà *et al.*, 2005; Dieterich *et al.*, 2005). Other efforts have emphasized the *de novo* discovery of unknown regulatory sequences through the use of motif discovery methods such as *Gibbs sampling* and more primitive approaches based on overrepresented *n*-mers (Lawrence *et al.*, 1993; Bailey and Elkan, 1995; Corà *et al.*, 2005). Recent work by (Corà *et al.*, 2005) has highlighted the usefulness of an integrative approach, in which expression data, motifs, and *Gene Ontology* classifications (Gene Ontology Consortium, 2000) may be combined to improve the accuracy of promoter prediction. Structural information has also been utilized, including such features as *DNA curvature* (e.g., Ohler *et al.*, 2001), *helix destabilization* (e.g., Burden *et al.*, 2005), and *palindrome structures* (e.g., Davuluri *et al.*, 2001).

While much progress has been made in recent years, the field of promoter prediction is still considered by many to be in its infancy (e.g., Rombauts *et al.*, 2003), with the accuracy of several well-known programs on particular test sets still not exceeding that expected by random guessing (Bajic *et al.*, 2004; but see also Bajic *et al.*, 2006).

12.4 Generative versus discriminative modeling

From the foregoing chapters of this book it should be clear that most practical gene finders in use today are based on probabilistic, Markovian models (i.e., HMMs, GHMMs, GPHMMs, PhyloHMMs, etc.). Such models are said to be *generative*, in the sense that they prescribe methods for generating sequences statistically similar to actual DNA sequences for a particular organism – e.g., via the transition and emission distributions of HMMs, or the substitution models of PhyloHMMs, etc. These generative models are then used, not for *generation* of sequences, but for parsing them into discrete, nonoverlapping elements such as exons and introns.

The accuracy of these systems is clearly dependent on their ability to *discriminate* between the genomic elements of interest. Thus, we see that in the majority of probabilistic gene-finding systems, *generative* models are used to drive *discriminative* parsing algorithms. This naturally leads to the question of optimality – i.e., whether a generative model trained via maximum likelihood will produce an optimal parser, or whether some other class of model (or training regime) might achieve greater parsing accuracy in practice.

The generative/discriminative dichotomy has received considerable attention in the field of machine learning, with the emphasis largely being on the problem of *classification*, though the issue has also received some attention in the parsing context within the field of natural-language processing (e.g., Chou *et al.*, 1992). In the case of optimal classification, the consensus seems to be that discriminative models – that is, systems designed specifically for maximum discrimination rather than maximal generative fidelity – tend in many cases to perform better than their generative counterparts (e.g., Tong and Koller, 2000; Bouchard and Triggs, 2004; Goutte *et al.*, 2004; but see also Ng and Jordan, 2002).

The issue has received less attention in the field of gene finding, though several groups have begun to investigate the use of discriminative models such as *conditional random fields (CRFs* – Culotta *et al.*, 2005; Vinson *et al.*, 2006), with initial results appearing very promising. Within the classification setting, methods for discriminative exon identification have been explored by (Jaakkola and Haussler, 1999) using a combination of generative and discriminative models. Maximum discrimination techniques have also been employed for other bioinformatic tasks such as protein classification (e.g., Eddy *et al.*, 1995).

The related issue of optimal training regimes for generative systems was briefly addressed in sections 6.9 and 8.6.2, where we considered discriminative training for HMMs and GHMMs. Though early work by Krogh (1997) showed that discriminative training could improve the accuracy of an HMM for gene prediction, little attention seems to have been given to the discriminative training of HMMs for gene finding by other groups. A crude method for discriminative training of GHMMs was described by Majoros and Salzberg (2004), though improvements to this method are clearly needed. To our knowledge, the discriminative training of PhyloHMMs has not yet been explored, though it would seem to be a promising avenue for future research. Along these lines, the work by Vinson *et al.* (2006) mentioned earlier showed that phylogenetic modeling within the discriminative CRF framework is also feasible.

Regarding the discriminative training of Markovian systems, it was pointed out earlier that the resulting systems typically serve as poor generative models of sequences, so that other applications of the model – such as posterior decoding – may suffer as a result. Indeed, in pursuing more effective discriminative training methods, some practitioners in the field of speech recognition have found some utility in relaxing sum-to-1 constraints on the probability distributions in their HMMs, so that their models are no longer strictly probabilistic (e.g., Johansen, 1996).

Other means of obtaining an accurate parser for DNA sequence may be conceived which do not begin with the training of a generative model. Interestingly, such methods were proposed relatively early in the history of computational gene finding (e.g., Snyder and Stormo, 1993; Snyder, 1994; Stormo and Haussler, 1994), but were later supplanted (for better or worse) largely by HMM-based methods. In the work by Stormo and Haussler (1994) a framework was outlined whereby multiple content sensors in each state may be combined via:

$$s(i, j, q) = \sum_{\zeta \in \Psi_q} w_\zeta f_\zeta(x_i \ldots x_j) \tag{12.2}$$

where the summation is over the sensors $\zeta \in \Psi_q$ associated with "state" q, and the weights w_ζ are to be determined during training (e.g., via gradient-ascent optimization of the parser's accuracy on the training set). The "states" of the model are simply labels chosen from $Q = \{exon, intron\}$. The parameters of the individual sensors may be defined naturally, such as via raw hexamer counts, or using (generative) probabilistic models such as Markov chains. The actual parsing of an input sequence using such a model may be carried out using a dynamic programming algorithm which is very reminiscent of Viterbi decoding for a GHMM:

$$\phi^* = \underset{\theta}{argmax} \sum_{(i,j,q)\in\phi} s(i, j, q) \tag{12.3}$$

where the optimal parse ϕ^* is defined as a partitioning of the input sequence into a series of nonoverlapping intervals (i, j) consistent with eukaryotic gene syntax (e.g., section 3.4), each having a label $q \in Q$.

It should be reasonably clear that the latter formulation is roughly a generalization of the approach outlined in sections 6.9 and 8.6.2 involving the use of discriminatively trained (G)HMMs with Viterbi decoding, so that the latter techniques effectively recapitulate the methods originally proposed by Haussler, Snyder, Stormo, and others in the early 1990s, but in a more probabilistic framework based very strongly on generative modeling. Whether the use of generative models in discriminative parsing systems is advantageous or disadvantageous remains an open question. To see that this question is not purely academic, we invite the reader to consider (for example) the popular use of phylogenetic modeling (section 9.6) in comparative gene finders, and whether accuracy gains in such systems may be obtained by relaxing strict probability constraints (i.e., sum-to-1 constraints) and instead viewing the phylogenetic component of a comparative gene finder as merely an evidence network in which the branch lengths and substitution matrices may be optimized so as to maximize the expected accuracy of the resulting gene finder. Indeed, the widespread use of so-called "fudge factors" in these and other comparative gene-finding systems (e.g., "coding bias" in ExoniPhy: Siepel and Haussler, 2004b; "conservation score coefficient" in N-SCAN: Gross and Brent, 2005; non-maximum-likelihood value for P_{match} in TWAIN: Majoros *et al.*, 2005b) suggests that precisely this type of "nonprobabilistic" manipulation of the model may be required in order to obtain optimal parsing behavior from these systems in practice. Further exploration of these issues is clearly needed.

12.5 Parallelization and grid computing

The commoditization of personal computing hardware makes parallel processing – in which a task is solved by partitioning the task into a number of sub-tasks which are then solved simultaneously on different processors – an increasingly viable alternative to traditional single-cpu computing. A popular form of multi-processor computing is so-called *grid computing*, in which a number of (typically) inexpensive computers are networked into a computing cluster to which computing jobs may be sent from an individual's personal workstation. Another emerging alternative is the use of multi-core processors, in which one processor contains multiple, near-identical computing cores; the more traditional multiple-cpu computers in which each computing core is contained within a separate cpu remains another effective means of parallel computing.

The challenge in using multiple processors to solve a single task is in finding a way to parallelize an algorithm so that its subproblems may be solved simultaneously rather than in strict sequence. In the case of gene finding, this may be done in a number of ways. The most obvious is to segment the genome of interest into overlapping or nonoverlapping sequences and to run identical copies of the gene finder on all processors in a system, with each copy responsible for processing a different set of segments, until the entire genome has been processed. Even this simplistic approach requires some work, however, to map the predicted gene coordinates – which the gene finder will give relative to the beginning of each segment – back to chromosome-level coordinates, so that the resulting genome annotation can be represented using a consistent coordinate system. One must also deal with the problem of genes that were inadvertently split because they straddle a segment boundary.

More fine-grained parallelization is possible as well, though with greater effort. For GHMM-based systems using the prefix sum arrays (PSA) approach to decoding, the prefix sum arrays for an entire genome may be computed in parallel using a segmented approach as described above, with the final decoding process being applied to the full arrays after they have been recombined; parallelization of the decoding phase can still be achieved at the chromosome or contig level, so that the problem of recombining partial genes does not arise as a practical complication (since the missing portions of any putative partial genes would not be present in the assembly).

Partitioning of prefix sum arrays can also be performed in the case of PhyloHMMs, since the $P(I^{(0)}, I^{(1)}, \ldots, I^{(n-1)}|S, q)$ terms computed for each column in the informant alignment are not dependent on the results of the $P(S|q)$ computations of the underlying GHMM decoder. Nonindependence of columns in the alignment, if modeled by the gene finder, produces only a minor complication at the beginning of each segment of the alignment; this can easily be addressed by allowing segments to overlap by n columns, where n is the order of dependence modeled in the PhyloHMM.

Parallelization of combiner-like approaches to comparative gene finding would seem to be a quite feasible undertaking, since the computation of evidence tracks may proceed independently on different nodes in a cluster. Because most combiner systems utilize standard machine-learning algorithms to combine evidence, known methods for parallelizing these machine-learning algorithms may be employed (e.g., Provost and Hennessy, 1994).

In the case of gene finding with an ORF graph, the genome may again be partitioned into reasonably long segments for the construction of the ORF graph, though some extra work would then be required to stitch together the graphs for the individual segments into complete chromosome or contig-level graphs. The decoding problem on a long ORF graph might be decomposed by noting *cut vertices* in the graph – i.e., vertices which if removed would segment the graph into distinct subgraphs having no edges between them (Cormen *et al.*, 1992) – and segmenting the graphs at these points, if they are common enough.

To our knowledge, none of these suggested methods of parallelization has been explored in practice. There is potentially much to gain, however, through the parallelization of gene finders, since the reduction in compute time for annotating an entire genome using current state-of-the-art gene-finding techniques would afford greater flexibility in exploring methods (such as modeling the secondary structure or chromatin state of the DNA) that may be computationally too expensive on a uniprocessing system.

Whereas muti-core and multi-cpu architectures benefit from the sharing of memory between all computing elements, memory banks remain physically separate in the case of grid computing, so that messages must be passed via network connections when one process (i.e., program solving a sub-task of the overall problem) needs to communicate with another process running on a different node in the cluster. A popular protocol for inter-process communication on clusters is *MPI*, the *message passing interface* (Gropp *et al.*, 1994); the interested reader is referred to the given references for more information.

EXERCISES

12.1 State and prove a theorem illustrating that SCFGs are strictly more powerful than HMMs.

12.2 Prove that any grammar $G = (V, \alpha, S, R)$ may be converted into Chomsky normal form (CNF) without affecting the language $L(G)$ defined by the grammar, as long as $\varepsilon \notin L(G)$.

12.3 Prove that, as a result of applying the CYK algorithm, grammar G is capable of deriving the given input string I iff $S \in V_{0,n-1}$.

12.4 Modify the CYK algorithm to compute the probability $P(I|G)$ of a sequence I given a grammar G.

12.5 Modify the algorithm derived in the previous exercise to find instead the highest-scoring parse.

12.6 The example grammar $\mathcal{G}$ of section 12.2 may be seen as a highly simplified and constrained model of stem–loop structures, with the loop being derived through L and the stem through the recursive productions involving S. Extend $\mathcal{G}$ to incorporate loops of arbitrary length.

12.7 Continuing with the previous exercise, extend $\mathcal{G}$ to incorporate mismatches and single-base insertions in the stem.

12.8 Continuing with the previous exercise, extend $\mathcal{G}$ to incorporate insertions of maximal length 4 in the stem. Enforce a geometric distribution on the length of insertions.

12.9 Extend $\mathcal{G}$ to allow loops of arbitrary size.

12.10 Extend $\mathcal{G}$ to allow loops of any size in the interval [5,8], where the size of a loop is defined as the number of residues comprising it.

12.11 Suppose that some SCFG G is capable of deriving only a finite number of different strings. Explain how an equivalent HMM may be obtained from G.

12.12 Explain how an HMM may be converted into an SCFG.

12.13 Give an algorithm for converting any CFG G into CNF, assuming that G cannot derive the empty string, ε (i.e., $S \not\Rightarrow^* \varepsilon$)

12.14 Extend the algorithm in the previous problem to convert (non-ε-deriving) SCFGs into CNF.

12.15 Assuming grammar $\mathcal{G}$ from section 12.2, compute the probability of each of the following sequences, assuming $P(S) = 0$ for sequences not derivable via the grammar:

(a) ACTACGTACGTACGT

(b) ACGTACGT

(c) TCTCGGGAGA

(d) TCTCGGGAGATCTCGGGAGA

(e) ATGATGCATGATGCATGATGCTGCAGCATCTAGCATCTAGCATCTA

APPENDIX

Online resources

A.1 Official book website

The official website for this book is located at:

http://www.geneprediction.org/book

Here you will find sample code, sample data, links to other online resources, information about other editions of the book, supplementary chapters, errata, and other information relevant to the book. In the near future we hope to add course materials such as slides and solutions to selected exercises from the book.

A.2 Open-source gene finders

As of the current printing of this book, the following gene finders are known to be available as open-source software. Note that all open-source programs are not governed by the same license; before modifying or redistributing any of this software, you should check the licensing agreement under which a specific program is distributed. Information about specific licensing agreements can be found at the web addresses given below, or by contacting the author(s) of the specific programs.

- Eugène: http://www.inra.fr/Internet/Departements/MIA/T/EuGene/
- GENEID: http://genome.imim.es/software/geneid/index.html
- GeneZilla: http://www.genezilla.org
- GlimmerHMM: http://www.cbcb.umd.edu/software/glimmerhmm/
- JIGSAW: http://www.cbcb.umd.edu/software
- Phat: http://bioinf.wehi.edu.au/Phat/
- SGP2: http://genome.imim.es/software/sgp2/index.html
- SNAP: http://homepage.mac.com/iankorf
- TWAIN: http://www.tigr.org/software/pirate
- TWINSCAN: http://genes.cs.wustl.edu/
- UNVEIL: http://www.tigr.org/software/pirate

A.3 Gene-finding websites

Currently, the most informative website dedicated to gene finding is geneprediction.org, which is found at the full address:

http://www.geneprediction.org

Much of the content of this site is mirrored at the alternate site:

http://www.genefinding.org

These websites feature links to a large number of gene finders and other external information related to gene finding. Submissions are welcome – visit the sites for information on submitting content or contacting the maintainers regarding suggestions.

A.4 Gene-finding bibliographies

A highly useful bibliography is currently maintained by Wentian Li and is available at:

http://www.nslij-genetics.org/gene/

References

Aha D.W. and Bankert R.L. (1996) A comparative evaluation of sequential feature selection algorithms. In Fisher D. and Lenz H.-Z. (eds.) *Learning from Data*, pp. 199–206. New York: Springer.

Aho A.V., Sethi R., and Ullman J.D. (1986) *Compilers: Principles, Techniques, and Tools*. Reading, MA: Addison-Wesley.

Allen J.E. and Salzberg S.L. (2005) JIGSAW: integration of multiple sources of evidence for gene prediction. *Bioinformatics* **21**:3596–3603.

Allen J.E., Pertea M., and Salzberg S.L. (2004) Computational gene prediction using mutliple sources of evidence. *Genome Research* **14**:142–148.

Allen J.E., Majoros W.H., Pertea M., and Salzberg S.L. (2006) JIGSAW, GeneZilla, and GlimmerHMM: puzzling out the features of human genes in the ENCODE regions. *Genome Biology* **7**(Suppl. 1):S9.

Alexandersson M., Cawley S., and Pachter L. (2003) SLAM: cross-species gene finding and alignment with a generalized pair hidden Markov model. *Genome Research* **13**:496–502.

Altschul S.F., Madden T.L., Schaffer A.A., Zhang J., Anang Z., Miller W., and Lipman D.J. (1997) Gapped BLAST and PSI-BLAST: a new generation of protein database search programs. *Nucleic Acids Research* **25**:3389–3402.

Anton H. (1987) *Elementary Linear Algebra*, 5th edn. New York: John Wiley.

Apweiler R., Attwood T.K., Bairoch A., Bateman A., Birney E., Biswas M., Bucher P., Cerutti L., Corpet F., Croning M.D.R., Durbin R., Falquet L., Fleischmann W., Gouzy J., Hermjakob H., Hulo N., Jonassen I., Kahn D., Kanapin A., Karavidopoulou Y., Lopez R., Marx B., Mulder N.J., Oinn T.M., Pagni M., Servant F., Sigrist C.J.A., and Zdobnov E.M. (2001) The InterPro database, an integrated documentation resource for protein families, domains and functional sites. *Nucleic Acids Research* **29**:37–40.

Attwood T.K., Bradley P., Flower D.R., Gaulton A., Maudling N., Mitchell A.L., Moulton G., Nordle A., Paine K., Taylor P., Uddin A., and Zygouri C. (2003) PRINTS and its automatic supplement, prePRINTS. *Nucleic Acids Research* **31**:400–402.

Azad R.K. and Borodovsky M. (2004) Effects of choice of DNA sequence model structure on gene identification accuracy. *Bioinformatics* **20**: 993–1005.

Bafna V. and Huson D.H. (2001) The conserved exon method for gene finding. *ISMB'2000*, **8**: 3–12.

Bahl L. R., Brown P. F., de Souza P. V., and Mercer R. L. (1986) Maximum mutual information estimation of hidden Markov model parameters for speech recognition. In *Proceedings of the IEEE International Conference on Acoustics, Speech and Signal Processing*, pp. 49–52.

Bailey T. L. and Elkan C. (1995) Unsupervised learning of multiple motifs in biopolymers using expectation maximization. *Machine Learning* **21**:51–83.

Bairoch A. and Apweiler R. (1996) The SWISS-PROT protein sequence data bank and its new supplement TrEMBL. *Nucleic Acids Research* **24**:21–25.

Bajic V. B., Seah S. H., Chong A., Zhang G., Koh J. L. Y., and Brusic V. (2002) Dragon promoter finder: recognition of vertebrate RNA polymerase II promoters. *Bioinformatics* **18**:198–199.

Bajic V. B., Tan S. L., Suzuki Y., and Sugano S. (2004) Promoter prediction analysis on the whole human genome. *Nature Biotechnology* **22**:1467–1473.

Bajic V. B., Brent M. R., Brown R. H., Frankish A., Harrow J., Ohler U., Solovyev V. V., and Tan S. L. (2006) Performance assessment of promoter predictions on ENCODE regions in the EGASP experiment. *Genome Biology* **7**(Suppl. 1):S3.

Baker J. K. (1979) Trainable grammars for speech recognition. In *Proceedings of the Spring Conference of the Acoustical Society of America*, Boston, MA, pp. 547–550.

Bartel D. P. (2004) MicroRNAs: genomics, biogenesis, mechanism, and function. *Cell* **116**:281–297.

Bateman A., Coin L., Durbin R., Finn R. D., Hollich V., Griffiths-Jones S., Khanna A., Marshall M., Moxon S., Sonnhammer E. L. L., Studholme D. J., Yeats C., and Eddy S. R. (2004) The Pfam protein families database. *Nucleic Acids Research* **32**:D138–D141.

Batzoglou S., Pachter L., Mesirov J. P., Berger B., and Lander ES. (2000) Human and mouse gene structure: comparative analysis and application to exon prediction. *Genome Research* **7**:950–958.

Baum L. E. (1972) An inequality and associated maximization technique in statistical estimation for probabilistic functions of Markov processes. *Inequalities* **3**:1–8.

Baum L. E., Petrie T., Goules G., and Weiss N. (1970) A maximization technique occurring in the statistical analysis of probabilistic functions of Markov chains. *Annals of Mathematical Statistics* **41**:164–171.

Beaudoing E., Freier S., Wyatt J. R., Claverie J.-M., and Gautheret D. (2000) Patterns of variant polyadenylation signal usage in human genes. *Genome Research* **10**:1001–1010.

Benson D. A., Karsch-Mizrachi I, Lipman D. J., Ostell J., and Wheeler D. L. (2005) GenBank. *Nucleic Acids Research* **33**:D34–D38.

Bernardi G., Olofsson B., Filipski J., Zerial M., Salinas J., Cuny G., Meunier-Rotival M., and Rodier F. (1985) The mosaic genome of warm-blooded vertebrates. *Science* **228**:953–958.

Besemer J., Lomsadze A., and Borodovsky M. (2001) GeneMarkS: a self-training method for prediction of gene starts in microbial genomes – implications for finding sequence motifs in regulatory regions. *Nucleic Acids Research* **29**:2607–2618.

Birney E., Andrews D., Caccamo M., Chen Y., Clarke L., Coates G., Cox T., Cunningham F., Curwen V., Cutts T., Down T., Durbin R., Fernandez-Suarez X. M., Flicek P., Gräf S., Hammond M., Herrero J., Howe K., Iyer V., Jekosch K., Kähäri A., Kasprzyk A., Keefe D., Kokocinski F., Kulesha E., London D., Longden I., Melsopp C., Meidl P., Overduin B., Parker A., Proctor G., Prlic A., Rae M., Rios D., Redmond S., Schuster M., Sealy I., Searle S., Severin J., Slater G., Smedley D., Smith J., Stabenau A., Stalker J., Trevanion S., Ureta-Vidal A., Vogel J., White S., Woodwark C., and Hubbard T. J. P. (2006) Ensembl 2006. *Nucleic Acids Research* **34**:D556–D561.

Blanco E., Parra G., and Guigó R. (2002). Using GENEID to identify genes. In A. Baxevanis (ed.), *Current Protocols in Bioinformatics*, unit 4.3. New York: John Wiley.

Boguski M. S., Lowe T. M., and Tolstoshev C. M. (1993) dbEST: database for expressed sequence tags. *Nature Genetics* **4**:332–333.

Borodovsky M. and McIninch J. (1993) GENMARK: parallel gene recognition for both DNA strands. *Computers and Chemistry* **16**:37–43.

Borodovsky M., Rudd K. E., and Koonin E. V. (1994) Intrinsic and extrinsic approaches for detecting genes in a bacterial genome. *Nucleic Acids Research* **22**:4756–4767.

Bouchard G. and Triggs B. (2004) The trade-off between generative and discriminative classifiers. In J. Antoch (ed.), *Proceedings of International Symposinm on Computational Statistics (COMPSTAT) 2004*, pp. 1–9.

Bray N., Dubchak I., and Pachter L. (2003) AVID: a global alignment program. *Genome Research* **13**:97–102.

Breiman L., Friedman J. H., Olshen R. A., and Stone C. J. (1984) *Classification and Regression Trees*. Monterey, CA: Wadsworth International.

Brejová B., Brown D. G., Li M., and Vinar T. (2005) ExonHunter: a comprehensive approach to gene finding. *Bioinformatics* **21**(Suppl. 1):i57–i65.

Brendel V. and Kleffe J. (1998) Prediction of locally optimal splice sites in plant pre-mRNA with applications to gene identification in *Arabidopsis thaliana* genomic DNA. *Nucleic Acids Research* **26**:4748–4757.

Brown R. H., Gross S. S., and Brent M. R. (2005) Begin at the beginning: predicting genes with 5′ UTRs. *Genome Research* **15**:742–747.

Bucher P. and Bairoch A. (1994) A generalized profile syntax for biomolecular sequence motifs and its function in automatic sequence interpretation. In *Proceedings of the 2nd International Conference on Intelligent Systems for Molecular Biology*, pp. 53–61.

Buratti E. and Baralle F. E. (2004) Influence of RNA secondary structure on the pre-mRNA splicing process. *Molecular Cell Biology* **24**:10505–10514.

Burden S., Lin Y. X., and Zhang R. (2005) Improving promoter prediction for the NNPP2.2 algorithm: a case study using *Escherichia coli* DNA sequences. *Bioinformatics* **21**:601–607.

Burge C. (1997) Identification of complete gene structures in human genomic DNA. Ph.D. thesis Stanford University, Stanford, CA.

Burge C. (1998) Modeling dependencies in pre-mRNA splicing signals. In Salzberg S., Searls D., and Kasif S. (eds.), *Computational Methods in Molecular Biology*, pp. 127–163. Amsterdam: Elsevier.

Burge C. and Karlin S. (1997) Prediction of complete gene structures in human genomic DNA. *Journal of Molecular Biology* **268**:78–94.

Burge C. B., Tuschl T., and Sharp P. A. (1999) Splicing of precursors to mRNAs by the spliceosomes. In Gesteland R. F., Cech T. R., and Atkins J. F. (eds.) *The RNA World*, 2nd edn., pp. 525–560. Cold Spring Harbor, NY: Cold Spring Harbor Laboratory Press.

Burges C. J. C. (1998) A tutorial on support vector machines for pattern recognition. *Data Mining and Knowledge Discovery* **2**:121–167.

Burks C., Cassidy M., Cinkosky M. J., Cumella K. E., Gilna P., Hayden J. E.-D., Keen G. M., Kelley T. A., Kelly M., Kristofferson D., and Ryals J. (1991) GenBank. *Nucleic Acids Research* **19**:S2221–S2225.

Burset M. and Guigó R (1996) Evaluation of gene structure prediction programs. *Genomics* **34**:357–367.

Burset M., Seledtsov I. A., and Solovyev V. V. (2000) Analysis of canonical and non-canonical splice sites in mammalian genomes. *Nucleic Acids Research* **28**:4364–4375.

Cai D., Delcher A., Kao B., and Kasif S. (2000) Modeling splice sites with Bayes networks. *Bioinformatics* **16**:152–158.

Carninci P., Sandelin A., Lenhard B., Katayama S., Shimokawa K., Ponjavic J., Semple C.A., Taylor M.S., Engstrom P.G., Frith M.C., Forrest A.R., Alkema W.B., Tan S.L., Plessy C., Kodzius R., Ravasi T., Kasukawa T., Fukuda S., Kanamori-Katayama M., Kitazume Y., Kawaji H., Kai C., Nakamura M., Konno H., Nakano K., Mottagui-Tabar S., Arner P., Chesi A., Gustincich S., Persichetti F., Suzuki H., Grimmond S.M., Wells C.A., Orlando V., Wahlestedt C., Liu E.T., Harbers M., Kawai J., Bajic V.B., Hume D.A., and Hayashizaki Y. (2006) Genome-wide analysis of mammalian promoter architecture and evolution. *Nature Genetics* **38**:626–635.

Cawley S.E., Wirth A.I., and Speed T.P. (2001) Phat: a gene finding program for *Plasmodium falciparum*. *Molecular and Biochemical Parasitology* **118**:167–174.

Choo K.H., Tong J.C., and Zhang L. (2004) Recent applications of hidden Markov models in computational biology. *Genomics, Proteomics, and Bioinformatics* **2**:84–96.

Chou W., Juang B.H., and Lee C.H. (1992) Segmental GPD training of HMM based speech recognizer. In *Proceedings of the IEEE International Conference on Acoustics, Speech, and Signal Processing*, vol. 1, pp. 473–476.

Chow C.K. and Liu C.N. (1968) Approximating discrete probability distributions with dependence trees. *IEEE Transactions on Information Theory* **14**:462–467.

Chuang T.-J., Lin W.C., Lee H.C., Wang C.W., Hsiao K.L., Wang Z.H., Shieh D., Lin S.C., and Chang L.Y. (2003) A complexity reduction algorithm for analysis and annotation of large genomic sequences. *Genome Research* **13**:313–322.

Chuang T.-J., Chen F.-C., and Chou M.-Y. (2004) A comparative method for identification of gene structures and alternatively spliced variants. *Bioinformatics* **20**:3064–3079.

Clark F. and Thanaraj T.A. (2002) Categorization and characterization of transcript-confirmed constitutively and alternatively spliced introns and exons from human. *Human Molecular Genetics* **11**:451–464.

Clote P. and Backofen R. (2000) *Computational Molecular Biology*. New York: John Wiley.

Cocke J. and Schwartz J.T. (1970) *Programming Languages and their Compilers: Preliminary Notes*, Technical Report. New York: Courant Institute of Mathematical Sciences, New York University.

Corà D., Herrmann C., Dieterich C., Di Cunto F., Provero P., and Caselle M. (2005) *Ab initio* identification of putative human transcription factor binding sites by comparative genomics. *BMC Bioinformatics* **6**:110.

Cormen T.H., Leiserson C.E., and Rivest R.L. (1992) *Introduction to Algorithms*. Cambridge, MA: MIT Press.

Cover T.M. and Hart P.E. (1967) Nearest neighbor pattern classification. *IEEE Transactions on Information Theory* **13**:57–67.

Culotta A., Kulp D., and McCallum A. (2005) *Gene Prediction with Conditional Random Fields*, Technical Report UM-CS-2005–028. Amherst, MA: University of Massachusetts.

Darwin C. (1859) *On the Origin of Species by Means of Natural Selection, or the Preservation of Favoured Races in the Struggle for Life*. London: John Murray.

Davuluri V.D., Grosse I., and Zhang M.Q. (2001) Computational identification of promoters and first exons in the human genome. *Nature Genetics* **29**:412–417.

Dawkins R. (1982) *The Extended Phenotype: The Long Reach of the Gene*. Oxford: Oxford University Press.

Dawkins R. (1997) Human chauvinism. *Evolution* **51**:1015–1020.

Dayhoff M., Schwartz R.M., and Orcutt B.C. (1978) A model of evolutionary change in proteins. *Atlas of Protein Sequence and Structure* **5**:345–352.

Delcher A.L., Kasif S., Fleischmann R.D., Peterson J., White O., and Salzberg S.L. (1999a) Alignment of whole genomes. *Nucleic Acids Research* **27**:2369–2376.

Delcher A.L., Harmon D., Kasif S., White O., and Salzberg S.L. (1999b) Improved microbial gene identification with GLIMMER. *Nucleic Acids Research* **27**:4636–4641.

Delcher A.L., Phillippy A., Carlton J., and Salzberg S.L. (2002) Fast algorithms for large-scale genome alignment and comparision. *Nucleic Acids Research* **30**:2478–2483.

Delphin M.E., Stockwell P.A., Tate W.P., and Brown C.M. (1999) Transterm, the translational signal database, extended to include full coding sequence and untranslated regions. *Nucleic Acids Research* **27**:293–294.

Dieterich C., Grossmann S., Tanzer A., Röpcke S., Arndt P.F., Stadler P.F., and Vingron M. (2005) Comparative promoter region analysis powered by CORG. *BMC Genomics* **6**:24.

Ding Y. (2006) Statistical and Bayesian approaches to RNA secondary structure prediction. *RNA* **12**:232–331.

Doudna J.A. and Cech T.R. (2002) The chemical repertoire of natural ribozymes. *Nature* **418**:222–228.

Down T.A. and Hubbard T.J.P. (2002) Computational detection and location of transcription start sites in mammalian genomic DNA. *Genome Research* **12**:458–461.

Dror G., Sorek R., and Shamir R. (2004) Accurate identification of alternatively spliced exons using support vector machines. *Bioinformatics* **21**:897–901.

Duda R.O., Hart P.E., and Stork D.G. (2000) *Pattern Classification*, 2nd edn. New York: Wiley-Interscience.

Dunteman G.H. (1989) *Principal Components Analysis.* London: Sage Publications.

Durbin R., Eddy S., Krogh A., and Mitchison G. (1998) *Biological Sequence Analysis.* Cambridge: Cambridge University Press.

Eddy S.R. (2002) Computational genomics of noncoding RNA genes. *Cell* **109**:137–140.

Eddy S.R. (2005) A model of the statistical power of comparative genome sequence analysis. *PLoS Biology* **3**:e10.

Eddy S., Mitchison G., and Durbin R. (1995) Maximum discrimination hidden Markov models of sequence consensus. *Journal of Computational Biology* **2**:9–23.

Edwards A.W.F. (1992) *Likelihood.* Baltimore, MD: Johns Hopkins University Press.

Eisen J.A., Coyne R.S., Wu M., Wu D., Thiagarajan M., Wortman J.R., Badger J.H., Ren Q., Amedeo P., Jones K.M., Tallon L.J., Delcher A.L., Salzberg S.L., Silva J.C., Haas B.J., Majoros W.H., Farzad M., Carlton J.M., Smith R.K., Garg J., Pearlman R.E., Karrer K.M., Sun L., Manning G., Elde N.C., Turkewitz A.P., Asai D.J., Wilkes D.E., Wang Y., Cai H., Collins K., Stewart B.A., Lee S.R., Wilamowska K., Weinberg Z., Ruzzo W.L., Wloga D., Gaertig J., Frankel J., Tsao C.C., Gorovsky M.A., Keeling P.J., Waller R.F., Patron N.J., Cherry J.M., Stover N.A., Krieger C.J., Del Toro C., Ryder H.F., Williamson S.C., Barbeau R.A., Hamilton E.P., and Orias E. (2006) Macronuclear genome sequence of the ciliate *Tetrahymena thermophila*, a model eukaryote. *PLoS Biology* **29**:**4**(9).

Fairbrother W.G., Yeh R.F., Sharp P.A., and Burge C.B. (2002) Predictive identification of exonic splicing enhancers in human genes. *Science* **297**:1007–1013.

Falconer D.S. (1996) *Introduction to Quantitative Genetics*, 4th edn. Englewood Cliffs, NJ: Prentice-Hall.

Fariselli P., Martelli P.L., and Casadio R. (2005) The posterior-Viterbi: a new decoding algorithm for hidden Markov models. *BMC Bioinformatics* **6** (Suppl. 4):S12.

Fausett L.V. (1994) *Fundamentals of Neural Networks*. Englewood Cliffs, NJ: Prentice-Hall.

Felsenstein J. (1981) Evolutionary trees from DNA sequences. *Journal of Molecular Evolution* **17**:368–376.

Felsenstein J. (1989) PHYLIP: phylogeny inference package (version 3.2). *Cladistics* **5**:164–166.

Felsenstein J. (2004) *Inferring Phylogenies.* Sunderland, MA: Sinauer Associates.

Felsenstein J. and Churchill G.A. (1996) A hidden Markov model approach to variation among sites in rate of evolution. *Molecular Biology and Evolution* **13**:93–104.

Fischer C.N. and LeBlanc R.J. (1991) *Crafting a Compiler with C.* Menlo Park, CA: Benjamin/Cummings.

Fisher R.A. (1936) The use of multiple measurements in taxonomic problems. *Annals of Eugenics* **7**:179–188.

Fletcher R. (1980) *Practical Methods of Optimization, vol. 1, Unconstrained Optimization.* New York: John Wiley.

Florea L., Hartzell G., Zhang Z., Rubin G.M., and Miller W. (1998) A computer program for aligning a cDNA sequence with a genomic DNA sequence. *Genome Research* **8**:967–974.

Florea L., Di Francesco V., Miller J., Turner R., Yao A., Harris M., Walenz B., Mobarry C., Merkulov G.V., Charlab R., Dew I., Deng Z., Istrail S., Li P., and Sutton G. (2005) Gene and alternative splicing annotation with AIR. *Genome Research* **15**:54–66.

Fogel D.B. (2005) *Evolutionary Computation: Toward a New Philosophy of Machine Intelligence*, 3rd edn. New York: Wiley-IEEE Press.

Foissac S. and Schiex T. (2005) Integrating alternative splicing detection into gene prediction. *BMC Bioinformatics* **6**:25.

Freund Y. and Schapire R.E. (1995) A decision-theoretic generalization of on-line learning and an application to boosting. In *Proceedings of the European Conference on Computational Learning Theory*, pp. 23–37.

Friedman J.H. (1989) Regularized discriminant analysis. *Journal of the American Statistical Association* **84**:165–175.

Galassi M., Davies J., Theiler J., Gough B., Jungman G., Booth M., and Rossia F. (2005) *GNU Scientific Library Reference Manual*, 2nd edn. Bristol, UK: Network Theory Ltd.

Gangal R. and Sharma P. (2005) Human pol II promoter prediction: time series descriptors and machine learning. *Nucleic Acids Research* **33**:1332–1336.

Gene Ontology Consortium (2000) Gene ontology: tool for the unification of biology. *Nature Genetics* **25**:25–29.

Gierasch L.M. (1989) Signal sequences. *Biochemistry* **28**:923–930.

Goldberg D.E. (1989) *Genetic Algorithms in Search, Optimization and Machine Learning.* Reading, MA: Addison-Wesley.

Gould S.J. (1994) The evolution of life on earth. *Scientific American* **271**:62–69.

Goutte C., Gaussier E., Cancedda N., and Dejean H. (2004) Generative vs. discriminative approaches to entity recognition from label-deficient data. In *Féme Journées Internationales Analyse Statistique des Données Textuelles (JADT 2004)*, pp. 1–10.

Gropp W., Lusk E., and Skjellum A. (1994) *Using MPI: Portable Parallel Programming with the Message-Passing Interface.* Cambridge, MA: MIT Press.

Gross S.S. and Brent M.R. (2005) Using multiple alignments to improve gene prediction. In *Research in Computational Molecular Biology (RECOMB'05)*, pp. 374–388.

Guigó R. (1998) Assembling genes from predicted exons in linear time with dynamic programming. *Journal of Computational Biology* **5**:681–702.

Guigó R., Flicek P., Abril J.F., Reymond A., Lagarde J., Denoeud F., Antonarakis S., Ashburner M., Bajic V.B., Birney E., Castelo R., Eyras E., Gingeras T.R., Harrow J., Hubbard T., Lewis S., Ucla C., and Reese M.G. (2006) EGASP: the human ENCODE genome annotation assessment project. *Genome Biology* **7**(Suppl. 1):S2.

Haas B.J., Delcher A.L., Mount S.M., Wortman J.R., Smith R.K., Hannick L.I., Maiti R., Ronning C.M., Rusch D.B., Town C.D., Salzberg S.L., and White O. (2003) Improving the *Arabidopsis* genome annotation using maximal transcript alignment assemblies. *Nucleic Acids Research* **31**:5654–5666.

Hall K.B., Green M.R., and Redfield A.G. (1988) Structure of a pre-mRNA branch point / 3′ splice site region. *Proceedings of the National Academy of Sciences of the USA* **85**:704–708.

Hannenhalli S. and Levy S. (2001) Promoter prediction in the human genome. *Bioinformatics* **17**:S90–S96.

Hasegawa M., Kishino H., and Yano T. (1985) Dating of the human–ape splitting by a molecular clock of mitochondrial DNA. *Journal of Molecular Evolution* **22**:160–174.

Hastie T., Tibshirani R., and Friedman J.H. (2003) *The Elements of Statistical Learning*. New York: Springer.

Heber S., Alekseyev M., Sze S.-H., Tang H., and Pevzner P.A. (2002) Splicing graphs and EST assembly problem. *Bioinformatics* **18**:S181–S188.

Heckerman D., Geiger D., and Chickering D. (1994) Learning Bayesian networks: the combination of knowledge and statistical data. In *Knowledge Discovery and Data Mining Workshop (KDD '94)*, pp. 85–96.

Henderson J., Salzberg S., and Fasman K. (1997) Finding genes in human DNA with a hidden Markov model. *Journal of Computational Biology* **4**:127–141.

Henikoff J.G., Greene E.A., Pietrokovski S., and Henikoff S. (2000) Increased coverage of protein families with the BLOCKS database servers. *Nucleic Acids Research* **28**:228–230.

Henikoff S. and Henikoff J.G. (1992) Amino acid substitution matrices from protein blocks. *Proceedings of the National Academy of Sciences of the USA* **89**:10915–10919.

Heylighen F. (1999) The growth of structural and functional complexity during evolution. In Heylighen F., Bollen J., and Riegler A. (eds.) *The Evolution of Complexity*, pp. 17–44. New York: Kluwer.

Hirschberg D. (1975) A linear space algorithm for computing maximal common subexpressions. *Communications of the Association of Computing Machinery* **18**:341–343.

Hofacker I.L., Fontana W., Stadler P.F., Bonhoeffer L.S., Tacker M., and Schuster P. (1994) Fast folding and comparison of RNA secondary structures. *Chemical Monthly* **125**:167–188.

Hopcroft J.E. and Ullman J.D. (1979) *Introduction to Automata Theory, Languages, and Computation*. Reading, MA: Addison-Wesley.

Hoskins R.A., Smith C.D., Carlson J.W., Carvalho A.B., Halpern A., Kaminker J.S., Kennedy C., Mungall C.J., Sullivan B.A., Sutton G.G., Yasuhara J.C., Wakimoto B.T., Myers E.W., Celniker S.E., Rubin G.M., and Karpen G.H. (2002) Heterochromatic sequences in a *Drosophila* whole-genome shotgun assembly. *Genome Biology* **3**:0085.1–0085.16.

Howe K.L., Chothia T., and Durbin R. (2002) GAZE: a generic framework for the integration of gene-prediction data by dynamic programming. *Genome Research* **12**:1418–1427.

Huang X., Adams M. D., Zhou H., and Kerlavage A. R. (1997) A tool for analyzing and annotating genomic sequences. *Genomics* **46**:35–45.

International Human Genome Sequencing Consortium (2001) Initial sequencing and analysis of the human genome. *Nature* **409**:860–921.

International Human Genome Sequencing Consortium (2004) Finishing the euchromatic sequence of the human genome. *Nature* **431**:931–945.

Jaakkola T. S. and Haussler D. (1999) Exploiting generative models in discriminative classifiers. *Advances in Neural Information Processing Systems* **11**:487–493.

Jeffreys A. J., Wilson V., and Thein S. L. (1985) Individual-specific "fingerprints" of human DNA. *Nature* **316**:76–79.

Jelinek F. (1997) *Statistical Methods for Speech Recognition.* Bradford, MA: Bradford Books.

Jelinek F. and Mercer R. L. (1980) Interpolated estimation of Markov source parameters. In *Proceedings of the Workshop on Pattern Recognition in Practice*, May 1980,

Jensen F. V. (2001) *Bayesian Networks and Decision Graphs.* New York: Springer.

Jenuwein T. and Allis C. D. (2001) Translating the histone code. *Science* **293**:1074–1080.

Johansen F. T. (1996) A comparison of hybrid HMM architectures using global discriminative training. In *Proceedings of the 4th International Conference on Spoken Language Processing*, pp. 498–501.

Jordan M. I., Ghahramani Z., Jaakkola T. S., and Saul L. K. (1999) An introduction to variational methods for graphical methods. In Jordan M. I. (ed.) *Learning in Graphical Models*, pp. 105–162. Cambridge, MA: MIT Press.

Jukes T. H. and Cantor C. R. (1969) Evolution of protein molecules. In Munro H. N. (ed.) *Mammalian Protein Metabolism*, pp. 21–132. New York: Academic Press.

Käll L., Krogh A., and Sonnhammer E. L. (2005) An HMM posterior decoden for sequence feature prediction that includes homology information. *Bioinformatics* **21** (Suppl. 1): i251–i257.

Kamal M., Xie X., and Lander E. S. (2006) A large family of ancient repeat elements in the human genome is under strong selection. *Proceedings of the National Academy of Sciences of the USA* **103**:2740–2745.

Karlin S. and Altschul S. F. (1990) Methods for assessing the statistical significance of molecular sequence features by using general scoring schemes. *Proceedings of the National Academy of Sciences of the USA* **87**:2264–8.

Kasami T. (1965). *An Efficient Recognition and Syntax-Analysis Algorithm for Context-Free Languages, Scientific Report AFCRL-65-758.* Bedford, MA: Air Force Cambridge Research Laboratory.

Katz S. M. (1987) Estimation of probabilities from sparse data for the language model component of a speech recognizer. *IEEE Transactions on Acoustics, Speech, and Signal Processing* **35**: 400–401.

Kaufman L. (1998) Solving the quadratic programming problem arising in support vector classification. In Scholkopf B., Burges C. J. C., and Smola A. J. (eds.) *Advances in Kernel Methods: Support Vector Learning*, pp. 147–167. Cambridge, MA: MIT Press.

Keilson J. (1979) *Markov Chain Models: Rarity and Exponentiality.* New York: Springer.

Kent W. J. (2002) BLAT: the BLAST-like alignment tool. *Genome Research* **12**:656–664.

Kent W. J., Sugnet C. W., Furey T. S., Roskin K. M., Pringle T. H., Zahler A. M., and Haussler D. (2002) The human genome browser at UCSC. *Genome Research* **12**:996–1006.

Kimura M. (1980) A simple method for estimating evolutionary rate of base substitutions through comparative studies of nucleotide sequences. *Journal of Molecular Evolution* **16**: 111–120.

Kingsbury N. G. and Rayner P. J. W. (1971) Digital filtering using logarithmic arithmetic. *Electronic Letters* **7**:56–58.

Kohavi R. and Sahami M. (1996) Error-based and entropy-based discretization of continuous features. In *Proceedings of the 2nd International Conference on Knowledge Discovery and Data Mining*, pp. 114–119.

Korf I. (2004) Gene finding in novel genomes. *BMC Bioinformatics* **5**:59.

Korf I., Flicek P., Duan D., and Brent M. R. (2001) Integrating genomic homology into gene structure prediction. *Bioinformatics* **17**:S140–S148.

Korf I., Yandell M., and Bedell J. (2003) *BLAST*. Sebastopol, CA: O'Reilly.

Krogh A. (1994) Hidden Markov models for labeled sequences. In *Proceedings of the 12th IAPR International Conference on Pattern Recognition*, pp. 140–144.

Krogh A. (1997) Two methods for improving performance of an HMM and their application for gene finding. In *Proceedings of the 5th International Conference on Intelligent Systems for Molecular Biology*, pp. 179–186.

Krogh A. (1998) An introduction to hidden Markov models for biological sequences. In Salzberg S. L., Searls D. B., and Kasif S. (eds.) *Computational Methods in Molecular Biology*, pp. 45–62. Amsterdam: Elsevier.

Krogh A. (2000) Using database matches with HMMGene for automated gene detection in *Drosophila*. *Genome Research* **10**:523–528.

Krogh A., Mian I. S., and Haussler D. (1994) A hidden Markov model that finds genes in *E. coli* DNA. *Nucleic Acids Research* **22**:4768–4778.

Koza J. (1992) *Genetic Programming: On the Programming of Computers by Means of Natural Selection*. Cambridge, MA: MIT Press.

Kullback S. (1997) *Information Theory and Statistics*. New York: Dover.

Kulkarni O. C., Vigneshwar R., Jayaraman V. K., and Kulkarni B. D. (2005) Identification of coding and non-coding sequences using local Hölder exponent formalism. *Bioinformatics* **21**:3818–3823.

Kulp D. and Haussler D. (1997) Integrating database homology in a probabilistic gene structure model. *Pacific Symposium on Bioinformatics* **2**:232–244.

Kulp D., Haussler D., Reese M., and Eeckman F. (1996) A generalized hidden Markov model for the recognition of human genes in DNA. In *Proceedings of the 4th International Conference on Intelligent Systems for Molecular Biology*, pp. 134–142.

Kurtz S., Phillippy A., Delcher A. L., Smoot M., Shumway M., Antonescu C., and Salzberg S. L. (2004) Versatile and open software for comparing large genomes. *Genome Biology* **5**:R12.1–R12.9.

Lam F., Alexandersson M., and Pachter L. (2003) Picking alignments from (Steiner) trees. *Journal of Computational Biology* **10**:509–520.

Lander E. S. and Waterman M. S. (1988) Genomic mapping by fingerprinting random clones: a mathematical analysis. *Genomics* **2**:231–239.

Landry J. R., Mager D. L., and Wilhelm B. T. (2003) Complex controls: the role of alternative promoters in mammalian genomes. *Trends in Genetics* **19**:640–648.

Lapedes A., Barnes C., Burks C., Farber R., and Sirotkin K. (1989) Application of neural networks and other machine learning algorithms to DNA sequence analysis. In Bell G. and Marr T. (eds.) *Computers and DNA: SFI Studies in the Sciences of Complexity*, pp. 157–182. Reading, MA: Addison-Wesley.

Larsen F., Gundersen G., Lopez R., and Prydz H. (1992) CpG islands as gene markers in the human genome. *Genomics* **13**:1095–1107.

Lawrence C. E., Altschul S. F., Boguski M. S., Liu J. S., Neuwald A. F., and Wootton J. C. (1993) Detecting subtle sequence signals: a Gibbs sampling strategy for multiple alignment. *Science* **262**:208–214.

Lee C., Grasso C., and Sharlow M. F. (2002) Multiple sequence alignment using partial order graphs. *Bioinformatics* **18**:452–464.

Lewin B. (2003) *Genes VIII*. New York: Prentice-Hall.

Lian Y. and Garner H. R. (2005) Evidence for the regulation of alternative splicing via complementary DNA sequence repeats. *Bioinformatics* **8**:1358–1364.

Lim L. P., and Burge C. B. (2001) A computational analysis of sequence features involved in recognition of short introns. *Proceedings of the National Academy of Sciences of the USA* **98**:11193–11198.

Liò P. and Goldman N. (1998) Models of molecular evolution and phylogeny. *Genome Research* **8**:1233–1244.

Loots G. G., Ovcharenko I., Pachter L., Dubchak I., and Rubin E. M. (2002) rVista for comparative sequence-based discovery of functional transcriptional factor binding sites. *Genome Research* **12**:832–839.

Lowe T. M. and Eddy S. R. (1999) A computational screen for methylation guide snoRNAs in yeast. *Science* **283**:1168–1171.

Lukashin A. V. and Borodovsky M. (1998) GeneMark.hmm: new solutions for gene finding. *Nucleic Acids Research* **26**:1107–1115.

Mackey A. (2005) GLEAN: improved eukaryotic gene prediction by statistical consensus of gene evidence. Poster presented at *Genome Informatics Conference*, October 28, 2005.

Maglott D. R., Katz K. S., Sicotte H., and Pruitt K. D. (2000) NCBI's LocusLink and RefSeq. *Nucleic Acids Research* **28**:126–128.

Majoros W. H. and Salzberg S. L. (2004) An empirical analysis of training protocols for probabilistic gene finders. *BMC Bioinformatics* **5**:206.

Majoros W. H., Pertea M., Antonescu C., and Salzberg S. L. (2003) GlimmerM, Exonomy and Unveil: three *ab initio* eukaryotic genefinders. *Nucleic Acids Research* **31**:3601–3604.

Majoros W. H., Pertea M., and Salzberg S. (2004) TIGRscan and GlimmerHMM: two open source *ab initio* eukaryotic gene finders. *Bioinformatics* **20**:2878–2879.

Majoros W. H., Pertea M., Delcher A. L., and Salzberg S. L. (2005a) Efficient decoding algorithms for generalized hidden Markov model gene finders. *BMC Bioinformatics* **6**:16.

Majoros W. H., Pertea M., and Salzberg S. L. (2005b) Efficient implementation of a generalized pair hidden Markov model for comparative gene finding. *Bioinformatics* **21**:1782–1788.

Manly B. F. J. (1994) *Multivariate Statistical Methods: A Primer*, 2nd edn. New York: Chapman and Hall.

Manning C. and Schütze H (1999) *Foundations of Statistical Natural Language Processing*. Cambridge, MA: MIT Press.

Marashi S. A., Goodarzi H., Sadeghi M., Eslahchi C., and Pezeshk H. (2006) Importance of RNA secondary structure information for yeast donor and acceptor splice site predictions by neural networks. *Computational Biology and Chemistry* **30**:50–57.

Markov K., Nakagawa S., and Nakamura S. (2001) Discriminative training of HMM using maximum normalized likelihood algorithm. In *Proceedings of the International Conference on Acoustics, Speech and Signal Processing*, pp. 497–500.

Matlin A.J., Clark F., and Smith C.W. (2005) Understanding alternative splicing: towards a cellular code. *Nature Reviews: Molecular Cell Biology* **6**:386–398.

McAuliffe J.D., Pachter L., and Jordan M.I. (2004) Multiple-sequence functional annotation and the generalized hidden Markov phylogeny. *Bioinformatics* **20**:1850–1860.

McCaskill J.S. (1990) The equilibrium partition function and base pair binding probabilities for RNA secondary structure. *Biopolymers* **29**:1105–1119.

Mealy G.H. (1955) A method for synthesizing sequential circuits. *Bell System Technical Journal* **34**:1045–1079.

Meyer I.M. and Durbin R. (2002) Comparative *ab initio* prediction of gene structures using pair HMMs. *Bioinformatics* **18**:1309–1318.

Meyer I.M. and Durbin R. (2004) Gene structure conservation aids similarity based gene prediction. *Nucleic Acids Research* **32**:776–783.

Mitchell T. (1997) *Machine Learning*. New York: McGraw-Hill.

Mitrophanov A.Y. and Borodovsky M. (2006) Statistical significance in biological sequence analysis. *Briefings in Bioinformatics* **7**:2–24.

Mizrahi A. and Sullivan M. (1986) *Calculus and Analytic Geometry*, 2nd edn. Belmont, CA: Wadsworth.

Moore E.F. (1956) Gedanken experiments on sequential machines. In Shannon C.E. and McCarthy J. (eds.) *Automata Studies*, pp. 129–153. Princeton, NJ: Princeton University Press.

Mouse Genome Sequencing Consortium (2002) Initial sequencing and comparative analysis of the mouse genome. *Nature* **420**:520–562.

Mural R.J., Adams M.D., Myers E.W., Smith H.O., Miklos G.L., Wides R., Halpern A., Li P.W., Sutton G.G., Nadeau J., Salzberg S.L., Holt R.A., Kodira C.D., Lu F., Chen L., Deng Z., Evangelista C.C., Gan W., Heiman T.J., Li J., Li Z., Merkulov G.V., Milshina N.V., Naik A.K., Qi R., Shue B.C., Wang A., Wang J., Wang X., Yan X., Ye J., Yooseph S., Zhao Q., Zheng L., Zhu S.C., Biddick K., Bolanos R., Delcher A.L., Dew I.M., Fasulo D., Flanigan M.J., Huson D.H., Kravitz S.A., Miller J.R., Mobarry C.M., Reinert K., Remington K.A., Zhang Q., Zheng X.H., Nusskern D.R., Lai Z., Lei Y., Zhong W., Yao A., Guan P., Ji R.R., Gu Z., Wang Z.Y., Zhong F., Xiao C., Chiang C.C., Yandell M., Wortman J.R., Amanatides P.G., Hladun S.L., Pratts E.C., Johnson J.E., Dodson K.L., Woodford K.J., Evans C.A., Gropman B., Rusch D.B., Venter E., Wang M., Smith T.J., Houck J.T., Tompkins D.E., Haynes C., Jacob D., Chin S.H., Allen D.R., Dahlke C.E., Sanders R., Li K., Liu X., Levitsky A.A., Majoros W.H., Chen Q., Xia A.C., Lopez J.R., Donnelly M.T., Newman M.H., Glodek A., Kraft C.L., Nodell M., Ali F., An H.J., Baldwin-Pitts D., Beeson K.Y., Cai S., Carnes M., Carver A., Caulk P.M., Center A., Chen Y.H., Cheng M.L., Coyne M.D., Crowder M., Danaher S., Davenport L.B., Desilets R., Dietz S.M., Doup L., Dullaghan P., Ferriera S., Fosler C.R., Gire H.C., Gluecksmann A., Gocayne J.D., Gray J., Hart B., Haynes J., Hoover J., Howland T., Ibegwam C., Jalali M., Johns D., Kline L., Ma D.S., MacCawley S., Magoon A., Mann F., May D., McIntosh T.C., Mehta S., Moy L., Moy M.C., Murphy B.J., Murphy S.D., Nelson K.A., Nuri Z., Parker K.A., Prudhomme A.C., Puri V.N., Qureshi H., Raley J.C., Reardon M.S., Regier M.A., Rogers Y.H., Romblad D.L., Schutz J., Scott J.L., Scott R., Sitter C.D., Smallwood M., Sprague A.C., Stewart E., Strong R.V., Suh E., Sylvester K., Thomas R., Tint N.N., Tsonis C., Wang G., Wang G., Williams M.S., Williams S.M., Windsor S.M., Wolfe K., Wu M.M., Zaveri J., Chaturvedi K., Gabrielian A.E., Ke Z., Sun J., Subramanian G., Venter J.C., Pfannkoch C.M., Barnstead M., and Stephenson L.D. (2002) A comparison of whole-genome shotgun-derived mouse chromosome 16 and the human genome. *Science* **296**:1661–1671.

Murphy P.M. and Aha D.W. (1994) *UCI Repository of Machine Learning Databases*, Irvine, CA: University of California, Department of Information and Computer Science. Available online at www.ics.uci.edu/~mlearn/MLRepository.html/

Murthy S.K., Kasif S., and Salzberg S. (1994) A system for induction of oblique decision trees. *Journal of Artificial Intelligence Research* **2**:1–32.

Needleman S. and Wunsch C. (1970) A general method applicable to the search for similarities in the amino acid sequence of two proteins. *Journal of Molecular Biology* **48**:443–453.

Ng A.Y. and Jordan M.I. (2002) On discriminative vs. generative classifiers: a comparison of logistic regression and naive Bayes. *Advances in Neural Information Processing Systems (NIPS)* **14**:841–848.

Normandin Y. (1996) Maximum mutual information estimation of hidden Markov models. In Lee C.-H., Soong F.K., and Paliwal K.K. (eds.) *Automatic Speech and Speaker Recognition*, pp. 58–81. New York: Kluwer.

Normark S., Bergstrom S., Edlund T., Grundstrom T., Jaurin B., Lindberg F.P., and Olsson O. (1983) Overlapping genes. *Annual Review of Genetics* **17**:499–525.

Ohler U., Stemmer G., Harbeck S., and Niemann H. (2000) Stochastic segment models of eukaryotic promoter regions. *Proceedings of the Pacific Symposium on Biocomputing* **5**:377–388.

Ohler U., Niemann H., Liao G., and Rubin G.M. (2001) Joint modeling of DNA sequence and physical properties to improve eukaryotic promoter recognition. *Bioinformatics* **17**:S199–S206.

Ohler U., Liao G., Niemann H., and Rubin G. (2002) Computational analysis of core promoters in the *Drosophila* genome. *Genome Biology* **3(12)**:r0087.1–r0087.12.

Ohler U., Shomron N., and Burge C.B. (2005) Recognition of unknown conserved alternatively spliced exons. *PLoS Computational Biology* **1**(2):e15.

Oliver J.L., Carpena P., Hackenberg M., and Bernaola-Galván P (2004) IsoFinder: computational prediction of isochores in genome sequences. *Nucleic Acids Research* **32**:W287–W292.

Osuna E., Freund R., and Girosi F. (1997) An improved training algorithm for support vector machines. In *Proceedings of the IEEE Workshop on Neural Networks for Signal Processing*, pp. 276–285.

Pachter L., Batzoglou S., Spitkovsky V.I., Banks E., Lander E.S., Kleitman D.J., and Berger B. (1999) A dictionary based approach for gene annotation. *Journal of Computational Biology* **6**:419–430.

Pachter L., Alexanderson M., and Cawley S. (2002) Applications of generalized pair hidden Markov models to alignment and gene finding problems. *Journal of Computational Biology* **9**:389–399.

Parra G., Agarwal P., Abril J.F., Wiehe T., Fickett J.W., and Guigó R. (2003) Comparative gene prediction in human and mouse. *Genome Research* **13**:108–117.

Patterson D., Yasuhara K., and Ruzzo W.L. (2002) Pre-mRNA secondary structure prediction aids splice site prediction. *Pacific Symposium on Bioinformatics* **7**:223–234.

Pavesi A., De Iaco B., Granero M.I., and Porati A. (1997) On the informational content of overlapping genes in prokaryotic and eukaryotic viruses. *Journal of Molecular Evolution* **44**:625–631.

Pearl J. (1991) *Probabilistic Reasoning in Intelligent Systems: Networks of Plausible Inference*, 2nd edn. Los Altos, CA: Morgan Kaufmann.

Pearson W.R. and Wood T.C. (2001) Statistical significance in biological sequence comparison. In Balding D.J., Bishop M., and Cannings C. (eds.) *Handbook of Statistical Genetics*, pp. 39–65. New York: John Wiley.

Pedersen J. S. and Hein J. (2003) Gene finding with a hidden Markov model of genome structure and evolution. *Bioinformatics* **19**:219–227.

Pertea M. (2005) The Glimmer HMM Home Page. Available online at: www.cbcb.umd.edu/software/GlimmerHMM

Pertea M., Lin X. and Salzberg S. L. (2001) GeneSplicer: a new computational method for splice site prediction. *Nucleic Acids Research* **29**:1185–1190.

Pertea M. and Salzberg S. L. (2002) Computational gene finding in plants. *Plant Molecular Biology* **48**:48–49.

Platt J. (1998) *Sequential Minimal Optimization: A Fast Algorithm for Training Support Vector Machines*, Microsoft Research Technical Report MSR-TR-98-14. Redmond, WA: Microsoft Corporation.

Pontius J. U., Wagner L., and Schuler G. D. (2003) UniGene: a unified view of the transcriptome. In McEntyre J. and Ostell J. (eds.) *The NCBI Handbook*, pp. 21–1–21–12. Bethesda, MD: National Center for Biotechnology Information.

Pop M., Salzberg S. L., and Shumway M. (2002) Genome sequence assembly: algorithms and issues. *IEEE Computer* **35**:47–54.

Potamianos G. and Jelinek F. (1998) A study of *n*-gram and decision tree letter language modeling methods. *Speech Communication*, **24**:171–192.

Powell M. J. D. (1981) *Nonlinear Optimization*. New York: Academic Press.

Pozo R. (1997) Template numerical toolkit for linear algebra: high performance programming with C++ and the Standard Template Library. *International Journal of Supercomputer Applications and High Performance Computing* **11**:251–263.

Press W. H., Flanner B. P., Teukolsky S. A., and Vetterling W. T. (1992) *Numerical Recipes in C: The Art of Scientific Computing*, 2nd edn. Cambridge: Cambridge University Press.

Provost F. J. and Hennessy D. N. (1994) Distributed machine learning: scaling up with coarse-grained parallelism. In *Proceedings of the 2nd International Conference on Intelligent Systems for Molecular Biology*, pp. 340–347.

Pruitt K. D., Tatusova, T., and Maglott D. R. (2005) NCBI Reference Sequence (RefSeq): a curated non-redundant sequence database of genomes, transcripts and proteins. *Nucleic Acids Research* 33:D501–D504.

Quinlan R. (1993) *C4.5: Programs for Machine Learning*. Los Altos, CA: Morgan Kaufmann.

Rabiner L. R. (1989) A tutorial on hidden Markov models and selected applications in speech recognition. *Proceedings of the IEEE* **77**:257–286.

Rätsch G, Sonnenburg S., and Schölkopf B (2005) RASE: recognition of alternatively spliced exons in C.elegans. *Bioinformatics* **21** (Suppl. 1):i369–i377.

Reese M. and Eeckman F. (1995) Novel neural network prediction systems for human promoters and splice sites. In Searls GSD., Fickett J., Noordewier M. (eds.) *Proceedings of the Workshop on Gene-Finding and Gene Structure Prediction*, Philadelphia, PA, pp. 311–324.

Reese M. G., Eeckman F. H., Kulp D., and Haussler D. (1997) Improved splice site detection in Genie. *Journal of Computational Biology* **4**:311–323.

Reese M. G., Hartzell G., Harris N. L., Ohler U., and Lewis S. E. (2000) Genome annotation assessment in *Drosophila melanogaster*. *Genome Research* **10**:483–501.

Reichl W. and Ruske G. (1995) Discriminative training for continuous speech recognition. In *Proceedings of the 4th European Conference on Speech Communication and Technology*, pp. 537–540.

Rissanen J. (1978) Modeling by shortest data description. *Automatica* **14**:465–471.

Ristad E. S. and Thomas R. G. (1997) Hierarchical non-emitting Markov models. In *Proceedings of the 35th Annual Meeting of the Association for Computational Linguistics and 8th Conference of the European Chapter of the Association for Computational Linguistics.*

Rivas E. and Eddy S. R. (1999) A dynamic programming algorithm for RNA structure prediction including pseudoknots. *Journal of Molecular Biology* **285**:2053–2068.

Rivas E. and Eddy S. R. (2000) Secondary structure alone is generally not statistically significant for the detection of noncoding RNAs. *Bioinformatics* **16**:583–605.

Rivas E. and Eddy S. R. (2001) Noncoding RNA gene detection using comparative sequence analysis. *BMC Bioinformatics* **2**:8.

Rombauts S., Florquin K., Lescot M., Marchasl K., Rouzé P., and Van de Peer Y. (2003) Computational approaches to identify promoters and *cis*-regulatory elements in plant genomes. *Plant Physiology* **132**:1162–1176.

Rosenblatt F. (1958) The Perceptron: a probabilistic model for information storage and organization in the brain. *Psychological Review* **65**:386–408.

Roth V. and Steinhage V. (1999) Nonlinear discriminant analysis using kernel functions. In *Proceedings of the 12th International Conference on Advances in Neural Information Processing Systems*, pp. 568–574.

Saetrom P, Sneve R., Kristiansen K. I., Snøve O. J., Grünfeld T., Rognes T., and Seeberg E. (2005) Predicting non-coding RNA genes in *Escherichia coli* with boosted genetic programming. *Nucleic Acids Research* **33**:3263–3270.

Saeys Y. (2004) Feature selection for classification of nucleic acid sequences. Ph.D. thesis, University of Ghent, Belgium.

Saeys Y., Degroeve S., Aeyels D., Rouzé P., and Van de Peer Y. (2004) Feature selection for splice site prediction: a new method using EDA-based feature ranking. *BMC Bioinformatics* **5**:64.

Saitou N. and Nei M. (1987) The neighbor-joining method: a new method for reconstructing phylogenetic trees. *Molecular Biology and Evolution* **4**:406–425.

Sakai M., Yoneda M., and Hase H. (1998) A new robust quadratic discriminant function. In *Proceedings of the 14th International Conference on Pattern Recognition*, pp. 99–102.

Salzberg S. L. (1999) On comparing classifiers: a critique of current research and methods. *Data Mining and Knowledge Discovery* **1**:1–12.

Salzberg S. L., Delcher A. L., Kasif S., and White O. (1998a) Microbial gene identification using interpolated Markov models. *Nucleic Acids Research* **26**:544–548.

Salzberg S. L., Pertea M., Delcher A. L., Gardner M. J., and Tettelin H. (1998b) Interpolated Markov models for eukaryotic gene finding. *Genomics* **59**:24–31.

Schadt E. and Lange K. (2002) Codon and rate variation models in molecular phylogeny. *Molecular Biology and Evolution* **19**:1534–1549.

Schlüter R., Macherey W., Müller B., and Ney H. (2001) Comparison of discriminative training criteria and optimization methods for speech recognition. *Speech Communication* **34**:287–310.

Schultz J., Milpetz F., Bork P., and Ponting C. P. (1998) SMART, a simple modular architecture research tool: identification of signaling domains. *Proceedings of the National Academy of Sciences of the USA* **95**:5857–5864.

Schwartz R. and Chow Y.-L. (1990) The *N*-best algorithm: an efficient and exact procedure for finding the *N* most likely hypotheses. In *Proceedings of the IEEE Conference on Aconstics, Speech, and Signal Processing*, pp. 81–84.

Seneff S., Wang C., and Burge C.B. (2004) Gene structure prediction using an orthologous gene of known exon–intron structure. *Applied Bioinformatics* **3**:81–90.

Servant F., Bru C., Carre S., Courcelle E., Gouzy J., Peyruc D., and Kahn D. (2002) ProDom: automated clustering of homologous domains. *Briefings in Bioinformatics* **3**:246–251.

Shannon C.E. (1948) A mathematical theory of communication. *Bell System Technical Journal* **27**:379–423, 623–656.

Shmatkov A.M., Melikyan A.A., Chernousko F.L., and Borodovsky M. (1999) Finding prokyarotic genes by the "frame-by-frame" algorithm: targeting gene starts and overlapping genes. *Bioinformatics* **15**:874–886.

Siepel A. and Haussler D. (2004a) Combining phylogenetic and hidden Markov models in biosequence analysis. *Journal of Computational Biology* **11**:413–428.

Siepel A. and Haussler D. (2004b) Computational identification of evolutionarily conserved exons. In *Research in Computational Molecular Biology (RECOMB'04)*, pp. 277–286.

Siepel A. and Haussler D. (2004c) Phylogenetic estimation of context-dependent substitution rates by maximum likelihood. *Molecular Biology and Evolution* **21**:468–488.

Siepel A. and Haussler D. (2005) Phylogenetic hidden Markov models. In Nielsen R. (ed.) *Statistical Methods in Molecular Evolution*, pp. 1034–1050. New York: Springer.

Simonoff J.S. (1996) *Smoothing Methods in Statistics*. New York: Springer.

Sinha S., van Nimwegen E., and Siggia E.D. (2003) A probabilistic method to detect regulatory modules. *Bioinformatics* **19**:i292–i301.

Smit A.F.A., and Green P. (1996) *RepeatMasker*. Available online at http://ftp.genome.waschington.edu/RM/ RepeatMasker.html/

Smith T.F. and Waterman M.S. (1981) Identification of common molecular subsequences. *Journal of Molecular Biology* **147**:195–197.

Sneath P.H.A. and Sokal R.R. (1973) *Numerical Taxonomy*. San Francisco, CA: W.H. Freeman.

Snyder E.E. (1994) Identification of protein coding regions in genomic DNA. Ph.D. thesis, University of Colorado, Boulder, CO.

Snyder E.E. and Stormo G.D. (1993) Identification of coding regions in genomic DNA sequences: an application of dynamic programming and neural networks. *Nucleic Acids Research* **21**:607–613.

Sokal R.R. and Rohlf F.J. (1995) *Biometry: The Principles and Practice of Statistics in Biological Research*. New York: W.H. Freeman.

Solovyev V.V. and Shahmuradov I.A. (2003) PromH: promoters identification using orthologous genomic sequences. *Nucleic Acids Research* **31**:3540–3545.

Solovyev V.V., Salamov A.A., and Lawrence C.B. (1995) Identification of human gene structure using linear discriminant functions and dynamic programming. In *Proceedings of the 3rd International Conference on Intelligent Systems for Molecular Biology*, pp. 367–375.

Sonnenburg S. (2002) New methods for splice site recognition. Diploma thesis, Humboldt University, Berlin, Germany.

Sonnenburg S., Zien A., and Rätsch G. (2006) ARTS: accurate recognition of transcription starts in human. In *Proceedings of the 14th International Conference on Intelligent Systems for Molecular Biology*, pp. 472–480.

Sorek R., Ast G., and Graur D. (2002) Alu-containing exons are alternatively spliced. *Genome Research* **12**:1060–1067.

Staden R. (1984) Computer methods to locate signals in nucleic acid sequences. *Nucleic Acids Research* **12**:505–519.

Stajich J. E., Block D., Boulez K., Brenner S. E., Chervitz S. A., Dagdigian C., Fuellen G., Gilbert J. G., Korf I., Lapp H., Lehvaslaiho H., Matsalla C., Mungall C. J., Osborne B. I., Pocock M. R., Schattner P., Senger M., Stein L. D., Stupka E., Wilkinson M. D., and Birney E. (2002) The Bioperl toolkit: perl modules for the life sciences. *Genome Research* **12**:1611–1618.

Stanke M. and Waack S. (2003) Gene prediction with a hidden Markov model and a new intron submodel. *Bioinformatics* **19**:II215–II225.

Stein L. (2001) Genome annotation: from sequence to biology. *Nature Reviews: Genetics* **2**: 493–503.

Stormo G. D. and Haussler D. (1994) Optimally parsing a sequence into different classes based on multiple types of evidence. In *Proceedings of the 2nd International Conference on Intelligent Systems for Molecular Biology*, pp. 369–375.

Suzek B. E., Ermolaeva M. D., Schreiber M., and Salzberg S. L. (2001) A probabilistic method for identifying start codons in bacterial genomes. *Bioinformatics* **17**:1123–1130.

Tikhonov A. N. (1963) Solution of incorrectly formulated problems and the regularization method. *Soviet Mathematics, Doklady* **4**:1035–1038.

Tipping M. E. (2001) Sparse Bayesian learning and the relevance vector machine. *Journal of Machine Learning Research* **1**:211–244.

Tong S. and Koller D. (2000) Restricted Bayes optimal classifiers. In *Proceedings of the 17th National Conference on Artificial Intelligence*, pp. 658–664.

Toutanova K., Mitchell M., and Manning C. D. (2003) Optimizing local probability models for statistical parsing. In *Proceedings of the 14th European Conference on Machine Learning*, pp. 409–420.

Tveter D. (1998) *The Pattern Recognition Basis of Artificial Intelligence*. Indianapolis, IN: Wiley-IEEE Computer Society Press.

Uberbacher E. C. and Mural R. J. (1991) Locating protein coding regions in human DNA sequences using a multiple-sensor neural network approach. *Proceedings of the National Academy of Sciences of the USA* **88**:11261–11265.

Usuka J., Zhu W., and Brendel V. (2000) Optimal spliced alignment of homologous cDNA to a genomic DNA template. *Bioinformatics* **16**:203–224.

Vapnik V. (1998) *Statistical Learning Theory*. New York: John Wiley.

Venter J. C., Smith H. O., and Hood L. (1996) A new strategy for genome sequencing. *Nature* **381**:364–366.

Venter J. C., Adams M. D., Myers E. W., Li P. W., Mural R. J., Sutton G. G., Smith H. O., Yandell M., Evans C. A., Holt R. A., Gocayne J. D., Amanatides P., Ballew R. M., Huson D. H., Wortman J. R., Zhang Q., Kodira C. D., Zheng X. H., Chen L., Skupski M., Subramanian G., Thomas P. D., Zhang J., Gabor Miklos G. L., Nelson C., Broder S., Clark A. G., Nadeau J., McKusick V. A., Zinder N., Levine A. J., Roberts R. J., Simon M., Slayman C., Hunkapiller M., Bolanos R., Delcher A., Dew I., Fasulo D., Flanigan M., Florea L., Halpern A., Hannenhalli S., Kravitz S., Levy S., Mobarry C., Reinert K., Remington K., Abu-Threideh J., Beasley E., Biddick K., Bonazzi V., Brandon R., Cargill M., Chandramouliswaran I., Charlab R., Chaturvedi K., Deng Z., Di Francesco V., Dunn P., Eilbeck K., Evangelista C., Gabrielian A. E., Gan W., Ge W., Gong F., Gu Z., Guan P., Heiman T. J., Higgins M. E., Ji R. R., Ke Z., Ketchum K. A., Lai Z., Lei Y., Li J., Li Z., Liang Y., Lin X., Lu F., Merkulov G. V., Milshina N., Moore H. M., Naik A. K., Narayan V. A., Neelam B., Nusskern D., Rusch D. B., Salzberg S., Shao W., Shue B., Sun J., Wang Z., Wang A., Wang X., Wang J., Wei M., Wides R., Xiao C., Yan C., Yao A., Ye J., Zhan M., Zhang W., Zhang H., Zhao Q., Zheng L., Zhong F., Zhong W., Zhu S., Zhao S., Gilbert D., Baumhueter S., Spier

G., Carter C., Cravchik A., Woodage T., Ali F., An H., Awe A., Baldwin D., Baden H., Barnstead M., Barrow I., Beeson K., Busam D., Carver A., Center A., Cheng M. L., Curry L., Danaher S., Davenport L., Desilets R., Dietz S., Dodson K., Doup L., Ferriera S., Garg N., Gluecksmann A., Hart B., Haynes J., Haynes C., Heiner C., Hladun S., Hostin D., Houck J., Howland T., Ibegwam C., Johnson J., Kalush F., Kline L., Koduru S., Love A., Mann F., May D., McCawley S., McIntosh T., McMullen I., Moy L., Moy M., Murphy B., Nelson K., Pfannkoch C., Pratts E., Puri V., Qureshi H., Reardon M., Rodriguez R., Rogers Y. H., Romblad D., Ruhfel B., Scott R., Sitter C., Smallwood M., Stewart E., Strong R., Suh E., Thomas R., Tint N. N., Tse S., Vech C., Wang G., Wetter J., Williams M., Williams S., Windsor S., Winn-Deen E., Wolfe K., Zaveri J., Zaveri K., Abril J. F., Guigó R., Campbell M. J., Sjolander K. V., Karlak B., Kejariwal A., Mi H., Lazareva B., Hatton T., Narechania A., Diemer K., Muruganujan A., Guo N., Sato S., Bafna V., Istrail S., Lippert R., Schwartz R., Walenz B., Yooseph S., Allen D., Basu A., Baxendale J., Blick L., Caminha M., Carnes-Stine J., Caulk P., Chiang Y. H., Coyne M., Dahlke C., Mays A., Dombroski M., Donnelly M., Ely D., Esparham S., Fosler C., Gire H., Glanowski S., Glasser K., Glodek A., Gorokhov M., Graham K., Gropman B., Harris M., Heil J., Henderson S., Hoover J., Jennings D., Jordan C., Jordan J., Kasha J., Kagan L., Kraft C., Levitsky A., Lewis M., Liu X., Lopez J., Ma D., Majoros W. H., McDaniel J., Murphy S., Newman M., Nguyen N., Nguyen T., Nodell M., Pan S., Peck J., Peterson M., Rowe W., Sanders R., Scott J., Simpson M., Smith T., Sprague A., Stockwell T., Turner R., Venter E., Wang M., Wen M., Wu D., Wu M., Xia A., Zandieh A., and Zhu X. (2001) The sequence of the human genome. *Science* **291**:1304–1351.

Voorhees E. M. (1986) Implementing agglomerative hierarchical clustering algorithms for use in document retrieval. *Information Processing and Management* **22**:465–476.

Vinson J., DeCaprio D., Luoma S., and Galagan J. E. (2006) Gene prediction using conditional random fields. In: *The Biology of Genomes*, Cold Spring Harbor Laboratory, New York, May 10–14, 2006 (abstract).

Viterbi A. (1967) Error bounds for convolutional codes and an asymptotically optimal decoding algorithm. *IEEE Transactions on Information Theory* **13**:260–269.

Von Hippel P. T. (2005) Mean, median, and skew: correcting a textbook rule. *Journal of Statistics Education* **13**.

Wain H. M., Lovering R. C., Bruford E. A., Lush M. J., Wright M. W., and Povey S. (2002) Guidelines for human gene nomenclature. *Genomics* **79**:464–470.

Watson J. D. and Crick FHC. (1953) Molecular structure of nucleic acids. *Nature* **4356**:737–738.

Wheelan S. J., Church D. M., and Ostell J. M. (2001) Spidey: a tool for mRNA-to-genomic alignments. *Genome Research* **11**:1952–1957.

Wheeler D. L., Barrett T., Benson D. A., Bryant S. H., Canese K., Church D. M., DiCuccio M., Edgar R., Federhen S., Helmberg W., Kenton D. L., Khovayko O., Lipman D. J., Madden T. L., Maglott D. R., Ostell J., Pontius J. U., Pruitt K. D., Schuler G. D., Schriml L. M., Sequeira E., Sherry S. T., Sirotkin K., Starchenko G., Suzek T. O., Tatusov R., Tatusova T. A., Wagner L., and Yaschenko E. (2005) Database resources of the National Center for Biotechnology Information. *Nucleic Acids Research* **33**:D39–D45.

Wiehe T., Gebauer-Jung S., Mitchell-Olds T., and Guigó R. (2001) SGP-1: prediction and validation of homologous genes based on sequence alignments. *Genome Research* **9**:1574–1583.

Wingender E., Kel A. E., Kel O. V., Karas H., Heinemeyer T., Dietze P., Knuppel R., Romaschenko A. G., and Kolchanov N. A. (1997) TRANSFAC, TRRD and COMPEL: towards a federated database system on transcriptional regulation. *Nucleic Acids Research* **25**:265–268.

Wojtowicz W. M., Flanagan J. J., Millard S. S., Zipursky S. L., and Clemens J. C. (2004) Alternative splicing of *Drosophila Dscam* generates axon guidance receptors that exhibit isoform-specific homophilic binding. *Cell* **118**:619–633.

Wortman J. R., Haas B. J., Hannick L. I., Smith R. K., Maiti R., Ronning C. M., Chan A. P., Yu C., Ayele M., Whitelaw C. A., White O. R., and Town C. D. (2003) Annotation of the *Arabidopsis* genome. *Plant Physiology* **132**:461–468.

Wu C. H., Yeh L.-S. L., Guang H., Arminski L., Castro-Alvear J, Chen Y., Hu Z.-Z., Ledley R. S., Kourtesis P., Suzek B. E., Vinayaka C. R., Zhang J., and Barker W. C. (2003) The Protein Information Resource. *Nucleic Acids Research* **31**:345–347.

Xu Y. and Uberbacher E. C. (1996) Gene prediction by pattern recognition and homology search. In *Proceedings of the 4th International Conference on Intelligent Systems for Molecular Biology*, pp. 242–256.

Yan J. and Marr T. G. (2005) Computational analysis of 3′-ends of ESTs shows four classes of alternative polyadenylation in human, mouse, and rat. *Genome Research* **15**:369–375.

Yang Z. (1994) Maximum likelihood phylogenetic estimation from DNA sequences with variable rates over sites: approximate methods. *Journal of Molecular Evolution* **39**:306–314.

Yeh R.-F., Lim L. P., and Burge C. B. (2001) Computational inference of homologous gene structures in the human genome. *Genome Research* **11**:803–809.

Yeo G. and Burge C. B. (2004) Maximum entropy modeling of short sequence motifs with applications to RNA splicing signals. *Journal of Computational Biology* **11**:377–394.

Yeo G. W., Van Nostrand E., Holste D., Poggio T., and Burge C. B. (2005) Identification and analysis of alternative splicing events conserved in human and mouse. *Proceedings of the National Academy of Sciences of the USA* **102**:2850–2855.

Younger D. H. (1967) Recognition and parsing of context-free languages in time n^3. *Information and Control* **10**:189–208.

Yu P., Ma D., and Xu M. (2005) Nested genes in the human genome. *Genomics* **86**:414–422.

Zar J. H. (1996) *Biostatistical Analysis*, 3rd edn. Englewood Cliffs, NJ: Prentice-Hall.

Zhang M. Q. (1997) Identification of protein coding regions in the human genome by quadratic discriminant analysis. *Proceedings of the National Academy of Sciences of the USA* **94**:565–568.

Zhang M. Q. (2003) Prediction, annotation, and analysis of human promoters. *Cold Spring Harbor Laboratory Symposium in Quantitative Biology* **68**:217–225.

Zhang M. Q. and Marr T. G. (1993) A weight array method for splicing signal analysis. *Computer Applications in the Biosciences* **9**:499–509.

Zhang H., Hu J., Recce M., and Tian B. (2005) PolyA_DB: a database for mammalian mRNA polyadenylation. *Nucleic Acids Research* **33**:D116–D120.

Zhang L., Pavlovic V., Cantor C. R., and Kasif S. (2003) Human–mouse gene identification by comparative evidence integration and evolutionary analysis. *Genome Research* **13**:1190–1202.

Zhao J., Hyman L., and Moore C. (1999) Formation of mRNA 3′ ends in eukaryotes: mechanism, regulation, and interrelationships with other steps in mRNA synthesis. *Microbiology and Molecular Biology Reviews* **63**:405–445.

Zien A., Rätsch G., Mika S., Scholkopf B., Lengauer T., and Muller K.-R. (2000) Engineering support vector machine kernels that recognize translation initiation sites. *Bioinformatics* **16**:799–807.

Zuker M., Mathews D. H. and Turner D. H. (1999) Algorithms and thermodynamics for RNA secondary structure prediction: a practical guide. In Barciszewski J. and Clark B. F. C. (eds.) *RNA Biochemistry and Biotechnology*, pp. 11–43. New York: Kluwer.

Index

Note: page numbers in *italics* refer to figures and tables

abstract syntax trees 346, *347*
accumulators, GHMM 237, 238
AdaBoost 359
adenine 2–3
adjacency matrix 71–72
AIR program 370–371
algorithms 35–38
  computational complexity 62–63
  efficiency *63*
  greedy 74
  searching 66–68
  sensitivity–specificity curve 117
  sorting 66–68
alignment, multiple 55–56
alignment scores *296*
  approximate *297*
  clustering strategies 110
  Needleman–Wunsch 296, 307
alignment-based prediction 277–281
alleles 7
alphabet 75
alternation 75
alternative polyadenylation 369–370
alternative promoter recognition 369–370
alternative splicing 24, 95, 101, 369–370, *370*, 373
  alternative donor/acceptor sites 369–370
  classification 372, 373
  exon skipping 351
  gene-finding systems 326–327
  parsing 372
  predictions 370–371
  suboptimal path identification 372
ambiguity codes 101–102, 127
amino acids, alignment 290–291
anchor signals 93–94, 235
annealing, simulated 263, 349
annotation of function 24–25
application programming interface (API) 258–259
approximate correlation 116–117
Arabidopsis thaliana genes
  GHMM-based gene finders *263*
  gradient ascent procedure 262
  hidden Markov model 148, *157*
  intron lengths 250
  maximal dependence decomposition 203
argmax
  of function 38–39, 54
  hidden Markov models 175
array literal 36
assembler program 83
assignment 36
associative arrays 38
AUGUSTUS gene finder 359
  isochore-specific submodel training 249
  short noncoding feature modeling 250
  single-strand 223
auto-annotation pipelines 319–320, 373
automata 76–77

automatic classification 325–328
average conditional probability (ACP) 116–117

back-off models, HMM 171–173
Backward algorithm 163–165
  Baum–Welch algorithm with scaling 165–169
  dynamic programming 163–164, *178*
  posterior decoding 177
bacterial gene finding 209–211
  open reading frames 210
  Viterbi algorithm 210
base-caller 16–17, 83
basis step 33–34
Baum-Welch algorithm 159, *168*
  emission probability 167
  log-likelihood changes 167
  maximum likelihood solution 165
  naive 163–165
  with scaling 165–169
  training 162–169
  transition probability 167
Bayes model, naive 329–330, 354
Bayes' rule 42
Bayesian inverse 330
Bayesian network 330–331, *331*, 332
  dependency graph 207
  maximal dependence decomposition 205
  re-rooted phylogenetic tree 304
  splice site prediction 332
Bernoulli trial 48–49, 56
BFGS algorithm 313
binary search 67–68
binary trees 70
binning 50–51
binomial coefficient 31, 38
binomial distribution 48
  hypothesis testing 55–56
bit (binary digit) 58
BLAST hits, gene finder training sets 105–106
BLAST programs 22–23, 24–25, 269
  pairHMM 285
BLASTN 22, 269, 287
BLASTX 22, 271–272, 274, 275
BLAT program 269
BLOSUM matrix 295–296
Boltzmann constant 349, 376
boolean algebra 34
boolean variables 42
boosting 358–359
bootstrapping 359–360
  gene finder training data 108
branch points 361, 364–365
break statement 37
Burge's noncoding predecessor theorem 226–228

C4.5 decision tree algorithm 105, 337–339
CAP site weight matrices 363–364
CEM (conserved exon method) 279–281
chaperones 11
character literals 37
χ^2 goodness-of-fit test 56–57, 174
χ^2 test of homogeneity, interpolated Markov models 174
χ^2 test of independence 57–58
  Bayesian networks 330–331
  maximal dependence decomposition 204
Chomsky hierarchy 76, 372, 376
Chomsky normal form 378–379
chromatogram 16
chromosomes
  heterochromatic regions 17
  human genome 2
clades 13, 300
classification
  accuracy 113, 194–195, *198*, *305*, 325
    genetic programming 346
  alternative splicing 372, 373
  automatic 325–328
  binary 343
  discriminative modeling 382–383
  DNA interval 354
  entropy 338
  exons *114*, *355*
  Gene Ontology 381–382
  generative modeling 382–383
  genetic programming 347–348, *348*
  minimum error 176

classification (*cont.*)
  paradigm *326*
  test case 325
  weight array matrices experiments 197, *198*
cloning operator 347
cloning vector 16
clustering strategies
  average link 110
  gene finders 109–110
coding bias 318, 384
coding predecessors *226*
coding segments (CDSs) 21–22, 23, 83–85, 86, 89
  alignment-based gene finder 278
  conservation 268
  DNA interval classification 354
  forwardstrand *89*
  frame progression *91*
  phase progression *91*
  repeats 367
  reverse strand *90*, 90
  signals delimiting parts *85*
coding–noncoding boundaries, signal sensors 200–201
codon(s) 9–10, *10*, 15, 89–91
  phases 89
  position modeling 150, 158
  usage scores 278
  usage table 274
  see also start codon; stop codon
codon bias 10, 85–86, 127–130
  exon finding 130–134
coefficient of variation 47
combiners 275–277, 373
  decision tree 276–277
  linear 276
  parallelization 385–386
  voting 276
communication, inter-process 386
complementary DNA (cDNA) 19
  full-length 268
  gene finder training sets 105–106
complementation operator 86
composite event, decomposition 43–44
computational complexity 62–63
conditional execution 36
conditional independence 42
conditional maximum likelihood (CML) 175–176
conditional random fields (CRFs) 383
conjugate gradients 352
connected component of *G* 69
consensus sequences 84–85
  canonical 193
  noncanonical 193
  signal sensors 193, 194
conservation 268
  score coefficient 318, 384
  sequences *270*
    Markov chain training 270–271
  tracks 22
conserved noncoding sequence (CNS) states 281–282
content sensors 184, 185–193
  feature length modeling 185
  generalized hidden Markov models 238
  Markov chains 185–188
    implementation 188
    improved implementation 188–190
    interpolated 191–192
    nonstationary 192–193
    three-periodic 190–191
  multiple 383–384
  prefix sum array α 228–229
context length, optimal predictive 61
context-free grammars (CFGs) 376–379
contigs 17–18, *18*, 83
  Genomicus simplicans genome 120–121, *121*, *122*
  identifiers 112–113
  ORF graphs *94*, *251*
contingency table *57*
contrapositive of implication 33
converse of implication 33
coordinate transformation 32
core promoters 380
  prediction 351
correlation coefficient 116
covariance 47

CpG islands 2, 361
  gene finding 363
  promoter prediction 381
CRASA program 269
crossover operator 347, 348
cross-validation 113–115
C-terminus 11
cumulative distribution function 48–49
cytogenetic bands 23
cytosine 2–3
  methylation suppression 363

database identifiers (IDs) 25
database matches 273
decision trees 70, *337*, 337–339
  bifurcation *338*
  C4.5 algorithm 105, 337–339
  classification entropy 338
  combiners 276–277
  entropy-based 354
  information gain *338*
  maximal dependence decomposition 201–202, *202*, *203*, 205, 337
  predicates 339
  recursive 338–339
  training set 339
  weight array matrices 339
decoding
  automata 77
  higher-order emissions 170–171
  Markov chains 186–187
  ORF graphs 253, 254–256
  pairHMM 282–285, 289
  PhyloHMM 300, 304–306
  posterior 177, *178*, 211, 255–256, 373
  TWINSCAN 270
  Viterbi algorithm 154, *179*, 210, 384
  see also generalized hidden Markov models (GHMM), decoding; hidden Markov models (HMM), decoding
dependency graph 206–207
  Bayesian networks 330–331
  probabilistic tree models 207
dependency structures
  Bayesian networks 330–331
  trees 331
descent with modification 267–268
deterministic finite automaton (DFA) 76, 77
  Markov chains 188–189
DFP (Davidson–Fletcher–Powell) algorithm 313
differencing method 352
dimensionality reduction 330, 343, 352–354
diploid organisms 2
directed acyclic graph (DAG) 69
  topological sort 72–73, 74
directed graphs 65
directionality bias 278
discrete distribution 48–49, 58–59
discriminant function analysis 340–342
discrimination techniques, maximum 383
discriminative modeling 382–384
discriminative training 365, 383
  generalized hidden Markov models 260–263, 326–327
  hidden Markov models 326–327
distribution 46, 48–54
  binomial 48
  continuous 49
  cumulative distribution function 48–49
  discrete 48–49, 58–59
  exponential 48–49
  geometric 48–49
  independent and identically 48
  normal 49
  probability 49–50
  uniform 49, 58–59
DNA 2–3
  5′-to-3′ direction 4–5
  coding *117*
  curvature 381–382
  hydrogen bonds 5
  interval classification 354
  noncoding *117*
  nucleotides 2–5
  RNA sequence template 7
  secondary structure 209
  shearing 16

DNA (*cont.*)
strands 86–87
structure *4*
DNA ligase 16
DNA sequence
GRAIL gene finding 337
signals 85
do loop 37
dot product 31–32
DoubleScan program 285–287, *287*
dynamic programming 63–66
alignment-based prediction 279
arrays 74
Backward algorithm 163–164, *178*
bottom-up 65
CEM program 280–281
Forward algorithm *143*, *178*
matrix *283*
memoization 65–66
OASIS program 294–295
pairHMM 282, 283–284
PhyloHMM 314
posterior decoding *178*
subproblems *280*
topological sort 72–74
Viterbi algorithm *143*
dynamic score propagation (DSP) 237–241
algorithm *238*, *241*, 241
example 244–246
PhyloHMM 304–306
PSA equivalence 242–244
shortcomings 246–247

early stopping technique 328
ECOPARSE bacterial gene finding 210
EGASP human gene-finding 277
EGC system 319–320
eigenvalues 32, 310
linear discriminant analysis 340–341
eigenvectors 32, 310
decomposition 312, 341
linear discriminant analysis 340–341
principal components analysis 353–354
emission probability, generalized hidden Markov models 230–231
emissions, higher-order technique 158
hidden Markov models 169–172
Ensembl system 319–320
entropy 58–59
conditional 60–61, *61*
cross 59–60
negative 130, 132
functions 59–61
joint 60–61
reduction 337–338
relative 59–60, 130
stop codon 59
estimation of distribution algorithms (EDA) 348–349
Euclidean distance 328
Euclidean norm, vectors 31–32
Eugene program 370–371
eukaryotes 2, 8
cell structure *3*
gene structure *84*
genomic annotation 19–20
selective pressures 268
Euler number 28–29
evaluation metrics 113–118
approximate correlation 116–117
average conditional probability 116–117
baseline performance 117–118
classification accuracy 113
correlation coefficient 116
F-measure 116, 118
granularity level 113
simple matching coefficient 113–115, 118
test substrate 115
evenness 58–59
event 41
mutually exclusive 41
probability 41–42
EvoGene 317
evolution 13–15
descent with modification 267–268
genetic programming 348
instantaneous rate matrix 309–310
models 306–310, 318
parameter estimation 312–316
phylogenetic tree *308*

random drift 268
substitution models 309–310
evolutionary computation, genetic programming 346–349
existential quantifier 33
exon(s) 7–8, 83–86
alignment fuziness *279*
assembly by combiner program 275
cassette 369–370
classification *114*, *355*
coding portion 85–86
data pooling *222*
deletions 286
entropy 60–61, *61*
false positive 115
final 88
finding 97
fusion *297*
generalized hidden Markov models 218, 221
Genomicus simplicans genome 120–121
GENSCAN modeling 208
high-confidence putative 272
individual identification 351
initial 88
insertions 286
internal 88
length distribution *50*, *51*, *216*, 216
likelihood ratios 236
optimism parameter 261–262
prediction
codon bias 127–130
bias 126–127
high-confidence 372–373
in low-coverage genomes 102
probability value for sequence 177–178
putative paired 278, 279
random prediction 121–126
sensitivity 115
single 88
skipping 351, 369–370, 371–372
specificity 115
splitting *297*
subgraphs 280–281
true positive 115
types *89*
exon finder, toy 120–134
acceptor site weight matrix *131*
codon bias 127–134
codon usage *128*
donor site weight matrix *131*
exon coordinates *122*
formulation 127
bias 126–127
nucleotide accuracy 125–126, 133–134
open reading frame 121–123, 124–125, 126
codon bias 127–130, 133
WMM score 132, 133
potential signals of sequence 123–124
pseudocode 123
random exon prediction 121–126
start codon weight matrix *131*
stop codon weight matrix *131*
WMM score 130–134
exon states 218, 221
following acceptor site 154
reading frame prefix sum array α 229
ExoniPhy 317
coding bias 318
tuning parameter 318
expectation 45
expectation maximization (EM) update equations 313
explicit length modeling *249*, *250*
exponential distribution 48–49
expressed sequence tags (ESTs) 19, 268
alternative splicing 370–371, 372
gene finder training sets 105–106
expression-based evidence 268, 269

FASTA files 83, *84*, 112
hidden Markov model 148
feature sensors 184–185, 208–209
DNA secondary structure 209
predicted RNA secondary structure 208–209
see also content sensors; signal sensors
Felsenstein's 4-parameter model 311, 313
Felsenstein's pruning algorithm 302–303, *303*, 316
FGENEH program 342
Fibonacci sequence 64–65, 66

finite automata 76–77
finite state transducer (FST), Markov chains 188–190
fitness 14, 347
fluorescent nucleotides 16
F-measure 116, 118
foreach statement 37
formal language theory 74–75
Forward algorithm 144–145
  Baum–Welch algorithm with scaling 165–169
  dynamic programming *143*, *178*
  posterior decoding 177
forward strand 87–88, *89*
  phases 90
  sequence 87
frame tracking 30–31
frames 90–92
frameshifts 19, 99–100
F-ratios 353
frequency 41
F-score *157*, *359*
fudge factors 318, 384
function calls 37–38
functional annotation 20
functions, mathematical 28–32
  argmax 38–39
  domain 29
  gradient 38–39
  inverse 29
  max value 38–39, 40
  maximization 38
  range 29
  sets 35

gaps 18
  characters 55–56
  closure 18–19
  positions 281
Gaussian distribution 49
GAZE program 277
GenBank database 21–22
gene(s) 1–2
  curated 19
  duplication 13
  expression 6–12
    patterns 12
    regular 92
  finding 83–86
  housekeeping 363
  identifiers 25
  isoforms 369
  loci 7
  nested 98
  noncoding 373–379
  overlapping 98
    prokaryote genes 209–210
  partial 98–99
  predicted 19
  promoter regions 12
  sensitivity 115
  structure *11*, 92
    prediction accuracy 113
  see also mutations
gene finders 21–22
  ab initio 86, 96–97
    accuracy 100
    alternative splicing 101
  accuracy 108, 115–116
  ambiguity codes 101–102
  bacterial 107
  baseline performance 117–118
  bias due to sampling error 113–115
  clustering strategies 109–110
  comparison 111–112
  contig identifiers 112–113
  deployment *105–107*
  discrimination power 254
  evaluation 104–118
  file formats 112
  {G,C} density 192–193
  generalization 104–105
  GHMM-based *263*, 263–264
  hash table *190*
  hidden Markov models 147–156
  hold-out set 108, 110–113
  isochore handling methods 248–249
  Markov chain content sensors 190
  ORF graph interface 257–258
  ORF use 107–108

overlapping genes 98
over-prediction 116–117
protocol testing 104–113
results interpretation 111–112
re-trainable *109*
sensitivity scores 116–117
sensitivity–specificity curve 117
specificity scores 116–117
split stop codons 101
submodels 184–185
test set 111, 118
test substrate 115–116
transducers *190*
under-prediction 116–117
whole-gene-level accuracy of predictions 262–263
gene finding
accuracy 365–366
bacterial 209–211
computational 326
cross-species 360
integrative 275
neural networks 336–337
parsing 92–97
phylogenomic 299–319
gene finding, comparative 267–268
alignment-based prediction 277–281
auto-annotation pipelines 319–320
combiners 275–277
computational advances *268*, *320*
informant techniques 269–275
pairHMM 279–280, 281–299
phylogenomic 299–319
techniques 22
topological sort 267
gene modeling, probabilistic techniques, conditional probability 42
gene ontology 24–25, 381–382
gene predictions
alignment-based 277–281
isoform prediction 24
noncoding genes 373–379
ORF graphs 225, 251–259
traditional machine learning 326–327
whole-gene-level accuracy 262–263
gene predictions, computational 21–22, 24, 83–102
alternative splicing 101
ambiguity codes 101–102
common assumptions 97–102
evaluation metrics 113–118
feature length constraints 100
frameshifts 99–100
frames 90–92
goal 84
haplotype 102
nested genes 98
noncanonical signal sequences 99
optimal parsing 100
orientation 86–88
overlapping genes 98
partial genes 98–99
phases 89–92
problems 25–26
relative direction terminology 87
selenocysteine codons 101
sequencing errors 99–100
simplifications 97–98
split start codons 100
split stop codons 101
test substrate 115
gene shadows 24
GENEID 272
GeneMark.hmm bacterial gene finding 210
General Feature Format (GFF) file 112
generalization, gene finder ability 104–105
generalized hidden Markov models (GHMM) 51, 175, 214
accumulators 237, 238, 244, 245
additional gene features 361–366, *366*
anchor signals 235
backward algorithm 255–256
content sensors 238
decoding 215, 223–228, *228*, 247
accumulators 237, 238, 244, 245
algorithm 270
coding predecessors 225
distance from prospective predecessor 230
dynamic score propagation 237–241, 244–246

generalized hidden Markov models (*cont.*)
    dynamic score propagation/prefix sum array equivalence 242–244
    eclipsed signals 233–234
    emission probability 230–231
    joint probability 300
    optimal predecessor selection 239–240, 243–244
    optimization function 231–232
    ORF graphs 253, 254–256
    ORFs 225–226
    posterior 255–256
    predecessor signals 232
    prefix sum arrays 228–237
    propagators 237–238, 239–240, 244–246
    putative signals 225
    signal sensor scoring 240
    time complexity bounding 224–225
    transition patterns 230
  discriminative training 260–263, 326–327
  double-stranded version 236
  duration
    distribution 214–215, 216–217
    functions 216–217
    term 215–216
  dynamic programming algorithm 255–256
  dynamic score propagation 237–241
    algorithm *241*, 244–246
    PSA equivalence 242–244
    shortcomings 246–247
  eclipsed signals 233–234
  emission
    probabilities 250
    terms 217
  exons 221
    length distribution 216
    states 218
  feature length modeling 185
  forward algorithm 255–256
  forward strand 222
  gene finders *263*, 263–264
  gradient ascent procedure 260, *262*, 262, 263
  higher-fidelity modeling 247–251
  introns
    content sensor 246
    length 250–251
    phase tracking 219–221
    states 218–219
  isochore modeling 247–249
  likelihood evaluation 184–185
  likelihood ratios 236
  log transformation 187
  log-likelihood ratio 236
  maximum likelihood estimation 259–260, 365
  metamodels 217
  noncoding feature lengths 224, 249–251
  noncoding predecessors 226–228, 245, 250–251, 252–253
  noncoding signal queues 250–251
  optimal predecessor selection 243–244
  ORF graphs 225
    building 252–253
    data interchange format 256–259
    gene prediction 225, 251–259
  pair 287–299
  parameter estimation 259
  parsing 194, 253, 254–255
  partial genes 98–99
  predecessor signals 232
  prefix sum arrays 228–237, 385
    algorithm 235–236, *241*
    dynamic score propagation equivalence 242–244
    shortcomings 246–247
  propagators 237–238, 239–240, 244–246
  putative donor sites 250–251
  putative exon probability 215–216
  reverse strand 222, 232, 235, 238
  signal queue 229–230, 233, 234, 237, 238, 240–241, 250–251
  signal sensor scoring 240
  single-strand 223
  sparse state topologies 223–224
  strand number 222–223
  submodel abstraction *217*, 217–218
  topologies 218–223
  traceback pointers 224–225

traceback procedure 234–235
training 259–263
discriminative 260–263
gradient ascent *262*
maximum likelihood 260, *262*
training data 221, *222*, 248–249
transition probability omission 216
trellis links 232–233
two-strand 222–223
variable state duration 218
Viterbi algorithm 223
Viterbi decoding 384
generative modeling 382–384
GeneSequer 269
GeneSplicer 201
genetic code *10*
genetic drift 14–15
genetic programming 346–349
abstract syntax trees 346
classification 347–348, *348*
cloning operator 347
crossover operator 347, 348
evolution 348
fitness function 347
mutation operators 346–347
mutations 346–347, 348
nonterminals set 346, 348
objective function 347
operator set 346
population 346
GeneZilla 361–362, *362*, 366
training 362
see also TIGRscan
GENIE program *219*, 274–275, 337
genome
annotation 19, *258*
assembly 15–19
see also human genome
genome browser 20–23
genome sequencing 15–19
BAC-by-BAC method 15–16
mRNAs 19
practical issues 17
reads 17
whole-genome-shotgun method 15–16
GenomeScan 269, 271–272
genomic annotation 19–25
genomic markers 23
genomic repeats 18, 20
Genomicus simplicans genome 120–121
acceptor site weight matrix *131*
codon bias 127–134
codon usage *128*
contig *121*, *122*
donor site weight matrix *131*
exon coordinates *122*
bias 126–127
nucleotide accuracy 125–126, 133–134
open reading frame 121–123, 124–125, 126
codon bias 127–130, 133
WMM score 132, 133
potential signals of sequence 123–124
pseudocode 123
random exon prediction 121–126
start codon weight matrix *131*
stop codon weight matrix *131*
WMM score 130–134
GENSCAN 21–22
exons
modeling 208
parse probability 255
score computing 177
{G,C} density 192–193
intron states 220–221
isochore definition 247
maximal dependence
decomposition 201–202, 208
nonstationarity 192–193
promoter modeling 208
recursion termination 205
signal sensing 207–208
state topology *220*
untranslated regions 220–221
geometric distribution 48–49
GFF files, hidden Markov model 148
Gibbs free energy 374–375
Gibbs sampling 381–382
Gist program 351
Glimmer microbial gene finder 107, 174,
210

GlimmerM gene finder 339
GNU Scientific Library 313, 341, 343
 hill climbing 352
gradient ascent procedure 38–39, 40, 260, *261*, 262, 263
 algorithm 344–345
 generalized hidden Markov models *262*
 hidden Markov models 176
 simulated annealing 263
gradient descent 38–39
GRAIL gene finding 336–337
graphs 68–74
 acyclic 330–331
 complete 69
 computer representation 71–72
 connected 69
 directed 65, *68*, 68–69
 edge 68–69, 93
 fully connected 69
 isomorphic 69
 splice 370–371
 state-transition *76*, 76
 topological sort 72–74
 tree 69, 70–71
 undirected *68*, 68–69
 vertices 68–69, 93
 weighted 68
 see also directed acyclic graph (DAG); open reading frame (ORF) graphs
grid computing 384–386
guanine 2–3
guanosine, mRNA capping 8–9

Hadamard matrix 314–315
haplotype 2, 102
hash table
 gene finders *190*
 Markov chains 188, *190*
helix destabilization 381–382
hidden Markov models (HMM) 77–78, *78*, 79, 93, 136–140
 ambiguous 140, 159–169
 Arabidopsis thaliana genes 148
 back-off models 171–173
 Backward algorithm 163–165, 177
 base frequency modeling at fixed positions 154–156
 baseline accuracy 149
 Baum–Welch algorithm 159
  naive 163–165
  with scaling 165–169
 Baum–Welch training 162–169
 case study 157–158
 codon position modeling 150, 158
 composite model definition 161
 conditional maximum likelihood 175–176
 decoding 140–145
  algorithms 137, 142, 144–145
  higher-order emissions 170–171
  posterior 177–179
 discriminative training 174–176, 326–327
 emission frequencies *146*
 emission matrix 139–140
 emission probability *138*, 162, 167
 exon probability value for sequence 177–178
 exon state 150, 151, 154
 formulations 136–137
 Forward algorithm 144–145, 165–169, 177
 gene finder 147–156
 generalization 214–218
 generalized pair 287–299
 generative 137
 gradient ascent 176
 higher-order emissions 158, 169–172
  accuracy results *172*
  decoding 170–171
  labeled sequence training 169–170
  sample size 171
 inductive score 143
 intergenic sequence generation 147–148
 interpolated 173–174
 intron phase tracking 152–153, 154
 intron state 147–148, 150, 151
 isochore prediction *248*
 Laplace prior 172
 log-likelihood changes 167
 maximum a posteriori parse 141
 maximum likelihood estimates 146–147, 174–175

maximum mutual information 176
metamodels 161
minimum classification error 176
minimum sample size rule 171–172
most probable path 140–143
notational conventions 137–139
nucleotide accuracy 156
numerical underflow 139
optimal accuracy decoder 178, *179*
pairHMM 281–287
parameter estimation 145
parameter tying 154
partial genes 98–99
phylogenetic 299–304
position-specific matrix 154
prediction accuracy 149–150
pseudocounts 168, 172
representation 139–140
sample size 171
self-transition 147–148
sequence 139, 143–145
simple pair *281*
smoothing 173
splice site modeling 150–154
start codon modeling 150–154
stop codons 150–154
submodels 159–162
test sets 148
traceback 143
training 149, *160*
training sequence labeling 145–147, 148, 150, 169–170
training sets 148, 175
transition
  definition 161–162
  frequencies *146*
  matrix 139–140
  probability 146, 167
variable-order 171–174
Viterbi algorithm 140–143, 149
  decoding 154, *179*, 210
  posterior 178–179, *179*
Viterbi training 159–160, *160*, 161
weight matrix 154, *155*
higher-order emissions
  hidden Markov models 169–172
  technique 158
highest scoring path algorithm *253*
hill-climbing 38–39, *39*, 40, 351–352
Hirschberg algorithm *286*, 286–287, *287*, 371
histograms
  binning 50–51
  regression 52–53
  smoothing 51–52
histone(s) 2
histone code 209
HKY (Hasegawa–Kishino–Yano) model 311–312
HMMgene 273–274
homologs 13
housekeeping genes 363
HSP graphs 297–298, *298*, 299
human gene structure *11*
human genome 2
  entropy 60–61
  intrinsic word length 61
  optimal predictive context length 61
  repeats 18
hydrogen bonds 5
hyperplanes, maximum margin *350*

in silico reassembly 17–18
incidence matrix 71–72
independence 42
induction, mathematical 33–34, 64
inductive hypothesis 33–34
inductive proof 33–34
inductive scores 64
inductive step 33–34
infinity 28
informants
  selection for PhyloHMM 318–319
  sequences 269
  techniques 269–275
information gain 353
  see also mutual information
information theory 58–62
in-frame stop codons 23
  intron straddling 154
Inside–Outside algorithm 379

integers
  interval over 35
  mod 3 operation 31
  non-negative 29
  rounding up/down 29–30
intergenic regions, entropy 60–61, *61*
inter-process communication 386
intron(s) 7–8
  acceptor sites 7, 85
  donor sites 7, 85
  entropy 60–61, *61*
  evidence track 277
  explicit length modeling *250*
  generalized hidden Markov models 218–221
  insertion into gene lineages *298*
  length 250–251
    human *224*
  retention 369–370, 371–372
  skipping 369–370
intron states, GENSCAN 220–221
isochores
  boundary identification 248, 362, 365
  differentiation 247–248
  {G,C} density 247
  gene finder methods 248–249
  modeling 247–249
  prediction *248*
iteration 36–37

JIGSAW program 275–277, 339
Jukes–Cantor model *305*

Kimura 2-parameter model 311
K-nearest neighbors 328–329
Kronecker delta function 30, 326
Kronecker match function 302–303
Kullback–Leibler divergence 59–60
kurtosis 47, 261–262

Lagrangian multipliers 351
languages 74–80
  ambiguity 92
  parsing 92–93
Laplace prior 172
lariat 7
law of large numbers 46
learning algorithm 326
learning problem, automatic classification 325
least-squares estimation (LSE) 54
ligation 7–8
likelihood 48
likelihood ratio
  exon models 236
  generalized hidden Markov models 236
  signal sensors 195, 200–201
  see also log-likelihood ratio
limit at infinity 34
linear discriminant analysis 340–341, *341*, 342
  F-ratios 353
local optimality criterion
  GeneSplicer 201
  signal sensors 198–200, 201
log odds of probability 48
log transformation 29, 30
logic 34
logical complementation 33
logical conjunction/disconjunction 33
logical implication 33
logical negation 33
logically equivalent values 33
logistic curve *344*
logistic regression 343–345
  regularized 345
  response variable 343–344
log-likelihood ratio 48, *188*
  changes in Baum–Welch algorithm 167
  generalized hidden Markov models 236
  Markov chains 187–188
longest increasing subsequence (LIS) algorithm 287

machine learning 325–355
  applications 354–355
  attribute values 327
  automatic classification 325–328
  Bayesian network 330–332
  boosting 358–359
  classification paradigm 326–327
  decision tree 337–339
  dimensionality reduction 352–354

estimation of distribution
algorithms 348–349
exemplar objects 328
feature selection 352–354
genetic programming 346–349
K-nearest neighbors 328–329
linear discriminant analysis 340–342
logistic regression 343–345
multivariate regression 343
naive Bayes model 329–330
neural networks 332–337
overtraining 327–328
quadratic discriminant analysis 342
simulated annealing 349
support vector machines 349–351
traditional 326–327
Mahalanobis distance 329
Manhattan networks 299
manual annotation 19
marginalization 42
Markov assumption 187
signal sensors 205–206
Markov chains
accuracy *192*
content sensors 185–188
implementation 188
improved implementation 188–190
interpolated 191–192
nonstationary 192–193
three-periodic 190–191
continuous-time 309–310
decoding 186–187
dependency graph 206
deterministic finite automaton 188–189
5th-order *190*
finite state transducer 188–190
1st-order 79, *206*
GeneSplicer 201
GENSCAN exon modeling 208
hash tables 188, *190*
implementation 188
improved implementation 188–190
inhomogeneous 191
interpolated 191–192
three-periodic 191
log-likelihood ratios 187–188
Moore machine 189
nonstationary 191, 192–193
2nd-order *188*
sequence conditional probability 187,
205–206
sharing 236
6th-order *79*
three-periodic 190–191
training 186–187, 270–271
transducers *190*
two-state *186*
weight array matrices 198
Markov models
0th-order 79
explicit duration 218
high-order emission probabilities 61
interpolated 173–174, 191–192
*n*th-order 79
see also generalized hidden Markov models
(GHMM); hidden Markov models (HMM)
Markov property 309
mate-pairs 18–19
matrix 32
max value of function 38–39, 40
maximal dependence decomposition (MDD)
Arabidopsis thaliana 203
Bayesian network 205
consensus bases 203
contingency table *204*
decision tree 201–202, *202*, *203*, 205,
337
dependence values 204–205
GeneSplicer 201
GENSCAN 208
recursion termination 205
signal sensors 201–205
training signals 203–205
weight array matrices 208
weight matrices 208
maximal spanning trees 331
maximum a posteriori (MAP) estimation 54
maximum a posteriori (MAP) parse 54
hidden Markov models 141
maximum discrimination techniques 383

maximum likelihood estimation (MLE) 54
  Baum–Welch algorithm 165
  generalized hidden Markov models 259–260, *262*, 365
  hidden Markov models 146–147, 174–175
maximum likelihood parse 54
maximum margin hyperplanes *350*
maximum mutual information (MMI), hidden Markov models 176
McPromoter 381
mean, standard deviation 46
mean of sample 45
median of sample 45
memoization 65–66
messenger RNA (mRNA) 6, 7
  capping 8–9
  computational gene prediction 25–26
  genome sequencing 19
  mature 9
  splicing *8*
metamodels 161, 217
methionine 11
mfold program 374
microRNA (miRNA) 21–22, 373–374
  gene finding 374
  secondary structure *375*
minimal spanning tree (MST) 70–71, *71*
minimum classification error (MCE) 176
minimum description length (MDL) 328
mode of sample 45
modular arithmetic 30–31
molecular biology, central dogma 1–6, *6*, 12
Moore machine 189
MPI (message passing interface) 386
multicollinearity 343
multi-core processors 385
multiplication 30
multivariate regression 343
MUMmer whole-genome alignment program 289, 290–291
mutation operators, genetic programming 346–347
mutations 11
  frameshifts 19, 99–100
  genetic programming 346–347, 348
  natural selection 14
  nonsynonymous 15
  point 347
  selective pressures 14–15
  silent 15
  subtree 347
  synonymous 15
mutual information 61–62
  Bayesian networks 330–331
  maximum in hidden Markov models 176
  pointwise 61–62
  probabilistic tree models 207
  redundancy 353
MZEF program 342

natural selection 14–15
Needleman–Wunsch alignment scores 296, *297*, 307
neighbor joining 307–308
  phylogenetic tree reconstruction *308*
neural networks 332–337
  activation function 334
  backpropagation 334–335
  error function 335
  error propagation 334–335
  feedforward 332, *333*
    propagation rules 333
    topology 333
  gene finding 336–337
  hidden layers 333
  input value 334
  learning rate 335
  paths *334*
  sigmoid activation function 335
  training 334–335, 336
neurons 332
  activation function 334
  activation value 332–333
  GRAIL gene finding 336–337
next statement 37
noncoding predecessors *226*
noncoding RNA (ncRNA) 373–374, 378
non-maximum-likelihood value 384
nonnegative integers 29

N-SCAN 317, 318
N-terminus 11
nucleotide triplets 9–10
nucleotides 2–5
  accuracy 114–115, 125–126, 133–134
  alignment 290–291, 297–299
  fluorescent 16
  F-score *157*
  orientation 4–5
  sensitivity 114–115
  simple matching coefficient 115
  specificity 114–115
  structure *4*
  substitution 304
NUCmer program 290–291
  alignments 292–293
  nucleotide alignment 297–299
  prefix sum arrays 295–296
null hypothesis 55
  power of test 55
numbers 28–32

O formalism 62
OASIS program 289–299
  alignment procedure *293*, 295–296, *296*
  dynamic programming 294–295
  Needleman–Wunsch alignment scores 296
  ORF graph 293–294
  prefix sum arrays 293, 295–296
  state-transition diagram *292*
  trellis 293–294, *294*
OC1 decision tree method 339
one-tailed test 56
one-to-one correspondence 35
ontogeny 1
open reading frame (ORF) 23, 91–92
  bacterial gene finding 210
  coding 354
  codon bias 127–130, 133
  DNA
    anonymous in gene finders 107–108
    random *361*
  {G,C} density 126
  generalized hidden Markov model
      decoding 225–226
  long 107–108, 360
  noncoding 354
  overlapping 125
  prokaryote gene finding 211
  random exon prediction 121–123, 124–125
  WMM score 132, 133
open reading frame (ORF) graphs 93–96
  alignment *291*
  alternative splicing 95, 370–371
  anchor signals 93–94
  application programming interface 258–259
  archiving 259
  Backward algorithm 255–256
  building 252–253
  contigs *94*, *251*
  data interchange format 256–259
  decoding 253, 254–256
  edge number 252–253
  edge types 93–95, 96
    coding/noncoding 258
  Forward algorithm 255–256
  gene-finding programs 257–258
  gene prediction 225, 251–259
  highest scoring path algorithm *253*
  OASIS 293–294
  parallelization 386
  parses 93–94
  parsing
    extraction of suboptimal 254–255, 257
    optimal 253
  phase constraints 95–96
  priority queue 252
  pruned 254–255, 256, 257–258
    GENIE program 274
  re-weighted 256, 257
  signal types 95
  target sequence 96
  TIGRscan 291–292
  universal decoder 256
  vertices 93, 94–95, 258
optimal accuracy decoder, hidden Markov
    models 178, *179*
optimal predecessor selection, generalized
    hidden Markov models 239–240
optimization 38–40

orthologs 13
  pairHMM 288–289
  prediction 296–297
Otto system 319–320
overlapping subproblem phenomenon 64–65
overtraining 104–105, 107
  classification accuracy 325
  early stopping technique 328
  gene finders 104–105, 107
  machine learning 327–328
  prevention techniques 328
  regularization method 328

pairHMM (PHMM) 279–280, 281–287
  computational costs 284–285
  decoding 282–285
  deletion states 282
  DoubleScan 285–287
  dynamic programming 282, 283–284
  gap positions 281
  gene predictions 285
  generalized 287–299
    accuracy *299*
    decoding 289
    homology information 299
    joint duration distribution 288–289
    known orthologues 288–289, *299*
    most probable parse 288
  heuristics 284–285
  insertion states 282
  match states 282
  recursion 283–284
  traceback pointers 284
  training 285
pairwise sequence alignment 281
palindrome structures 381–382
parallelization 384–386
paralogs 13
parameter estimation 54
parameter mismatching 108
parsing 74–80, *96*
  alternative splicing 372
  gene finding 92–97
  heuristic 96–97
  languages 92–93
  maximum a posteriori 141
  optimal 100
  phase constraints 95–96
  techniques 92–93
PASA program 269
Pearson correlation coefficient 47, 330–331
phase-tracking 19, 30–31
phenotype/phenotypic effects 11–12
phosphodiester bonds 5
phylogenetic footprinting 381–382
phylogenetic modeling 32, 299–304
phylogenetic trees 13, *14*, 300, *301*
  conditional probabilities 301–302
  evolution *308*
  genomic sequences *308*
  gradient evaluation 313–316
  informant species 301
  likelihood *303*
  neighbor joining *308*
  reconstruction 307–309, *308*
  re-rooting *301*, 301, 304
phylogenomic gene finding 299–319
phylogeny 13, 300
PhyloGHMM 317
PhyloHMM 299–304
  coding elements *305*
  conditional likelihood 303
  decoding 300, 304–306
  decomposition by state 304
  descent base conditioning 317–318
  discriminating power enhancement 318
  dynamic programming 314
  dynamic score propagation 304–306
  evolution models 306–310
  evolutionary parameter estimation 312–316
  expectation maximization update
      equations 313
  higher-order dependency modeling 316–318
  informant selection 318–319
  instantaneous rate matrix 309–310
  interpolation 317–318
  maximum a posteriori parse 304
  neighbor joining 307–308
  noncoding elements *305*
  phylogeny reconstruction 307–309

prefix sum array 304–306, 385
rate matrix 310–312, 313, 314–315
recursion 302–303
reversibility 310, 311–312
substitution models 309–310, 311
substitution rate matrix 306
training run *316*
transition–transversion modeling 310, 311–312
tree likelihood gradient evaluation 313–316
tree topology 306
tuning parameters 318
PhyloIMM 317–318
pointer representation 71–72
Poisson distribution 49
poly-A tail 8–9
polyadenylation, alternative 369–370
polyadenylation signal 8–9, 361, *370*, *371*
gene finding 363
polypeptide(s) 6, 9
polypeptide chain 10–11
polypyrimidine tracts (PPTs) *364*
population, sampled 53
position-specific matrix (PSM) 154, 195–198
posterior decoding 373
Powell algorithm 313
predicted RNA secondary structure 208–209
predictive accuracy 54
training set size *359*
prefix sum arrays (PSA) 229, 293, 295–296
algorithm 235–236, *241*
alignment scores *296*
dynamic score propagation equivalence 242–244
generalized hidden Markov model decoding 228–237, 385
parallelization 385
partitioning 385
PhyloHMM 304–306, 385
putative exon emission probability *231*
reverse-strand states 229
shortcomings 246–247
principal components analysis (PCA) 330, 353–354
priority queues 70
probabilistic tree models (PTM)
dependency graph 206–207
mutual information 207
signal sensors 205–207
probability 40–48
conditional 42, 43, 48
density 49
discrete distribution 44
distribution 49–50
event 41–42
joint 43–44
log odds 48
marginal 42
posterior 42
Projector program 287
prokaryote gene finding 209–211
open reading frames 211
posterior decoding 211
sliding window 210
weight matrices 210–211
prokaryotes 8
PROmer program *290*, 290–291
alignments 292–293, *293*
prefix sum arrays 295–296
promoter elements 361
promoter modeling, GENSCAN 208
promoter prediction 379–382
promoter recognition, alternative 369–370
promoter state modeling 363–364
propagators 237–239
generalized hidden Markov models 238, 239–240
protein(s) 1–2
domains 12
families 24–25
folding 11
hydrophobicity 11
synthesis *9*
proteome 12
pseudocode 35–38
pseudocounts 168, 172
pseudogenes 24
retrotransposed 8
pseudosignals, signal sensors 194

quadratic discriminant analysis 342
quadratic programming 350–351
quick sort algorithm 67

random drift 268
random gene predictor 117–118
random prediction, exons 121–126
random variable 40–41
  coefficient of variation 47
  correlation 47
  covariance 47
  distribution 46
  entropy 58–59
  expected value 45
  mutual information 61–62
  variance 45–46
rate matrix parameterization 310–312
receiver operating characteristic (ROC) 117
recurrence relation 64–65
  dependency structure 65
  evaluation order 65
  memoization 65–66
recursion 64
  pairHMM 283–284
  PhyloHMM 302–303
  termination 205
redundancy of mutual information 353
regression 52–53
regular expressions 75–76
regularization 328, 345
relative frequency 41
relevance vector machines 381–382
repeat sequences, masking 366–367
RepeatMasker program 22, 366
repeats, low-complexity 366, 367
reproducing kernel Hilbert spaces 351
restriction enzyme 16
retrotransposed pseudogenes 8
revcoord 88
reverse complement 5–6, 86, 87, 88
reverse strand 87–88, *90*
  generalized hidden Markov models 222, 232, 235, 238
  phases 90
  sequence 87
reversibility modeling 310, 311–312
ribosomal binding sites (RBS) 210–211
ribosomal RNA 373–374
ribosomes 9, 10–11
ribozymes 8
risk minimization 350–351
RNA 2–3
  5′-to-3′ direction 4–5
  ensembles 375–376
  hydrogen bonds 5
  microstates 375–376
  protein synthesis *9*
  secondary structure
    free energy 374–375
    prediction 208–209, 374–375, *375*, 379
    probability 376–379
    thermodynamic method of prediction 374–376
  stochastic context-free grammars 376–379
RNA polymerase 7
RNAfold program 374
ROSE program 289–299
ROSETTA program 278

sample space 40–41
  subset 41, 42
sample variance 46
sampling error 14, 53–54
  bias 113–115
scaffolds 17, 18–19
scalar 32
scaling, Baum–Welch algorithm 165–169
searching 66–68
selective pressures 14–15, 268
selenocysteine codons 101
semi-Markov models 218
sensitivity–specificity curve 117
  algorithms 117
  gene finder training 117
sensor windows, nonoverlapping *232*
sequence conservation 13, 289
sequence motifs 12
sequencing errors 99–100
sequencing machine *17*

sets 34–35
cardinality 34
cartesian product 34–35
countable 35
functions 35
membership 34
power of 35
subtraction 34
universal 35, 42–43, 44
SGP-1 program 278–279
SGP-2 program 269, 272–273
Shine–Delgarno sequence, prokaryotic gene finding 210–211
signal identification sensitivity/specificity 99
signal peptides 361, 364
signal queue 229–230, 233, 234, 237
generalized hidden Markov models 238, 240–241
signal sensors 184, 193–208
application 193–194
boosting 358–359
classification accuracy 194–195
coding–noncoding boundaries 200–201
consenus sequence 193, 194
dependency graph 206–207
fixed-length state matches 229
GeneSplicer 201
GENSCAN 207–208
likelihood ratio 195, 200–201
local optimality criterion 198–199, *199*, 200
Markov assumption 205–206
maximal dependence decomposition 201–205
probabilistic tree models 205–207
pseudosignals 194
putative signal removal 198–200
sliding window *194*
training 194–195
weight array matrices 197–198
weight matrices 195–198
signal sequences, noncanonical 99
signal thresholding 229
sim4 19
simple matching coefficient (SMC) 113–118
simulated annealing 349, *350*
single nucleotide polymorphisms (SNPs) 7
haplotype 102
single-stranded DNA (ssDNA) 16
skewness 45, 47, 261–262
sliding window
prokaryotic gene finding 210
promoter prediction 381
signal sensors *194*
smoothing 51–52, 173
SNAP gene finder, single-strand 223
sorting 66–68
speciation 13, 14–15
species separation *341*, 341–342
spectral decomposition 309–310
splice graphs 370–371
splice sites
coding–noncoding boundaries 200–201
modeling 150–154
prediction 332, 348–349
ROSETTA program 278
spliced alignment program 19
spliceosome 7, 24
splicing 7
mRNA *8*
patterns 369
see also alternative splicing
standard deviation 46
distribution 46
of mean 46
standard error 46
start codon(s) 11, 19–20
modeling 150–154
prokaryote genes 209–211
split 100
translation of mature transcript 84
Venn diagram *44*
weight matrix *195*
statistical hypothesis testing 55–58
rejection criteria 55, 56
Steiner graphs 299
step size 39
stepping-stone algorithm *287*, 287
stochastic context-free grammars (SCFGs) 376–379

stop codon(s) 11, 19–20
  coding–noncoding boundaries 200–201
  entropy 59
  in-phase 89–90, 91
  intron straddling 154, 234
  modeling 150–154
  prokaryotic gene finding 210–211
  split 101
  translation
    mature transcript 84
    termination 89
  valid 92
  see also in-frame stop codons
string(s) 74–75
  alternation 75
  concatenation 75
  deterministic acceptors 77–78
  exponentiation 75
  Kleene closure 75
  nondeterministic generators 77–78
  parsing 92
  positive closure 75
  regular expressions 75–76
  substrings 37
  syntactic structure modeling 76
string literals 37
subset operators 35
substitution matrix 23, 300
substitution models 309–310, 311
substrings 37
summation 30
superset operators 35
support vector machines 349–351
  dot product 350–351
  kernel function 350–351
  risk minimization 350–351
SVMlight 351
synapses 332, 333, *334*
syntax, stochastic models 77–80
syntax trees 346, *347*
systems biology 12

target genome 269
TATA-box 363–364, 381
taxa 13
Taylor expansion 309–310
TBLASTN 22
TBLASTX 22, 272–273, 280, 290–291
termination site 11
test case
  classification 325
  computational gene finding 326
test statistic 55
thymine 2–3
TIGRscan 291–292
  see also GeneZilla
TNT (Template Numerical Toolkit) 312, 341
topological sort 72–73, *73*, *74*, 74, 267
toy genes/genome 120–121
TOYSCAN 127, 133–134
trace file 16
traceback pointers
  generalized hidden Markov models 224–225
  pairHMM 284
training *105–107*
  Baum–Welch algorithm 162–169
  gene finders *105–107*
  generative modeling 383–384
  GeneZilla 362
  Markov chains 186–187, 270–271
  neural networks 334–335, 336
  pairHMM 285
  PhyloHMM *316*
  sensitivity–specificity curve 117
  signal sensors 194–195
  Viterbi 159–161
  see also discriminative training; generalized hidden Markov models (GHMM), training; hidden Markov models (HMM), training; overtraining
training data 53–54
  accuracy measurement inflation 107
  bootstrapping 108, 359–360
  checks 106
  collection 106–110
  small sets 106–108, 110
  see also generalized hidden Markov models (GHMM), training data
training problem, automatic classification 325

training sets 105–106, *107*, 108
  bias 109
  BLAST hits 105–106
  cDNA 105–106
  decision trees 339
  dimensionality 326
  ESTs 105–106
  genes 105
  partitioning 110–113
  predictive accuracy *359*
  redundancy elimination 108–113
  see also hidden Markov models (HMM), training sets
transcription 6–9
transcription factor(s) 380
transcription factor binding sites 210–211, 380–381
  library 363
transcription start sites (TSS) 380
  alternative 381
transcriptome 12
transcripts 7, 84
transducers *190*
TRANSFAC library 83–86, 363, 380–381
transfer RNA (tRNA) 9, 10–11, 373–374
  codon reading 89–90
transition–transversion modeling 310, 311–312
translation 9–11, 84–85
  initiation site 11
  start/stop sites 84–85
transposable repeat elements 366
transposons 367
tree(s) *69*
  maximal spanning 331
  minimal spanning 70–71, *71*
  traversal 70
  see also decision trees; phylogenetic trees; syntax trees
tree graphs 69, 70–71
trellis *232*, 232–233, 252
trellis links 232–233, 254–255
  OASIS program 293–294, *294*
trinucleotide frequency, codon bias 85–86
tuning parameters 318
TWAIN program 289–290, *290*, 299
  sequence conservation modeling 289
TWINSCAN 269–271
  decoding algorithm 270
two-tailed test 56
Type I and Type II errors 55

UCI machine-learning repository 354–355
UCSC Genome Browser *21*, 21–22, 372–373
UNCOVER program 371–372
underflow 30
uniform distribution 49, 58–59
unimodal distribution 45
universal quantifier 33
untranslated regions (UTRs) 21–22, 85
  GENSCAN 220–221
UNVEIL gene finder *158*, *179*
  accuracy results *159*, *172*, 263–264
  case study 157–158
UPGMA algorithm 306, 307–308
  phylogenetic tree reconstruction *308*
uracil 2–3
UTR models 362

variable number of tandem repeats (VNTR) 366
variance 45–46
vector machines 381–382
vectors 31–32
VEIL gene finder 157–158, *158*
Venn diagram 35, *43*, *44*
Viterbi algorithm 77–78, 140–143, 149
  bacterial gene finding 210
  decoding 154, *179*, 210
  dynamic programming *143*
  generalized hidden Markov models 223
  hidden Markov models *160*, *179*
  posterior 178–179, *179*
Viterbi decoding 384
Viterbi training 159–161

Watson–Crick complementarity 2–3, 5, 10–11
  genome sequencing 16
weight array matrices (WAM) 197–198
  classification experiments 197, *198*
  decision trees 339

weight array matrices (WAM) (*cont.*)
GeneSplicer 201
likelihood ratios 197
Markov chains 198
maximal dependence decomposition (MDD) 208
windowed 198
weight matrices (WMM) 195–196, *196*, 198
CEM program 280
classification accuracy *198*
exon finding 130–134
hidden Markov model *155*
maximal dependence decomposition 208
method 130–134, 154
pictograms 195
prokaryotic gene finding 210–211
promoter prediction 380–381
sequence conditional probability 195–197
start codons *195*
TATA-box 363–364
while loop 36–37
whole-genome alignment program 289, 290–291
word length, intrinsic 61

zero-based indices 29